T0261126

Megafauna

Life of the Past

JAMES O. FARLOW, EDITOR

Indiana University Press   BLOOMINGTON & INDIANAPOLIS

MEGAFAUNA

GIANT BEASTS OF PLEISTOCENE SOUTH AMERICA

RICHARD A. FARIÑA

SERGIO F. VIZCAÍNO

GERRY DE IULIIS

This book is a publication of

Indiana University Press
Office of Scholarly Publishing
Herman B Wells Library 350
1320 East 10th Street
Bloomington, Indiana 47405 USA

iupress.indiana.edu

Telephone orders   800-842-6796
Fax orders   812-855-7931
Orders by e-mail   iuporder@indiana.edu

© 2013 by Richard A. Fariña, Sergio F. Vizcaíno,
and Gerardo De Iuliis

All rights reserved

No part of this book may be reproduced or
utilized in any form or by any means, electronic
or mechanical, including photocopying and
recording, or by any information storage
and retrieval system, without permission in
writing from the publisher. The Association
of American University Presses' Resolution on
Permissions constitutes the only exception to
this prohibition.

♾ The paper used in this publication meets
the minimum requirements of the American
National Standard for Information Sciences—
Permanence of Paper for Printed Library
Materials, ANSI Z39.48-1992.

Manufactured in the
United States of America

Library of Congress
Cataloging-in-Publication Data

Megafauna : giant beasts of Pleistocene South
America / Richard A. Fariña, Sergio F. Vizcaíno,
and Gerry De Iuliis.
    p. cm. — (Life of the past)
Includes bibliographical references and index.
ISBN 978-0-253-00230-3 (cloth : alk. paper) —
ISBN 978-0-253-00719-3 (eb)
1. Mammals, Fossil—South America.
2. Paleobiology—South America. 3. Geology,
Stratigraphic—Pleistocene. I. Fariña, Richard
A. II. Vizcaíno, Sergio F. III. De Iuliis, Gerardo,
[date]
  QE881.M475 2012
  569.098—dc23
                                    2012017801

1 2 3 4 5   17 16 15 14 13

To past and current researchers of South American fossil mammals.
Those of the past are an endless source of inspiration,
those still current of intellectual motivation.

To the memory of Mirta Tosar, my mother,
who taught me to be bold and love animals as a person,
and to Neill Alexander, who encouraged me in the same way as a scientist.

R.A.F.

To the memory of my parents, Eric and "Negra,"
who instilled in me the value of hard work and honesty.

To Susi, Rulo, Leo, Tano, Guille, and Nestor,
the people with whom I share every day the joy of doing this job.

S.F.V.

To my family and the memory of my father and father-in-law,
whose sacrifices allowed me the luxury of doing what I love.

To Charles "Rufus" Churcher,
who instilled in me the intellectual discipline to carry it out.

G.D.I.

The number of the remains embedded in the grand estuary deposit which forms the Pampas and covers the granitic rocks of Banda Oriental, must be extraordinarily great. I believe a straight line drawn in any direction through the Pampas would cut through some skeleton or bones. Besides those which I found during my short excursions, I heard of many others, and the origin of such names as "the stream of the animal," "the hill of the giant," is obvious. At other times I heard of the marvellous property of certain rivers, which had the power of changing small bones into large; or, as some maintained, the bones themselves grew. As far as I am aware, not one of these animals perished, as was formerly supposed, in the marshes or muddy river-beds of the present land, but their bones have been exposed by the streams intersecting the subaqueous deposit in which they were originally embedded. We may conclude that the whole area of the Pampas is one wide sepulchre of these extinct gigantic quadrupeds.

Charles R. Darwin, November 26, 1833

Contents

The first reports, during the late 1700s and early 1800s, of the fossil remains of South America's magnificent Pleistocene beasts, so fantastically bizarre, immediately caused a stir among the general public and, in particular, the European scientific community. The first notices of their discovery described them as monsters, firing the imagination and interest of several eminent scientists and politicians, and leading some of them to believe that these great beasts still wandered among the unknown (for Europeans, at any rate) reaches of the New World. The fossils helped usher in a new episode among the fledgling nations of both South and North America, striving then for recognition and validation in the eyes of the established European powers: finally they had something of their own that rivaled the great treasures of the Old World. Eventually, the fossils contributed significantly to the establishment of new scientific institutions and traditions as the New World countries took hold of their destinies and exploration of their territories.

The fossil mammals of both North and South America began to reveal an unimagined chapter in the history of mammals, based as it then was mainly on knowledge unearthed from European deposits, but it was those from South America that were most strikingly different and garnered much of the early attention. Perhaps because of this distinctness, largely as a result of the long, past isolation of South America from other continental landmasses, they played crucial roles in the development of modern biological thought. We may note as examples of their scientific achievements that a South American fossil mammal (*Megatherium americanum*, a giant fossil sloth) was the first fossil to be formally described and named scientifically, and its skeleton was the first to be mounted in a lifelike pose. The sharp mind of Georges Cuvier, the great French comparative anatomist, forged the concept of extinction (in the modern sense of this word) based on this fossil sloth (as well as on North and South American remains of fossil elephant relatives). Perhaps most significantly, it was the giant sloths, the giant armadillo-like glyptodonts, and the majestic and ponderous toxodonts (among other South American fossil remains) that struck most fervently upon the fertile mind of the young Charles Darwin, both during and after his famous voyage aboard the HMS *Beagle*, as he worked out his ideas on evolutionary theory.

Despite the relative isolation of the new South American countries, these ideas greatly affected scientists and intellectuals on both sides of the Río de la Plata, several of whom (such as the Ameghinos) took on the void created by Darwin's return to England and restarted the study of the South American fossil mammals with renewed enthusiasm. Such was their influence that it even affected the excited atmosphere of the newly born city of La Plata, which inspired an illustration that adorns one of the entrances of the División Paleontología de Vertebrados of the Museo de La Plata. As explained in Chapter 8, it was conceived at the end of the nineteenth century, just after the city's founding, as an artist's rendition of a plan to embellish the gardens between the museum and the neighboring zoo, two of the proud city's new jewels. In these gardens, visitors could stroll between the institutions that housed the living and the long dead and have a sense of those extinct beasts, brought back to life in the form of life-sized sculptures.

Despite such an early and auspicious beginning, the study and our understanding of South America's extinct mammals has generally lagged behind those from most other continents. In part this is certainly due to their distinctness, generally leading scientists to either consider them too odd to expend much energy on, or regard them as somewhat inferior variants of more typical mammals, as was once done for dinosaurs. In accepting such views, the past was condemned to be more or less like the present and the magnificent mammals of South America relegated to being antiquated curiosities of "better" and "more modern" mammalian designs.

In this book we endeavor to reveal that such views are erroneous. Despite their differences and the fact that no living analogues exist for many native South American mammals, thus making comparisons difficult, we show that these mammals and the environments in which they lived and evolved were likely not merely slight variations of those that exist today. Their study is thus entirely worth the effort. By combining a wide variety of techniques from several biological disciplines, researchers over the past decade or so have begun bringing these beasts back to life. The picture that has begun to emerge is that of a marvelous biota that resists being pigeonholed, one that at once enlightens our concept of mammalness and enhances our knowledge of the past.

We begin our story in Chapter 1 with an introduction to the splendor of the South American megafauna, the assemblage of large-sized and gigantic beasts (or megamammals) that are the main subject of this book. In addition, we provide an analysis of how paleontology (and science in general) works, discussing its place among the evolutionary sciences, and consider other relevant background topics such as the role of fossilization, geology, and biological classification. This sets the stage for the more detailed discussions of the mammals and their environments provided in later chapters. We continue in Chapter 2 by outlining the historical contributions to the study of South America's fossil mammals: a pantheon of researchers whose efforts illuminate our present knowledge, for these are the giants on whose shoulders modern paleontologists stand. Chapter 3 provides more detailed discussions, begun in Chapter 1, of the geological and ecological contexts in which this fauna evolved. These historical contexts are indispensible, for they allow us to understand the natures of and differences between the Pleistocene and modern South American fossil assemblages. This theme is continued in Chapter 4, which focuses particularly on the North American fauna and its relationship with that of South America—although we are getting ahead of ourselves; much of the Pleistocene faunas, and to an even greater degree modern South American faunas, are ultimately of North

American origin, and thus the northern stock plays an integral role in our story.

In Chapter 5 we bring together the diverse threads of the previous chapters as we discuss the history of the native South American mammals, the geological processes leading to the emergence of the Isthmus of Panama, finally ending South America's long isolation, and the episode that followed: the extensive intermingling of the North and South American faunas known as the Great American Biotic Interchange and its consequences. As well, we consider the habitats present in the Pleistocene, which served as the essential backdrop to the events occurring among the mammals. Chapter 6 rejoices in the wonders of this splendid assemblage by focusing on many of the main large-bodied species, noting their relationships, main characteristics, and roles as members of the fauna. Chapters 7 and 8 outline our more recent attempts at bringing these mammals back to life—or at least in viewing them as having once been real flesh-and-blood beasts rather than the dry and scattered bones that we have left of them. We provide explanations of some of the methods used to study the paleobiology and paleoecology of these animals, particularly of the giants, and present the results of such research into their habits, ecology, metabolism, appearance, lifestyles, and diets.

As is generally known, the magnificent megafauna of South America (and several other continents) is no more. Chapter 9 presents our attempts to explain the demise of these beasts. The extinction of the megafauna is a touchy subject: whereas we are generally content to explain away other extinction events (of the nonavian dinosaurs, for example) as due to any, all, or some combination of environmental, tectonic, catastrophic, and climatic factors, the disappearance, only some 10 thousand years ago, of the South American Pleistocene megafauna occurred in the presence of humans and debate rages on the role, if any, that humans played. Our treatment of the subject is set within the framework of the competing hypotheses with all their scientific and ethical implications. Last, the Appendices provide guidance on the basic anatomy required to interpret our discussions and descriptions of the mammals, and more in-depth explanations of some of the methods described in this book's chapters.

The publication of this book is due to the efforts and resources of many people and institutions, and so we have many to thank. Foremost among these, we acknowledge the generosity and graciousness of Edmundo Canalda. In addition to his much-appreciated encouragement, Edmundo (and Fin de Siglo) kindly allowed us to borrow liberally from *Hace Sólo Diez Mil Años* (Fariña and Vizcaíno, 1995). Though briefer in scope and less sophisticated and detailed in treatment of the themes presented in the current book, the similarities between them are unmistakable. In addition, we thank Ángeles Beri for help with and information on Cenozoic plants, Sebastián Tambusso and Néstor Toledo for many of the illustrations, Eva Fariña for recreating some of the medieval bestiary figures in Chapter 6, Signe Haakonsson for reviewing the spelling in some of the references in Danish, Guillermo Cassini for help with Fig. 3.14, Mauro Muyano for many illustrations, Jonathan Perry for reading parts of this book and making many good suggestions, Dino Pulerà for kind permission to reproduce several anatomical illustrations, Celestino De Iuliis for reading and commenting on the proposal and initial drafts of the earlier chapters, Peter Reali for help with photography and image reproduction, Silvia Ametrano, director of the Museo de La Plata, for permission to reproduce images of publications, specimens, and art of this institution, Susi Bargo for collaborating with the three of us on many aspects of this book and much of the original research, Eduardo Mizraji and Andrés Pomi for their help in finding Vilardebó's image, Antonio Carlos Sequeira Fernandes and Deise Henriques for their help with the photograph of Paulo Couto, Jhoann Canto and the Eberhardt family for their help with the photograph of Captain Eberhardt, Juan Carlos Fernicola for helpful discussions on glyptodonts and the image of H. G. Burmeister, the American Museum of Natural History (New York) and the Field Museum of Natural History (Chicago) for permission to reproduce images and the staff who facilitated our requests, the many other colleagues who provided original images and whom we acknowledge individually in the figure captions, colleagues too numerous to list who provided original information, shared their views, and collaborate with us in the ongoing process of learning about South American fossil mammals, our graduate students who have developed their research under our supervision (and the innumerable other tasks that they inevitably take on), Marco Tosi for a long-ago evening in Rome that helped launch G.D.I.'s long and fruitful relationship with South American researchers. S.F.V. acknowledges the contributions of CONICET (Consejo nacional de Investigaciones Científicas y Ténicas), Universidad Nacional de La Plata and the Agencia Nacional de Promoción Científica y Tecnológica, for these agencies' continued support for much of his research activities. Last, a huge thanks to our spouses, Ángeles, Miriam, and Gina, and children, Eva, Josefina, Julieta, Theodore, and Jacob, for their encouragement and tolerance throughout the duration of this project.

Megafauna

Paleontology and Science: What Is Science?

South America, the southern half of the pole-to-pole landmass named, according to the usual attribution, after the Italian merchant and cartographer Amerigo Vespucci — or, as convincingly argued by Lloyd and Mitchinson (2008), after the wealthy Bristol merchant Richard Ameryk, a main investor in Giovanni Caboto's second transatlantic voyage — remains a territory full of interest, intrigue, and biological treasures. Artificially severed from the northern half by the Panama Canal, it extends from the tropics, where the marvelous Amazonian rain forest offers its biological diversity and chemical riches of trapped carbon, to the elevations and endless steppes of Patagonia, its tapered south that points at and nearly touches frozen Antarctica. From west to east, the assortment of landscapes includes the soaring Andes, followed in places by the Altiplano that so aroused past greed for silver and gold, and then descends into the low-lying eastern plains, where the fossils discussed in this book have mainly been found.

Despite this variety in habitats, latitudes, and altitudes, the great naturalist Georges-Louis Leclerc, better known as Comte de Buffon, claimed that South America lacked enough vital energy to yield true giants among its fauna. Indeed, the present-day fauna boasts no true megamammals, i.e., those for which body mass is given in tonnes (or megagrams, hence the term). Impressive as it is for a rodent, the capybara, at 60 kg, can hardly lay claim for membership in the same category as elephants and rhinos. Also of similar size are two xenarthrans (a group of mammals particularly characteristic of South America), the giant armadillo *Priodontes maximus* and the giant anteater *Myrmecophaga tridactyla* (both, sadly, of dwindling populations), for which the adjective applies only in comparison with their close kin, but not if placed side by side with hippos and giraffes. The largest South American mammal is the tapir, whose bulk (approximately 300 kg) is less than striking compared with bison. Among carnivores, the jaguar is the largest, but it cannot match other top predators, such as lions and tigers, respectively two and three times its size.

It was Charles Darwin himself who corrected Buffon: rather than absence, the reality for huge South American mammals is recent demise. Ten thousand years ago, an instant in terms of geologic time, South America was inhabited by a mammalian fauna so large, diverse, and rare that today's African national parks would pale in comparison. Bears, sabertooth cats, and elephant-like gomphotheres lived alongside much-larger-than-extant 150-kg capybaras and oversized llamas. Horses roamed these lands and went extinct thousands of years before Spanish explorers reintroduced the domestic species. In addition to these relatively familiar mammals, there were also bizarre creatures only distantly related to modern forms, such as terrestrial sloths several meters in height when standing bipedally; completely armored glyptodonts, hippo-sized animals related to armadillos; the camellike macrauchenids; and the rhinolike toxodonts, weird yet strangely familiar in echoing the ungulate, or hoof-bearing, types common in North America, Africa, and Asia.

Thanks to the efforts of many people, both collectors who found and prepared the material, often in decidedly uncomfortable and in some cases outright dangerous conditions, and scholars who interpreted those remains, we have now resurrected them for our contemplation and awe.

Traveling today through the sparsely populated regions of South America, it boggles the mind to think that the plains and low hills were once filled by a fauna so grand that it was rivaled only by the dinosaurs in magnificence. But unlike these long-extinct creatures of the Mesozoic, the South American fauna is much closer to us in time and in the phylogenetic (or genealogical) relationships to living mammals, as well as being more abundantly preserved. It is therefore easier to understand the biology and infer the ways of life of these magnificent mammals; further, the extraction of genetic material from the more recent fossils is already being accomplished, in contrast to the currently impossible science fiction world of Jurassic Park. Perhaps even more intriguing is that some of these mammals coinhabited these lands with early Americans, who may, regrettably, have had a hand in their demise.

In this book, we deal with the life and times of these remarkable beasts, the adaptations they evolved, their origin and journeys, the ecology of those (not so) long ago times, and the possible reasons for their extinction. All these subjects are dear to paleontology, a discipline that claims its own place within science. Most people think of paleontologists as adventurous individuals dashing through remote areas of the world to find and dig out dinosaur bones. Although it is true that paleontologists need fossils to understand life of the past, their scope of knowledge and skills is broader than that required to unearth old dead things. Paleontologists usually receive broad training in both biology and geology, and they are generally equally comfortable in the field, deciphering the geology of the area under investigation, and in the lab, researching and describing the material they have collected. The latter in particular requires a wide knowledge of living organisms—how else could we hope to understand the way extinct animals appeared and behaved from the meager scraps we have left of them? This requires, among other things, an extensive knowledge of anatomy—it is no coincidence that many paleontologists also teach comparative anatomy in universities and human anatomy in medical schools.

Paleontology and science

Paleontology falls under the aegis of evolutionary biology, the scientific field that deals with the remarkable diversity of the organic world, past and present. The theory of evolution, the cornerstone of modern biology, includes the set of ideas explaining how and why evolution occurs. It is one of science's most robust theories, supported by an impressive array of evidence. Surprisingly, it also happens to be among those scientific theories that the general public is least confident in. A recent poll among Britons, for example, indicated that fewer than half of respondents accept the theory of evolution as the best explanation for the development of life, with nearly 40% opting instead for some form of creationism—the set of ideas that rely on some sort of supernatural creator or designer as an explanation for the origin and diversity of life.

The situation in the United States is pretty much the same as in the United Kingdom. It seems incredible that one of science's strongest theories should be viewed with such skepticism in such a scientifically and technologically advanced nation. Then again, some have cast doubt on the theory of evolution and voiced support for the teaching of intelligent design theory in U.S. public schools. This position was soundly routed in U.S. district judge John E. Jones's 2005 ruling against Pennsylvania's Dover area school board's decision to insert intelligent design into the public school science curriculum. In one of the most stinging criticisms of creationism to date, Judge Jones noted the overwhelming evidence that establishes intelligent design as a religious view, a mere relabeling of creationism, rather than a scientific theory, and he condemned the "breathtaking inanity" of the Dover board's policy, accusing some of the board members of lying to conceal their true motive of promoting religion.

Much of the public's suspicion of evolution is the result of a consistent battle waged by a small, but politically active and well-connected fundamentalist religious faction that has tried to disparage evolution while trying to impose its own viewpoint on U.S. public schools; that is, on students during what is probably their most intellectually vulnerable stage. Originally a concern mainly in the United States, creationism has taken root in other Western nations over the last two decades. Its onslaughts on science have both contributed to and played upon, in synergistic fashion, the general public's misperception of what science is. It is one thing when people who have not received a higher level of education are swayed by creationists, especially as many people start out their school careers already inculcated with some worldview that includes elements of creationism. It is quite another when the more highly educated members of society (U.S. polls indicate about 40% of college graduates do not believe that humans evolved from some other mammals) and indeed a former leader of the nation—a Yale University graduate with any number of highly qualified advisors supposedly at his disposal—misunderstand science. It is even worse when many professional academics venture into a field not their own and, with an embarrassingly limited understanding of the subject (remedied by even a first-year university-level course), decide they are qualified to hold forth on the perceived inadequacies of evolutionary theory. To place this sort of dabbling into perspective, image an evolutionary biologist trying to set a nuclear physicist straight on real or imagined inconsistencies in the theory of quantum mechanics due to interpretations over the value of particle accelerators. It is so preposterous an idea that we wouldn't even take its proposition seriously. Yet we must wonder why so few eyebrows are raised when (rather than if) the reverse happens.

There seem to be two main reasons why so many people have an aversion to evolution. One is that evolution happens to touch on aspects of humanity dear to many of us: our place in this universe, and the meaning and purpose of our lives. The centrality of these facets leads many of us to assert that our personal convictions are as equally valid explanatory tools as evolutionary theory in dealing with the diversity, development, and history of life on this planet. However, this is simply not true. Evolutionary theory analyses the physical and behavioral changes in the forms of life over time and provides explanations for how those changes occurred. That such changes have occurred is beyond any reasonable doubt. Opinions on such questions are not equally valid, any more so than is any particular person's opinion on whether matter is really composed of atoms or the earth revolves around the sun. There may still be people somewhere in the world that believe in a flat earth. Their opinion, bluntly put, simply does not count. Evolutionary theory does not deal with, nor does it purport to deal with, questions of spirituality, which, for reasons explained below, are outside the realm of science. Many scientists, including a good number of evolutionary biologists, such as Francisco J. Ayala (Fig. 1.1), have reconciled religion and science, maintaining their spiritual faith while recognizing the fact of evolution.

Another reason why people mistrust evolution has to do with how they view science. This calls into question the effectiveness and role of science education in much of Western society. Science education has made great strides in the last half century or so, particularly in conveying scientific knowledge. Yet we have to wonder whether our emphasis on presenting the results and successes of scientific investigation has overshadowed the more basic principles of how we arrive at that knowledge, on what actually qualifies as science. If we have done our job properly, then by the time a student graduates from high school—given that we start science education in the elementary grades—the student would have a perfectly good grasp of what science is and what it is meant to do, and there would be little if any reason for debate on whether we can accept evolution as scientific or whether it is true or not because "it is only a theory, after all." When almost half of college graduates—in the United States, at least—are confused, it might be a sign that we are not doing all we can as educators. Given the current level of misunderstanding about science in general, and to place the information of this book in context, it is worth spending a few pages to consider what science is.

Science is so important to modern society that rarely does a day go by without a media report on some new discovery. It is perhaps because science is so ubiquitous that we have begun to take it for granted and fail to reflect on just what it is that's behind the headlines. This is certainly one reason why science is so misunderstood, but the fact that so many people complete their formal education without understanding the fundamentals of science means that we may not be instructing students properly—or at least not emphasizing the essentials.

What is it about science that makes it so special? Most people view scientists as diligently working away to fill in some piece of a giant puzzle, a great edifice of knowledge that, once completed, would represent the truth about the way the nature works. This picture is misleading. Scientists are not out to seek some universal truth—at least, not in the way most people understand truth. At best, we can say that scientists are out to describe the way that nature works, and that the best we can usually do is to arrive at some approximation of the truth.

Science provides us with tentative answers. Any scientist worth his or her salt realizes that any answer is provisional and can expect to be corrected or revised (Tattersall, 2002). Eventually a larger understanding

1.1. Francisco J. Ayala in 2008, former Dominican priest and famed geneticist and evolutionist. Although he does not discuss his personal opinions, his views represent an appropriate way of dealing with science and religion.

Image courtesy of F. J. Ayala, University of California, Irvine, USA.

of nature's workings is achieved, but this knowledge is always based on our current abilities to perceive nature. If new techniques or conceptual frameworks provide us with novel ways of analyzing phenomena, then we can expect our conceptualization of nature to change. This is precisely what happened when we switched our picture of matter and motion from a Newtonian to a quantum perspective.

This means that there is no giant puzzle to fill in. A puzzle is static, whereas science is a dynamic process, constantly rearranging its view of how nature works (Tattersall, 2002), with scientists accepting new ideas and rejecting old ones, as the need arises. The fact that there is change does not mean that the methods used to discover the knowledge are invalid or somehow inferior. Indeed, we have made enormous strides over the past 300 hundred years in understanding the natural world using these methods, which have been far more successful than any other method humankind has devised. The success of science lies in its methodology. Science, as summarized in Fig. 1.2, is factual knowledge of the natural world gained through an unbiased procedure that is generally labeled as the scientific method. That is, we can think of science as consisting of a body of information or knowledge (facts, in most people's estimation)—for example, an atom of hydrogen has a single proton in its nucleus. As well, science is a human endeavor; it also includes the way (by using scientific methodology) we obtain that knowledge.

In considering how scientists carry out their activities, we can begin to explore the essence of what makes science so successful at revealing the workings of the natural world. Through prior knowledge and observations, scientists propose ideas about some particular facet of nature. Why do they do this? In part, it is because, like most people, scientists are inquisitive beings. Further, scientists are generally driven to understand their particular corner of the universe; it is simply part of who they are. However, many people have ideas, so simply coming up with ideas isn't what sets science apart. Once a scientist formulates an idea, it is evaluated by putting it through some sort of test, then published so that it is available to the scientific community (though usually only a small number of scientists are interested in or qualified to evaluate it). There, in the scientific arena, it is open to further scrutiny and testing. Other scientists can check this idea, evaluating whether it might indeed be a valid description of nature. How do scientists do this? By banding together and trying to prove it, as many people might believe? No, it is precisely the opposite: scientists try to knock it down; they try to show that it is not correct. In short, they try to falsify it. Scientists are not striving to prove something, because they realize that certainty is unattainable.

So how does science advance our knowledge? Before we get to this, let's consider why other scientists would want to test another's ideas. There can be many reasons. Some scientists may disagree, on the basis of their own observations and research. They may have competing ideas that they either have already proposed or are in the process of developing. They may simply doubt a scientist's results based on their understanding of the way nature works. Being human, scientists also sometimes dislike one another, they may be envious of each other, or they may be competing for a limited pool of research funds.

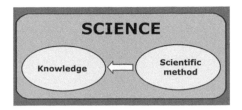

1.2. Science is a human activity that is both a body of knowledge (scientific knowledge) and a way (scientific method) of investigating natural phenomena. The method is peculiar to science and results in the knowledge we consider scientific. As explained in the text, there isn't a single "scientific method," although most people are unaware of this. The important aspect of scientific methodology is that it provides us with an objective or unbiased way to investigate natural phenomena.

Whatever the reasons (and we hope that they are more often than not of the nobler variety), the point is that ideas are tested. When scientists formulate ideas, they are careful to state them in such a way that they are open to testing or to being falsified. If a proposal is not open to falsification, then it is not considered scientific. This is why creationism (whether labeled as scientific creationism or intelligent design theory) is not and can never be scientific. These attempts begin by asserting that creationism and its tenets are true based on faith. If there is evidence that refutes any part of creationism (say, that the earth is only a few thousand years old), then it is the evidence that is wrong, because creationism must be true. Faith, by definition, is not open to testing, otherwise it wouldn't be faith. Science accepts only natural explanations because supernatural explanations are not testable. The instant we resort to a supernatural explanation, we have stopped doing science altogether.

Science is grounded in doubt, rather than faith (Tattersall, 2002), and so ideas must be testable. When a testable idea, technically termed a hypothesis, resists falsification, scientists' confidence that it is an accurate description of nature is strengthened. We can reformulate the sequence of events given above into the more formal (and stereotypical) flowchart (Fig. 1.3) that most students are exposed to during the primary and secondary school years.

This is the classic scientific method that so many of us have grown up with. This stereotype, however, is yet another hurdle to overcome, for students come away with the idea that it is *the* scientific method. Educators rarely go beyond it to explain that this method, the experimental method, is not the only one. It is critically important in many scientific fields, particularly those amenable to investigating phenomena in artificial environments, such as a laboratory, where the researcher has control over which and how many variables (or factors) are manipulated. Examples are much of physics and chemistry. However, it is unsuitable for investigating many scientific questions, such as those posed by much (but clearly not all) of cosmology, biology, geology, and meteorology — and paleontology.

These disciplines, which rely on other equally valid methods, are usually considered historical sciences because they investigate phenomena resulting from past events that cannot be recreated in a laboratory setting. Although aspects of these sciences are important components of school curricula, it is the experimental method (which is inappropriate for these sciences) that is typically presented to students. If students fail to make the link between the material they are learning and the method they are told is used to investigate scientific questions, we should not be surprised — the correlation does not exist! A volcanic eruption or the movement of the earth's crust cannot be recreated in a laboratory; certainly, they may be modeled, but that is not the same thing, as explained in Chapter 9. Neither can the beginning of the universe, nor the interactions among the factors involved in ecological settings. For example, in many cases there is no way to experiment with an ecological system that does not also alter the system. If we wanted to know what would happen to a system if one element — say, a top carnivore, were eliminated — we could not do this experimentally. It would be neither ethical nor desirable to exterminate the species in

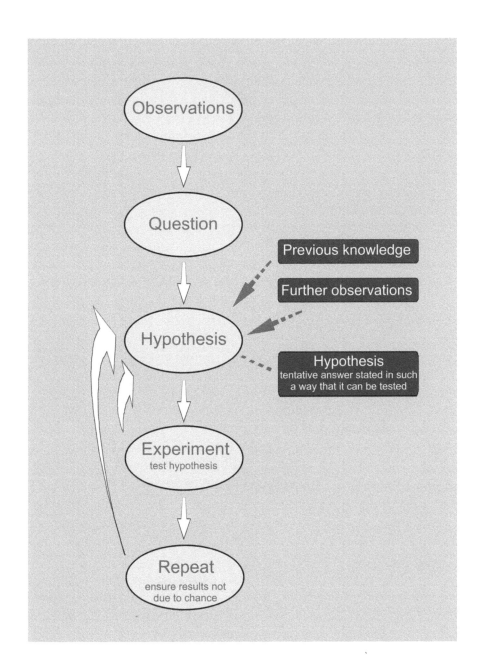

1.3. A flow diagram outlining the steps in the experimental method of scientific investigation. This represents the stereotypical view of how science is carried out, but in reality this is the general pattern followed in the experimental method, where a hypothesis is amenable to being tested by experiment. However, not all scientific hypotheses can be tested by conducting laboratory experiments, in which variables can be strictly controlled. Such testing is not easily applicable to scientific investigations of historical phenomena because the conditions that produced the phenomenon under investigation cannot be artificially duplicated. The appropriate methodology applied to the investigation of historical phenomena, where the "experiments" have been already conducted by nature, is the comparative method.

question to determine the outcome of its absence. In such realms of study, scientists are presented with outcomes that are the result of historical and therefore nonrepeatable events. The experiment, so to speak, has already been performed, in this case by nature, and it is the task of the scientist to unravel the conditions and factors responsible for producing the "results." Scientists involved in such questions use a methodology called the comparative method.

The comparative method, then, is used to investigate situations—or natural experiments—where variables cannot be manipulated because they occurred in the past, because they are impossible to manipulate, or because it is unethical to do so. It is termed comparative because the researcher investigates by comparison of situations that differ in the variable of interest. It is often necessary to set up classification systems to accomplish this. If we are interested, for example, in comparing habitats, then categories such as "top predators" and "physical factors" must be defined and described so that similarities and differences may be recognized in the first place. The researcher proceeds by observing the outcomes of the natural experiment based on the varying components of the different systems under study—in other words, by reconstructing what must have been the conditions that produced the outcomes. Let's consider a simplified scenario to illustrate how the comparative method may be used to test hypotheses. In our example, we are investigating an ecosystem that contains a top predator such as a tiger and a potential prey species such as an antelope. We have noticed a dramatic decline in the numbers of antelope and hypothesize that the tigers must be responsible. How can we test this hypothesis? One way is to seek another ecosystem where tigers are scarce or absent. If the antelope are in decline there as well, it suggests our hypothesis is wrong. If the antelope population is in good shape, however, it suggests (but does not prove) that our hypothesis might be correct.

Although many aspects of biology and other sciences rely on the comparative method, this does not mean that all activities in these fields do. For many biological disciplines, such as physiology, anatomy, and molecular biology, the experimental method is clearly applicable. Others, such as ecology, systematics, and paleontology, normally rely on the comparative method. Many of the results discussed in this book, therefore, were obtained by means of the comparative method. In some cases, though, a combination of comparative and experimental methods were used, such as in the analysis of functional anatomy.

Before we leave the subject of the nature of science, we must address yet one more aspect that causes confusion, and this is the difference between fact and theory, and what scientists mean by theory. Many people equate facts with certainty but are less confident about theories. Most believe that there is a gradation between fact, an incontrovertible truth, and theory, as though a theory were not quite as good as fact, but something that is still awaiting proof (Gould, 1983).

A fact is an observable (keeping in mind that observation includes more than just sight) attribute of nature. Let's consider gravity as a familiar example. If an object were dropped from a height, it is possible that it would fall up, but we recognize that the innumerable observations made about what happens to dropped objects makes the probability of such an event infinitesimally small, and we recognize that gravity exists; it is a fact (Gould, 1983). The *theory* of gravity is an attempt to explain why it exists and how it functions, and is decidedly not a discourse on whether gravity exists. Gravity would exist regardless of why we think it exists. Similarly, that evolution has occurred and continues to do so is indicated by numerous observations from many areas of knowledge. The theory of evolution is an explanation of why and how evolution occurs. It may be that some aspects of our ideas on how evolution occurs may be wrong, which is something that only scientific methodology could reveal and correct. More to the point here, however, is that our being wrong would have absolutely no effect on evolution's factuality, which is supported by overwhelming evidence. As another example, consider that when, some four centuries ago, the religious establishment debated leading philosophers on whether to accept or reject the heliocentric explanation for the motion of celestial bodies in our solar system, the planets took no mind and continued to do what they had always done; they did not care in the least what humans thought.

A theory in science has a precise meaning that is distinct from its colloquial use. In everyday language, a theory is something akin to an opinion, as when, in a murder-mystery TV show, one detective asks another for his theory on what happened. In science, a theory is an explanation or a set of explanations for some attribute of the natural world that scientists have come to accept as being the best explanation at a particular time. Scientists do not accept theories because they have been proven to be correct—recall that scientists do not strive to provide proof (although scientists, being human, sometimes slip up in everyday talk and refer to the proof of one thing or another)—but because they have resisted repeated attempts to disprove them. Generally, theories start out as hypotheses that are tested over and over again. When scientists are sufficiently confident of their explanatory power, they come to be considered theories. A further caveat about scientists is necessary. They often speak about their own ideas as theories, when in fact they should refer to them as hypotheses—unless of course, the ideas have been accepted by other scientists as the best possible explanation. To start applying these concepts to the disciplines that study the life of the past, the raw materials of paleontology, the fossil remains of once-living organisms, are discussed in the following section.

Who has not marveled at the discovery of something ancient, or come across dusty family albums in the attic, the staid expressions of our great-grandparents staring back at us, as though wondering what planet they were being observed from? Who has not stood in wonder at discovering an ancient object casually found in a field and reflected on the hand that gave it form? Human beings, like many other mammals, are curious creatures, and questions on origins, on what once was and now is no longer, often stand uppermost in our thoughts.

Fossils and taphonomy

This is probably the main reason why there are professional students of the past, men and women who earn their living scrutinizing the past through the years, centuries, millennia, or eras, according to their own personal interests. In particular, the historian of ancient life is designated by a wonderful word of Hellenic elegance: paleontologist. To decipher its meaning, let's examine its component parts: *palaios* (παλαιός in Greek) means "old," *ontos* (ὄντος) refers to existing things and thus is a roundabout way of referring to life, and *logos* (λόγος) means a treatment or discussion, and by extension the body of knowledge associated with it—or in other words, the corresponding branch of science. Thus, paleontology is the scientific field that is concerned with ancient life.

The paleontologist lives and breathes fossils. The etymology of the word *fossil* is ambiguous. It is certainly derived from Latin, but its precise meaning is not easy to translate. Probably the most appropriate translation is "dug up," referring to the fact that fossils are usually buried and that a pick is the favored, or at least stereotypical, tool of the trade. Actually, in centuries past, the term *fossil* was used to describe a wide array of things that were dug up, such as gems and interesting concretions, as well as objects that we now recognize as fossils (Rudwick, 1985). As we shall see, one South American mammal in particular was important in helping establish what fossils, in the modern usage of this term, represent.

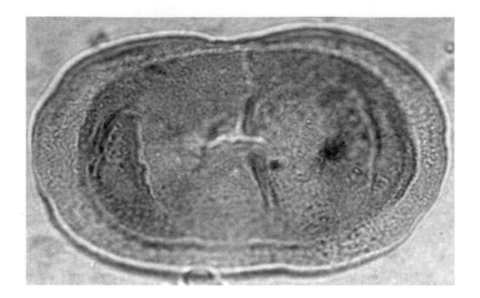

1.4. Fossil pollen: reproductive structure of a 250-million-year-old gymnosperm (cone-bearing plant). Width approximately 40 μ.

Image courtesy of Ángeles Beri.

The erstwhile pick is the tool of choice for most of the cases we discuss in this book, but not all fossils are the same. Most people think of fossils as the lithified remains of bones, teeth, and shells. A good number are, but fossils include much more than animal remains that have been turned to stone. A fossil is defined as any trace, impression, or remains of a once-living organism. Thus, a set of footprints preserved along an ancient mudflat is a fossil, as are tracks, trails, and burrows made by crawling invertebrates, or even more esoterically, boreholes made by predatory organisms in the shell of clams—two fossils in one! Some fossils may be microscopic (Fig. 1.4). For example, the deposition of plant pollen, as with dead plankton in general that sinks to the bottom of a body of water, is almost imperceptible.

The fragmentary remains of small mammals and birds are surrounded by resistant matrix. The use of a pick to recover such fossils, given the good fortune that they are preserved at all, is akin to using a shotgun to scatter flies. Instead, the remains are removed along with a good deal of the surrounding rock (Fig. 1.5)—a rock saw often comes in handy—and then packaged in a protective jacket, usually made of plaster, for safe transfer back to a laboratory. But even when digging out the bones that are normally thought of as fossils, we face a common dilemma: the bones, large or small, are often fragile. Their extraction requires great patience and the delicacy of a brush and needle, rather than brute force; the surrounding matrix must be removed, often grain by grain (in the laboratory, various techniques may be used), and the fossil itself must be coated with frequent applications of a dilute cohesive that penetrates its substance (Fig. 1.6). The old saying that paleontology could not exist without glue is an exaggeration, but it is not too far off the mark.

However, before the wondrous moment when the remains of an animal or plant from an age long past are extracted, a seemingly miraculous series of events must occur. The scientific field that studies these events is called taphonomy, which refers, according to the paleontologist I. A. Efremov (Fig. 1.7; also a great novelist), who coined the term, to the laws of burial or, more expansively, the study of the transition of remains from the biosphere (the realm of living organisms) to the lithosphere (rocks or the

1.5. Recovery of a fossil. (A) Fossil remains being extracted along with the surrounding rock from the early Miocene Santa Cruz Formation along the Atlantic coast of Argentine Patagonia. (B) The block (including the fossil and surrounding rock) is then packaged in a protective jacket for safe transport back to a laboratory, where further preparation is carried out.

inanimate world). If something is to be transformed into a fossil, then the natural recycling of material must be interrupted at some point. Usually, nature recycles all the materials previously incorporated into once-living organisms (Fig. 1.8) because the earth's resources are limited and we cannot count on meteors and comets for perpetual deliveries of extraterrestrial supplies. Even though these resources exist in great quantities, they would be exhausted with the inexorable passing of time. This does not occur, though, because the lithosphere receives continuous transfer of material from the biosphere.

Fossilization may appear to be as nothing less than miraculous, the outcome of an improbable series of events. It occurs, however, because of the uncountable number of opportunities that are available without pause during the extremely long periods of time that nature takes in going about its business. For an organism to become fossilized, it must first die in such a way that it avoids being completely devoured by another animal (not such an easy trick in the real world), and then further hope that its more resistant parts are not feasted on by putrefying bacteria. One common way

1.6. In the laboratory, preparator Leonel Acosta of the Museo de La Plata (Argentina) patiently picks away the matrix surrounding the fossil remains. Note the bottle of glue in the right foreground of the image.

Courtesy of the Museo de La Plata, Argentina.

for this to happen is to be buried quickly, but even in such cases, not all bones that survive postmortem events reach a sedimentary environment conducive for preservation—in other words, it depends on where they are finally laid to rest. Fluvial sediments are often a good bet for a bone bent on becoming a fossil. On the other hand, for example, we have a relatively poor record of forest-dwelling animals because the acidic soils of forests lead to disintegration of buried bones.

The different parts of an organism have different potentials for fossilization. Fortunately for us, we share a common interest with most organisms. They happen to require protective and supportive structures, which, as a result of the nature of their functions, are quite resistant. We, on the other hand, can discern a great deal from such structures, when and if we can find them. Protective and skeletal structures are formed by a few primary materials common in the biological world. The shells of mollusks, for example, are made of calcium carbonate, a material with great potential for fossilization.

It is not mere coincidence that the bones of vertebrates are those most commonly associated with the word *fossil*. The calcium phosphate that forms them is stupendously well preserved. Similarly, this occurs with teeth, which are the hardest elements of the vertebrate skeleton. The most common examples of fossils are often formed by petrification, which results in stony or lithified remains, and may occur by two processes. Permineralization occurs when soft structures decay while water containing dissolved minerals works its way into every nook and cranny of the hard structures.

Minerals are then deposited in these cavities and pores, resulting in a fossil that still contains a good deal of its original hard parts. Replacement occurs when water dissolves the original material forming the hard structures and replaces them with minerals. This may happen so slowly that the new minerals duplicate almost exactly the shape and structure of the original organic material (Ausich and Lane, 1999).

Other readily fossilized substances include silicon, which forms the coverings of diatoms and other microscopic organisms; chitin, present primarily in the covering of insects and other arthropods, but also in lichens, earthworms, and other invertebrates; and cellulose, which provides rigidity to the cell walls of plants. Once in a while, the impossible happens: soft tissues, the daily bread of bacteria and predators, escape decomposition and present us with precious information that is almost always missing in the fossil record, as can be seen in the skin and feces of a ground sloth in Fig. 1.9 (see Chapters 5 and 7). Even more remarkably, once in a long while, an organism may be so completely preserved that its last meal is too, as occurs in deposits from near Messel, Germany.

The process of decomposition may be interrupted or altered in the absence of oxygen. This is the case in natural mummification, when the environment is too dry and hot for bacteria to carry on their activities. One example is the famous mammoths of Siberia, some so well preserved in ice that a banquet, with a main course of mammoth trunk, was served at an academic gathering. At least that's the way this story is usually told. However, it may be apocryphal and based on the assumption that the mammoths were frozen extremely rapidly in some sort of extraordinary climatic catastrophe. The famous paleontologist Björn Kurtén suggested that what probably happened is that one member of the mammoth-hunting expedition made a heroic attempt to consume a bit of mammoth meat, but could not quite get it down—it seems that the carcasses had been partly decomposed before freezing (Farrand, 1961). Still, the flesh was so well preserved that someone might actually have tried.

Soft tissues may also be preserved by substitution of the organic matrix by minerals (Fig. 1.10). At times, this happens in so perfect a manner that the nature of the original structures may be studied microscopically. Wonderful examples of this sort of preservation occur in lower Cretaceous rocks (dated at about 125 Mya) of the Yixian Formation of Liaoning Province, China. These deposits preserve not only such delicate structures as skin and insect wings, but also feathered dinosaurs, findings that have revolutionized our idea of how some dinosaurs looked and lived.

Fossil tree trunks, which are not overly rare, may also be subject to such fine preservation. In Patagonia, there is a petrified forest, declared a national monument, that preserves the trunks, branches, and even fruit in exquisite detail. There is also a petrified forest in Arizona, and the conifer trunks in Yellowstone Park are petrified.

Further, fecal material may be fossilized under the right conditions. These fossils, termed coprolites, are invaluable because they can reveal dietary information, such as pollen, fiber, bones, and shells. The coprolites of several fossil sloth species have been recovered (Fig. 1.9).

Paleontologists' traditional ideas of how fossilization occurs involve the replacement of the original tissue, including soft tissues, by minerals to

1.7. Ivan Antonovich Efremov (1907–1972), pioneer in the field of taphonomy, during an expedition to Mongolia in 1949.

Image from the Paleontological Institute of the Russian Academy of Sciences, courtesy of Sergey Rozhnov and Valeriy Golubev.

1.8. (A) The recently dead remains of a guanaco from the steppes of Argentine Patagonia. The flesh and most of the skin have already been re-cycled by nature, but the skeleton is almost intact. Many people believe paleontologists routinely find such complete and well-preserved skeletons as fossils. However, such finds are exceedingly rare. (B) Much more commonly, the bones of animals are disarticulated, scattered about, eaten, or otherwise destroyed, as is shown in this image of a fossil sloth from the Pleistocene of Neuquén, Argentina.

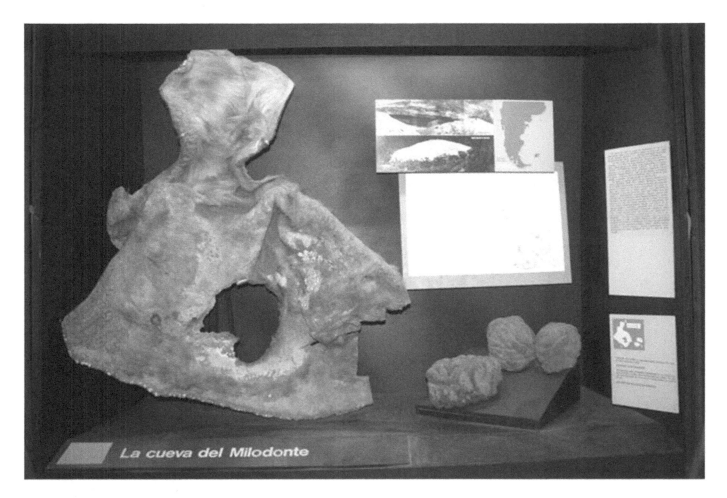

La cueva del Milodonte

produce a hard or lithified representation of the original. Several remarkable recent discoveries hint at other methods of preservation that suggest that fossilized remains need not be restricted to lithification of structures. Remains of what seem to be Mesozoic soft tissues of dinosaurs have been reported, essentially like those of modern species (Asara et al., 2007; Schweitzer et al., 2005, 2007). In addition to preservation of tissue, scientists have uncovered preserved protein sequences—these are real molecules present in the organism during the time it was alive! Fossilization of such materials was once thought to be impossible, as scientists could not conceive of a way that such fossilization could occur. Although the fossilization method is still unclear, it appears, at least in the case of tissue preservation, to have something to do with postmortem formation of polymers, long-chained molecules that become inert and thus resistant to further chemical changes.

But it does not end here. Fossilization requires still further luck. Once buried, structures still face the possibility of destruction from the pressure of overlying sediments. Finally, there is erosion, the paradoxical ally of the paleontologist. Erosion exposes a fossil originally buried deep within sediments, but there is only a small window of opportunity, for if the remains are not recovered, the same forces will destroy them.

A fragment of an organism is exposed to and must survive all of these risks before it may finally rest in relative peace in the drawers of a collection. Then it must be studied, but here the adventure becomes, to use Einstein's definition of science, one of thought. These adventures are referred to

1.9. A portion of the hide, including some fur, and dung of *Mylodon darwinii* from Última Esperanza cave, Chile, exhibited at the Museo de La Plata, Argentina.

Courtesy of the Museo de La Plata, Argentina.

1 mm

1.10. A portion of lithified pebblelike skin of an embryonic sauropod dinosaur from Auca Mahuida, Neuquén, Argentina.

Image courtesy of Rodolfo Coria, Museo Carmen Funes, Plaza Huincul, Neuquén, Argentina.

below and in Chapter 2, exemplified through the activities of the people who have dealt with the fossil animals that are the subject of this book.

Linnaeus and classification

Taxonomy is the branch of the biological sciences that establishes the rules for and carries out the classification of living organisms. Classifying things, be they animate or inanimate, is an essential practice that helps humans keep track and make sense of the numerous objects and concepts that we must deal with daily. We can classify most inanimate objects fairly easily on the basis of their unchanging features. Nearly anyone living in a modern industrialized society would be able to pick out a van from an SUV, pickup, jeep, convertible, sedan, hatchback, station wagon, and so on. The reason is that we all agree on the features that distinguish these different subcategories of the category motor vehicles. If we were to switch languages, say from English to Spanish, then we would simply have to learn a new vocabulary (well, and grammar too . . .).

Living things also need to be named. For many purposes, their names are not terribly important—it doesn't much matter if people refer to their local species of chickadee as the black-capped chickadee, the boreal chickadee, or simply call it a chickadee, especially if it's the only one in the area. For formal scientific purposes, however, the naming of organisms is critically important and not as straightforward. Further, as we shall see, we cannot use characters to define groups of organisms.

Scientists face two problems. One is naming all the different units or kinds of life, technically known as species, and making sure that no two closely similar species have the same name. When scientists from different parts of the world communicate about a species, they need to be sure that they are talking about the same thing. It may sound trivial, but the vast diversity of life forms can lead to considerable confusion. In some cases, a single species may have many informal local or common names (not due solely to the fact that scientists from different countries speak distinct

CAROLVS LINNÆVS KNIGHT OF THE POLAR STAR FIRST PHISICIAN TO THE KING PROFESSOR OF BOTANY IN THE UNIVERSITY AT UPSAL &c &c

1.11. Portrait of Carolus Linnaeus (Carl von Linné, 1707–1778), by Swedish artist Alexander Roslin, held in the Nationalmuseum, Stockholm. His system for naming organisms revolutionized the organization of living things and remains highly influential. His most lasting legacy is the practice of the use of two names—thus the term *binomial nomenclature*—to designate a species.

Courtesy of Wellcome Library, London.

languages), and, at the other extreme, different species may have the same common name. The second problem for scientists is how to organize the bewildering number of living and extinct species so that a broader understanding of life, including interactions and relationships among its components and how it arose and diversified, can be achieved.

Successful scientific resolutions to these problems began about 250 years ago with the work of the Swedish naturalist Carl von Linné (1707–1778; Fig. 1.11). Better known to us as Carolus Linnaeus, he served stints as army doctor and physician to the king, but his main calling, during and after duty to king and country, was as a botanist at the universities of Stockholm and Uppsala. He also served as president of the Academy of Sciences.

Linnaeus solved the vexing problem of naming species, bringing stability to what had been a chaotically confusing profusion of names, by devising a system based on binomial nomenclature. Since then, each species has been christened according to this system, with names rigorously derived from Latin or latinized colloquial words. (More recently, this rule has been relaxed, and names need no longer conform to Latin.) In Linnaeus's system, the name of each species has two parts (hence, binomial nomenclature): a generic name and a specific epithet (or trivial name). The first letter of the generic name is capitalized, and both the generic name and specific epithet are italicized or underlined to set them off from surrounding text. For example, two closely similar species are the wolf, known scientifically as *Canis lupus*, and the coyote, known as *Canis latrans*. These species are in the same genus but have their own specific epithets. Although the naming of species, or taxonomy, is governed by a complex set of rules and requires knowledge of a specialized vocabulary, the important point is that binomial nomenclature allows unambiguous communication among scientists.

A second concern for biologists is that simply naming individual species is not sufficient; it is not of much help in organizing the vast diversity of life forms. The discipline that deals with the more encompassing aspects of organisms' biology is termed systematics. It is distinct from taxonomy, which concerns itself with the naming of organisms. Systematists are therefore concerned with all biological aspects of a group of organisms, including the relationships among them and their way of life. Linnaeus realized this as well and tackled the problem by grouping species into higher categories on the basis of similarities and by assigning groups a rank as indication of the degree of distinctness from other groups. For example, he included such mammals as cats, hyenas, dogs, wolves, jackals, bears, and weasels in the Carnivora and gave this group the rank of Order. In doing so, he implied a degree of difference from mammals in other orders, such as the Artiodactyla (though we tend these days not to use this term as a formal name), which includes cows, sheep, goats, pigs, deer, moose, and camels. All the different groups of mammals were collected together under the Class Mammalia. Class is a higher level rank than Order, signaling that groups of organisms in different classes were separated by a greater degree of difference than those included as orders of the same class. As an example, crocodiles, turtles, lizards, and snakes were grouped into Class Reptilia, implying a markedly different form and way of life than those classified under Mammalia. Similarly, the Class Pisces, containing all the fishes, emphasized the distinctness of its members from those of other classes. This method led naturally to the use of characters to define species or higher categories.

Linnaeus's resulting classification, the Systema Naturae (System of Nature), organized the diversity of life into a hierarchical pattern of decreasing inclusiveness: the further down the hierarchy, the fewer the members. Linnaeus's methods met with resounding success, mainly because his system was vastly superior to those proposed by other naturalists. It remained essentially intact until just a few decades ago, even surviving Darwin's intellectual upheaval. Most readers are familiar with the primary categories in common use since the early 1800s: kingdom, phylum, class, order, family, genus, and species. In practice, biologists have availed themselves of intermediate categories by using prefixes (e.g., super- and subfamilies) or entirely new categories (e.g., tribe, cohort).

Linnaeus's efforts were deeply grounded in the Western cultural and social fabric of the day: decidedly Christian and carried out to demonstrate

God's greater glory and wisdom. He was, however, an excellent biologist, and in most cases, he was able to identify the important similarities that distinguished groups of organisms from each other. With the near-universal acceptance of evolution among biologists soon after the publication of Darwin's (1859) *The Origin of Species by Means of Natural Selection, or The Preservation of Favoured Races in the Struggle for Life*, Linnaeus's groupings remained largely unaltered, but the cause behind the hierarchical pattern of diversity was very different indeed. Among Darwin's revelations was that all life shared a common descent and that the diversity was due to successive branchings of species over time (although we also recognize that in some cases new species may be created by combining two existing species). The hierarchical pattern of Linnaeus's classification was now considered to reflect an overall similarity that was due to the common ancestry of each of the different groups. Groups of organisms that shared a relatively recent common ancestor and tended to resemble each other closely fell together in a lower rank, whereas those sharing fewer similarities had a more remote common ancestry and fell together only in a higher rank. Thus, *Canis lupus* and *Canis latrans*, which are similar to each other, share a relatively recent common ancestor and are in the same genus. *Canis* and *Felis*, the genus including several smaller cats, are less similar. They share a common ancestor in the more distant past and are grouped together in the Order Carnivora. The Carnivora and the Order Rodentia (rodents) are even less similar, share an even more remote common ancestor, and are united in the Class Mammalia.

Within the last few decades, however, the classification of organisms has undergone a major restructuring, mainly as a result of the relatively recent methodology of recognizing evolutionary (or genealogical) relationships among organisms known as phylogenetic systematics. At first glance, much of the old system seems to have survived this recent intellectual development. For example, many of Linnaeus's groups and their names and ranks seem the same. However, there are several key differences.

One concerns the method by which organisms are classified. As noted above, the older classifications were based essentially on general overall similarity. However, it turned out that the use of overall resemblance often led to misleading interpretations of the evolutionary relationships among organisms. Specific features were identified as more important than others, but decisions on which were important enough were usually left to an authoritative researcher regarded as an expert on a particular group of organisms. As might be expected, this was a rather subjective way of going about things.

Biologists began to grasp that organisms cannot be defined, in contrast to inanimate things, on characters. Instead, groups of organisms are defined on ancestry. A meaningful classification can only be based on ancestry, because characters evolve and some members of a group may lack one or more defining characters. For example, Tetrapoda (or Stegocephalia, according to several researchers; De Iuliis and Pulerà, 2010) includes all vertebrates with four limbs. If we were to adhere strictly to this definition of this group, then snakes could not be included. In practice, biologists have long realized that snakes share ancestry with other reptiles and have been classified with them and other vertebrates as tetrapods. Biologists do, of course, make use of characters, especially to infer what the evolutionary relationships are (and to diagnose—as opposed to define—groups, which helps us recognize them), but it took some time before they understood that the only characters useful in delineating evolutionary relationships are those that are modified in an ancestor species and then inherited by its descendant species. Unmodified characters present in the ancestor and passed on to descendants are not useful.

For example, dogs, horses, and kangaroos all possess hair, a universal mammalian condition presumably present in and inherited from the earliest mammalian ancestor. One task of biologists is to understand the evolutionary relationships among organisms. At this level, the presence of hair tells us that these three mammals share a common ancestor and belong in the same group. What about the relationships among them? A biologist would want to know whether the dog is more closely related to the horse or the kangaroo. In this situation, the presence of hair is of no help. A useful character would be one that is present in only two of the three—say the dog and horse, but not the kangaroo. Such a character would be evidence that the dog and horse share this particular feature because it was derived from a more recent common ancestor. Scientists, not surprisingly, refer to these sorts of characters as shared derived characters. A derived character is technically termed an apomorphy, and a shared derived character is technically termed a synapomorphy. The more general presence of hair is a primitive (or, more appropriately, an ancestral) character, technically termed a plesiomorphy. A shared plesiomorphy is a symplesiomorphy. A group that includes an ancestor and all of its descendants is a natural or monophyletic group (also termed a clade). When scientists present classifications, they are actually proposing hypotheses on the phylogenetic (or genealogical) relationships among the organisms under study. Further data (for example, the discovery of new fossils) can cause a revision of phylogeny. The critical point is that in carrying out phylogenetic studies, scientists are attempting to reconstruct the unfolding of life. Over the course of evolution, new species have typically come into being through descent with modification from ancestors – an ancestral species split to produce new species. We are trying to reconstruct or discover the pattern of this descent. As this pattern of descent is a historical process, it happened only one way. For example, if we are dealing with three species, A, B, and C, then the possible relationships among them are that species A and B are derived from species C, or species A and C are derived from species B, or species B and C are derived from species A. However, only one of these possibilities actually did occur – another way of thinking about this is that the Tree of Life only occurred once. A monophyletic group is therefore not an arbitrary construct—it represents what we accept (until we can demonstrate otherwise) as the true pattern of the unfolding of life. Two groups, be they species or some higher and more inclusive taxa, that are each other's closest relatives (because they share a common ancestor to the exclusion of other species) are termed sister groups.

The reconstruction of genealogical relationships among organisms, or phylogeny, along these guidelines led to a clear idea of how to represent these relationships in the classification of organisms: any grouping of organisms must reflect the genealogical history of its members. All the organisms included within a group must be more closely related to one another than to organisms in other groups. Another way of saying this is that a group must include an ancestor and all of it descendants. This may seem obvious, but classifications began to apply this principle consistently only beginning about 30–40 years ago. Before this, much of our classification was based on symplesiomorphic rather than on synapomorphic characters, or on some combination of the two.

An example of this older method of classifying organisms is that the Class Reptilia encompassed a wide array of organisms including not only the living turtles, lizards, snakes, crocodiles, and their extinct relatives (dinosaurs, pterosaurs, and plesiosaurs, among others), but also the extinct relatives of mammals. Most of the latter, traditionally termed the mammal-like reptiles, were almost certainly more reptilelike in their overall appearance and way of life, but they nonetheless share a more recent common ancestry with the mammals than with the other reptiles. Not only this, but the older system excluded some of the descendants of the reptiles: birds, certainly sharing common ancestry with turtles, lizards, snakes, crocodiles, dinosaurs, pterosaurs, and plesiosaurs, were grouped on their own in the Class Aves; and mammals, sharing ancestry with the mammallike reptiles, were likewise separated as the Class Mammalia. For both of these groups, the ancestor was not grouped with them. To rectify this, the mammallike reptiles were removed from the Reptilia and grouped with their living and extinct mammalian descendants in the group Synapsida, and the Aves were included within the larger group of reptiles.

Such groups as the old concept of Reptilia is an example of an artificial group—it is not a natural group because it does not include an ancestor and all of its descendants. In this situation (when a group includes an ancestor but not all of its descendants), the group is described as paraphyletic. Another kind of artificial group is termed polyphyletic. In this case, the group includes organism that have more than one ancestral source. Polyphyletic groups place together organisms on the basis of a few characteristics that were derived independently (rather than inherited). An example is an older classification that recognized elephants, rhinoceroses, and pigs (among others) as belonging to a group – Pachydermata – based on the presence of a thick skin. This sort of similarity, in which organisms share features that are independently or convergently evolved, is termed homoplasy, and it can often be a source of confusion in reconstructing phylogeny.

One final, and more recent, development is that the Linnaean practice of assigning formal rankings to groups is beginning to be abandoned. The reason for this is simple: if the history of life proceeds by successive branchings of species to produce a tree of life, then there is no objective way to equate one group of organisms in one part of the tree with any other group in another part of the tree. Thus, there is no point in trying to say that Mammalia and Reptilia are equivalent and deserve special recognition as classes. Systematists are therefore starting to simply refer to groups of organisms by name, without any attempt at recognizing rank.

Paleontology is the intersection between the areas of knowledge that study life and the earth. A substantial part of that confluence, needed to understand the relationships between rocks, organisms, and time, is stratigraphy.

Steno and stratigraphy

This branch of geology deals with strata—those sedimentary rocks that lie in beds or layers. Sedimentary rocks are rocks formed by the settling, consolidation, and lithification of sediments (particles that are carried or suspended by wind or water). Sediments are formed by the weathering of previously existing rocks, minerals precipitated from solution, or minerals used by organisms to form skeletons. There are different aspects of stratigraphy, such as those concerned with the rocks themselves and their features, called lithostratigraphy (λίθος, "stone" or "rock"), those that study the correlations established by the shared fossil contents of rock bodies, or biostratigraphy (βιος, life), and those that have to do with the time in which the rocks were formed, or chronostratigraphy (χρόνος, "time"). Correlation is a crucial concept, essential to this whole discipline. It can be defined as the establishment of spatial and, especially, temporal correspondence between separate stratigraphic units. Although it can be said that the whole earth is stratified, in the case of rocks, the strata are recognized on the basis of discontinuities due to changes in the conditions in which they were deposited, as well as the effects of erosion.

The principles of stratigraphy were proposed by the Danish anatomist, geologist, and bishop Niels Stensen, better known, as noted by Rudwick (1985), by his literary name Stenonis, anglicized as Nicolaus Steno (1638–1686; Fig. 1.12), and they remain the fundamental concepts of the discipline (Ausich and Lane, 1999):

> The principle of original horizontality. As most sedimentary particles settle out from fluids by gravity, the sediment is deposited in layers that are nearly horizontal and parallel to the surface on which they are accumulating.
> The principle of strata superposition. In a sequence of undisturbed strata (or layers), the oldest layer is lowest in the sequence (at the bottom) and successively higher layers are successively younger.
> The principle of lateral continuity. Strata extend in all directions until they terminate by thinning out at the edge of a basin, end against some former barrier to deposition, or grade into a different kind of sediment; according to this, if a layer does not continue, we may assume that displacement of strata or erosion has occurred, and that the strata can be followed at some other location, as on, for example, the other side of a gorge. Clearly, the disruption of strata must occur after the deposition of strata.

To the modern reader, these three principles may seem trivial, but in Steno's time, they were revolutionary ideas and formed the basis of the theoretical framework that began allowing geologists to unravel the geologic history of the earth. In most cases, strata are not neatly arranged. Rather, they are often folded and distorted; sometimes a sequence may be entirely overturned. However, by following Steno's principles, geologists could begin to work out the historical sequence of events that must have occurred to produce a particular configuration of strata. This practice began in earnest during the nineteenth century, when geology was the science in vogue, and continued into the twentieth century, but two other theoretical concepts were necessary before geologists could understand the sequence over which life on earth had unfolded. (We choose the word *unfolded* carefully, rather than saying *evolved*, because most geologists, particularly during much of the nineteenth century, did not accept biological evolution.)

1.12. Portrait of Nicolaus Steno by J. P. Trap. His pioneering efforts in recognizing the fundamental principles of how rock layers are formed paved the way for the geologic sciences.

1.13. Portrait of James Hutton by Sir Henry Raeburn (before 1797) in the Scottish National Portrait Gallery. The Scottish physician and geologist viewed the earth's history as a dynamic, ongoing process, in contrast to the then commonly held view that the earth had remained essentially unchanged since its creation by the Creator. Hutton's genius lay in understanding that the application of the same processes that occur in the present hold the key to unraveling the historical development of the earth. His methodology, termed *uniformitarianism* or *actualism*, has been much maligned recently, as it smacks of gradualism, and many events that occur today and contribute to the geologic structures of the earth are anything but gradual—a volcanic eruption, landslide, or asteroid impact, for example. However, these sorts of episodes are also ongoing process and so can still be accommodated under Hutton's methods.

One was the principle of uniformitarianism or actualism, formulated by the Scottish physician and geologist James Hutton (1726–1797; Fig. 1.13), who viewed the earth as a dynamic, ever-changing place: new rocks, lands, and mountains rose as a balance against destruction by erosion and weathering. Hutton's method was to explain the history of the earth by observing the present, reasoning that the same processes that occur in the present also occurred in the past and could be used to explain how the earth's geologic features had come into being. Incidentally, a much more famous geologist, Charles Lyell (1797–1875; Fig. 1.14), made his mark largely by championing and amplifying Hutton's ideas, and his work was among the influences on Charles Darwin's development of evolutionary theory (see Chapter 2). Though Lyell, also Darwin's friend, was able to accept a changing and ancient earth, he did not accept biological evolution until later in his life. At any rate, Hutton's dynamic view of a continuously cycling series of processes that saw the buildup and destruction of geologic features led to his famous remark about the earth, in which "we find no vestige of a beginning, no prospect of an end" (Hutton, 1788). Among his contributions was his ability to correlate similar rock units at different localities; that is, he could say that such strata had been deposited at the same time. However, he had trouble determining whether dissimilar-looking strata were equivalent in age.

1.14. Portrait of Charles Lyell by G. J. Stodart. Lyell became the main champion of Hutton's methods, applying uniformitarianism to explain a wide body of geologic phenomena. He wrote several important and highly influential works, among them *The Principles of Geology* (1830–1833), a three-volume set that was read by and greatly influenced Charles Darwin during his voyage aboard the HMS *Beagle*.

1.15. Portrait of William Smith by the French painter Hugues Fourau. Although Smith worked mainly as a canal builder, he observed that fossil assemblages are specific and succeeded each other in predictable ways, allowing him to establish the idea known as the principle of biologic succession. This was a key insight that allowed geologists to extend the correlation of rock units across great distances. The ability to correlate rock units from far-flung localities greatly enhanced geologists' attempts in arranging the earth's rock units into a global chronologic succession.

The resolution to this last problem was supplied by William Smith (1769–1839; Fig. 1.15), an English surveyor, engineer, and canal builder. The link between canals in England and unraveling the history of life may seem elusive, but before the advent of the steam locomotive and development of the railroads, transportation in England was accomplished mainly by water routes, so that the country was crisscrossed by extensive canal networks (Cadbury, 2000). It was thus Smith's business to know the kinds and thicknesses of rock to cut through to build canals. He realized that stratified rocks occur in a particular sequence and could be identified on the basis of their lithology and the fossils they contain—that is, certain rocks units could be identified because they contained particular assemblages of fossils. Smith was thus able to correlate rock units over large distances. The reason for Smith's success is known as the principle of biological (or fossil) succession: the life forms of earth's history are unique for particular periods of time, and they succeed one another in a regular and predictable way. Such fossil assemblages permit geologists to recognize contemporaneous deposits all over the world. Combined with the principles of stratigraphy, they could be used to arrange the rock sequences in different parts of the world into a chronologic sequence. This combination is known as biostratigraphy (Fig. 1.16).

Armed with these concepts, geologists went to work on piecing together the relative ages of outcrops all over the world by determining the sequence of strata in one place and then correlating it with sequences at other locations. They began to subdivide time on the basis of such sequences, with a sequence in one part of the world serving as a basis for comparison (thus established as the type locality) to strata in other parts of the world. Eventually these nineteenth- and early twentieth-century geologists constructed, in piecemeal fashion, the geologic timescale (Plate 1), which arranged the rocks on earth into their relative temporal sequence or geochronology (Box 1.1). Much of this work, especially the broad subdivisions of earth's history, was completed by the beginning of the twentieth century. Finally, we were able to understand the progress of life, but only in a relative sense. With relative dating, we could tell what came before what, but could not say in absolute terms how much time had elapsed since any particular event. Although several attempts were made to determine absolute time, they turned out to be generally quite off the mark, particularly the deeper in time we went.

Attaching a value in real time to events in earth history had to await the discovery of radioactivity in the final years of the nineteenth century and then the development of techniques allowing the application of this concept to the dating of rocks. Although it had been shown in the first decade of the twentieth century that it was possible to date radioactive rocks, it was not until the 1940s and 1950s that methods were developed to allow the principle of radioactivity to be used in a consistent and systematic manner. Essentially, there were two developments during this period. One was that allowing organic matter to be dated directly on the basis of the radioactivity of an isotope of carbon; the other, that radioactivity of several other elements in igneous rocks—those formed by the solidification of molten rock or magma—would provide a reliable way of dating the formation of the rock itself.

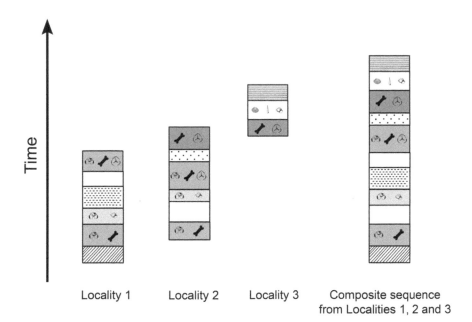

Locality 1   Locality 2   Locality 3   Composite sequence from Localities 1, 2 and 3

1.16. Schematic diagram showing how biostratigraphy is used to piece together local sequences of rock units into composite sequences encompassing greater amounts of time. Fossil-bearing units are represented by symbols, indicating specific fossil assemblages. The units without symbols represent units that lack fossils. Units may vary in thickness, reflecting differential deposition or erosion in different localities. By comparing the various sequences in Localities 1–3, geologists can learn much about overall time and the events that occurred in specific places. For example, by comparing the sequences in Localities 1 and 2, we note that two non-fossil-bearing strata are present between the top and middle fossil-bearing units in Locality 1. In Locality 2, however, the non-fossil-bearing units are missing between the two middle fossil-bearing units. We may surmise from this that the time during which the missing strata were deposited is not represented at Locality 2. If we only have knowledge of Locality 2, we might mistake the absence of the units as indicating that no time had elapsed between the deposition of the two middle fossil-bearing units. However, we know better because of the evidence from Locality 1. We can explain the absence in Locality 2 by supposing either that deposition did not occur in Locality 2, or that the deposition did occur but the strata were subsequently eroded.

Radioactive dating works on the facts that some isotopes of certain elements decay and that the rate of decay is constant and can be measured. An isotope is a naturally occurring form of an element—for example, carbon exists in various forms, the most frequent being ^{12}C and ^{14}C. Some, but certainly not all, isotopes decay—that is, they change into a stable isotope of the same element or another element altogether. The common ^{12}C is not radioactive and is therefore stable; it remains ^{12}C. But ^{14}C is radioactive, and so it decays over time, eventually changing into ^{14}N. We say that the parent (unstable) isotope decays into its daughter (stable) isotope. Radioactivity refers to the fact that decaying isotopes emit energy as they change from one substance into another (because, in the process, matter is converted to energy), but this is incidental to our discussion. We are interested in how the phenomenon can be used to date rocks and organic material.

The rate of decay for any particular isotope is constant and is termed its half-life. A half-life is the time required for half of a parent isotope to decay into its daughter isotope. Thus, if we start at time zero with a sample containing 100% of the parent isotope, after one half-life, there will be 50% parent and 50% daughter isotopes. After another half-life has passed, there will be 25% parent and 75% daughter isotopes, and so on. Dating involves calculating the ratio of parent to daughter isotopes, except in the case of carbon. An important condition is that we need to be sure that the sample being dated was composed initially of the parent isotope only, and for this reason, the rocks most used in estimating ages for fossil remains are those of igneous origin. The volatile nature of molten rock disrupts any parent–daughter isotope association, and they go their separate ways. When molten rock begins to cool and solidify, any coalescing radioactive isotope will consist entirely (or almost so) of parent isotope. This is time zero, and we can be assured that the dating of such a sample also represents the time of solidification, and hence formation, of the rock. The reason this is worth remembering is that we can date any rock fragment that contains radioactive isotopes, but that gives us the time elapsed since the particular rock fragment was formed. The fragment can, at some later time, become

Table 1.1. Examples of parent–daughter isotope series

Parent–Daughter Isotope	Half-life	Materials
^{238}U-^{206}Pb	4.47×10^6 Ma	Minerals (zircon, monacite, rutile)
^{87}Rb-^{87}Sr	48.8×10^9 Ma	Whole rock
^{40}K-^{40}Ar	1.25 Ma	Minerals (mica, clay) and whole rock
^{40}Ar-^{39}Ar	1.25 Ma	Minerals (mica, amphibole, clay)
^{147}Sm-^{143}Nd	106 Ma	Whole rock
^{14}C-^{14}N	5730 years	Organic matter, skeletons with carbon

incorporated into another rock, such as, for example, a layer of sandstone. But the age of the rock fragment does not provide us with the age of the sandstone. If we are interested in any fossils in the sandstone, which were deposited as the sandstone was being formed and thus are as old as the sandstone, then we have to find a way of estimating the age of the sandstone. The most straightforward way of doing this is to find a layer of igneous rock above and another below the sandstone, and date these two layers. As the two igneous layers bracket the sandstone, the age of the sandstone will be between those of the igneous layers. Of course, if fossils happen to be preserved in an igneous layer, then dating the layer also gives the date of the fossils. There are several parent–daughter series that are commonly used, each having a different half-life and thus applicable to different rocks and time, as shown in Table 1.1.

The ages of many of the fossil mammals we discuss in this book are based on carbon dating, which dates organic material rather than rock. This particular method is based on the fact that living organisms incorporate carbon into their component parts (such as the skeleton) in the same ratio as present in the atmosphere. Thus, the carbon in the bones of a mammal will be composed of ^{12}C and ^{14}C in the same ratio in which these isotopes are found in the environment. As soon as the mammal dies, it stops incorporating any new carbon (and anything else, of course). The ^{12}C remains stable, but the ^{14}C, being radioactive, starts to decay, so that over time the ratio of these two isotopes in its bones will differ from the ratio in the atmosphere. On the basis of the difference in the ratios, and given that we know the half-life of ^{14}C, we can determine the time that has passed since the mammal died. Note that in this particular method, the important ratio is not between parent and daughter isotopes—^{14}C decays into ^{14}N—but between the two carbon isotopes. The half-life of ^{14}C is about 5700 years, which is relatively short and limits the reliability of dates obtained by this method to about 50,000 years. The reason is that after several half-lives have passed, there would be so little of the parent isotope remaining that determining accurate amounts is difficult, thereby increasing the likelihood of error. There are other limitations to carbon dating, such as the rather likely possibility of contamination of a sample by other organic material, such as vegetation, and the fact that it is now clear that the ratio of ^{12}C to ^{14}C is not constant but can and has varied over the time for which the method is used. Scientist make allowances for such factors, but other methods have also been devised for dating organic remains, such as electron spin resonance.

The last point to consider with absolute dating is the apparently odd values given to divisions of earth history. Many readers may have wondered, in gazing at Plate 1 of the geologic timescale, why scientists have chosen to use such inconsistent time intervals for the divisions. For example, the boundary between the Eocene and Oligocene occurred approximately 33.7 million years ago; that between the Miocene and Pliocene about 5.3 million years ago. Why not simply use convenient multiples of, say, 5 or 10, which would make remembering the dates much easier? The reason is that the divisions were essentially established by the beginning of the twentieth century, whereas absolute dating of these divisions got underway more than half a century later. As we noted earlier, when a sequence of strata was first described, geologists established a particular example of that sequence as the type sequence, and all other similar sequences were compared to the type. In other words, the boundary between any two particular time units is defined on rocks in a particular location on earth, and *that* is where the division is. If we want to know the absolute time of the boundary, we must go to the type locality and date the particular layer that geologists have determined as representing the boundary.

There is also another reason. Each of the divisions is meant to represent a particular time interval in which the earth, and especially the life it bore, was marked by particular features. It is not by chance that most mass extinctions occurred precisely at the end of periods or even eras (Benton, 2003).

Stratigraphic units

Box 1.1.

LITHOSTRATIGRAPHIC UNITS

According to the hierarchy from more general to more specific, they are as follows: supergroup, group, formation (the basic unit), member, and bed. The basic unit is the formation, which is defined as a body of rock with particular lithological features that can be mapped at a scale of 1:10,000 or higher.

BIOSTRATIGRAPHIC UNITS

These are units of rock characterized by their fossil content, called zones or biozones. This concept is related to the existence of some fossils, called index fossils, that have a short biochron (or life span) and a wide distribution.

CHRONOSTRATIGRAPHIC AND GEOCHRONOLOGIC UNITS

The chronostratigraphic units are the body of rock formed during a certain time interval. The units of geologic time during which those rocks were formed are called geochronologic units. Both units have a strict correspondence (Box Table 1.1).

Chronostratigraphic Unit	Geochronologic Unit
Eonothem	Eon (e.g., Phanerozoic)
Erathem	Era (e.g., Mesozoic)
System	Period (e.g., Jurassic)
Series	Epoch (e.g., Lower Jurassic)
Stage	Age (e.g., Toarcian)
Substage	Subage

Box Table 1.1. Corresponding chronostratigraphic and geochronologic units

CUVIER

DARWIN

D'ORBIGNY

LAMARCK

LINNEO

Distinguished Paleomammalogists

2

In any human activity, there are people who distinguish themselves, whether through bold ideas or the pioneering character of their efforts, but always on the basis of hard work, such as those commemorated in the niches of the facade of the Museo de La Plata (see opposite page). For the paleontology of South American mammals, there are a few investigators to be included in the pantheon, and their feats with pick and pen are recounted and closely imitated. The following roster includes some of the people that most current South American mammalian paleontologists would include in their short lists. Some of them never set foot in South America but worked on material sent back to their home institutions (e.g., Cuvier and Owen), whereas others, foreigners or native South Americans alike (e.g., Darwin, Lund, the Ameghinos, Kraglievich), spent time there in excavating fossils, publishing on them, or both. By and large, the nonnative researchers often arrived with the then most current conceptual and methodological tools, whereas many of the natives had to overcome both intellectual and institutional isolation (an interesting though incidental parallel to the conditions of the South American mammals themselves) in which they had to work. Florentino Ameghino was clearly an exception, in large part because he had managed to travel to Europe to collaborate with some of the leading scientists of the day. Although sometimes disparaged for having invaded and carted away precious fossils, the foreigners were instrumental both in studying South American fossils, thus making them known to the rest of the scientific community, and in inspiring or directly assisting in the formation of a paleontological research community in South America.

Cuvier

Baron Georges Cuvier (1769–1832; Fig. 2.1), to give just two of his numerous names, was undoubtedly the greatest anatomist of his day—and one of the best of all time. In addition to his numerous scientific contributions, the French scientist (though he was born to Swiss parents and completed his formal education in Stuttgart; Hallam, 1983) is generally recognized as the founder of the sciences of vertebrate paleontology and comparative vertebrate anatomy by virtue of his *Recherches sur les ossemens fossiles: où l'on rétablit les caractères de plusieurs animaux dont les révolutions du globe ont détruit les espèces* (Cuvier, 1812, 1825). He was not a main player in the story of South American mammals, but his dabbling produced two important milestones.

One is that he described, in his first publication on fossil vertebrates (Cuvier, 1796a), one of the largest late Pleistocene South American giant ground sloths, *Megatherium americanum*, on the basis of a specimen recovered from Luján in 1788 (Simpson, 1984) or 1789 (Novoa and Levine, 2010). Luján is a small city in Buenos Aires Province, Argentina, that has

2.1. Portrait of Georges Cuvier by Mathieu-Ignace Van Brée. Cuvier, one of the greatest anatomists of all time, is generally acknowledged as the founder of vertebrate paleontology and comparative vertebrate anatomy. Among his earliest publications was the description and naming of *Megatherium americanum.*

become famous for its fossil treasures. There's been some confusion as to the subject of Cuvier's initial foray into fossil mammals, with his (Cuvier, 1796b) report of fossil proboscideans (elephants and their kin) sometimes considered first; but the report on *M. americanum* was published before that on proboscideans (Simpson, 1984). Because this was Cuvier's first paper on fossil vertebrates, a legitimate case might be made for regarding it as the seminal paper of this biological discipline, and thus for *M. americanum* as the subject of this important landmark. Incidentally, before Cuvier described this specimen, it had been sent to the Real Gabinete de Historia Natural in Madrid, where it was prepared by Juan Bautista Bru de Ramón (1784–1786), who was employed by the museum in a position akin to that of the modern preparator or conservator (Simpson, 1984; Rudwick, 1997). Further, Bru mounted and illustrated the skeleton in a lifelike position, the first fossil skeleton to be so mounted (Simpson, 1984; Fig. 2.2).

The pose given by Bru was incorrect, but the exhibit has remained nearly unchanged (although it was remounted at least once), thus preserving the original attempt at reconstructing the life pose of an extinct vertebrate. Bru was apparently set to publish his own description of the skeleton,

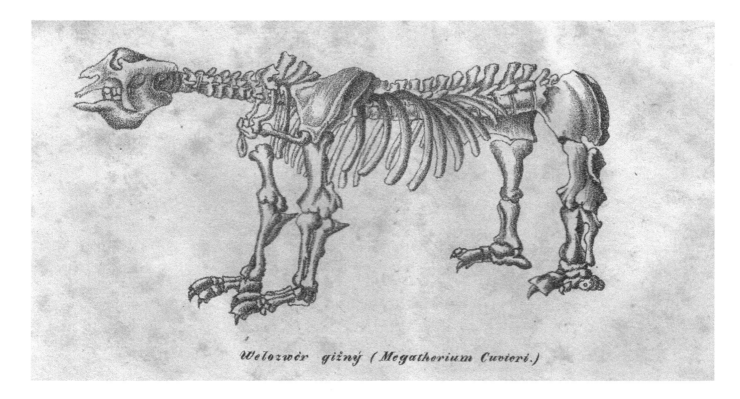

Welozwèr giżny (Megatherium Cuvieri.)

2.2. The original quadrupedal pose given to *Megatherium americanum* by Juan Bautista Bru de Ramón. This was the first fossil skeleton to be mounted in a lifelike pose. Although our ideas on its posture have changed, the skeleton remains approximately as originally mounted, reflecting its historic importance. The image used here comes from the Czech edition of the *Discours . . .* (Cuvier, 1834).

but his as-yet-unpublished illustrations fell into the hands of Cuvier, who hurried into print in 1796a (López Piñero, 1989).

In any case, Cuvier considered the giant sloth important for other reasons. He recognized that nothing like it existed in modern times—in other words, it had become extinct, which was a revolutionary idea at the time. During the few decades before and after the turn of the nineteenth century, biology, then known as natural history, was in the midst of a conceptual transformation, moving gradually away from a theological mind-set for explaining the pattern of life on earth toward more modern evolutionary ideas. At the beginning of this period of change, what a fossil represented was still not entirely clear. As noted in Chapter 1, the word *fossil* referred to items such as gems and odd concretions, as well as the lithified remains or traces of once-living organisms that are now gathered unambiguously under the rubric of paleontology. Among the more amusing examples were the natural internal casts of certain brachiopods (a kind of shelled marine animal) that bear an uncanny resemblance to female genitalia and were hence termed hysteroliths, or womb stones. The mystery of their formation was resolved by the middle of the eighteenth century, just a few decades before *M. americanum* was described (Gould, 2000).

Cuvier reasoned that if analysis of fossil bones by "the powerful new methods of comparative anatomy, could prove that they had belonged to species distinct from any known alive, the reality of extinction would be proved almost beyond dispute"; this conclusion was first suggested to Cuvier by the *Megatherium* remains (Rudwick, 1985:107). Although the importance of extinction to evolutionary theory as understood by modern science was still in the future, coming with the publication in 1859 of Charles Darwin's *The Origin of Species by Means of Natural Selection, or The Preservation of Favoured Races in the Struggle for Life*, Cuvier claimed that extinction was a reality and opposed the idea that fossil animals had been transformed into living species. The latter theory, which virtually denied the existence of extinction, was gaining in popularity during the first few decades of the nineteenth century and was favored by such great biologists as Jean-Baptiste de Lamarck. Although Cuvier has thus often been cast in a negative light as being an antievolutionist, his work greatly contributed to the proper understanding of fossils as representing forms of life that are, generally but not exclusively, no longer in existence (Gould, 1983).

It is unfortunate that our modern image of Cuvier has suffered ignoble slings and arrows. But this is almost entirely of our own making, for posterity has often judged him on the basis of a modern understanding of natural phenomena, rather than appreciating his efforts and ideas in the context of his own life and times. Indeed, scientists and historians of science tend to do this rather frequently, which may say more about modern thinkers than their intellectual forebears. Typically for the particular period of biology in question, the benchmark of glory is usually the acceptance of evolution, and in particular evolution by natural selection, or Darwinism. Those who were sufficiently capable intellectually of steering clear of prejudices to perceive objective truth that was based on empirical evidence are deemed worthy. The others are bit players, although some did have a few interesting things to say.

In Cuvier's case, there are two sources of supposed error that contribute to his present stature: his beliefs in the nonchanging nature of species and in the geologic concept of catastrophism (Gould, 1983). His adherence to these ideas has generally been viewed as due to the unyielding strength of his conventional theological convictions: that species had become extinct by successive and divinely induced catastrophic revolutions of the globe, and new species had been created to replace them. However, this is incorrect. Cuvier did not consider biblical scripture to be particularly reliable when it came to explaining the diversity and history of life; he believed, for example, that the earth was ancient and that empirical evidence was required to decide issues of earth's history. One of Cuvier's main goals was to show that the earth was indeed very old by using successive fossil assemblages. To accomplish this, he needed to show that the organisms embedded in rocks were restricted to (i.e., that they had lived and gone extinct during) particular periods of time (Gould, 1983).

The methods of comparative anatomy alluded to above were pioneered by Cuvier. He believed he had uncovered a general biological law, the correlation of parts, that allowed the prediction of the form of parts of an organism from other parts. Parts, or structures, of an organism were designed to perform specific functions, and their form was such that they were compatible with each other, working harmoniously to allow the organism to function properly and survive. From this, it follows that if one structure were changed, the parts would no longer mesh together to produce a functional organism, and the organism would fail, and so, Cuvier believed, species could not change. Although in hindsight we consider his view on species incorrect, it was just what Cuvier needed for explaining the broader perspective of earth history. If species evolved, as suggested by Lamarck (and others), then they merely changed over time rather than becoming extinct, precluding their use to calibrate past time and thus unravel geologic history. Cuvier's view that geologic change was explained mainly by episodes of catastrophic events was replaced during the 1800s by the idea of uniformitarianism or actualism (see Chapter 1), which maintains that geologic history could be explained by the same sorts of processes that are still occurring today. One need look no further than, for example, present-day river deltas or lava flows, acting over immense periods of time, to infer how thick rock sequences might have accumulated. Although catastrophism as advanced by Cuvier has not had many followers, at least in the scientific community, we have begun to appreciate that catastrophic events can play a major role in biological history. For example, that the demise of the dinosaurs and many other creatures at the end of the Cretaceous period was triggered by the impact of a huge asteroid has largely become accepted by most evolutionary biologists.

Cuvier's comparative anatomical methods, if not his correlation of parts, remain among his more lasting legacies. In subjecting *Megatherium* to his analytical methods, Cuvier clarified how it was to be classified, for he realized that *M. americanum*, aberrant as it must have seemed, was a giant ground sloth and that its closest living relatives are tree sloths. That Cuvier was able to deduce this from Bru's crude illustrations is testament to his genius as an anatomist. However, Cuvier was able to go further: he reconciled the odd anatomy of the gigantic beast and demonstrated that its parts contributed to a harmonious whole, in contrast to the opinions of other naturalists who viewed the anatomy of *M. americanum* as proof that it was ill adapted for life and therefore led a miserable existence. Honoring Cuvier's legacy, we will deal repeatedly with this amazing beast in this book.

Although Baron Cuvier was the first of the great scientists to have studied an important South American fossil mammal, and although other finds were reported by South American scientific pioneers, the great developments in this field had to await Charles Darwin's (1809–1882; Fig. 2.3) arrival to the austral shores of this land. When Captain Fitz-Roy of the HMS *Beagle* raised anchor on 27 December 1831 from Devonport, he was probably completely unaware of his contribution in setting in motion the greatest of conceptual revolutions in biological thought. Darwin would later become one of the greatest academic celebrities ever, but who was then still a young and somewhat aimless gentleman, was chosen by Fitz-Roy as the onboard naturalist and to collaborate with him on the *Beagle*'s mission to circumnavigate the world and, especially, explore and map the southern end of South America.

In addition to this assignment, the ensuing voyage of 5 years included stops in the Galápagos, Hawaii, Australia, New Zealand, the Cape of Good Hope, and a plethora of other places. Nothing escaped Darwin's razor-sharp eye, and he filled his travel diary with observations on animals and plants, the morals of Tahitians, the virtues of the gauchos, the colonization of Australia, as well as the slavery in Brazil, which he found most disgusting:

> I may mention one very trifling anecdote, which at the time struck me more forcibly than any story of cruelty. I was crossing a ferry with a negro who was uncommonly stupid. In endeavouring to make him understand, I talked loud, and made signs, in doing which I passed my hand near his face. He, I suppose, thought I was

Darwin: traveler and collector

2.3. Charles Robert Darwin. Darwin, who remains among the most important scientists in history, provided evidence that convinced contemporary biologists of the fact of evolution. His voyage to South America on the HMS *Beagle* provided much of the evidence that convinced Darwin himself of the truth of evolution.

Courtesy of the Wellcome Library, London.

in a passion, and was going to strike him; for instantly, with a frightened look and half-shut eyes, he dropped his hands. I shall never forget my feelings of surprise, disgust, and shame, at seeing a great powerful man afraid even to ward off a blow, directed, as he thought, at his face. This man had been trained to a degradation lower than the slavery of the most helpless animal. (Darwin, 1839:24–25)

On the South American rulers of the day, we may note his assessment of Juan Manuel de Rosas, the political leader of Buenos Aires, who ruled the country as dictator until 1852. Darwin recounted in a letter to his sister, Caroline, of Rosas's ruthless "bloody war of extermination against the Indians" (Darwin, as quoted in Novoa and Levine, 2010:19). On the other hand, although it no consolation, Rosas did "donate war-surplus horses for Darwin's exploration of the interior" (Novoa and Levine, 2010:19).

Yet he still found time to collect great quantities of Pleistocene fossil mammals—and others—which he sent to England to be studied primarily by the great anatomist and his later adversary on evolutionary thought, Richard Owen. Many of the great mammals we discuss were first found by Darwin, or became better known through his efforts. Of particular interest for the significance of evolutionary theory were the remains of giant quadrupeds preserved in the cliffs near Bahía Blanca on the coast of Argentina. These made a great impression on Darwin—most certainly the first great influence of facts observed for the later development of his theory. In his own words, from his book *The Voyage of the Beagle,* written in San Julián on 9 January 1834: "This wonderful relationship in the same continent between the dead and the living, will, I do not doubt, hereafter throw more light on the appearance of organic beings on our earth, and their disappearance from it, than any other class of facts" (Darwin, 1839, chap. 8); and in his notebook: "To my view, in S. America parent of all armadilloes might be brother to Megatherium—uncle now dead" (Darwin, 1837:216). Hence, these fossils had convinced Darwin that the similarities between extinct and living forms should be explained by the existence of common ancestors, and, even more, the transformation of species to a large degree was not vertical in a ladderlike sequence, as Lamarck had proposed, but a tree with asymmetric branches. The strong impressions that Lyell's *Principles of Geology* (1830–1833) left on Darwin's mind are well known. However, there were differences in their points of view; Lyell, as well as other geologists of the day, followed the ideas of the eighteenth-century Scottish natural philosopher James Hutton, author of the rather impenetrable *Theory of the Earth* in 1795. Lyell stated that the Huttonian principle of uniformitarianism could easily explain the past processes of the earth and the whole evolution of the planet, a concept elegantly expressed by Geikie (1905:299) as the present being the key to the past. Therefore, in Lyell's eyes, Lamarck and the concept of biological evolution were overrated because it was not possible to demonstrate it in terms of current processes, and hence Lamarck's views could not be accepted as anything more than pure speculation. Fortunately enough, the fossils that Darwin, an adherent of Hutton's and Lyell's uniformitarianism, was collecting convinced him that organic evolution was a fact. Hence, what originally was only appealing reading material (Darwin confessed that Lamarck's work was among the few books he found really interesting during his time in Cambridge) turned into a decisive influence for his own creation (Howard, 1982).

Less well known is that when Darwin boarded the *Beagle*, he was already convinced of another important biological issue: adaptation. Paradoxically, this idea did not come from scientific readings but from the study of Paley's (1802) *Natural Theology, or Evidences of the Existence and Attributes of the Deity collected from the Appearances of Nature*, when he was in Cambridge and his greatest ambition was to become a rural pastor. Natural theology, supported by the Anglican Church and championed by Paley, argued that every organism was perfectly designed to its particular life conditions, that the perfection of the biological design was intentional, and that the designer was, of course, God. Thus, by the time the *Beagle* departed for South America, Darwin was already aware of two of the main concepts that were to become hallmarks of his theory of biological evolution: evolution itself, and adaptation (although he was not yet an evolutionist). The third pillar, natural selection, would await interpretation of the Galápagos data, Malthus, and years of collecting evidence.

Nevertheless, it was not only the idea, means, and processes of evolution that impressed Darwin, but also many aspects of what is today considered essential to the realm of paleontology. For instance, the idea of extinction, and even of mass extinction, is foreshadowed in Darwin's words. During his journey between Buenos Aires and Santa Fe, he wrote, "We may therefore conclude that the whole area of the Pampas is one wide sepulchre for these extinct quadrupeds" (*Voyage of the Beagle*, chap. 7, October 1833). By that time, his sharp mind had already noted some ecological paradoxes of great interest in reconstructing past interactions: "That large animals require luxuriant vegetation has been a general assumption, which has passed from one work to another. I do not hesitate, however, to say that it is completely false; and that it has vitiated the reasoning of geologists, on some points of great interest in the ancient history of the world" (*Voyage of the Beagle*, chap. 5, August 1833; "The Red Notebook of Charles Darwin" in Herbert 1980:54). Moreover, in January 1834, Darwin collected remains of the extinct litoptern *Macrauchenia* in Patagonia near San Julián, in what today is Santa Cruz Province. He made inferences about the environment in which this beast lived and reflected on its extinction:

> Mr. Owen . . . considers that they form part of an animal allied to the guanaco or llama, but fully as large as the true camel. As all the existing members of the family of Camelidae are inhabitants of the most sterile countries, so we may suppose was this extinct kind . . .
>
> It is impossible to reflect without the deepest astonishment, on the changed state of this continent. Formerly it must have swarmed with great monsters, like the southern parts of Africa, but now we find only the tapir, guanaco, armadillo, capybara; mere pigmies compared to antecedent races . . . Since their loss, no very great physical changes can have taken place in the nature of the Country. What then has exterminated so many living creatures? . . . We are so profoundly ignorant concerning the physiological relations, on which the life, and even health (as shown by epidemics) of any existing species depends, that we argue with still less safety about either the life or death of any extinct kind. (*Voyage of the Beagle*, chap. 9, January 1834)

In this way, Darwin not only triggered studies on the genealogical interpretation of the fossil mammal lineages, but also opened a door to research on their behavior, physiology, and extinction—what we call today their paleobiology. Phylogenetic studies, combined with morphological and

2.4. Portrait of Dámaso Antonio Larrañaga. Active in many aspects of Uruguay's early history, Larrañaga was also among the first naturalists born in South America.

taxonomic analyses, flourished in South America, particularly beginning during the latter half of the nineteenth century, as a result of the intellectual influence of Florentino Ameghino, but paleobiology has become a much more recent line of work, in apparent relation to innovations in methodology and technology (which are described in Chapter 8). Indeed, as Darwin noted, Pleistocene South American fossil mammals show a greater morphological diversity than their living counterparts; they include representatives of great body size and with peculiar features. Their peculiarity and general lack of modern analogues have encouraged creative paleobiological approaches that will be extensively treated later in this book. We may claim, with pride, that our approach is in line with the questions and remarks on the paleobiology, paleoecology, and extinction of the South American Pleistocene fossil mammals that the young Darwin himself made when he first collected them.

Many of the fossils collected by Darwin are still housed in the collections of British natural history museums. Sadly, however, some material, including a complete glyptodont carapace and many other fossils, sent to the Hunterian Museum (Plate 2) of the Royal College of Surgeons, was lost when that institution was bombed during World War II. Some of that specimen was mixed with the rubble until it was taken to the Natural History Museum of London, where it still waits, Humpty Dumpty–like, for its pieces to be put back together again.

There are many people who complain that this material, which belonged to South Americans, was taken so far away, and speak of Darwin in somewhat less than flattering terms. However, we should consider the other side of the coin. In contrast to the obelisk of Luxor, which now adorns the Parisian landscape thanks to Napoleon, or the tombs that were carted off and now reside in London, these fossils would have been lost to us through the effects of erosion had they not been collected and taken to Europe. Neither South Americans nor Europeans, much less the scientific community, would have known of their existence for decades, at least until the development of independent scientific research and institutions in South America, a story recounted below.

Presbyterian Dámaso Antonio Larrañaga

Born in Montevideo, later capital of Uruguay, Dámaso Antonio Larrañaga (1771–1848; Fig. 2.4), a man of the cloth, was a pioneering scientist in the Spanish colony. He was also active in the political turmoil of the early nineteenth century, as an army chaplain in the task force that took back Buenos Aires from the British in 1806, and later during the war of independence against Spain. Further, he was active in the founding of the new country: he was a representative in the early revolutionary congresses and a member of the first senate of the Republic of Uruguay, among many other assignments. He somehow also had time to create the national library, create the natural history museum, and introduce a bill for the creation of a university.

Here, however, we emphasize his efforts as one of the first naturalists born in this part of America, recognized by his acceptance, after having been proposed by Étienne Geoffroy Saint Hilaire, as a full member by France's Societé d'histoire naturelle. His scientific interests ranged widely: botany, zoology, astronomy, paleontology. In his *Diario de Historia Natural,* still largely unpublished (Fig. 2.5), Larrañaga studied 200 species of plants. Moreover, he wrote a treatise in Latin on the living mammals of Uruguay, and his zoological endeavors led to the description of 216 species of insects (including 19 ant species). Larrañaga also produced one of the first figures of *Megatherium americanum* (Fig. 2.6), as cited by Ramírez Rozzi and Podgorny (2001: fig. 1).

It was Larrañaga who in 1814, although indirectly, broke the first scientific news about a great armored extinct animal, when Darwin was 5 years old and the former Spanish colonies were in the midst of trying to establish themselves as independent republics. This fossil material, including a femur, carapace fragments, and a caudal tube, was assigned to "*Dasypus* (*Megatherium* Cuv.)" by Larrañaga in his *Diario* and was included as such in Cuvier's second edition of *Recherches* in 1824. This subgeneric assignment was based on the fact that the original material of *Megatherium americanum*—the very specimen sent to Spain and handled by Bru—was collected in association with carapace fragments (as noted by Méndez Alzola, 1950). The fact that Larrañaga's opinion was published in Cuvier's (1824) hugely influential work misled several later authors (e.g., Weiss, 1830; Clift, 1835) into recognizing armored megatheres.

It fell to Richard Owen to set matters straight. Owen (1839) founded the genus *Glyptodon* on material collected near Buenos Aires and shipped to London by the English secretary, Woodbine Paris. A year later, Owen reviewed all references to armored megatheres and established that the big armored animals were actually glyptodonts, a group related to armadillos, and that megatheres did not have a carapace.

2.5. A page from the manuscript of Larrañaga's *Diario de Historia Natural*, kept in the Archivo General de la Nación, Uruguay.

Many of the pioneering efforts of some early South American researchers have been largely forgotten, now treated essentially as historical footnotes. However, the early descriptions and functional studies by Teodoro Vilardebó and Bernardo Berro (see Chapter 6) in Uruguay, and especially Francisco P. Muñiz in Argentina, are of particular interest for their role in the emergence and credibility of natural history studies produced in South America.

Muñiz (1795–1871; Fig. 2.7), who was born in Monte Grande, Argentina, devoted his intense life to his country as a soldier, politician, and scientist. At the age of 11, he enlisted in the army and wounded his leg defending the city of Buenos Aires against the British. He studied medicine in the Military School of Medicine, which became the School of Medicine of the University of Buenos Aires in 1821, a year before he graduated. As a scientist, his work encompassed subjects as varied as medicine, topography, zoology and, of course, paleontology, and he is considered to be the first naturalist of Argentina. Although his contributions to paleontology are now generally neglected, his efforts and inspiration were acknowledged by his contemporaries and subsequent researchers. An example of such tributes is that of Florentino Ameghino in a letter, dated 20 January 1886, to F. Lajouane, the editor of Sarmiento's (1885) compilation of Muñiz's work (*Vida y escritos del coronel doctor Francisco Javier Muñiz*): "Él se ocupó de las mismas ciencias que constituyen mis estudios predilectos, vivió 15 años en donde yo pasé mi niñez, y explotó los mismos yacimientos fosilíferos que yo debía remover 30 años después . . . y los recuerdos de sus hallazgos,

Francisco Javier Muñiz

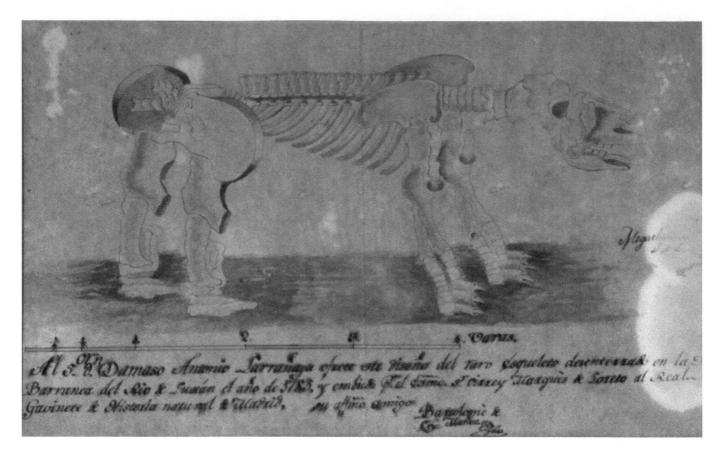

2.6. Larrañaga's drawing of the skeleton of *Megatherium americanum*. A translation of the caption, handwritten in Spanish below the figure, states, "Mr. Dámaso Antonio Larrañaga offers this drawing of the strange skeleton unearthed in the cliffs of the Luján river in the year of 1785 and sent by his Highness the Viceroy Marquis of Loreto to the Real Gabinete de Historia Natural in Madrid. Your most loyal friend Bartolomé de Muñoz."

Photograph courtesy of the Museo de La Plata, Argentina.

vueltos populares en Luján, no contribuyeron poco a que me lanzara tras de él a las mismas investigaciones." ["He concerned himself with my own chosen field of study and lived for 15 years in the place where I spent my childhood, exploring the same fossiliferous beds that 30 years later I would come to excavate . . . and the news of his discoveries, common knowledge at that time in Luján, contributed in no small way in drawing me toward the same endeavors."] This excerpt from Ameghino's letter appeared in the prologue of subsequent editions of the book.

Muñiz was no mere amateur fossil collector, but a well-rounded scientist who described and named his fossils, and postulated several sound hypotheses on aspects of their paleobiology based on his knowledge of anatomy. His first collections were made in 1825 from the environs of Chascomús (Buenos Aires Province). His collections from Luján represent a sample of the most characteristic mammals that inhabited the pampas during the Pleistocene, including megatheres and other ground sloths, proboscideans, macrauchenias, glyptodonts, and bears. The influential Buenos Aires newspaper *Gaceta Mercantil* published an annotated list of these specimens in 1841. Muñiz presented this collection (unwillingly, in the minds of many, but apparently with the hope of stirring interest in the establishment of a national natural history museum; Novoa and Levine, 2010) to Don Juan Manuel de Rosas, then the general governor (or, better yet, dictator) of Buenos Aires Province, only to have Don Juan turn around and give the collection to the commander of the French navy in the Río de La Plata. This collection and the fossils that Muñiz himself sent off to institutions in Europe enhanced his stature as a scientist and spurred the study of his fossils by such contemporaneous luminaries as Henri Gervais, of the

2.7. Portrait of Francisco Javier Muñiz at the DPV MLP. Another of the pioneering South American naturalists, Muñiz's efforts helped establish the scientific credibility of the fledgling republics.

Courtesy of the Museo de La Plata, Argentina.

Muséum national d'Histoire naturelle of Paris, who described the bear "*Ursus*" *bonaerensis* (later *Arctotherium*) and the ground sloth *Lestodon*.

In 1857, Muñiz donated many specimens to the Museo Público de Buenos Aires that now enrich the exhibits of this institution's present incarnation as the prestigious Museo Argentino de Ciencias Naturales "Bernardino Rivadavia," including a skull of *Toxodon*, a fossil horse, and a complete sabertooth. The skull of *Toxodon* and the horse were studied by the director of the museum, the Prussian-born naturalist Hermann Burmeister. The sabertooth, which Muñiz first (and rather egocentrically) named *Muñifelis bonaerensis* and later correctly assigned to the genus *Smilodon*, represents Muñiz's highest formal acknowledgement as a paleontologist. He had discovered, collected, and prepared the specimen, described it with the precision of an accomplished anatomist, and proposed hypotheses on its life habits that to a large extent have been supported by more modern studies. Muñiz compared its skeleton with that of present felids and concluded that the animal was in general more developed than the lion, most particularly in its forelimbs, and therefore that it did not hunt for prey by long-distance chasing, but by ambushing and seizing prey with its powerful forelimbs.

Owen: a tarnished reputation

Richard Owen (1804–1892; Fig. 2.8) became one of Darwin's fiercest rivals, although in many senses the tangle was mainly between Owen and T. H. Huxley, a great biologist himself, certainly at the forefront of paleontology and often remembered as Darwin's bulldog. Unfortunately, it is for these unpleasant facts that Owen is most often remembered, while his numerous contributions to science have largely been forgotten. In fact, Richard Owen was so brilliant and so meticulous an anatomist and paleontologist that he quickly assumed Cuvier's mantle. Indeed, in his prime, during much of the 1830s to the 1850s, he was known as the British Cuvier, although this nagged him because he would rather have been recognized on his own merits as the Owen of England.

It is no surprise that Darwin's South American fossil mammals were primarily entrusted to Owen for description and interpretation; at the time, there was no one more capable, and there was as yet no enmity between the two (Box 2.1). Although Darwin found them, it was Owen who made these great beasts known to science. Many of them were entirely new—*Toxodon*, *Macrauchenia*, *Scelidotherium*, *Glossotherium*, *Mylodon*, *Glyptodon*, all integral to the story told here, were christened by Owen. Characteristically, these were all handled with Owen's trademark descriptive style, methodical and thorough. He named one of the large ground sloths *Mylodon darwinii*, obviously in tribute to the greatest of evolutionary thinkers.

Box 2.1

Owen (1840) on Darwin

The osseous remains of extinct Mammalia, rank amongst the most interesting results of Mr. Darwin's researches in South America. The abundance and variety of the osseous remains of extinct Mammalia in South America are amply attested by the materials for the following description, collected by one individual, whose sphere of observation was limited to a comparatively small part of South America; and the future traveller may fairly hope for similar success, if he bring to the search the same zeal and tact which distinguish the gentleman to whom Oryctological Science is indebted for such novel and valuable accessions.

In addition, he produced important works on already known fossil mammals. Owen's (1851–1860) magnificent descriptions and illustrations of *Megatherium americanum* (Fig. 2.9), published as a series over several years as a result of funding constraints (Desmond, 1982; Rupke, 1994), were among the most extensive and comprehensive studies of a fossil vertebrate at the time. The complete work became a standard for the presentation of information on fossils, and remains, a century and a half later, an authoritative text on the skeletal anatomy of *M. americanum*. Owen's work gained enough public expression that he was even caricaturized in his time (Fig. 2.10).

Owen remained a respected and celebrated scientist for much of his long career. Perhaps his most lasting accomplishment was the lead role he played in establishing the British Museum of Natural History, now the Natural History Museum, thus realizing his lifelong dream of a separate institution to house the empire's natural history collections. The museum opened in 1881, and Owen served as its first director.

From his impressive early papers, such as that on the chambered nautilus, his ascendancy to the upper echelons of British science was swift. He happened upon his well-deserved fame at a time when the framework for doing science was undergoing a profound shift, from one where scientists were self-supporting aristocratic gentlemen to one where science became a professional discipline. Owen has often been criticized for belonging to or siding with the former, although in truth he came from a modest background. His family was not wealthy, and his father died while Owen was a young boy. He worked his way up, with some luck but also with considerable talent and enormous effort, to become a truly great scientist.

Fortune of a sort positioned him to step into the role of his supervisor, William Clift, who happened to be both the curator, as we would consider the position today, of the Hunterian Museum and Owen's prospective father-in-law. Owen began his career as a physician, completing his studies at the Royal College of Surgeons in London. By all accounts, he was a competent physician and an even more impressive anatomist—so much so that the director of the college and the Hunterian Museum made him his prosector and offered him a position at the museum, which housed the vast natural history collection amassed by the great surgeon John Hunter. This collection, which numbered approximately 18,000 specimens of human and other preserved, including fossil, vertebrate remains was purchased by the British crown and handed over to the college with the proviso that it be made accessible to the public and medical community alike through lectures and the establishment of a museum. Owen was assigned the task of partially fulfilling this mandate by organizing, cataloging, and publishing on the collection—functions that had been in a relatively catastrophic (and indeed fraudulent) state for the previous few decades. Owen attacked this task with zeal. His enthusiasm and industriousness on this collection, his popularity as a lecturer, and his anatomical publications on other vertebrates caught the eye of the scientific elite and his star rose quickly—all talent and hard work so far.

During this period, Owen served as one of Clift's assistants, the other being Clift's son, William Jr. Naturally, the son was set to take over for his father. Clift (the father) also had a daughter, Caroline. She and Owen soon fell in love and were engaged to be married. With Owen slated to be his son-in-law, Clift used his influence to help Owen further his own career; largely freed of administrative duties, Owen was able to focus on science (McGowan, 2001). As fate would have it, tragedy struck William (the son). A cab in which he was riding overturned, and he struck his head. Ironically, it was Owen who attended to him at the hospital, but there was nothing that could be done about his fractured skull (McGowan, 2001). With William's death, Owen moved up.

In addition to his numerous mainly descriptive works, Owen made many primarily theoretical contributions. For example, he showed exceptional foresight in delineating larger taxonomic groupings, and he is well remembered for his recognition and naming of Dinosauria, and for predicting that large flightless birds had once lived in New Zealand, based on a scrap of bone sent to him by a missionary. He was right, and he later fully described the moa on more complete remains.

2.8. Portrait of Richard Owen in 1856. Although a prickly character, he was the foremost anatomist and paleontologist of his day, and he studied the fossil mammals that Charles Darwin found in South America.

Courtesy of the Hunterian Museum, Royal College of Surgeons of England.

2.9. The skull and mandible of *Megatherium americanum* that Charles Darwin recovered from South America. This figure was produced for Owen's (1856: pl. 21) landmark study of this giant ground sloth. Length of skull approximately 85 cm.

RIDING HIS HOBBY.

2.10. Richard Owen riding a *Megatherium americanum* skeleton. This caricature by Frederick Waddy (from Anonymous, 1873) appeared in *Vanity Fair*, which described him as a wicked, simple-minded creature. As noted in the text, he was far from being the latter, and perhaps less wicked than usually portrayed.

Image courtesy of the Wellcome Library, London.

Further, he introduced several terms and concepts that are still in use and central to our modern understanding of evolutionary biology. Among these are *analogy* and *homology*, terms defined technically in 1843 that express the concepts of similarity of parts in different organisms based on functional requirements (analogy) and the same structure or part in different organisms appearing in all manner of form and function—that is, possibly appearing different but having a common underlying structural plan and similarity in arrangement and relationship of parts (homology). Owen then related these concepts to what may be his most famous and enduring scientific accomplishment: the vertebrate archetype (Fig. 2.11).

The archetype was formulated by Owen to represent the ideal or fundamental pattern on which the vertebrates had been constructed, the simplest or most basic notion of what a vertebrate is and from which all of the different vertebrates could be derived through varying degrees of differential modification of its parts. Conversely, the modifications in any vertebrate could be traced back to the basic parts present in the archetype. Although it has been considered a Platonic expression of the basic vertebrate form, and indeed Owen came to as advance it as such, it was not quite so. It was the simplest, least perfected representation of vertebrateness (with "man," of course, being the most perfect)—the lowest common denominator—in

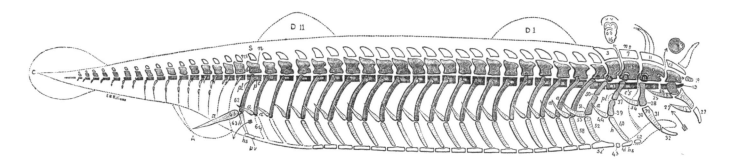

contrast to Plato's sense of the ideal as the highest or perfect metaphysical expression or form of any particular thing.

Importantly, Owen's concept was not evolutionary in a Darwinian sense; Darwin's *Origin* was more than a decade in the future. Instead, Owen's archetype eventually came to be conceptualized as "a divine forethought, a blueprint of design for animal life" (Rupke, 1993:245), and it fulfilled the biblical notion that the knowledge of the possibility of man must have existed before the appearance of man. Once Darwin's ideas on evolution appeared in print in 1859, Owen's lowest common denominator was easily and simply transformed into a generalized common ancestor from which subsequent vertebrate evolution proceeded.

Why is it, then, that Owen occupies such a lowly place in the history of the biological sciences? The answer to this question is not straightforward. Clearly, the fact that he was wrong about evolution is a factor in how his efforts have been judged. However, to make an assessment from our current perspective, in retrospect, would be erroneous and unfair; we should not stigmatize a scientist's entire career on this basis. In any case, we maintain a status of greatness for many scientists whose work and ideas have been subsequently ruled as off the mark.

To this knock against the value of his scientific contributions has been added his rather prickly character. In his professional life, Owen is often portrayed as having been selfish, bickering, egotistical, jealous, and arrogant, and much of this seems to be accurate. However, one can walk into almost any university department today and find that the same holds true for much of the faculty, as indeed it did for most of Owen's contemporaries. He has also been criticized for having failed to adequately acknowledge that some of his ideas were built partly on the work of others, such as Gideon Mantell's on dinosaurs or Carl Gustav Carus's on the archetype. There is some truth in this, but to Owen's mind, it was not so much a matter of priority per se. Rather, he felt little need to praise the work of others if it was incorrect or incomplete. To paraphrase Desmond (1982), it is not easy to love a man so full of himself. Further, he has been accused of hindering or impeding the career and work of others. In many cases at least, this seems to be a distortion of reality. It is clear that Owen was helpful in trying to further the careers of several researchers. For example, he was initially friendly with Huxley, the man who would later do his utmost to demolish Owen's career work, and accommodating in supplying him with a steady stream of glowing references (Desmond, 1982).

In any event, the sort of unpleasant behavior characterizing Owen has clearly not prevented others from being exulted. For example, one of

2.11. Richard Owen's archetype. The great anatomist saw this as the fundamental pattern or blueprint in the mind of the Creator from which all other vertebrates could be brought into existence. Within the framework of Darwin's descent with modification, the archetype became the common ancestor. From Owen (1848).

science's most revered icons, Sir Isaac Newton, was a decidedly nasty fellow, as described by Stephen Hawking in *A Brief History of Time* (1988). His career was marked by a series of bitter disputes, in which he brazenly wielded his power and influence to obtain what he wanted. In one case, Newton went so far as attempting to have another scientist's data seized so that his rival could publish it, an act that was prevented only by court order. More famous were his shenanigans in cheating (albeit temporarily) Gottfried Leibniz from the latter's rightful claim to having independently invented calculus. Indeed, Newton rejoiced at hearing of Leibniz's death. To be fair, though, Leibniz wasn't entirely blameless in their feud (Hellman, 1998).

As Desmond (1982) and Rupke (1994) have pointed out in Owen's case, much of the rancor seems to be the result of his having been targeted as the enemy of evolution (or transmutation, as it was then called) by Darwin's circle of supporters, and his subsequent vilification as the antievolutionary force impeding scientific progress (and of course that of the scientists who were competing for recognition). Topple Owen, who represented the scientific pinnacle of the religious and political establishments, and the whole edifice might just follow. Certainly, Owen's image hasn't been helped by the fact that it was Huxley's circle that rewrote much of Victorian science history (Huxley and Kettlewell, 1965; Desmond, 1982). However, as Owen himself tried to point out, and as anyone who has read Owen's work can attest, he was not an antievolutionist (he clearly was earlier in his career, but so was just about everybody else, including Darwin). He bristled, for example, over Darwin's labeling him an antievolutionist in the first edition of the *Origin*, an oversight that Darwin rectified in later editions. He had been struggling for years to come up with an explanation that would accommodate transmutation into the then-current (Western) worldview, one that would still allow man's place in the universe to remain uncontested. Owen wrote often of the progression of life on earth as being due to secondary causes (though he was often muddled on just what these were), while never doubting the primary cause—that is, a divine or supernatural force. Indeed, initially he was guardedly favorable toward natural selection, one of Darwin's main contributions to evolutionary theory, as one such possible secondary cause, until he found himself attacked on several fronts. After that, there was no turning back.

To be fair to Huxley and Darwin—and history has been more than that—these scientists sought natural rather than supernatural causes to explain the patterns and diversity of past and present life. For Owen, the secondary causes were equivalent to laws as set out by an omnipotent and omniscient creator—this Darwin and Huxley could not accept—and as such could not provide explanations for evolution. Owen certainly did not have Darwin's clarity of vision and courage in the realm of evolutionary thought, but he was a truly great anatomist, and his numerous contributions to comparative anatomy and paleontology are slowly but surely regaining the luster they deserve.

Lund: in blissful exile

Peter Wilhelm Lund (1801–1880; Fig. 2.12) is a figure often overlooked by students of vertebrate paleontology, despite having his biography turned into a marvelous novel in 1981 by his fellow Dane, Henrik Stangerup, in

Vejen til Lagoa Santa, the English translation of which was published as *The Road to Lagoa Santa* in 1987. However, he made an immense contribution to our knowledge of late Pleistocene South American mammals and is generally regarded as the father of Brazilian paleontology by virtue of his pioneering efforts in that country. His sustained, systematic effort of collecting in a limited region was among the first such paleontological undertakings, certainly so in South America. Lund was born in Copenhagen, Denmark, and graduated from the University of Copenhagen, where he studied medicine, philosophy, and music, but like Darwin, he decided instead to pursue the study of natural history. He also seems to have shared the great naturalist's empathy for fellow human beings; Lund used large amounts of money to buy and free slaves at a time when no one else in Brazil was campaigning for abolition (Cartelle, 2002).

Lund's change in plans was in large part dictated by his experiences during his first trip to Brazil, begun in 1825. Lund, as the story is usually told (Simpson, 1984), chose Brazil for its climate in the hope of escaping the fate of his two brothers, who had died from tuberculosis. The trip, lasting 3 years, resulted in a large collection of extant animals, most of which were sent back to Copenhagen's Royal Museum of Natural History. Lund returned to Denmark in 1829 and then traveled widely in Europe, visiting collections in the major European centers of study and meeting many of the day's famous natural historians, including Cuvier (Cartelle, 2002). Encouraged to return to Brazil to continue his research, and probably also influenced by the death of his mother as well as Brazil's favorable climate, Lund left again in late 1832, never to return to his native country.

Lund began his second sojourn by studying the botanical wonders of Brazil, but soon after meeting fellow Dane Peter Claussen, probably in 1835, and learning of the bones being recovered from caves, his focus and subsequent course in life were profoundly altered. Lund and Claussen met probably in Curvelo, a town about 170 km north of the major city of Belo Horizonte (Cartelle, 2002), in the state of Minas Gerais, which owes its name to its abundant mineral resources. It is in this general area that Lund spent nearly the rest of his life. This region of Brazil is dotted with limestone caves, some so large and resplendently beautiful that they have become famous tourist destinations. More importantly for our story, however, is that many of the caves yield abundant fossils. It was in these caves that Lund was to search systematically for the next decade, recovering by torchlight and pack animals a treasure of fossils, numbering over 12,000 specimens, from about 800 localities generally in the valley of the Rio das Velhas. Add to this the monumental task of preparing and cataloging the specimens, of sending them back to Europe, and of writing about them for scientific publication (often accompanied by exquisite illustrations; see Plate 3 for an example) from Lagoa Santa, a town far from any center of study and thus lacking laboratory and library facilities. Among his published work was the first report of South American cave art.

Despite the promise of further finds and the possibility of returning to Europe to further his study of the collection, Lund suddenly stopped his research in 1846. A short time later, once the last of the collection had been shipped to Copenhagen, he settled into a life of relative idleness in his adopted town. There are a number of factors that might have caused

2.12. Portrait of Peter Wilhelm Lund. Acknowledged as the founder of Brazilian paleontology, his systematic collecting from caves in Minas Gerais, Brazil, was among the earliest long-term paleontological expeditions.

2.13. Karl Hermann Konrad Burmeister in 1856. An anti-Darwinist, Germán (as he preferred to be called) was often at odds with Florentino Ameghino.

Hermann Burmeister

him to abandon his investigations with such finality. Some were apparently related to economics, his funding having been curtailed, and his health, which seems to have been delicate from the start. Also, his enthusiasm seems to have been tempered by the repetitiveness of the fossils he was finding. Most important, perhaps, was Lund's unshakable Protestant faith, which likely conditioned him to be a strict adherent of Cuvier's concept of catastrophism. He believed, at least initially, that the fossils of the extinct mammals he had been finding were the remains of those creatures that had been wiped out during a global flood, a commonly held view at the time. However, Lund began to make some troubling finds. In some cases, he recovered fossils of species that were still living. More serious were the remains that indicated the contemporaneous existence of humans, supposedly God's most recent creative enterprise, and species that had become extinct. This discovery was noted in a paper published in 1844, shortly before Lund retired from his scientific career.

Accompanying the first specimens shipped back to Copenhagen was a note from Lund to the Danish monarch, King Christian VIII, with instructions that the collection be used to its fullest potential because of its significant scientific value. Sadly, Lund died before fulfillment of this request was even begun. Another Danish scientist, Herluf Winge (1857–1923), spent a good part of his professional life studying and publishing on Lund's collection. Unfortunately, both Lund and Winge published in Danish, which was then, as it is now, little read by the scientific community. For a long time the work of these men was known by only a handful of people, until Winge's work was translated into English and published in 1941 and Lund's into Portuguese in 1950. Lund's enduring legacy is plainly evident, however, in the work of subsequent Brazilian paleontologists. In particular, first Carlos de Paula Couto, who spent most of his career at the Museu Nacional in Rio de Janeiro, and then Cástor Cartelle, currently of the Pontifícia Universidade Católica de Minas Gerais, followed Lund's example of systematic and sustained excavations. Cartelle's efforts in particular have produced a spectacular collection of fossils over the past two decades from caves in the state of Bahia, Brazil.

When Karl Herrmann Konrad Burmeister (or Carlos Germán Conrado, as he called himself; 1807–1892; Fig. 2.13) left Germany and arrived for the first time in South America in 1850, he was already an acknowledged scientist, having earned doctorates in philosophy and medicine from the University of Halle and the University of Greifwald in 1829. His socialist political leanings got him into trouble, so he left Europe following in the footsteps of Alexander von Humboldt—another of the great scientists, commonly acknowledged as a founder of geography—to study the natural history of the Brazilian states of Minas Gerais and Rio de Janeiro from 1850 to 1852. Once back in Europe, new political problems at the University in Halle forced his withdrawal from German academic circles, and Burmeister returned to South America in 1856 to study the natural history of Argentina and Uruguay. In 1858, he conducted a series of geologic and paleontologic investigations on the cliffs of the Paraná River. He presented himself to the authorities of Argentina for the post as director of the Museo Público de

Buenos Aires (now Museo Argentino de Ciencias Naturales "Bernardino Rivadavia") in Buenos Aires in 1862, received the position, and worked there until he died, 30 years later, after an accident in the museum: he was apparently sufficiently self-reliant to scorn the help of the museum employees and climbed a ladder without assistance (at the age of 85!) to try to close a window. Unfortunately, he fell from the ladder and died a few weeks later. His contribution and dedication to this institution were considered so important that his earthly remains were transferred to the Museo Argentino de Ciencias Naturales "Bernardino Rivadavia" in May 1967, as acknowledgment on the part of the Argentinian government to the foreign scientists who had dedicated their careers to their adopted land. Burmeister was chosen for this honor because he was among the most eminent of this group. His remains are not on public display, but a monument in his tribute, created by the sculptor Richard Aigner, was erected in the Parque Centenario near the museum (Fig. 2.14).

Burmeister was an indefatigable worker (Birabén, 1968). His broad encyclopedic knowledge and expertise included entomology, ornithology, geology, meteorology, and paleontology, including descriptions of several fossil mammals published, in those less specialized times, in the journal of the Sociedad Argentina de Farmacia (Burmeister, 1863–1864). Among his most notable contributions to paleontology are that, as director of the museum, he transformed a haphazard hodgepodge of specimens, artifacts, and curiosities into organized collections accessible for scientific study (in principle, at any rate—Florentino Ameghino would have told us a different story), improved the museum's holdings with new specimens, such as Muñiz's *Smilodon*, and created the first paleontological society in South America, the Sociedad Paleontológica. He also initiated the publication of the *Anales del Museo Público de Buenos Aires* in 1864, the first volumes of which were devoted to the study of the highly significant fossil mammals of the Pleistocene Pampean Formation, mostly massive sediments that underlie vast extensions of pampas soil, accompanied by his own exceptional illustrations (Fig. 2.15). In the *Anales*, Burmeister (1864:7) stated that the collection of fossil mammals is the richest part of the museum and that the province is their most abundant deposit: "los esqueletos más curiosos y completos de animales antidiluvianos, que se ostentan en los Museos de Londres, París, Madrid, Turín, etc., todos han salido de la provincia de Buenos Aires" ["the most curious and complete skeletons of antediluvian animals, on display in the Museums of London, Paris, Madrid, Turin, etc., are all from the province of Buenos Aires"]. His work on glyptodonts (Burmeister, 1874) is still quoted, and that on the Pampean fossil horses was selected by the government to represent Argentina at the Exposition of Philadelphia in 1876.

Burmeister was a determined anti-Darwinist, and this made him one of Florentino Ameghino's most significant rivals. Indeed, one of the most bitter and acrimonious disputes in paleontology occurred between these two men. Their bickering had been ongoing for years, but their scientific differences, ironically, boiled over and became overtly personal over the interpretation of a megathere that Ameghino had named after Burmeister. The insults flew fast and furious. The latter's intellectual position on evolution and adversarial relationship with Florentino, whose stature was

2.14. The monument to Burmeister by Richard Aigner, erected in the Parque Centenario near the Museo Argentino de Ciencias Naturales "Bernardino Rivadavia," Buenos Aires, Argentina.

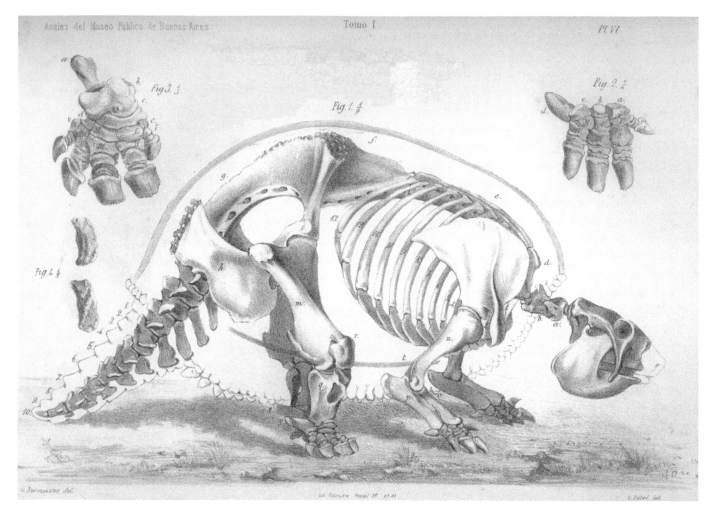

2.15. Among Burmeister's many contributions to paleontology is his work on glyptodonts, for which he provided his own high-quality illustrations, such as that reproduced here of *Glyptodon* (Burmeister, 1874: pl. 6).

subsequently raised to near legendary proportions in Argentina, have probably done much to unjustly relegate the memory of Burmeister's collective contribution to relative obscurity.

The Ameghino brothers: patriots and tireless researchers

Sons of humble Italian immigrants, the Ameghino brothers' body of work on South American paleontology is probably unrivaled. Indeed, the framework for understanding both the singular nature of South America's Cenozoic fossil mammals and the basic subdivisions of Cenozoic time were essentially established more than a century ago by the combined efforts of Florentino and Carlos Ameghino. It would not be too great an exaggeration to suggest that much of our subsequent efforts have gone to elaborating and filling in the gaps in their knowledge.

Florentino Ameghino (1854–1911; Fig. 2.16) was born perhaps in Liguri, Italy, possibly on the ship that bore his parents across the Atlantic, or probably more likely in Luján, Buenos Aires Province. He was conditioned for his profession by the city of his childhood, Luján, where news of Francisco Muñiz's discoveries had ignited the public's imagination. Luján is a sanctuary for Catholics, adorned by the impressive Basílica Nacional Nuestra Señora de Luján, the site of untold pilgrimages to which supplicants make their way on bended knee, or are meant to anyway, as part of the pact for having their prayers answered. But Luján is also a place held sacred by paleontologists. It was here where the *Megatherium americanum* specimen described by Cuvier was recovered and where Muñiz had collected fossils a few decades earlier. This locality has given us the most representative assemblage of fossils of the late Pleistocene, so that the age itself takes its name.

Florentino's younger brother, Carlos Ameghino (1865–1936; Fig. 2.18), born certainly in Luján, was an invaluable ally who began to help Florentino with his fieldwork in 1876 at the age of 11. His first forays were in the late Pleistocene deposits of Río Luján and Arroyo Frías (between Mercedes and Luján, Buenos Aires Province). Carlos was a tireless collector who crisscrossed Patagonia, often alone (although sometimes he hired workers or asked natives for help), in his search for fossils near the turn of the century. Anyone who has had the opportunity to travel through the desolate Patagonian landscape cannot help but shudder at the thought of what his life was like more than a century ago. But it was there, in so inhospitable a desert yet a garden of delight for fossil lovers, that Carlos was most at home. One particularly chilling anecdote was recorded by a naturalist making his way through the vast expanses of Patagonia's Pampa del Castillo. At a certain point, the naturalist was startled to find a pile of chests covered by a thick layer of snow, which had been falling for the previous few days. Searching through the pile, he realized there was a man bundled up beneath it. Surprise turned to astonishment when the bearded figure turned out to none other than Carlos, who had been forced to hole up there for 3 days because the vehicle that had been sent to fetch him and his packages had been delayed by the storm.

There was a middle Ameghino as well: Juan. This brother apparently did not share the same feverish enthusiasm for fossils as his brothers, but he nonetheless played an important role in the equation that allowed the continued scientific efforts of his brothers. Although it may seem incredible, most of the Ameghinos' efforts in collecting and describing fossils were not funded by a governmental institution. Except for a relatively short stint at the Museo de La Plata, and a somewhat longer period at the then Museo Nacional de Historia Natural de Buenos Aires—but the latter near the end of their active careers—the Ameghinos paid for much of their own research. Some of the expenses were covered from the sale of fossils; the rest derived from two bookstores that the brothers ran, where they sold items such as paper and rubber stamps in addition to books. The first of these, in Buenos Aires, was named El Glyptodon after the extinct, large, and tanklike relative of armadillos. It is of passing interest to note that a bookstore bearing the same name exists still in Buenos Aires, but we have been unable to verify whether its owner meant to honor the Ameghinos' first store. In any event, the bookstore was eventually moved to La Plata, under a different name, where Juan was the main shopkeeper while Florentino researched and published on the fossils that Carlos sent back from the field, as described in letters such as the one reproduced in Box 2.2.

2.16. Florentino Ameghino. The great Argentinian scientist was among the foremost paleontologists of the latter half of the nineteenth century and remains a revered figure in his native country and to all students of South American fossil mammals.

Courtesy of the Museo de La Plata, Argentina.

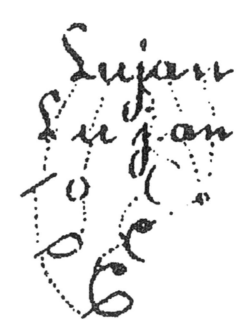

2.17. Example of Ameghino's stenographic system. The word and its shorthand forms is *Luján*. Although his system was concise, it never really caught on.

Box 2.2

Letter from Carlos Ameghino to his brother, Florentino (translated by the authors)

Luján, March 30, 1886
Mr. F. Ameghino—Buenos Aires

Dear Brother:

I did not get back to you promptly to tell you that I had received your package because I had wanted to know the results of our excavation of Olivera's Panochthus, which I had planned for and completed by Sunday, the day before yesterday, so that I could let you know about both things.

I left home early in D. Demetrio's cart. As soon as I had arrived, I began by clearing off all the earth that had entombed the great piece of carapace, with the idea of leaving it exposed to dry in the sun so I could glue it and carry it off in one piece. Unfortunately, however, the afternoon turned cloudy. Not only did the carapace remain wet, but so much water flowed into the excavation hole since we'd wrapped the fossil that it covered over and completely impregnated the fossil.

So as not to waste the trip and the day's work, I lit a thistle wood fire on it to see if that would help. It did dry off enough for me to harden it with glue, so that I could move it whole onto a bed of hay prepared beforehand in the cart.

All for naught, as it turned out. Having spent all day on the carapace, night came upon us and we were caught in darkness on poor and unknown roads. With the gates of the fields already shut, we had to rush off, and the subsequent jolting of the cart broke the carapace. It arrived in several large pieces, which, fortunately, can now easily be put back together.

As for the other parts of the animal that are in the bit of earth I was able to clear, well, they're no big deal, amounting only to a few other pieces of the carapace, some ribs, and two or three caudal vertebrae, which makes me think that the tail is probably not far behind

Anyway, what I've already got is a fine sight, with one of the larger carapace fragments preserving, for example, a good part of the cephalic notch, which is magnificent.

As I just mentioned, Olivera's gate was locked when we returned at night in the cart, so that I had to go to the nearest worker's quarters to ask to be let in. The *puestero* was the same peasant that saw us taking the *Mylodon* from the river. Well now, before opening the gate he warned us that it would be a wise thing, for him as well as for us, to fill in the holes that we had dug in the river. I asked him why and he replied that, besides preventing an animal from being killed, should it happen and Olivera find out about it, he would surely order him to bar anyone from entering the fields, and so he'd be obliged to prevent us from taking any fossils we might find. Of course, worried that he'd leave me to sleep out in the open, I promised to have the holes filled in. I don't think we'd lose much by this, as it wouldn't take long to fill them with a worker helping, and then I'd have the rest of the day to pick away at the "Panochtus" to look for its tail and go to Mastodont Creek to see if there's anything else from the Scelidotherium; and if not, then to the other Glyptodon carapace.

I'll await your instructions on what to do: should I fill them in or not? Let me know. We're preparing for the journey, and we'll leave as soon as possible. Regards to you, Leontina, and Juan from Mom and me.

Carlos Ameghino

Carlos's contributions are often overshadowed by Florentino's accomplishments, so that the former is generally, though incorrectly, viewed as the latter's sidekick, essentially the one who did the dirty work that Florentino polished and published. Without doubt, Carlos Ameghino was not

only a collector but also a great scientist in his own right, methodologically rigorous and able to express and formulate original ideas. Although he did not publish a great number of articles, his qualities and abilities are clearly demonstrated through his observations and comments in the abundant correspondence with his brother. In 12 subsequent trips to Patagonia, Carlos discovered four major successive faunas that reveal the general aspects of mammalian evolution of Patagonia from the early Eocene to the early Miocene. During these trips, Carlos made highly significant geologic observations and interpretations, including maps and profiles of the main stratigraphic units of the Cenozoic strata of Patagonia. He also recognized local and regional uncomformities, marine transgressions, and numerous other geologic features. This information permitted Florentino to synthesize a complete stratigraphic scheme for Patagonia that remains essentially valid (Ameghino, 1906). This bible of Cenozoic geology and paleontology of Patagonia could not have been completed without Carlos's fieldwork, ideas, and interpretations. There is also an incident in which Florentino ignored Carlos's interpretation and thereby erred. Indeed, when Carlos found *Argyrolagus* in Chapadmalal, he sent it to Florentino with the suggestion it was a very odd marsupial. Florentino instead described it as a rodent, but he later had to retract his opinion.

2.18. Carlos Ameghino. He never received the same recognition as his older brother, Florentino, but it was largely on Carlos's dedication, perseverance, and abilities as a geologist and collector in the field that Florentino's reputation was made. Image possibly taken in the old building of the Museo de Buenos Aires.

Courtesy of the Museo de La Plata, Argentina (donated by the Parodi family).

In 1913, Carlos was appointed chief of the paleontological section of the Museo Nacional de Historia Natural de Buenos Aires and in 1919 as director of the museum. After Florentino's death, Carlos dedicated himself to the organization of the immense collections he had accumulated and the training of a group of young disciples, including Lucas Kraglievich, Alfredo Castellanos, Carlos Rusconi, and Lorenzo Parodi.

Florentino was as tireless as his brother, but his strengths lay in analysis and synthesis. Although he died at the relatively young age of 56, he nevertheless managed to describe and study all of the important groups of South American mammalian faunas. He became entangled in bitter arguments with several of his colleagues. The dispute with Hermann Burmeister is mentioned above. In addition, he clashed with British naturalist Richard Lydekker and with Francisco Moreno, the director of the Museo de La Plata, a wonderful late nineteenth-century museum set in a wooded backdrop that remains still a beacon of light in the study of South American fossil mammals. Florentino was offered a position at the museum by Moreno in 1886 in exchange for exhibiting his fossils (most of which had been collected by Carlos, who was given a position as a traveling naturalist). Florentino's bitter tangle with Moreno led to his resignation (and eventually Carlos's) from the museum in 1888. He was appointed director of the Museo Nacional de Historia Natural de Buenos Aires but maintained his residence in La Plata until his death. In addition to being a great paleontologist, Florentino was talented in other ways. For example, he devised a system of stenography (or shorthand), as shown in Fig. 2.17.

A confirmed evolutionist, one of his works is entitled *Filogenia*, or *Phylogeny*, a monumental work published in 1884. Another of his great works concerns fossil mammals of Argentina (Ameghino, 1889) that won a gold medal in the Paris Exposition of that same year. He was a member of the main international scientific societies of his time. Although he could not be accused of being mean-spirited, his fierce first-generation patriotism occasionally led him astray—or at least that's how we view some of his ideas, rather unfairly because we have much more information than was available to him.

Even so, it is worth analyzing the errors of great thinkers because their mistakes can sometimes teach us as much or more than their accomplishments. It is not the minor prejudices that muddle their insight, but the larger assumptions firmly embedded in the cultural environment of a particular time and place. Ameghino stubbornly insisted on a South American origin for the Pleistocene fauna, as well as for humans. In the first case, modern investigators are reconsidering a South American origin in the broader context of the development of the southern continents themselves. With regard to human origins, however, the reinterpretation of American remains and the discovery of African hominids have firmly proved Ameghino wrong. However, such was his influence that even as late as 1978, a participant at the First Latin American Congress of Paleontology continued to defend his views.

Often, however, the criticisms leveled at the Ameghinos, especially Florentino's interpretation of an older age for the strata and fossils in South America, are completely out of context. One of the main mistakes made by Florentino is the great antiquity that he attributed to some of the *pisos* (fossiliferous layers) in Patagonia. He believed that mammals coexisted alongside dinosaurs in what are actually early Tertiary levels, but to which he assigned a late Cretaceous age on the basis of the apparent presence of dinosaurs (because all nonavian dinosaurs became extinct at the end of the Cretaceous).

Rather than dismissing Florentino's ideas as the result of an overzealous patriotic streak, we should take into account several factors in trying to make sense of Florentino's error and several of his subsequent ideas on mammalian evolution. For example, when Carlos began collecting in Patagonia, there were essentially no previous geologic studies to guide him; just about everything was totally new to the scientific eye. Carlos had certainly found dinosaurs in undoubtedly Cretaceous-aged levels. From Tertiary-aged deposits, he had recovered mammalian remains together with small, serrated teeth that were then known only from dinosaurs. These, too, were sent back to Florentino. Armed with this knowledge, the evolution of Florentino's thinking becomes obvious: because dinosaurs were Cretaceous, the levels from which the mammals came were of the same age. As a consequence, these mammals, more advanced in their evolution—because they were (supposedly) from the Cretaceous—than those from other continents, could be considered ancestral to the latter, with the concomitant evolutionary and paleobiogeographic implications that this entailed.

As time passed, researchers began to report the discovery of numerous new finds of the serrated dinosaur-like teeth in undoubted Tertiary-aged strata. Suspicion briefly fell on Carlos: it was claimed that he had mixed up specimens from different fossil layers. The mystery was solved much later, in the 1930s, when G. G. Simpson found remains of an unusual crocodile in early Tertiary beds. It had a dinosaur-like skull and serrated teeth similar to those of carnivorous dinosaurs. Carlos was vindicated. The teeth were exactly like those he had recovered, and they were from the same levels where he claimed he had found them. We now know that the early Tertiary of South America saw the evolution of Sebecosuchia, a group of crocodiles with dinosaur-like teeth that cohabited with, and probably fed on, a varied mammalian assemblage.

Between 1896 and 1899, John Bell Hatcher (1861–1904; Fig. 2.19), then curator of the department of vertebrate paleontology of Princeton University, planned and conducted three memorable expeditions to the essentially uncharted interior of Patagonia. These journeys are considered to be among the most important in the improvement of our knowledge on fossils from South America. By the close of the nineteenth century, many researchers had begun to cast a long and mainly critical gaze at the discoveries and investigations of Florentino and Carlos Ameghino, who were laboring under the relative geographical and intellectual isolation of southern South America. Hatcher was also inspired by the Ameghinos' work: his expeditions were conceived mainly in the belief that many of the challenging hypotheses set forth by Florentino Ameghino would be invalidated under the scrutiny of skilled and trained observers from the Northern Hemisphere. Of course, this feeling was not restricted to Hatcher, as observed by Wallace (2004:85): "Argentinean claims to mammalian priority disturbed North

John Bell Hatcher

2.19. John Bell Hatcher. His expeditions to the remote and isolated regions of Argentine Patagonia produced one of the most important South American fossil collections.

Americans even more than Europeans. Manifest Destiny was still in the air, and empire on the horizon." On the other hand, Hatcher was a bit more prickly than most, and he crossed the bounds of diplomacy (granted, not necessarily of much concern in many scientific endeavors), stating in print (as recounted by Wallace, 2004:28) that he doubted whether even someone as capable as Florentino could "determine the exact sequence of strata in Patagonia from the window of his study, situated in La Plata or Buenos Aires." In this Hatcher not only insulted Florentino unjustly (he knew full well that Carlos was the main source of information on stratigraphy), for there is nothing wrong with collaboration, but he also disparaged, perhaps even more unjustly, Carlos's astute abilities on geologic matters.

Hatcher's enormous collection was sent to Princeton University, but it now resides largely in Yale University's Peabody Museum. The scientific success of Hatcher's work is reflected in the *Reports of the Princeton University Expeditions to Patagonia, 1896–1899*. This monumental, lavishly illustrated work includes five volumes, published between 1903 and 1928. Hatcher (1903) wrote the first volume, *Narrative of the Expeditions: Geography of Southern Patagonia*, which was republished as *Bone Hunters in Patagonia* (1985) and is thus readily available. The other volumes represent the efforts of some of the foremost specialists at the time. For example, W. B. Scott (1903–1904), one of the main U.S. paleontologists, analyzed the xenarthran remains (such as those of the sloth *Hapalops* and the glyptodont *Propalaehoplophorus*, Figs. 2.20, 3.13), and W. J. Sinclair (1906) studied the marsupials.

The first expedition extended from 1 March 1896 to 16 July 1897, the second from 7 November 1897 to 9 November 1898, and the third from 9 December 1898 to 1 September 1899. Each of the three expeditions was a major undertaking, given the distance and geographical isolation of that region. Although Hatcher was accompanied by either O. A. Peterson or A. E. Colburn as assistants, he spent long stretches of time alone overcoming numerous obstacles either peculiar to Patagonia, such as the harsh climate with its fogs, driving rains, snowstorms, and blizzards—let's keep in mind that he overwintered in Patagonia—or those caused by physical ailments, such as a head wound inflicted by his horse and the persistent rheumatism that kept him in his tent for 6 weeks in the middle of the winter.

His strength, determination, and capacity to work in solitude were crucial for the success of the expeditions, but his solitude also reflects, we suppose, his inability to understand the cultural mix in that part of the world, so distant from his home. In his *Narrative*, he made it obvious that he was comfortable with the *estancieros*, progressive hard-working and mainly British farmers who had brought many of their customs and traditions from the old country. He was not so kindly disposed toward the hired help, individuals usually of combined Spanish and Amerindian heritage, that he must have come across on numerous occasions, yet whom he mentioned only sporadically, and in derogatory terms at that; or toward the gauchos, repeatedly referred to as the Argentinian equivalents of North American cowboys—a distortion of their true social status, which Hatcher should have grasped given the length of his stays—whom he treated with contempt; and the Tehuelche, the last descendants of a once great race already in serious

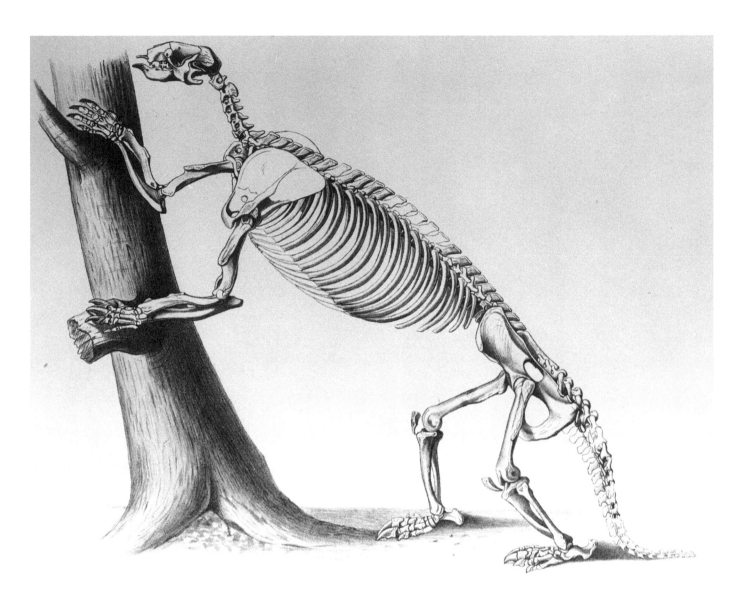

decline and condemned to extinction, whom he mentioned merely as objects of study.

Hatcher's *Narrative* is a wonderful record of the rigors of fossil collecting more than a century ago in a remote and desolate part of the world, and we recommend it wholeheartedly for anyone interested in paleontology. We may gather from it that in many ways, little has changed. Except for a few bits of modern technology—a winch, a rock saw, walkie-talkies—the fieldwork itself and the sensations derived from being in the remoteness of Patagonia and the abundance of Santa Cruz fossils (Fig. 2.21) are essentially the same as he portrayed them.

The *Narrative* is also filled with movingly vivid and descriptive passages, so that one might almost imagine and feel an affinity for the lonely figure in his tent, bent over the day's field notes after having beaten the odds, filling page after page, recording his impressions and thoughts and musings so that one day they might be published—granted, a romantic image, but probably one that most paleontologists would want for themselves. Oddly, there seems no room in the *Narrative* for comments that might reflect the joys, surprises, and nuances of daily life, other than those produced by the work itself. One wonders whether these omissions are merely

2.20. Among the spectacular fossils recovered by Hatcher's expeditions were those of xenarthrans. This illustration of the sloth *Hapalops* is from Scott's (1903–1904: pl. 30) report.

2.21. One of the authors (S.F.V.) collecting along the Atlantic coast near Rio Gallegos, Argentine Patagonia. Although modern paleontologists have a few modern amenities, such as vehicles instead of pack mules, the solitude and rhythm of the work and the beauty of the cliffs and ocean remain as they were during the times of Hatcher and the Ameghinos.

a reflection of Hatcher's personality. He does, however, reveal something of his empathy and sensitivities, as when he had to kill, skin, and collect the skeleton of extant mammals to fulfill the scientific requirements of his expeditions. He wrote of having just shot two guanacos:

> As I approached they stretched their long necks toward me, while at the same time from their strikingly large and beautiful eyes they looked up, as though imploring mercy. I almost wished I had not fired the fatal shots. Since, however, I had been the cause of their destruction, I would not be unnecessarily cruel and leave them to a lingering death, so, reaching for my knife, I drove the glistening blade between the atlas and occiput quite through the spinal cord and deep into the cranial cavity of each.

On the other hand, we may also gather that he was not above sticking his nobler faculties in his back pocket when it suited him. He was all too pleased at having managed to circumvent authorities and regulations in sending a load of fossils on their way home, on "account of certain laws prohibiting the exportation of fossils from Argentine territory" (Hatcher, 1903:85, in Hatcher, 1985). We will end Hatcher's story with his own words, for further comment would only spoil their wonderful imagery:

> Ours was a remarkable and interesting experience, as patiently pursuing our work, we sat on the surface of the talus-covered slope at a safe distance from the waters that dashed furiously beneath, with a enormous wall at ours backs, rising perpendicularly to a height of more than four hundred feet, while over the sandstones of the beach from which but a few hours previously we had been excavating the remains of prehistoric animals, there now rolled a sea sufficiently

deep for the safe navigation of the largest transatlantic liner. What a remarkable change, in so short a time! Not many experiences even among those of my childhood, that most impressionable period of our lives, have left themselves so indelibly engraved upon my mind. At this somewhat distant perspective, when not engaged in any particular line of thought, as some times happens, I frequently detect myself in the midst of a most vivid mental picture drawn from some particularly interesting point along this coast. Here seated in the foreground on some convenient ledge intently listening to the deep sonorous music of the waters beneath, as the crest of each great tidal wave comes rolling in from its long journey across the eastern seas and breaks with thundering force upon the rocks below once again, I am for a moment transported to some favorite spot and busily engaged in my chosen work, only to awaken almost immediately from my reverie to a painful consciousness of the delusion and to wonder whether or not the desire was father to the dream. (Hatcher 1903:74, in Hatcher 1985)

Kraglievich, a genius lost

Lucas Kraglievich (1886–1832; Fig. 2.22), born in Balcarce, Buenos Aires Province, was sufficiently talented and industrious to have become one of the all-time greats, which in many ways he was, except he died young—before his 46th birthday. In addition, his experience demonstrates that the scientific community is not any better or nobler than society at large. His work is impressive, especially because his university training was in mechanical engineering and he was essentially a self-taught paleontologist, although under the supervision of Carlos Ameghino.

A prodigious publication record is not necessarily a reliable guide of ability, because the quality of the written work must be considered as well. But when massive output is matched by quality, a scientist's stature may be judged by the quantity of publications. In relatively young scientific disciplines, there is considerable value in first descriptions of new information. This is particularly relevant here because paleontology in South America was at just such a stage in Kraglievich's day, so we can gauge the scope of his work by how many new species and higher categories he created. He named 28 families or subfamilies and more than 80 genera and subgenera, and he created or transferred 250 species or subspecies. In addition to this unrelenting activity, he directed his intellectual talents toward challenging theoretical questions and had begun proposing insightful hypotheses just as he was reaching his peak as a mature scientist. We can only speculate what more he might have produced had his life not ended prematurely.

However, hidden envy, as well as mediocrity and lack of courage on the part of those in positions to defend him as he deserved, forced Kraglievich into exile in Uruguay, where he pioneered the study of paleontology in that country. As Camacho (2006) noted, after many years of intense duty in the Museo de Ciencias Naturales, he obtained a modestly paid position as a technical assistant of paleontology from 1919 to 1929. Between 1925 and 1929, with Carlos Ameghino seriously ill, Kraglievich was temporarily appointed to be in charge of the museum. Carlos Ameghino retired in 1930. Even though Kraglievich had been acting admirably as deputy director since 1925, he was not offered the position as the next director, despite the unanimous opinion of his colleagues and even of Don Carlos himself. Instead, it was suggested that he accept a position as a traveling naturalist. Of course, he considered this decision unfair, given his achievements and hard work, and he tendered his resignation by the end of 1930. A few years before,

2.22. Lucas Kraglievich. An indefatigable researcher and writer, Kraglievich had the skill and production to have assumed directorship of the Museo Argentino de Ciencias Naturales, a position that was unfairly denied to him. He subsequently left Argentina and established paleontology as a professional discipline in Uruguay.

in 1927, he had visited Uruguay as part of an Argentinian–Uruguayan study committee, and he had taken advantage of the occasion to make several interesting discoveries, such as that of dinosaur remains that suggested the existence of Cretaceous sediments; and thus began his study of Uruguayan paleontology. Consequently, and because of the warm welcome he had received, he settled in Uruguay in the beginning of 1931. Lamentably, he was able to devote only a few hard-working years to his Uruguayan studies. Had he been allowed a few more years, paleontology in Uruguay would no doubt have developed more quickly.

Such is destiny. In compensation, however, his name and legacy are secure, whereas those who were much less capable and made his life impossible have been relegated to a well-deserved oblivion.

Simpson: all over the landscape

The U.S. paleontologist George Gaylord Simpson (1902–1984; Fig. 2.23) was a scientist of the highest international stature at the forefront of both paleontology and evolutionary biology. On strictly paleontological terms, he can fairly be regarded, along with Alfred S. Romer, as a giant of the twentieth century: it is no coincidence that the highest annual award offered by the Society of Vertebrate Paleontology (of which Simpson, along with Romer, was a founding member) is the Romer–Simpson medal. These two scientists were perhaps the last of a breed, both in terms of scope and production. They remain the ideal that many of us strive to emulate, although today's degree of specialization would seem to prevent any one person from achieving such widespread and lasting impact.

Simpson, by the scope and influence of his work and ideas, both theoretical and empirical, was a true colossus, rated by Gould (1985:229) as "the most important paleontologist since Georges Cuvier" in part because he "established and inspired nearly all the fruitful directions of modern paleontology." In his nearly 800 published works (Laporte, 2000), both popular and academic, Simpson displayed a breadth of ability and expertise that will likely never be matched. Indeed, his career really was all over the landscape, as he noted about himself as a young student field worker in a letter to one of his sisters (Laporte, 2000).

As a vertebrate paleontologist, Simpson was accomplished in the field, excellent at prospecting and collecting, and a brilliant researcher and writer. He led numerous successful field expeditions in several countries, including two early year-long trips to Patagonia, experiences that supplied material for one of his first popular books, the entertaining *Attending Marvels: A Patagonian Journal* (1934). He continued regular fieldwork until a near-fatal accident in the Brazilian Amazon in 1956 left him lame, curtailing further serious field efforts.

An expert in nearly all mammalian groups (as well as some nonmammalian taxa), his synthetic efforts on their systematics produced several landmark works, including *The Principles of Classification and a Classification of Mammals* (Simpson, 1945). In addition to such comprehensive coverage, he produced important monographs of particular groups, including Mesozoic (Simpson, 1928, 1929) and South American mammals (Simpson, 1948, 1967). His relevance to our story here is already obvious. Simpson distinguished himself by his dedication to the South American fossil fauna, on account of which he traveled repeatedly to this continent (or subcontinent, as South Americans prefer to recognize it). Indeed, through his many publications, he is the paleontologist who has made the greatest contribution to knowledge of South American fossils. In addition to the typical paleontological endeavors, such as describing and naming fossil remains, we owe to him the concept of faunal strata (discussed in Chapter 5), which clarified our understanding of the sequence of faunal compositions between North and South America before and during the Great American Biotic Interchange. Further, Simpson worked out the principles of historical biogeography that help explain these distributions.

As alluded to above, much of what made Simpson famous, at least among biologists, transcended his role as a traditional paleontologist. His theoretical contributions in classification, systematics, historical biogeography, and evolutionary biology were at the forefront of such efforts. Simpson's concepts on methods of classifying and naming animals (formalized in the above-mentioned *The Principles of Classification and a Classification of Mammals* and in *The Principles of Animal Taxonomy*, 1961) were hugely influential. Although the methodologies have been largely superseded by phylogenetic systematics (or cladistics; see Chapter 1) some of his views on mammalian relationships are still occasionally considered in the work of modern researchers.

More lasting was his work on variation in fossil taxa. Building on the work of some of his mentors, such luminaries as H. F. Osborn, W. K. Gregory, W. D. Matthew, and R. S. Lull, Simpson was among the first to consistently view fossils as once-living, complex biological entities. That this should be so may seem absurd, but much of paleontology during the early part of the twentieth century had come to concern itself essentially with empirical morphological description and its change and sequence through strata (Laporte, 2000). Simpson viewed fossils as real flesh-and-blood beings rather bits and pieces of broken bone. One of the factors that allowed this was Simpson's application of statistical concepts to fossil remains. This was in stark contrast to the common practice of viewing a single specimen, the type, as the all-important signature specimen of a species, the one that all others were meant, somehow, to be like. Simpson, however, viewed all individuals of an interbreeding species as equal members of the species, and he thought that a proper understanding of any species, particularly when working with the fossil record, requires population (as opposed to typological) thinking. That is, for many characteristics, such as size, a biological species has a range of variation and includes more typical individuals—those that cluster closely about a mean—as well as those that are farther removed from the mean and those at the extreme ends of the range. Simpson was adamant that we take such biological facts into account and set out to introduce statistical rigor into systematics. Much of his thinking about statistical principles as applied to zoological inquiry was formalized in *Quantitative Zoology* (1939), coauthored with Anne Roe, his second wife and a renowned psychologist (Laporte, 2000).

We have already noted many of Simpson's important contributions, but there is yet another, one that is universally recognized as his greatest contribution: his theoretical efforts in evolutionary biology (Gould, 1985). Together with several other biologists during the 1930s and 1940s,

2.23. George Gaylord Simpson. Certainly the greatest vertebrate paleontologist of the twentieth century and probably the most important since Cuvier, Simpson's ideas revolutionized paleontology and contributed enormously to our modern understanding of evolution. This photograph, courtesy of L. Laporte, was taken in 1933 or 1934 during Simpson's fieldwork in Patagonia. The photograph was given to Professor Laporte with "G. Patagonia 33–34" written on the back in Simpson's handwriting.

Simpson was a key founder of the modern synthesis, the essential paradigm of modern evolutionary theory, through which Darwin's theory of natural selection was integrated with the discoveries of genetics and ecology, disciplines that expanded enormously during the twentieth century, to provide a more complete understanding of the workings of nature. Simpson's main contribution was *Tempo and Mode in Evolution* (1944; and an expanded, more detailed treatment of the subject in *The Major Features of Evolution*, 1953a). Simpson supplied an essential pillar to the synthesis "by arguing that the grand sweep of life's history could be rendered consistent with the genetics of Darwinian processes in populations, thus forging a unified theory applicable to all scales" (Gould, 1985:229).

By using fossils in doing so, Simpson accomplished something else: he single-handedly brought "paleontology into the mainstream of biological research by validating the use of fossil evidence for solving evolutionary questions" (Laporte, 2000:2). This may seem odd, for today most people would naturally assume (in large part because of Simpson) that fossils provide some of the most important evidence for evolutionary theory, but as noted above, during the early years of the twentieth century, paleontology was concerned mainly with descriptive morphology. Gould (1985:231) suggested that the field had become "either backward or uninvolved [in evolutionary theory] right to the dawn of the synthesis—resolutely empirical, antitheoretical, traditionally focused on the narrow specialization of a chosen time, place and taxon"—somewhat exaggerated, perhaps, but in the main not far off the mark. By way of further demonstrating Simpson's influence, however, we note here that some theoretical interpretations had been postulated by paleontologists, but these were largely inconsistent with or contradicted the mechanisms of evolution as worked out through research in biological fields such as genetics and ecology; this only heightened the estrangement between paleontology and the rest of biology. These earlier paleontological explanations, such as inertia and momentum, racial senescence, aristogenesis, and orthogenesis, relied on inherent or internally directed forces as factors directing evolutionary change. In *Tempo and Mode in Evolution*, Simpson, a paleontologist, debunked such explanations as inconsistent with genetic theory (Laporte, 2000).

So far this account of Simpson paints a decidedly glowing picture of the man as a scientist; but this is not to say that his word is received gospel. He was wrong in many aspects, but this is only to be expected in science. As an example, his work on historical biogeography was undertaken in large part as a reaction to the earlier ideas of continental drift (see Chapter 3). Simpson held on to the concept of fixed continents, in agreement with most North American geologists of the time, until insurmountable evidence in the 1960s and 1970s made it (scientifically) impossible to do so. In fact, Simpson's work on fossil distributions had done much to sway the argument against moving continents; he showed that Tertiary fossil mammals, at least, could not be used to support drift. In this respect, despite being obviously wrong about drift, Simpson was partly right: much of the historical biogeography of Tertiary mammals does not depend on moving continents. Some of his mechanisms for explaining distribution (such as island hopping and rafting; see Chapter 4) are still useful and play a role in the story of South American mammals. As further examples, we may note that some of Simpson's ideas on the relative importance of several evolutionary mechanisms, such as an overly selectionist position and the dominance of slow, gradual phyletic change in evolutionary history, have been challenged.

For all of his scientific prowess, as a person, Simpson seems to be remembered as having been a difficult man. For example, Laporte (2000:3) noted that few were ever able to get emotionally close to him, even those with whom he had worked for a long time; he was sharply critical; he did not make friends easily; and perhaps most damning to this aspect of his reputation, "those whom he referred to as good friends in his autobiography were surprised to be so considered." Similarly, Gould (1983) noted that he was not an easy man to like, that he couldn't bear disagreement, and that he took offense easily. Certainly there are many stories about his disagreements with others, particularly in situations where he thought he had been unjustly treated—for example, the debacle that led to his resignation from the American Museum of Natural History, where he had worked for so long and established his reputation (and enhanced that of the museum). The impression is of an aloof, bitter, selfish, cranky, perhaps egotistical individual, one who recognized his own marked intellectual superiority over those around him and distanced himself. If he was guilty of this, then he is not much different from many of the individuals portrayed here, as well as from others who do not merit even the least mention. But we wonder whether Simpson deserves such a reputation. It is true, as Gould (1983) noted, that in the long run, it really doesn't matter, because it is the power and impact of his science that will last.

He was remembered rather differently by E. C. Olson, a renowned paleontologist who knew Simpson for nearly 50 years. We will end this brief biographical sketch with Olson's (1991:332–333) words:

> In those early years, I—who was very young, along with many others who were not—found myself awed and tongue-tied in his presence. This reticence in turn affected George, who, misunderstanding it, acted withdrawn and taciturn, confirming our expectations. I felt he deemed us not quite up to his advanced level of reason and knowledge—a fact that was certainly true but was not, I believe, a correct assessment of his reactions. Looking back this now seems very mixed, but it certainly seemed real at the time and continued to affect George's relations with others for years to come.

After Peter Lund's pioneering efforts during the mid-1800s, paleontology in Brazil fell into a lull. Certainly there were several, mainly foreign-trained and nonnative Brazilian researchers such as Hermann von Ihering (1850–1930) and Émil Goeldi (1859–1917), who continued to study and publish on Brazilian fossils, but none of them established a long-term research program involving the training of graduate students, hence future paleontologists, so as to carve out a tradition of paleontological research in Brazil. An analogous, though less severe, situation occurred in Argentina, following the efforts of the Ameghino brothers and Lucas Kraglievich.

It was into this relative vacuum that Carlos de Paula Couto (1910–1982; Fig. 2.24) burst onto the scene in the 1940s, going on to become the most important Brazilian mammalian paleontologist of the twentieth century. That he emerged seemingly from nowhere is not an exaggeration. Born in

Carlos de Paula Couto: paleontology in Brazil

2.24. Carlos de Paula Couto. A giant of twentieth-century mammalian paleontology in Brazil, his efforts produced a long line of intellectual descendants, ensuring a healthy future for paleontology in Brazil.

Courtesy of Departamento Nacional de Produção Mineral, Brazil.

the southern Brazilian state of Rio Grande do Sul, Paulo Couto never obtained a degree in geology or paleontology, because in those days advanced courses in these fields were not offered in the state's higher institutions. Rather, he obtained his formal education at a military school in his native city, Porto Alegre. By his early teens, he had acquired that passion for fossils that marks a paleontologist for life.

At the age of 21, he decided, perhaps impetuously and much to the chagrin of family and friends, on an uncertain future in science rather than a stable life in the military, and he continued in earnest the task, mainly through locally available resources, of becoming a self-taught paleontologist and geologist. Paula Couto worked at various jobs during his early adulthood, including a stint as an usher at a local movie theater—a position that he must have enjoyed, given his love of films. He settled into a more serious career, from 1936 to 1944, in the local offices of the national treasury (Simpson, 1984). He had already become a competent paleontologist by his late 20s, when, in 1937, the first of his many scientific publications appeared. He had managed nearly 30 by 1944.

An opportunity presented itself in that year, when the Museu Nacional, affiliated with the Universidade do Brasil (today the Universidade Federal do Rio de Janeiro), posted a position as naturalist. Paula Couto's application drew high praise, and he was offered the position. He remained at the museum for nearly 30 years and served as its director for a period. He participated in numerous field trips, augmenting fossil collections, continued his prolific publishing, and soon gained an international reputation. A particularly important episode in his career occurred during 1950–1951, when, as a John Simon Guggenheim fellow, Paula Couto was able to visit several museums in the United States, chiefly the American Museum of Natural History, where a long and close friendship began with George Gaylord Simpson. The two paleontologists, who remained great friends for the rest of their lives, collaborated and published together on a variety of fossil mammals. After his retirement from the museum, Paula Couto returned in the early 1970s to his native city, where he continued to work and publish on fossil mammals, established a postgraduate program in paleontology at the Universidade Federal do Rio Grande do Sul, and served as president of the state's Fundação Zoobotânica.

Paula Couto has left a grand legacy. Among his many accomplishments is the vast contribution he made to our knowledge of mammals through fieldwork and more than 200 scientific publications, in which he described many new taxa and provided synthetic analyses of many mammalian groups. Indeed, his contributions are of such scope that it is difficult even today to conduct research on nearly any fossil Brazilian mammal group without consulting or citing his work. Most of his publications were typical of articles published by paleontologists, but at least two merit special mention. In 1950, Paula Couto published *Memórias sobre a Paleontologia Brasileira*, a Portuguese translation of Lund's articles in Danish. This monumental effort, for it was both a translation and commentary, has had a huge and lasting impact, as it made Lund's work accessible to many modern paleontologists—for most of us, it is much easier to read Portuguese than Danish—and Paula Couto's numerous annotations provided a modern perspective on many of Lund's ideas. The second of Paula Couto's landmark publications came out in 1979, near the end his career and life. *Tratado de Paleomastozoologia*, a magnificent tome on the evolution and paleontology of mammals—and that's all mammals, not just Brazilian—may justly be considered Paula Couto's masterpiece. For its time, it was the most detailed and up-to-date treatment of Mammalia. Although its systematics is now dated, it remains a useful account of many mammalian groups, and its scope, diversity, and detail have not yet been equaled.

Another of Paulo Couto's legacies is his involvement in the preservation and protection of his country's national inheritance. His suggestions to a federal minister resulted in the drafting of a proposal that became a federal law (known as the Paula Couto Law; Simpson, 1984) setting out restrictions for the excavation of Brazil's fossil treasures. That this occurred relatively early in his life—before his formal career in paleontology, when he was still employed by the national treasury—makes it all the more exceptional.

His most lasting and arguably most important influence, however, was the establishment of a paleontological research tradition in Brazil. He recognized the value of training graduate students who would go on to continue formal study as independent researchers. This has led to a proliferation of paleontologists who have in turn established centers of mammalian paleontology in various parts of Brazil and continue to produce graduate students. As a result, the state of paleontology in Brazil is decidedly healthy (Box 2.3).

Cástor Cartelle	Box 2.3

Among many of Paula Couto's intellectual descendants, either directly or once removed, are most of present-day Brazilian paleomammalogists. Among them, we might single out here the efforts of Cástor Cartelle (Fig. 2.25). Born in Puente Barjas, Orense, Spain, in 1938, during the Spanish civil war, and currently well into his second career, the first being that of Jesuit priest, Cartelle arrived in Brazil in 1957 and began working on fossil mammals under the supervision of Paula Couto. Cartelle focused his attention on the caves of Minas Gerais, as had Lund, and Bahia, among others. From his base in Belo Horizonte as professor in both the Pontificia Universidade Católica de Minas Gerais (PUCMG) and the Universidade Federal de Minas Gerais, he amassed over the course of approximately 30 years one of the largest collections of sloth remains in the world, most of which (overall, 70,000 specimens) are housed in the Museu of PUCMG. In many ways, we might consider him a modern-day Lund. His systematic and sustained efforts have also produced the largest collection anywhere of a single sloth species from a single locality, which has helped enormously in understanding the variation and population structure in giant ground sloths. Indeed, anyone wishing to undertake serious systematic studies of sloths is well advised to visit these collections, as one of the current trends among some sloth researchers is the description of new taxa on woefully inadequate tidbits, a practice more in keeping with the paleontology of over a century ago than the modern biological discipline it has become.

We cannot end this narrative without mentioning another of Paula Couto's attributes, remarked on by anyone who met and knew the man: that of his character. Quite in contrast to many who reach a certain status during their lifetime, Paula Couto was and remained a sincere, quiet, and pacific person who treated students and colleagues with respect and dignity, regardless of possible conflicts in matters of scientific interpretation. There was none of the belligerence, acrimony, and conflict all too often evident in professional practitioners of the fossil trade, including several who have been written about here. He earned the respect of all those with whom he interacted. A particularly telling anecdote is that he occasionally expressed to students and colleagues the possibility of one day obtaining a formal advanced degree, given his position (a circumstance that was not uncommon in those days, but is almost unheard of now—and probably quite rightly—at major internationally recognized research institutions). The response he always received was something along the lines of, "And just where would we find someone sufficiently qualified to supervise you that hasn't already been trained and supervised by you?"

Robert Hoffstetter: the flavor of France

French science has a rich tradition in the study of South American fossil mammals, beginning of course with the landmark efforts of Georges Cuvier. He was followed by a string of researchers, less well known but who nonetheless made important contributions in the nineteenth and early twentieth centuries. Among these we may mention André Tournouër, who made impressive collections and published on parts of them; Albert Gaudry, who published on many of the important remains collected by Tournouër; and Pierre Marcellin Boule.

All but the last mentioned scientist focused their efforts on Argentinian fossils, which is not surprising because nearly all of the early systematic prospecting activities for South American fossils were directed toward Argentinian deposits, in large part, though not solely, because of the spectacular successes of the Ameghinos. This is reflected by the fact that the type localities for nearly all the South American land mammal ages (see Chapter 3) are in Argentina. Fossils had been recovered from other parts of South America, but (with the major exception of those of Lund in Brazil) these were largely sporadic finds that were not followed by organized prospecting and collecting. The fossils recovered from near Tarija, Bolivia, represented a bit of an anomaly, but even in this case, a major fossil collection amassed by Luis Echazú was mainly from local and decidedly amateur collectors. Two significant portions of this private collection were sold to important institutions, one to the Muséum national d'Histoire naturelle (MNHN) in Paris, where many of the fossils were described and published by Boule (with Armand Thévenin as coauthor), and the other to the Field Museum of Natural History in Chicago, but these fossils have been neither extensively studied nor published. The situation, then, was that the history of South American fossil mammals was based mainly on remains from Argentina, but with important though secondary and geographically and geologically (i.e., Pleistocene) much more restricted sites in Brazil and Bolivia.

2.25. Cástor Cartelle, one of Carlos de Paula Couto's intellectual descendants, has continued in the footsteps of both Lund and Paula Couto and has amassed a huge collection of mammalian fossils. This image was taken during his speech in 2010 at the ceremony celebrating him as an Illustrious Citizen of Minas Gerais.

Courtesy of Cástor Cartelle.

This focus on Argentina began to change in an important way only toward the middle of the last century, mainly through the early and sustained efforts of Robert Hoffstetter (1908–1999; Fig. 2.26), certainly the most influential twentieth-century French paleomammalogist, and a researcher who figured prominently in opening up numerous sites in several northwestern South American countries. It is fossils from these areas that have gained much attention in the last few decades. Further, we should not neglect the work of Ruben Stirton, a U.S. paleontologist who established Colombia as a major source of fossil mammals in the 1940s. Hoffstetter's efforts, however, were considerably more far-reaching, at least so far as South America is concerned.

As usual for paleontologists, Hoffstetter early developed a love of fossils. A graduate in natural sciences, he began his paleontological studies on the invertebrate fauna and stratigraphy of the Cenomanian (a stage of the late Cretaceous) of northern France while employed as a secondary school teacher in the city of Lyon. In association with the natural history museums of Paris and Lyon, he extended his research to squamates, the vertebrate group including lizards and snakes. In the 1940s, Hoffstetter turned his prodigious talents to mammals and South America during his tenure, between 1944 and 1952, as professor of paleontology and biology in the institute of Biological Sciences of the Escuela Politécnica Nacional in Quito, Ecuador. Over this period, he prospected more than 40 Quaternary sites. The abundant material he collected formed the basis for *Les mammifères pléistocènes de la République de l'Équateur* (1952), a classic monograph on Ecuadorian fossil mammals.

Although Hoffstetter returned to France and a position as scientist in France's Centre National de la Recherche Scientifique (CNRS) at the paleontological department of the MNHN (and indeed was appointed as director of both institutions in 1960 and 1972, respectively), his principal focus continued to be in South America. Through both extensive fieldwork and a prolific publication record, he forged a well-deserved reputation as an international expert on South American fossil mammals.

2.26 Robert Hoffstetter, wearing his famous field cap. The foremost French paleomammalogist of the twentieth century, his work emphasized fossiliferous localities in South American countries other than Argentina and Brazil, where most paleontologists had traditionally explored. This opened new windows onto the past life of the continent. Photograph taken in 1980 by Philippe Taquet in Tiupampa. No mammals were then known in that locality, only chelonians and crocodiles. The first mammalian remains were found in 1982.

Courtesy of C. de Muizon.

After his early successes in Ecuador, Hoffstetter extended his efforts to Bolivia, Peru, and Colombia, where he worked numerous new localities. Many of these were from considerably older (Oligocene to Pliocene) deposits than the more typically known Pleistocene sites of these countries. The fossil faunas yielded by these sites have allowed us a much broader and more comprehensive understanding of the history of mammals in South America than had hitherto been possible based mainly on Argentinian rock sequences and the fossils these contained. In Bolivia, for example, he uncovered about 15 new localities, including some of the earliest deposits yielding remains of sloths. Hoffstetter published on and made important contributions to several groups of mammals from the countries he actively prospected as well as from others (such as Argentina). These include, among others, primates (he described the oldest South American primate, *Branisella boliviana*), marsupials, and rodents.

Despite such expertise in so broad an array of mammalian groups, Hoffstetter is probably best remembered for his contributions to our knowledge of xenarthrans (one of the principal groups with which this book is concerned), to which he devoted 40 publications between 1948 and 1986. Though many of these are descriptive paleontological accounts of new taxa or revisions of taxa already known, he produced some important synthetic analyses of the group as a whole, such as *Phylogénie des Édentés Xenarthres* (1954b), *Remarques sur la phylogénie et la classification des Édentés Xénarthres (Mammifères) actuels et fossiles* (1969), the chapter on xenarthrans (1958) in Jean Piveteau's monumental *Traité de Paléontologie*, and in particular *Les Édentés Xenarthres, un groupe singulier de la faune néotropicale (origines, affinités, radiations adaptatives, migrations et extinctions)* (1982).

He officially retired in 1977 but continued to be active. Indeed, he wrote some of his most important synthetic analyses during his retirement. Among Hoffstetter's lasting contributions is the influence he has had on the career paths of subsequent French students in their choice of research topic. Several generations of graduate students, though some were not directly under his supervision, have followed his lead on xenarthran studies, and France continues to produce worthy additions to our knowledge of these mammals, particularly sloths. Among these students, some of whom remain active are Christian de Muizon, Christian Guth, Jean Bocquentin, Pierre-Antoine Saint-André, and François Pujos.

Born in Mendoza, Argentina, in 1925, Rosendo Pascual (Fig. 2.27) has already celebrated his 87th birthday and is still carrying the torch. Although primarily a geologist whose doctoral dissertation dealt with the geology of the high cordillera near his place of birth, he is responsible for having set in motion and nurtured the process of developing the rich and wide-ranging research program on vertebrate paleontology in the Museo de La Plata. Indeed, he was appointed chief of the División Paleontología Vertebrados in 1957. More than half a century later, this academic unit has contributed impressively to this field, being by far the most prominent institution in this area of research in South America and among the most active in the world.

Rosendo Pascual: torchbearer

This was quite a turnaround, and no small feat. After the work of the Ameghinos and Lucas Kraglievich, paleontology fell into a bit of a lull. To be sure, work in Argentina continued and at a high level; we might single out the work of such researchers as Ángel Cabrera and Jorge Lucas Kraglievich (Lucas's son). However, none of these scientists was able to organize a system to produce a steady supply of new workers. When Rosendo arrived at the Museo de La Plata, he found that little was going on, and the collections of this illustrious institution were falling into disrepair. Given his prodigious talents, industriousness, and intellectual ability, Rosendo decided to study everything. He also began to train graduate students, and as they arrived, he would hand off a chunk to each. For example, Zulma Brandoni de Gasparini began studying reptiles and Eduardo Tonni the birds. Eventually, Rosendo helped produce specialists in several vertebrate groups, and these students went on to become independent researchers who multiplied themselves, so that several generations of paleontologists have emerged from Rosendo's program. Currently the group comprises more than 15 researchers who have developed research programs producing new investigators and covering most subjects in vertebrate paleontology: paleoichthyology, paleoherpetology, paleornithology, paleomammalogy, biostratigraphy, paleoclimatology, and paleobiology (functional morphology and biomechanics). The presence of this research contingent and its scientific production, in addition to the museum's collections and history, make the División a sort of Mecca for anyone interested in the paleontology of South America.

However, being the director of such a considerable team, in which freedom of thinking was always highly valued and achieving international recognition persistently encouraged, is not the only activity for which Rosendo deserves recognition. Besides being generous with his ideas, always freely

2.27. Rosendo Pascual. His efforts at the Museo de La Plata helped revitalize paleontology in Argentina. His many students continue the rich tradition of paleontology in that country. Here Rosendo stands beside the portrait of Florentino Ameghino in his office at the MLP.

Courtesy of Ricardo Pasquali.

shared with those who showed an interest in developing them (as indicated by the 14 doctoral students he supervised and the many other researchers he trained), Rosendo set the standard for quality and hard work for his beloved subjects, the late Mesozoic and early Tertiary mammals. Among his many contributions, the descriptions of primitive mammals of Gondwanian origin and of the first South American platypus stand as landmarks among his many contributions, including books and more than 160 papers (and counting).

Further, Rosendo set out to upgrade the museum's collection, both in terms of organizing and modernizing the older collections and increasing its holdings. He established his own regular field seasons, prospecting and collecting numerous specimens, as did his students once they had become independent researchers. A measure of the value of his efforts may be drawn by growth of the collection since he took over leadership of the division: the collection of vertebrate paleontology, started in 1877 when Francisco P. Moreno donated his 15,000 specimens (both paleontological and archeological), amounts today, greatly as a result of Rosendo and his intellectual descendants, to 140,000 specimens, of which about 10% correspond to type material.

We may further note Rosendo's character and personality as important elements in the development of paleontology in La Plata over the past half century. Rosendo,* a kind and enthusiastic individual, was able to forge productive and amiable relationships with internationally renowned researchers and students alike. During his tenure, he fostered an environment of independent thought, backed of course by rigorous investigation, that brought a sense of excitement into intellectual discourse and spurred his students to achieve their own scientific goals. He was particularly interested in young students during their formative intellectual development, realizing that this stage was critical to their progression toward becoming independent thinkers. This aspect was clearly demonstrated to one of us (G.D.I.) on a memorable evening in Rome during the Fourth International Theriological Congress in 1989. Several of the students could not afford the formal banquet, so G.D.I. organized an impromptu gathering at the home and with the help of one of his Italian friends, Marco Tosi. The gathered included paleontology students as well as other student and nonstudent friends, but all were between 20 and 30—except Rosendo, of course, who became the life of the party. This was understandable from the paleontology students' point of view: we were aware of his stature. However, the other guests soon realized they were in the presence of a special man—but not because he spoke to them about himself or paleontology. Rather, the focus was on whomever Rosendo was conversing with, and their particular thoughts, goals, aspirations, and direction in life.

We'd like to end this short tribute by explaining one of Rosendo's famous sayings, first uttered in English: "Yes, for sure—maybe!" Many of us have heard this phrase, and it seems to apply so perfectly to just about any facet of paleontology—whether the weather will cooperate on a particular field day, whether fossils will be found, whether a particular idea will withstand the test of more evidence and time—that it seemed Rosendo came up with it for just that purpose. But it turns out that its origin is only tangentially related to paleontology. Well into his 80s now, Rosendo has slowed down a bit and only occasionally comes to his office to work. During one of these sessions, we were fortunate enough to sit down with him and get him to reminisce about his earlier days, and he told us about this phrase. Rosendo first blurted it out in exasperation rather than through paleontological inspiration. During the 1960s, Rosendo conducted extensive fieldwork with Alfred Romer (one of North America's leading paleontologists) in some of Argentina's more remote reaches. Despite this, Romer was persistent every morning in trying to secure a proper (i.e., North American) breakfast of bacon and eggs, and every morning, Rosendo did his best to accommodate his friend. One day, as Rosendo was trying to coordinate the day's activities in a particularly remote area, Romer came up and asked, "So, Rosendo, do you think we can get some eggs this morning?" They were, as Romer knew full well, in the middle of nowhere, and securing a few eggs would waste the better part of the day. With a sigh, Rosendo turned to his colleague and said, "Yes, for sure," then added ". . . maybe." Needless to say, they did not find eggs that day.

*With great sadness, we inform our readers that Rosendo passed away on December 23, 2012, after the book had gone to print.

3.1. Polar view map showing Gondwana, the supercontinent of the south, as it appeared in the late Mesozoic, when it started to split up into the current landmasses.

Illustrated by Sebastián Tambusso.

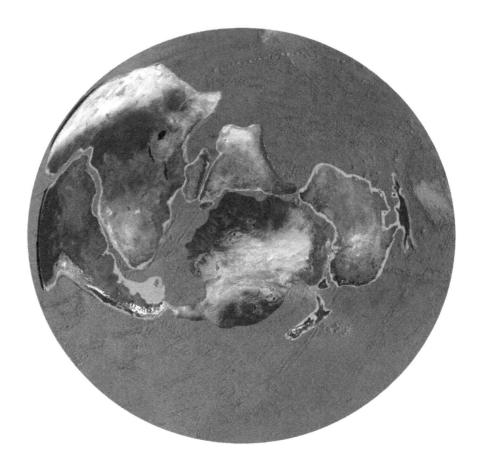

Geological and Ecological History of South America during the Cenozoic Era

3

In this chapter, we provide a broad outline of the tectonic, climatic, and biotic changes that occurred in South America over the course the Cenozoic, focusing on the mammals, given that they have served as the main basis for establishing the biostratigraphic framework in South America. Our story will extend only through to the Pliocene (because the changes in South America before this time were essentially self-contained, given its long isolation from North America and, through it, other landmasses) so that we may consider, in Chapter 4, the North American mammals and their fate during this earlier period. With the stage thus set for subsequent events, we return, in Chapter 5, to the thread of the South American climate and biota, consider the events and conditions that led to the formation of the Isthmus of Panama, and explore the changes that occurred once the faunas from the two continents began to intermingle during the Pliocene and, especially, the Pleistocene.

This earth that we live on, the world we are comfortable with when it can be considered fixed and unchangeable, is our point of reference. Once in a while, reports of an earthquake, tsunami, or other calamity remind us of just how fragile this stability is. In fact, not only is it fragile, it is totally illusory, because, as we have known for decades, the continents move or drift, sliding heavily over deeper layers of the earth at the scanty pace of a few centimeters per year. Over millions of years, which are mere moments in the immense dimension of geologic time, these few centimeters are transformed into hundreds of kilometers. The continents thus unite and separate, like pieces of a giant puzzle.

South America and Africa were united until 180 million years ago (Mya), and the outlines of their respective Atlantic coastlines reveal their former union. The meteorologist Alfred Wegener, who first proposed a scientific theory of drifting continents at the beginning of the last century (the idea of moving continents has roots stretching back to the 1500s) and was ridiculed by the contemporary scientific establishment for his efforts, presented this superposition as among his primary evidence. India was also united to Africa. In separating and migrating north, much of India's former territory was wrinkled against Asia to form the mighty Himalayas. These three landmasses, plus Australia, New Zealand, and Antarctica, were all united until 180 Mya and formed the supercontinent of the south, termed Gondwana (Fig. 3.1).

When Gondwana existed, there was a time during the late Paleozoic in which the South Pole was situated near the south of modern-day Brazil, and gigantic glaciers dominated the regions that today are the prairies bordering

Continents on the move: plate tectonics

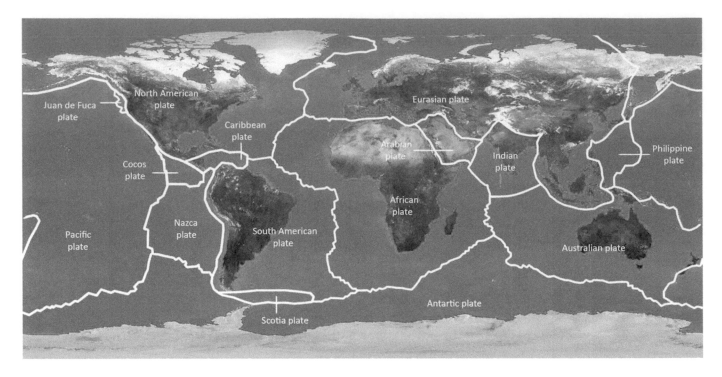

3.2. The major plates of the earth's lithosphere.

Illustration by Sebastián Tambusso.

the Río de la Plata. The northern continents were also united, forming the supercontinent of Laurasia. Both supercontinents, representing nearly all the exposed landmasses, were united into a single unit, termed Pangaea, which started to break apart about 200 Mya.

This pattern of union and separation may have occurred several times in the earth's 4500 million (or 4.5 billion) year history. Clearly, each union and separation greatly affected life on the continents as well as life in the oceans as these were created or connected. Although here we focus on the fauna that lived only hundreds of centuries ago and no appreciable change has occurred since then in this regard, various critical aspects of this history cannot be easily understood except in the context of moving continents.

We have been speaking so far about the movement of continents. However, the continents, the earth's main current portions of land, are really only the emergent parts of larger components of the earth's surface. These components are known as plates, and it is the plates that move (hence the term *plate tectonics*), with the continents going along for the ride, in response mainly to seafloor spreading. A plate is a large, mobile slab of rock and may be oceanic, topped with oceanic rocks submerged to form the seafloor, or continental, formed of both oceanic and continental rock (although some of the smaller plates are made entirely of continental rock). Fig. 3.2 shows the earth's major plates.

Adjacent plates move in different ways relative to each other (Fig. 3.3). At a divergent boundary, the plates move away from each another, whereas at a convergent boundary, they move toward each other. At a transform boundary, the plates slide past each other. Many of the geologic events that make the news, such as earthquakes and volcanic eruptions, usually occur at plate boundaries, as do less well-known events, such as mountain building. The events occurring at the boundary between adjacent plates depends on the types of plates. For our story, we will be concerned with convergent plates, and specifically their oceanic portions, so we

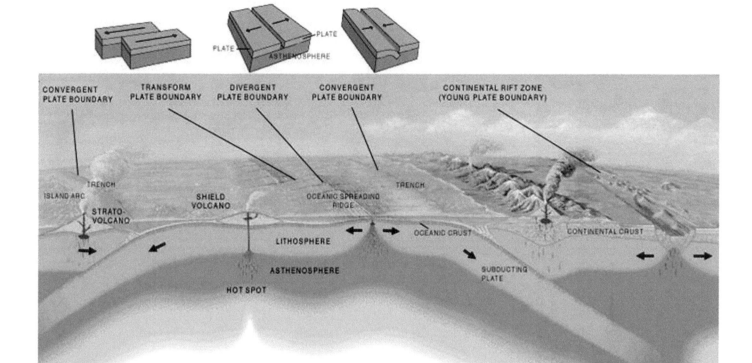

CONVERGENT PLATE BOUNDARY TRANSFORM PLATE BOUNDARY DIVERGENT PLATE BOUNDARY CONVERGENT PLATE BOUNDARY CONTINENTAL RIFT ZONE (YOUNG PLATE BOUNDARY)

PLATE PLATE ASTHENOSPHERE

TRENCH TRENCH
ISLAND ARC SHIELD VOLCANO OCEANIC SPREADING RIDGE CONTINENTAL CRUST
STRATO-VOLCANO OCEANIC CRUST
LITHOSPHERE SUBDUCTING PLATE
ASTHENOSPHERE
HOT SPOT

will discuss the processes occurring at the boundary between convergent oceanic plates.

Before continuing, there are a few further terms to consider for the benefit of those not familiar with geology. The interior of the earth is divided into three zones, in order from surface to interior: the crust, mantle, and core. The crust is the thinnest layer and includes the rocks that form the continents and underlie the oceans. The mantle, the middle layer, is much thicker and is considered to be mainly solid, although the rock forming it appears to be mostly different from that of the crust, on the basis of the way certain seismic waves travel through these layers. The crust and the uppermost part of the mantle are together regarded as the lithosphere, a zone that generally varies between 70 and 125 km in thickness. It is the lithosphere that is broken up into the various plates. The next more interior layer, termed the asthenosphere, is entirely mantle and is about 100 km thick (although there is disagreement among geologists on this point). The rocks of the asthenosphere are apparently soft and perhaps even partially melted, because they are closer to their melting points as a result of the great pressure at such depths. It is this slushy quality of the asthenosphere that, geologists believe, allows the plates of the lithosphere to move.

Let's return to what happens at the boundary of convergent oceanic plates. When two such plates converge, one plate subducts, which means that it slides under the other, with the subducting plate usually bending downward into the asthenosphere, as shown in Fig. 3.3. An oceanic trench, curved toward the subducting plate, is formed where the two plates meet. An accretionary wedge accumulates on the nonsubducting plate, adjacent to the trench, formed by faulted portions of the plate and material scraped from the subducting plate. As the latter descends and reaches a depth of at least 100 km, partial melting of the asthenosphere overlying the plate

3.3. Schematic diagram showing the types of movement possible between adjacent plates.

Courtesy of the United States Geological Survey.

occurs (apparently due to dehydration of the descending oceanic crust), and magma, or molten rock, is eventually generated. The magma begins to migrate toward the surface, where it erupts as an island arc, a line of volcanoes that forms a series of islands parallel to the trench. The region between the island arc and trench is the forearc basin, whereas that on the opposite side of the arc is the backarc region. The relevance of these processes and structural features will become apparent later during discussion of the formation of the Panamanian land bridge.

Paleoenvironmental change during the South American Cenozoic

Today, South America occupies nearly 18 million km^2 (nearly 7 million square miles). A large portion of it (more than 80%) lies in the intertropical region, while the southernmost part of its mainland reaches about 55°S, i.e., far from the polar circle. This uneven distribution along the latitudinal, and hence climatic, gradient decisively influences its varied landscapes. Another important factor is the high elevations of the Andes, a sort of vertebral column along all the western region of the continent, which not only shapes the biomes distributed among its different altitudes, but also acts as a barrier to humidity in certain regions.

Biomes, the world's major communities, are classified according to the predominant climate and vegetation and are characterized by adaptations of organisms to that particular environment (Campbell, 1996). The main biomes generally recognized in present-day South America are shown in Plate 4.

The extremes in the climate, and consequently the flora and fauna, of these regions are striking. In the west, the narrow, mixed mountain systems along the Andes are flanked on the west by an elongated strip of arid land nearly 1000 km in length. This is the Atacama Desert, a virtually rainless plateau between the Andes to the east and the western coastal mountains, which is one of the driest places on earth. To its north are tropical forests, and to its south are savannas, deciduous forests, and woodlands. East of the Andes, tropical and subtropical forests dominate in the north, reaching their fullest development in the dense Amazon rain forest, with its varied tall and continuously distributed tree canopy. The tropical rain forest exceeds all the other biomes in the diversity of its flora and fauna. This dominant biome largely gives way southward to tropical and temperate grasslands, the latter particularly in southern Brazil, Uruguay, and Argentina, and these in turn are replaced farther south by the warm to cold deserts of Argentinean Patagonia that extend to the southernmost tip of continental South America.

This complex modern mosaic of biomes is considerably different from those that were in place during earlier intervals of Cenozoic time. The history of these earlier, shifting environments and biotas up to about the end of the Pliocene will be treated in this chapter. The history of South America during the subsequent Pleistocene will be deferred to Chapter 5, for reasons that will become clear once we consider the dramatic effects of the emergence of the Isthmus of Panama. In any case, Ortiz Jaureguizar and Cladera (2006) concluded that from the early Paleocene (and up into the late Pleistocene), South American climatic conditions, especially in the better-known region south of 15°S, changed to colder, drier, and

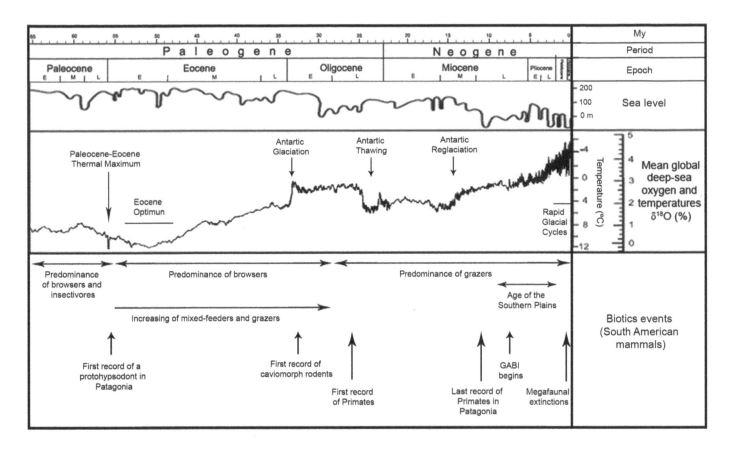

3.4. Changes in sea level, temperature, and ratio of oxygen-stable isotope levels during the Cenozoic, correlated with biotic and abiotic events.

Modified by Sebastián Tambusso based on Ortiz Jaureguizar and Cladera (2006).

seasonal from warm, wet, and nonseasonal conditions. As a consequence of this, biomes changed from tropical forests to steppes, across a sequence of subtropical forests, woodland savanna, park savanna, and grassland savanna. During the Cenozoic Era, the glacial cycles produced cold and dry conditions, interspersed with warmer and wet periods. Accordingly, several pulses of expansion and retraction of steppes (and, concomitantly, advances and retreats of the northern tropical forests) are recorded along this 65-million-year time interval.

These patterns of change contributed to the provincialism that characterizes South American biotas through time, along with changing connections with other landmasses (Antarctica in the early Tertiary, North America in the last few million years) and the lifting of the Andes during the middle Miocene. This orogenic (mountain building) event caused many transformations in Patagonian environments, as the rising Andes began to act as an effective barrier to humid westerly winds. The Patagonian climate then became increasingly arid, and the temperature dropped as a result of the separation of Antarctica and consequent glaciation in that polar continent. The forests retreated toward the cordillera, leaving the Patagonian plains as the steppes they currently are, ending the so-called Age of the Great Plains (Pascual and Bondesio, 1982). As indicated in Fig. 3.4, the present is the coldest time since the end of the Cretaceous. Temperatures were much higher during much of the Cenozoic, and in addition, sea level rose as much as 200 m above present. These conclusions are drawn from a variety of sources, biotic and otherwise. In the following pages, biotic evidence will be considered.

Evolution of the paleofloras and climates during the Cenozoic

Modern ecosystems are dominated by the flowering plants (angiosperms or Magnoliophyta). On the basis of the discovery of *Archefructus* in China (Sun et al., 2002), their origin is proposed to have occurred 125 Mya ago during the Barremian (early Cretaceous). This solved what Darwin, frustrated with the fossil record of flowering plants and what he perceived as an abrupt origin and highly accelerated rate of diversification of flowering plants in the mid-Cretaceous (Friedman, 2008), called an "abominable mystery" in a letter written in 1879 to Sir Joseph Dalton Hooker, his close friend and a great botanist. During the early Cenozoic, following the K-T crisis that ended the reign of the nonavian dinosaurs, several angiosperm groups are frequently recorded: Palmae (the palm trees), Myrtaceae (to which the ubiquitous *Eucalyptus* belongs), and Nothofagaceae (the southern beeches so common in the southern continents, including, as fossils, Antarctica) were already present in the Paleocene (Stewart, 1994; Schönenberger and Conti, 2003; Li and Zhou, 2007). True grasses, today's most widespread group, apparently arose at the Paleocene–Eocene boundary (Crepet and Feldman, 1991). On the other hand, Compositae (the daisies, which comprise the richest and most diverse group of angiosperms) appeared nearly at the same time, 47 Mya, as demonstrated by the discovery near the beautiful Patagonian city of Bariloche, Argentina, of impressive remains that include both flower and pollen grains (Barreda et al., 2010).

As usual with living organisms, the evolution and distribution of plants depend on various general factors, such as tectonics, geography, and climate. In the case of the Southern Hemisphere, the most decisive phenomenon in this regard was the fragmentation of Gondwana. According to Palazzesi and Barreda (2007), the Cenozoic events that determined the major vegetation trends in Patagonia were variations of atmospheric CO_2, the separation of South America and Antarctica (resulting in the establishment of the circumpolar current), the lifting of the Andes (which act as a barrier to humidity in the western regions of the continent), marine transgressions that affected temperature and humidity, and, finally, volcanic activity that affected the composition of the atmosphere and soils. In southern South America, more specifically in Patagonia, a rich record of both plant micro- and megafossils allows the reconstruction of the history of vegetation, and hence of the climate, during the Cenozoic (Barreda and Palazzesi, 2007; Palazzesi and Barreda, 2007; Prámparo et al., 2007; Barreda et al., 2007).

The succession of environments in which the South American fauna, including the megafauna, occurred may be summarized as four major paleofloras as described for the southern part of South America (Frenguelli, 1928; Hinojosa and Villagrán, 1997; Menéndez, 1971; Romero, 1978, 1986; Troncoso and Romero, 1998). The spatial and temporal succession of these paleofloras appears to have been closely related to tectonic and climatic events during the Cenozoic (Hinojosa and Villagrán, 1997; Hinojosa, 2005; Barreda et al., 2007).

First, the neotropical paleoflora developed mainly during the Paleocene when South America, Antarctica, and Australia were still linked geographically and warm climatic conditions extended to at least 50°S. The mean annual temperatures and rainfall oscillated between approximately 2 and 25°C, and 1500 and 2000 mm, respectively (Hinojosa, 2005). During the Paleocene–early Eocene, the rain forest dominated in Patagonia

(Barreda and Palazzesi, 2007), with a great diversity of species suited to high temperature and humidity. However, the presence of such groups as Anacardiaceae (the cashews) and some Leguminosae (whose well-known fruits are called legumes, such as peas, beans, and vanilla) also suggests the existence of more arid zones.

This flora was replaced during the Eocene and early Oligocene by a mixed paleoflora, coincident with a decrease of temperature levels and mean annual precipitation. According to Hinojosa (2005), the mean annual temperature was between 17 and 20° C, and the mean minimum precipitation was around 570 mm. The mixed paleoflora is characterized by species distributed today in the tropics and the austral–Antarctic territories, and endemic forms distributed at present in tropical and subtropical regions of the continent. The presence of *Nothofagus* forests are recorded during the middle Eocene–early Oligocene. This genus, including the southern beeches, is an important component of today's Andean forests and indicates colder climates. Toward the early Oligocene, a temperate forest developed containing elements of Gondwanian lineage, such as Nothofagaceae, and the conifer (cone bearing) clades Podocarpaceae (a characteristic group of Southern Hemisphere conifers, such as the yew plum pine, and a classic member of the Antarctic flora) and Araucariacea (to which the distinctive monkey puzzle tree belongs), without evidence of species indicative of warmer temperatures.

By the late Eocene–early Oligocene until the early Miocene, global temperatures fell, influenced by the glaciations of eastern Antarctica as a consequence of the separation of Australia from Antarctica–South America. Accordingly, the mixed paleoflora was largely replaced by an Antarctic paleoflora, characterized by the abundance of forms from temperate–cold environments. The notable increase of the annual thermal amplitude would favor the development of taxa adapted to colder conditions (mean annual temperature approximately 15°C) and extreme temperatures. Annual mean rainfall increased to 870 mm by the late Oligocene and to 1120 mm by the early Miocene (Hinojosa, 2005).

During the late Oligocene–early Miocene of Patagonia, the climate turned warmer, allowing the southward dispersion of neotropical elements—that is, those belonging to tropical South America and neighboring areas, such as the palm trees that are added to the Gondwanian local flora. The expansion of herbaceous–bushy plants adapted to more arid (xerophytes) and saline (halophytes) environments during the late Oligocene–early Miocene initiated the formation of current vegetation patterns. The more xerophytic communities possibly occupied the coastal salt environments and gallery forests occurred in extra-Andean areas, but there is no evidence of well-established grasslands.

The subtropical flora became established as a consequence of global warming that characterized the middle Miocene. Mean annual temperatures ranged between 21 and 26°C (Hinojosa, 2005). According to Zachos et al. (2001), the climatic optimum occurred around 15 to 17 Mya. After the late early Miocene climatic optimum, temperature and rainfall decreased (430 mm), related to an increase of the thermal difference between extreme temperatures, producing conditions for the development of arid-adapted subtropical floras.

3.5. Timescale showing various subdivisions of the Cenozoic Era in millions of years (Mya). The South American Land Mammal Ages or stages/ages are indicated in the right column. Faunal units are also shown.

Illustrated by Sebastián Tambusso based on several sources.

Since the middle Miocene–Pliocene, the establishment of the circumpolar and Humboldt oceanic currents (resulting from the separation of Antarctica and South America) produced higher temperatures and increased aridity in the subtropics. The final elevation of the Andes caused a rain shadow effect that, by the Plio-Pleistocene, resulted in the fragmentation of the subtropical paleoflora and in the spread of taxa adapted to arid environments along the region called the arid diagonal, which extends, up to the present day, from the southeastern tip of the continent, across the Andes near central Chile, and continues north along the Pacific coast to near the equator (Hinojosa and Villagrán, 1997; Villagrán and Hinojosa, 1997). In the middle to late Miocene of Patagonia, there is a greater diversity and abundance of xerophytic-adapted taxa, such as Compositae (daisies, as noted above), Chenopodiaceae (the group including spinach) and Convolvulaceae (the morning glories). From the late Miocene on, there is a great expansion of the grass-dominated communities in southern South America. The scarce record of Podocarpaceae (yew plum pine) and Nothofagaceae in the extra-Andean region suggests that since the late Miocene the distribution of vegetation was similar to that of today, i.e., with the steppe expanding across non-Andean Patagonia and the forests restricted to the western part where rainfall is abundant (Barreda and Palazzesi, 2007; Palazzesi and Barreda, 2007; Prámparo et al., 2007; Barreda et al., 2007).

The main characteristics of the environmental changes and conditions during the subsequent Pleistocene Epoch will, as noted, be deferred to Chapter 5, once the emergence of the Isthmus of Panama and its consequences have been considered.

Stratigraphy and sequence of vertebrate faunas over the Cenozoic

Patagonia is a privileged region from the paleontological point of view. Rarely, if anywhere, in the world can such a succession of units of fossil-bearing strata be found. They are known from the pioneering efforts of some of the personalities mentioned in Chapter 2 down through the present day. In the last 50 years, this wealth of Cenozoic fossil mammals has been complemented by finds in other parts of Argentina, notably in the northwestern region, as well as in other South American countries and in Antarctica. This diversity has been summarized in proposals such as that of the faunistic cycles, an approach based on the affinities of successive South American Cenozoic faunas first advanced by Kraglievich (1930) and more recently complemented by Ortiz Jaureguizar and Pascual (1989) by means of multivariate statistics.

The stratigraphical scheme of the South American Cenozoic is given in Fig. 3.5. In the past, informal time units, the South American Land Mammal Ages (SALMAs), were adopted in the establishment of a biostratigraphic framework for the continent (Pascual et al., 1965), but later they were formalized into the more widely accepted and technically rigorous stage–age concept (Cione and Tonni, 1995a), which revives Ameghino's (1889) view to some extent. Florentino Ameghino established the standard reference scale for the continental Cenozoic of southern South America and proposed a sequence of stages (*pisos*) grouped into higher-order units (formations = *formaciones*, which were chronostratigraphic and not lithostratigraphic units; e.g., Ameghino, 1889). Later, it was extended to

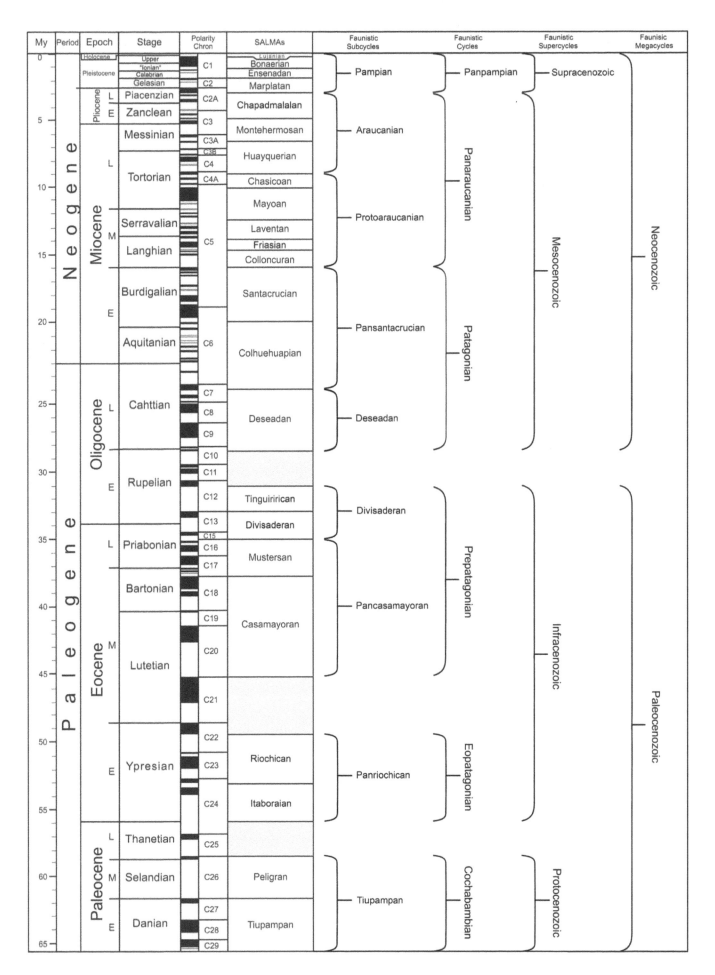

My	Period	Epoch	Stage	Polarity Chron	SALMAs	Faunistic Subcycles	Faunistic Cycles	Faunistic Supercycles	Faunisic Megacycles
0		Holocene	Upper	C1	Lujanian				
		Pleistocene	"Ionian"		Bonaerian	Pampian	Panpampian	Supracenozoic	
			Calabrian		Ensenadan				
			Gelasian	C2	Marplatan				
		Pliocene L	Piacenzian	C2A	Chapadmalalan				
5		E	Zanclean	C3	Montehermosan	Araucanian			
			Messinian	C3A					
				C3B	Huayquerian		Panaraucanian		
	Neogene		L Tortorian	C4					
10				C4A	Chasicoan				
		Miocene			Mayoan				Neocenozoic
			M Serravalian		Laventan	Protoaraucanian		Mesocenozoic	
15			Langhian	C5	Friasian				
					Colloncuran				
			Burdigalian		Santacrucian				
			E	C6		Pansantacrucian			
20			Aquitanian		Colhuehuapian		Patagonian		
25			L Cahttian	C7 / C8 / C9	Deseadan	Deseadan			
30		Oligocene		C10 / C11					
			E Rupelian	C12	Tinguiririrican	Divisaderan			
35				C13	Divisaderan				
			L Priabonian	C15 / C16	Mustersan		Prepatagonian		
				C17					
	Paleogene		Bartonian	C18				Infracenozoic	Paleocenozoic
40				C19	Casamayoran	Pancasamayoran			
		Eocene	M	C20					
45			Lutetian	C21					
50				C22					
			E Ypresian	C23	Riochican	Panriochican	Eopatagonian		
55				C24	Itaboraian				
			L Thanetian	C25					
60		Paleocene	M Selandian	C26	Peligran		Cochabambian	Protocenozoic	
				C27		Tiupampan			
65			E Danian	C28 / C29	Tiupampan				

the rest of South America (Pascual et al., 1965). Ameghino's (1909) scale of stages, largely based on biostratigraphy, remains valid even today in the chronologic local standard scale of at least southern South America (Nabel et al., 2000).

As such, we will refer to these intervals as ages. We emphasize this point because SALMAs are still in widespread use, although "ages" is the more correct concept. Pascual (1984a,b) noted that the limits of these ages apparently correspond with changes in climatic, environmental, and tectonic conditions. This observation led to a new formulation of Kraglievich's (1930) approach of faunal cycles, but from more recent multivariate statistical techniques that grouped the SALMAs (or, more properly, ages), on the basis of their contained taxa, into hierarchically higher categories termed faunal units (Ortiz Jaureguizar, 1986, 1988). According to the resulting scheme (Fig. 3.5, right side), the two higher levels, faunal megacycles and supercycles, are defined on the basis of wide-scale, global events, whereas the lower levels, faunal subcycles and cycles, reflect climatic–environmental changes at more restricted regional scales.

The earlier megacycle, the Paleocenozoic, begins with the start of the Cenozoic and ends in the early Oligocene. It is characterized by humid, sylvan environments as inferred from the low-crowned teeth of the native ungulates, among other evidence (Pascual and Ortiz Jaureguizar, 1990). Also, the mammals found in the Protocenozoic Supercycle are peculiar, representing old Gondwanian lineages, such as gonwanatheres and monotremes, implying an Australian connection through Antarctica. The later Infracenozoic Supercycle begins after a hiatus in the record of terrestrial faunas. It exhibits a rather high systematic diversity of marsupials and native ungulates, but they are of restricted morphological diversity compared with that of the taxa present in the following (and younger) cycles.

The Neopaleozoic Megacycle begins with the late Oligocene, a time marked by global climatic cooling that saw the first permanent glaciation event in Antarctica and that was correlated by Pascual and Ortiz Jaureguizar (1990) with the North American terminal Eocene event (Prothero, 1994) and the European La Grande Coupure, first observed by the Swiss paleontologist Hans Georg Stehlin (1910). It is divided in two supercycles, the earlier of which is the Mesocenozoic, encompassing all but the last few million years of the Cenozoic, i.e., the Pleistocene. The earlier part of the Mesocenozoic is represented especially in Patagonia and was hence named the Patagonian Cycle, while in the younger Panauracanian Cycle, following one of the phases of the rise of the Andes (termed the Quechua phase; Charrier and Vicente, 1970), mammal-bearing sedimentary rocks start to be found farther north than in the previous records (Pascual and Ortiz Jaureguizar, 1990). Finally, the Supracenozoic Supercycle includes only one cycle, the Panpampian Cycle, which evidences the trend to less mild global and South American climates, and in consequence, the mammal-bearing sediments are found in northern latitudes, with Patagonia progressively becoming deprived of its once diverse mammalian assemblages.

This scheme was further refined as new SALMAs or ages were proposed (Ortiz Jaureguizar and Pascual, 1989; Pascual and Ortiz Jaureguizar, 1990, 2007; Pascual et al., 1996), and it was compared with the North American mammalian faunas (Pascual and Ortiz Jaureguizar (1992), an approach that proved to be productive (Ortiz Jaureguizar et al., 1999; Ortiz Jaureguizar and Posadas, 1999, 2000).

Below, we describe the ages and provide a synopsis of the mainly mammalian faunal characteristics for each arranged in the scheme of faunal units just described. The dates for the ages mainly follow Flynn and Swisher (1995), and several of their characteristics are taken from Aceñolaza and Herbst (2000), Vucetich et al. (2007), and Cione et al. (2007).

Paleocenozoic Megacycle

Protocenozoic Supercycle
  Cochabamban Cycle
This cycle is the only one included in the Protocenozoic Supercycle (Fig. 3.5). It includes the Tiupampan Age and the Peligran Age (Pascual and Ortiz Jaureguizar, 2007). The first, named after the Tiupampa locality in Bolivia, corresponds to the very first part of the early Paleocene, when the world had just been deprived of the dominant dinosaurs, between approximately 64.5 to 63 Mya. Its fauna was recovered from the Santa Lucía Formation, deposited in channels of meandering rivers on a then flat alluvial plain (Gayet, 2001), long before the dramatic uplift of the Andes transformed that region into a 3000-m-high plateau. This fauna is so peculiar that it was earlier considered to be of Cretaceous age (Marshall and Muizon, 1988). It includes one Multituberculata, an ancient, mainly Mesozoic mammalian group; several marsupials, such as *Pucadelphis andinus* (Fig. 3.6; belonging to Didelphidae, the group of modern opossums), among others; and primitive ungulates of uncertain relationships with other placentals, as such the long-surviving Notoungulata and the primitive Condylarthra and Proteutheria (Box 3.1). The xenarthrans (armadillos, sloths, and anteaters) had not yet appeared, which is rather surprising, given the great diversity later achieved in this clade. Some of the Tiupampan mammals are named in homage to prominent scientists and science funding institutions, such as the marsupial species *Roberthoffstetteria nationalgeographica*, belonging to the marsupial clade Caroloameghininae.

The Peligran Age is represented by an early Paleocene fauna (62.5 to 61 Mya), found in the Salamanca Formation, the so-called Banco Negro Inferior in Punta Peligro, San Jorge Gulf, Chubut, southern Argentina. It includes, among other surprises, one of the most astonishing finds in mammalian paleontology: a tooth belonging to a fossil platypus, *Monotrematum sudamericanum*, the first of its kind outside Oceania and suggestive of a connection (already disrupted by the Peligran) to a then not yet frozen region in Antarctica. Because living adult platypuses lack teeth and bear instead a ducklike beak, it should be noted that the tooth is similar to those of the genus *Obdurodon* (Fig. 3.7), found in the Oligocene–Miocene of Australia.

Also present is Dryolestida, an extinct clade of mammals related to the living marsupials and placentals (formerly known as Pantotheria or Eupantotheria). The teeth of this group, which is considered a survivor of a Mesozoic Gondwanian radiation, lack the typical three-cusped or tribosphenic pattern (Appendix 1) of mammals (Gelfo and Pascual, 2001). The Peligran fauna is regarded as a mixture of Gondwanian elements and those from the

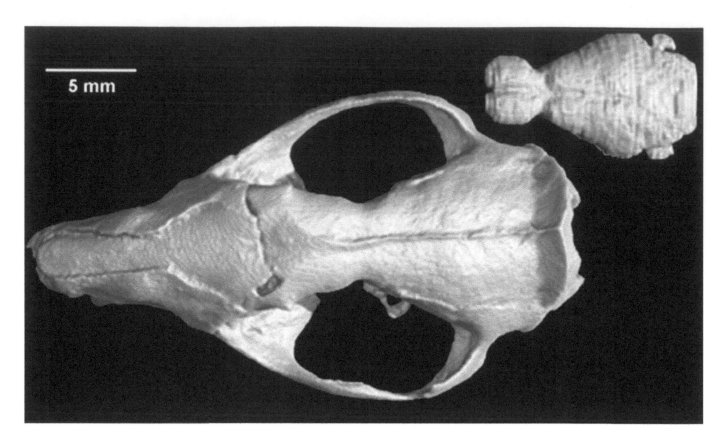

3.6. Dorsal views of the skull of *Pucadelphis andinus* and of the digital endocast (a mold of the cranial cavity).

Modified from Macrini et al. (2007), by kind permission of the Society of Vertebrate Paleontology.

northern lands (gathered in the supercontinent of Laurasia). Among the former group are anurans (frogs) and chelonians (turtles), well known from the Gondwanian Cretaceous, as well as the gondwanatherid *Sudamerica* and the already mentioned monotreme *Monotrematum*. Among the forms of Laurasian origin are alligatorid crocodiles and members of the placental ungulate clade Mioclaenidae, also known from the early and middle Paleocene of North America, although there are opinions in favor of the Gondwanian origin of ungulates (Prasad et al., 2007, 2008).

Pascual et al. (1996) suggested that this Cretaceous–early Paleocene Patagonian biota evolved in isolation from the rest of South America because Patagonia was then a region forming a separate biogeographic province, without ready access to the more northerly areas. As described by Ortiz Jaureguizar and Cladera (2006), the absence of large continental topographic barriers allowed a widespread Atlantic transgression (a landward migration of the shoreline so that once exposed land becomes inundated by sea), the island-spangled Salamancan Sea that may have had a buffering influence on climate, to cover the major part of southern South America, from Patagonia in the south to Bolivia and Peru in the north (Fig. 3.8).

Slightly younger is the Cerrejón Formation in Colombia (Jaramillo et al., 2007), from which the gigantic boid snake *Titanoboa cerrejonensis*, whose body mass was estimated as about 1100 kg, was recovered (Head et al., 2009). As the maximal size of poikilothermic animals (i.e., whose internal temperature varies with the ambient temperature) at a given temperature is limited by metabolic rate, a snake of this size would require a minimum mean annual temperature of 30–34°C to survive. This temperature is consistent with hypotheses of hot Paleocene neotropics with high

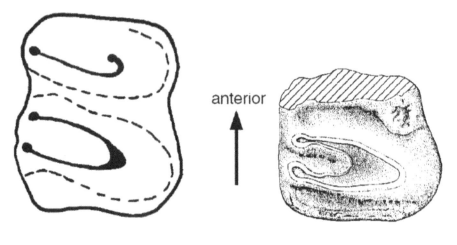

anterior

3.7. Occlusal view of the incomplete second upper molar of *Monotrematum sudamericanum* (right), compared with a schematic view of the first lower molar of the Australian ornithorhynchid *Obdurodon dicksoni* (left). Not to scale.

Modified by Sebastián Tambusso based on Pascual et al. (2002), by kind permission of Acta Palaeontologica Polonica.

concentrations of atmospheric CO_2 based on climate models, and a steep temperature gradient during the early Paleogene greenhouse.

Paleocenozoic Megacycle

Infracenozoic Supercycle
    Eopatagonian Cycle
       Panriochican Subcycle

In the first part of this supercycle, land-mammal communities were dominated by marsupials, particularly didelphimorphians and polydolopimorphians (Pascual and Ortiz Jaureguizar, 2007), while later the presence of native ungulates becomes more conspicuous. Already in the Eopatagonian Cycle, gondwanatheres and monotremes are no longer present, an initial step toward the establishment of the native fauna that was to dominate South America through much of the Cenozoic. The Panriochican Subcycle includes two ages, the Itaboraian and the Riochican, summarized below.

In São José of Itaboraí, near the glamorous and violent city of Rio de Janeiro, Brazil, an extraordinary Paleocene outcrop was found in the 1940s and studied by the great Brazilian paleontologist Carlos de Paula Couto. The fossils, of fragmentary nature, are preserved in channels and carbonate sediments. The fauna of São José of Itaboraí forms the basis of the Itaboraian Age, between 56 to 53 Mya, which corresponds to the earliest Eocene. Although invertebrates are also found, it is composed mainly of vertebrates, which are similar to those found in the previous ages but already with a strong signal of South American endemism. About two-thirds of the species of mammals found in this age are marsupials. However, the ungulates are more abundant in terms of specimens, including clades that survived until the Pleistocene—Litopterna and Notoungulata, as well as the more widely distributed Condylarthra and the mysterious Xenungulata and Astrapotheria (Box 3.1). In this age, the xenarthrans, later to become major players in the drama depicted in this book, are found for the first time among the South American mammals; and thus the three faunal elements—marsupials, native ungulates, and xenarthrans—that characterized South America for much of the Cenozoic were present for the first time.

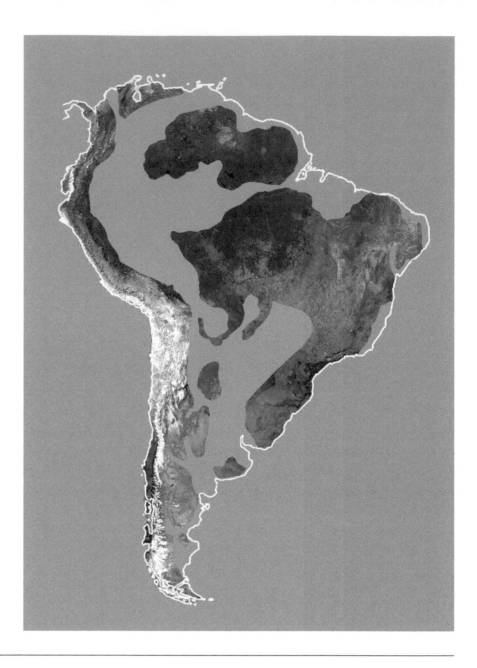

3.8. Map of South America showing the extent of the Atlantic marine transgression (gray regions within South American outline) during the Peligran South American Land Mammal Age.

Modified by Sebastián Tambusso from Ortiz Jaureguizar and Cladera (2006: fig. 2).

Box 3.1   Native South American ungulates

The native South American ungulates have been compared to living herbivorous analogues (as summarized by Croft, 1999) and fall into five groups: the astrapotheres, pyrotheres, notoungulates, litopterns, and xenungulates (Simpson, 1980; Marshall and Muizon, 1988). The phylogenetic relationships of these groups are unclear (Cifelli, 1985, 1993). They were all once united in a single taxon, Meridiungulata, originally founded on the idea that all endemic South American ungulates were monophyletic (McKenna, 1975). However, the term *ungulate* as used here does not imply that the ungulate groups endemic to South America and modern Ungulata (Perissodactyla and Cetartiodactyla) share a most recent common ancestor, or even form a single clade themselves.

Astrapotheres (including trigonostylopids) were rhinoceros-like mammals (Cifelli, 1985) found in deposits of Paleocene to Miocene age. They attained maximal diversity during the early Miocene Colhuehuapian and Santacrucian ages (Marshall and Cifelli, 1990; Johnson and Madden, 1997). Astrapotheria means "lightning beasts," yet these large mammals were anything but fast or quick. As Scott (1937) noted, this is an unfortunate misnomer, but it assumes that Burmeister intended the name as a descriptor of physical attribute. However, Burmeister (1879) chose the name to complement that of a North American group to which he believed they were closely allied, the brontotheres (or titanotheres, which are members of Perissodactyla), or "thunder beasts." Thunder and lightning do go together, but not in this case, as their supposed relationship is no longer considered likely. Astrapotheria is thus still an inappropriate designation, but we're stuck with it, and in our opinion, it's a good thing, because it remains a wonderful expression. Astrapotheres had large, ever-growing, possibly sexually dimorphic tusks. The best-known species is *Astrapotherium magnum* ("great lightning beast"), which was a large mammal nearly 3 m in length whose remains were found first in the Santa Cruz Formation. It probably had a trunk and was likely semiaquatic, given its relatively slender limbs (Naish, 2008).

Pyrotheres were elephant-like but were never as diverse, nor did they cover as great a span of time as the astrapotheres. The name Pyrotheria, "fire beasts," refers to their occurrence in volcanic ash deposits. They are among the most mysterious ungulates known, largely because their remains are sparse—indeed, the skull is known only from a single specimen of *Pyrotherium* (Billet, 2010). Their remains are known only from the middle Eocene (Casamayoran) through the late Oligocene (Deseadan). *Pyrotherium* was a large beast, with bilophodont cheek teeth, a short skull, two pairs of broad upper and a lower pair of tusklike incisors, and probably a proboscis (Benton, 2005; Billet et al., 2009).

Notoungulates are by far the most diverse and abundant lineage of South American ungulates (nearly 140 species in 13 families; Croft, 1999). This group includes animals similar to rhinoceroses, hippopotami, rabbits, and rodents. Other notoungulates do not closely resemble any living mammal. Many have high-crowned cheek teeth, and a couple of genera reach the late Pleistocene (see Chapter 6).

Litopterns were the second most successful group of South American ungulates in terms of diversity and longevity, spanning from the late Paleocene (Itaboraian) to the late Pleistocene (Marshall and Cifelli, 1990). They include forms similar to antelopes, horses, and camels, all with relatively low-crowned cheek teeth. *Macrauchenia patachonica* belongs to this group and is typical of the late Pleistocene, so it will be further dealt with in Chapter 6.

Xenungulates, "strange ungulates," are primitive, poorly known tapir-like mammals, restricted to the Paleocene deposits of Brazil, Argentina, and Colombia (Gelfo et al., 2008). They are best known from the remains of the genus *Carodnia*, of which *C. vieirai* (Gelfo et al., 2008) had large canine teeth (Muizon and Cifelli, 2000) and was relatively large. It reached the size of a tapir and apparently had a gait similar to the living African elephant (Ávilla and Bergqvist, 2005).

From the Riochican Age to the end of this supercycle, notoungulates replaced marsupials as the dominant mammals (Pascual and Ortiz Jaureguizar, 2007). Mammals of Riochican Age were discovered and first studied by the Ameghinos. This fauna, found in sediments deposited between 53 and 50 Mya (early Eocene), also comes from the basin of the San Jorge Gulf, Chubut, Argentina, but is known as well from the northern Argentinian province of Tucumán, 2000 km away. It comprises some 40 species, especially ungulates (Box 3.1). In late 1898 and early 1899, Carlos Ameghino recognized, in what Florentino had called "beds with *Pyrotherium*," two quite distinct faunas. The older was called "bed with *Notostylops*," and the younger kept the name of "bed with *Pyrotherium*." Later, Florentino created a zone, the older of them termed "basal Notostylopan." This name was inappropriate because in those sediments there were no remains of *Notostylops brachycephalus*, a basal early Eocene rabbitlike notoungulate. The current name of Riochican for the basal Notostylopan was suggested by Simpson (1933, 1935), in reference to the Río Chico of Chubut. This author proposed a Tertiary age for these sediments, previously regarded as Cretaceous as a result of the supposed presence of dinosaur remains that were finally understood to represent a group of dinosaur-like toothed crocodiles, Sebecosuchia (see Chapter 2).

Among the vertebrates are teleosts and amphibians, including the leptodactylid frog *Caudiverbera amazonensis* and the bufonid toad *Xenopus pascuali*, a genus also present in the Cretaceous and Miocene of Africa. Turtles and snakes phylogenetically close to those of the North American Cretaceous are also found, while others (i.e., *Crassochelys corniger*) belong to a clade recorded also in the Pleistocene of Australia. Another spectacular creature is the 12-m-long boa *Madtsoia bai*, one of the longest snakes known, living or extinct, which was discovered in 1931 by Simpson, and known as well from the upper Cretaceous of Madagascar and the Casamayoran Age. Some ratite birds (the group comprising, e.g., the emu in Australia, the ostrich in Africa, and the ñandu in South America), are also present. The mammalian fauna consists of ñandú, xenarthrans, and an array of native ungulates.

Paleocenozoic Megacycle

Infracenozoic Supercycle
    Prepatagonian Cycle
        Pancasamayoran Subcycle

Over the course of the Prepatagonian Cycle, the younger of the Infracenozoic Supercycle, the native South American marsupials and ungulates reached their radiation climax (Pascual and Ortiz Jaureguizar, 2007). The mammal-bearing sediments of the Prepatagonian Cycle are also recorded in northwestern Argentina, as well as Central Patagonia (Pascual et al., 1990; Pascual and Ortiz Jaureguizar, 2007). The Pancasamayoran Subcycle witnessed a dramatic global drop in ocean temperature (Savin et al., 1975).

The oldest age is the Casamayoran, which succeeds the Riochican. It was deposited during the middle Eocene, approximately 41.6 to 39.0 Mya. The name is derived from Punta Casamayor, also located in the richly fossiliferous region of the San Jorge Gulf, but in this instance from Santa Cruz Province. It was proposed by Albert Gaudry, professor at the Muséum national d'Histoire naturelle in Paris, based on fossils collected at the top of the sequence by André Tournouër beginning in 1898. The fauna found in these sediments (and in Patagonia and northern Argentina) is similar to that of the Riochican. For instance, the already mentioned *Madtsoia bai* is also present. The presence of this snake and of other species, like chelonians (turtles and tortoises) and crocodiles, as well as palm trees, indicates that the climate of Patagonia was then subtropical with high humidity.

The strange-looking astrapotherians were also present, as well as carnivorous bearlike marsupials, such as the impressive *Callistoe vincei* (Fig. 3.9; Babot et al., 2002). Another Casamayoran marsupial was the opossum-like, frugivorous *Caroloameghinia*. The xenarthrans were represented by early armadillos; and among the ungulates, *Notostylops* was beaver sized. There are two different assemblages among Casamayoran mammals: the primitive Cañadón Vaca fauna, and the more derived Gran Barranca fauna. Other fossils found in Casamayoran sediments are reptiles and birds, as well as insects, teleosteans (advanced bony fishes), and dipnoans (lungfish) of the extant genus *Lepidosiren*, crocodiles, chelonians, and lizards. Again, the inferred climate is tropical to subtropical.

Such conditions also favored the presence of a rather rich fauna and flora in the Antarctic peninsula during this age. Indeed, the James Ross Basin, in King George and Seymour Islands, among others, includes the La Meseta Formation, of Eocene age and with fossils of terrestrial mammals, invertebrates, and plants in poorly consolidated siliciclastic (made of silica particles) sediments indicative of a deltaic–estuarine environment (Reguero and Gasparini, 2006).

In many cases, the fossils from this and several succeeding ages are known also from the Sarmiento Formation, outcropping in the locality of the Gran Barranca (Fig. 3.10) the southern cliff surrounding Lake Colhué-Huapi, discovered by Tournouër and later thoroughly worked by Carlos Ameghino in 1896. The Sarmiento Formation is remarkable in that it comprises a continuous sequence from the Eocene to the early Miocene, including at least eight biostratigraphical units with evidence of important events, such as after beginning of the development of hypsodonty (increase in tooth height, particularly among ungulates), temperature drop in the Eocene–Oligocene, and the probable arrival of rodents in South America (Vucetich et al., 2007).

The next younger beds deposited during the late Eocene, from 35.3 to 28.8 Mya, yielded the Mustersan Age fauna. The sediments, outcropping in several places in Patagonia above the Casamayoran formations (including the impressive Gran Barranca from 38.7 to 37.0 Mya; Fig. 3.10), were termed the *Astraponotus* zone, after the astrapotherian *Astraponotus assymetrum*, by Florentino Ameghino. The name Mustersan, proposed by Kraglievich (1930), is derived from outcrops of this age found in the cliffs along Lake Musters, in southern Chubut Province, Argentina. Incidentally, the lake, in turn, had been named after George Chaworth Musters (1841–1879), an English sailor, explorer, and writer perhaps best known for his travels through the Patagonian interior in the company of the indigenous Tehuelches (Box 3.2).

3.9. Skull and jaw of the carnivorous bearlike marsupial *Callistoe vincei* from the Casamayoran Lumbrera Formation of Salta, Argentina. Scale bar = 10 cm.

Photograph courtesy of Judith Babot.

Box 3.2

George Musters

Box Fig. 3.2. George Chaworth Musters was an English explorer famous for having endured an epic journey accompanying a tribe of Patagonian natives during the mid-nineteenth century.

Born in Naples to traveling British parents, George Musters's (Box Fig. 3.2) adventurous life included military action in the royal navy and exploration of Patagonia. He wrote about those experiences in a book awarded a gold watch in 1872 by the Royal Geographical Society, *At Home with the Patagonians: A Year's Wanderings over Untrodden Ground from the Straits of Magellan to the Rio Negro* (1871; 2nd ed., 1873), which also dealt with Tehuelchean habits, curiosities, rituals, and vocabulary, and included observations on the geology, orography, hydrography, and fauna and flora of the uncharted territories. He belonged to a family, the Chaworth Musters, that, though cast in a minor role, was an interesting clan. George's uncle, Charles Musters, was as a boy a volunteer first class aboard the HMS *Beagle* (Keynes, 2001). A favorite among the officers and crew, young Charles accompanied Charles Darwin on several exploratory strolls in South America. Unfortunately, the young lad died in 1832, probably of malaria. Darwin (1839) records his death: "Poor little Musters; who three days before his illness heard of his Mothers death." The latter, Mary Ann, was a distant cousin and an apparent early love interest of Lord Byron, the poet (Musters, 2001).

Later, a rich outcrop of Mustersan mammals was found in the interior of the town of Antofagasta de la Sierra, Catamarca Province, Argentina. Its fauna differs markedly from that of the Casamayoran Age, as some of the most prominent members of the South American fossil fauna appear: the glyptodonts, although of reduced size compared to their later relatives, and sloths of the clade Megalonychidae. They are accompanied by abundant and diversified notoungulates of the clades Notohippidae and Hegetotheriidae. The notohippids ("southern horses"), as their name implies, superficially resembled horses in overall shape of the snout, enlargement of the incisor teeth for cropping vegetation, and complex and high-crowned cheek teeth, which formed efficient grinding batteries. Hegetotheres demonstrate enlargement of the first incisors and greatly simplified cheek teeth. Some of them exhibit extreme convergence with modern rabbits in their cranial and postcranial adaptations (Cifelli, 1985). In addition, the astrapothere *Astraponotus*, a genus considered typical of this age, was present. Other ancient lineages of peculiar South American ungulates make their last appearance, such as the notostylopid and notopithecine notoungulates and the trigonostylopid astrapotheres. Marsupials are also known from northern Argentina.

Paleocenozoic Megacycle

Infracenozoic Supercycle
    Prepatagonian Cycle
        Divisaderan Subcycle

The traditionally recognized and rather long hiatus (Pascual et al., 1965) in continental faunas between Mustersan and Deseadan times has begun to be filled by the work of several recent authors. The ages so far identified as partially filling this gap are included in the Divisaderan Subcycle of the Prepatagonian Cycle. The earlier time period is represented by a relatively modest but important assemblage of mammals, known since the work of Minoprio (1947), from the Divisadero Largo Formation, named after a hill in Mendoza Province in western Argentina, called Cerro Divisadero, from which the natives kept watch over passing guanaco herds for hunting purposes. This Divisaderan Age fauna has been dated as between 35 and 33 Mya, which represents a transitional position between the late Eocene and the early Oligocene (Kay et al., 1999; Prothero, 2006). Among the mammals found here are litopterns, notoungulates, and Groeberiidae, remarkable marsupials with an extreme rodentlike appearance probably reflecting a dietary preference for hard items in their seasonally marked environment (Pascual et al., 1994). Moreover, herbivores continued the trend toward the development of hypsodont or high-crowned teeth, an achievement that far predated its development on other continents in other mammalian groups, suggesting that the climate in the Eocene was much drier than in other continents.

The gap mentioned above was further filled by the discovery in the 1980s by North American paleontologists (Wyss et al., 1993) of a rich fauna in the sediments of the Tinguiririca Valley in the Andes of central Chile that is dated as between 33 and 31.5 Mya, known as the Tinguiririican Age. The oldest record of South American rodents was found here. The material, an incomplete lower jaw, has teeth with the distinctive pattern of five crests, which clearly reveals its affinities with African rodents, as North American species had only four. It is remarkable that most low-crowned groups that predominated in the South American Eocene faunas are absent. They are replaced in this age by the first community of high-crowned mammals, including the groeberiid (Fig. 3.11) marsupials, already noted as present in the Divisaderan Age, and indicative of the presence of grass and hence of drier, more seasonally defined times than occurred in the tropical Eocene.

3.10. The Gran Barranca in Patagonia, preserving the most complete sequence of middle Cenozoic paleofaunas in South America, is the only continuous continental fossil record of the Southern Hemisphere between 42 and 18 million years ago, when climates at high latitudes transitioned from warm humid to cold dry conditions.

Photograph courtesy of Richard Madden.

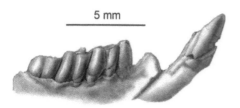

3.11. The Tinguirirican Age groeberiid *Klohnia charrier.*

From Flynn and Wyss (1999), with the kind permission of the Society of Vertebrate Paleontology.

Mesocenozoic Supercycle
  Patagonian Cycle
    Deseadan Subcycle

In a period spanning some two million years just before the end of the Oligocene, global sea level dropped from about 280 m above the present level to 150 m below it (Prothero, 2006). This sudden drop coincides with a remarkable change in the mammalian fauna of Patagonia, characterized by the extinction of the typical early Oligocene species. This is also the start of a new megacycle, the Neocenozoic Megacycle, which Pascual and Ortiz Jaureguizar (1990) correlated with the North American terminal Eocene event (Prothero, 1994) and in Europe with La Grande Coupure, as noted above.

This also marks the start of the Mesocenozoic Supercycle, which extended until the beginning of the Supercenozoic Supercycle. In the latter, the pampas take over as the region on which the ages are defined. The Patagonian Cycle is characterized by the first major faunistic change leading ultimately to a modernization of the mammals (Pascual et al., 1985), with the extinction of many lineages of marsupials and of native ungulates, as well as of utaetine armadillos.

Only one age is recognized in the Deseadan Subcycle, the late Oligocene Deseadan (28 to 24 Mya). In this age, the composition of mammalian assemblages in widely separated regions (Patagonia, Bolivia, Peru, Uruguay) became more uniform, and temperate–warm climates have been proposed to have reached as far south as modern Santa Cruz Province (Pascual, 1984a,b). Grazing herbivores were dominant, which suggests also the predominance of grass prairies. However, some beds in Chubut Province contain abundant petrified wood, evidence of the reappearance of forests after the late Paleocene in Patagonia.

The name Deseadan comes from the presence of sediments of that age along the Deseado River, derived itself from the nearby port (Puerto Deseado), which in turn was originally named Port Desire by the English privateer Thomas Cavendish in 1586 after his flagship, then later translated into Spanish. It is located in northern Santa Cruz Province, Argentina, and was known since the Ameghinos' time as the zone with *Pyrotherium*, a large beast with long, tusklike incisors belonging to (and indeed the last representative of) Pyrotheria (Box 3.1). This assemblage marks the beginning of a change in faunal composition to one that was to remain dominant for the rest of the Tertiary, with some taxa, particularly among marsupials and xenarthrans, attaining very large sizes.

In the Deseadan, novelty among mammals is reflected by new lineages of notoungulates (Typotheria and homalodotheriid Toxodonta, such as the horned-nose *Leontinia*), xenarthrans (the mylodontid sloths, which continued into the Pleistocene with great diversity), and the platyrrhine monkey *Branisella boliviana*, representing our own clade Primates. Although the hystricomorph rodents are known already in the Tinguirirican (and perhaps before, according to Vucetich et al., 2007), they here show the beginnings of what would become a great degree of taxonomic, ecologic, and size diversity. Deseadan birds are also notable, particularly in the initial

3.12. The terror bird *Phorusrhacos* was, as its name implied, a fearsome sight. It stood nearly 3 m in height.

Drawing by Charles R. Knight in Lucas (1901).

diversification of the impressive, large-sized, carnivorous phorusrhacids (Fig. 3.12).

Neocenozoic Megacycle

Mesocenozoic Supercycle
  Patagonian Cycle
    Pansantacrucian Subcycle

This subcycle includes the Colhuehuapian and the Santacrucian ages. The oldest mammals belong to the late Oligocene–early Miocene Colhuehuapian Age (24 to 20 Mya), named after Lake Colhué Huapi in Chubut Province, Argentina. This name is derived from the native language, Mapuche, and is composed of the words *colhué* ("reddish place") and *huapi* ("island"). Carlos Ameghino, again, collected most of the Colhuehuapian fauna along the cliffs on the southern end of this lake. As usual, the specimens were studied by his brother, Florentino, who referred to the fossil-yielding sediments as the zone with *Colpodon*. *Colpodon propinquus* is a large notoungulate first found in the Chubut Valley and described by Burmeister (1885). The presence of platyrrhine monkeys (such as *Dolichocephalus gaimanensis*, not far removed from its African ancestors) indicate that warm and humid climates, with well-developed forests, had returned to Patagonia. Also, the first anteaters are found here, and the native ungulates are mesodontic (i.e., with teeth less high crowned than in the hypsodont condition) in a higher proportion than in the previous subcycle.

The Santacrucian Age (19.3 and 16.2 Mya) follows, with deposits from the late early Miocene. It is represented by the sediments of the Santa Cruz Formation, which outcrops extensively in continental Patagonia. These sediments, some of which yield beautifully preserved fossils, were first observed by Francisco P. Moreno (a prominent figure of his time as naturalist, explorer, and politician, as well as occasional rival of Florentino Ameghino) in the valley of the Santa Cruz River. In 1887, Carlos Ameghino conducted his first field expedition as a traveling naturalist of the Museo de La Plata to Patagonia along the banks of the Santa Cruz River, from which he recovered an important collection of Santacrucian mammals that was studied by Florentino. *Homunculus patagonicus*, a platyrrhine monkey that lived in the Santacrucian of Patagonia, was described by F. Ameghino (1891a).

The presence of primates during the last part of the early Miocene in Patagonia indicates that there were forests and warmer climatic conditions than currently exist in this region. Monkeys later disappeared from Patagonia and became restricted to subtropical areas farther north. The phorusrhacids, the already-mentioned cursorial carnivorous birds related to the seriemas and coots, that lived in South America from the Eocene to the Pliocene (and perhaps even the Pleistocene) were more diverse than in previous and later faunas. The first remains of a phorusrhacid, an incomplete mandibular symphysis, were recovered by Carlos Ameghino in 1887 from the Santacrucian of Santa Cruz Province. Florentino Ameghino (1889) named it *Phororhacos longissimus*, but because the specimen was enormous and robust, he described it as belonging to an edentulous xenarthran, such as an anteater. Phorusrhacids attained a height of between 50 cm to over 2 m. Their powerful hind limbs bore three anterior digits and one posterior digit. The wings, so reduced as to be unsuitable for flight, must have been effective in balance during running, much as in living rheas.

The Patagonian environment consisted mainly of wooded intertropical savannas, with reptiles such as iguanas and lizards. As noted above, the phorusrhacids reached their peak diversity at this time, and other clades of birds made their appearance (Rheiformes, Anseriformes, Accipitriformes, and Galliformes). The Santacrucian included several herbivorous and predatory marsupials, xenarthrans (sloths such as *Hapalops*), glyptodonts such as the sheep-sized *Propalaehoplophorus* (Fig. 3.13), and armadillos, anteaters, rodents, and ungulates (such as the notoungulate toxodonts and typotheres, and litopterns). *Nesodon imbricatus*, a toxodont the size of a tapir, was among the most characteristic of faunal elements (Fig. 3.14).

The proterotheriid litopterns were highly diversified, with mono- and tridactyl (single- and three-toed) forms. Interestingly, the proterotheriids achieved the characteristic equid foot morphology millions of years earlier than did horses—indeed, because their side toes were even more reduced and splintlike than in horses (Appendix 1), they were the most monodactyl of all mammals (Simpson, 1980). Macrauchenid litopterns included

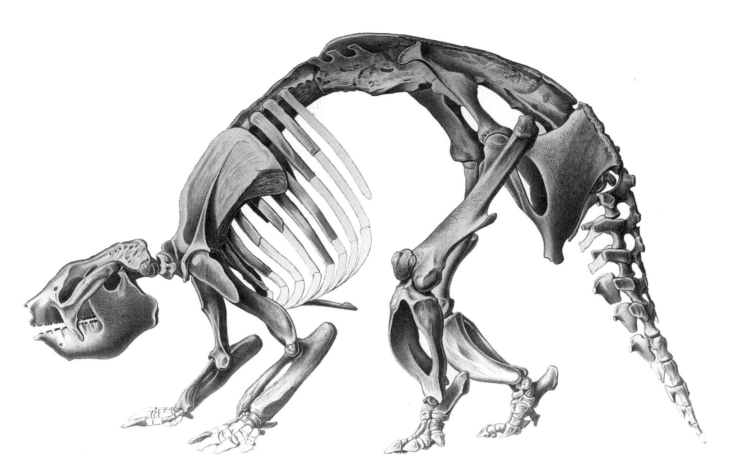

Theosodon, which resembled the llama. The enigmatic molelike mammal *Necrolestes* (Fig. 3.15) and giant astrapothere *Astrapotherium* were the sole representatives of their respective groups.

Neocenozoic Megacycle

Mesocenozoic Supercycle
   Panaraucanian Cycle
      Protoaraucanian Subcycle

In middle Miocene sediments, mammals typical of tropical or subtropical warm climates are no longer recorded from the austral region of South America. Indeed, the Colloncuran Age (15.5 to 14 Mya) marks a dramatic shift in the composition of the mammalian communities. The age owes its name to fossiliferous outcrops near the Collón Cura River, in the southern part of Neuquén Province, Argentina. Here the Panaraucanian Cycle begins. In this and the following Pampian Cycle, the mammals show evidence of preferring extensive open-country environments, from wetter subtropical savannas to cold–temperate, steppelike habitats.

This fauna has received less attention than others, but it is noteworthy that the first Megatheriinae are recorded here (Scillato-Yané and Carlini, 1998). Other Colloncuran vertebrates are anteaters, platyrrhines (Kay et al., 1998), and porcupines (Candela and Morrone, 2003). The Ostracoda (small, shelled crustaceans) recovered from the lower beds of the Collón Cura Formation suggest a paleotemperature near 20°C for the body of water in which Colloncuran sediments were deposited (Bertels-Psotka, 2000).

3.13. Propalaehoplophorines are the only group of glyptodonts represented during Santacrucian times. The estimated body mass of the only specimen of this glyptodont preserving an almost complete skeleton, *Propalaehoplophorus australis* (Scott, 1903–1904: pl. XXII), had values between 74 and 93 kg (Vizcaíno et al., 2011). *Propalaehoplophorus* was a highly selective feeder in moderately open habitats (Vizcaino et al., 2011a,b).

3.14. Skull and mandible of the late Oligocene-Miocene notoungulate toxodont *Nesodon imbricatus* (AMNH 15256; photo by Guillermo Cassini, reproduced with kind permission of the American Museum of Natural History). This animal reached a size of approximately 500 kg and had a complete dentition without diastema and large masseteric and temporal muscle attachment areas, as well as lateral incisors and high-crowned cheek teeth of finite growth. Madden (1997) proposed that the complex molar crowns would have provided considerable shearing ability among toxodontids, allowing them to break down grasses more easily. However, Townsend and Croft (2008), based on microwear analyses of three Santacrucian notoungulate genera, disagreed with this dietary inference, and suggested instead that *Nesodon imbricatus* was a leaf browser which focused more on hard browsing, possibly including bark consumption. Cassini et al. (2011) and Cassini and Vizcaíno (2012), using ecomorphological, ontogenetic-allometric and biomechanical approaches, concluded that *Nesodon* was a generalized herbivore (i.e., not a specialized grazer or browser) capable of achieving large masticatory forces, suggesting a diet rich in hard-objects, such as bark, as first proposed by Townsend and Croft (2008). Also, its hypertrophied lateral incisors and an unusually robust muzzle suggest potentially aggressive behaviors, in addition to obvious defensive uses (Cassini and Vizcaíno 2012).

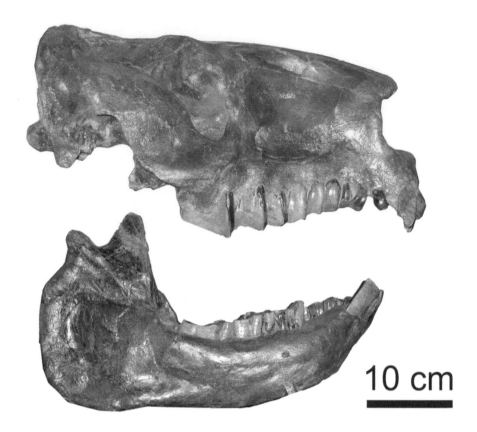

10 cm

Apart from the Patagonian location, Colloncuran mammals are also known from Arica, northern Chile (Bond and García, 2002).

As just noted, the rise of the Andes and subsequent glaciation in Antarctica during the middle Miocene caused the Patagonian climate to turn colder and more arid. Forests then gave way to the steppes that are still present there today. Many Santacrucian mammals became extinct, and monkeys retreated from the high latitudes to more northerly locations, where they are still found. The fauna of the Araucanian Subcycle arose in the Colloncuran Age, but reached its climax in the later Chasicoan (middle Miocene) and Huayquerian (middle Pliocene) and continued until the Montehermosan (middle to late Pliocene) ages. At that point in earth history, the time of the Great Plains came to an end when Pliocene diastrophism (deformation of the earth's crust) raised and fractured the regions they had occupied, restricting them to the Chaco-Pampean. The Friasian community was transitional. In it, the currently known features of the neotropical faunas began to be established. The typical megatheriines and the large leptodactylid *Gigantobatrachus* are found in rocks of this age, which also marks one of the last appearances of astrapotheres.

The subsequent Laventan Age (13.8 to 12 Mya) owes its name to the site of La Venta, Colombia. It was proposed by Madden et al. (1997), and it is based on the mostly endemic taxa described from the Villavieja Formation in Colombia: tortoises, alligators, primates, glyptodonts, armadillos, and notoungulates are found in this unique tropical fauna, as well as a surviving astrapothere (Fig. 3.16).

Another fauna of this age is that of Quebrada Honda in southern Bolivia. Its higher-level taxonomic affinities are closer to those of the slightly older high-latitude fauna of Collón Cura than to the low-latitude fauna

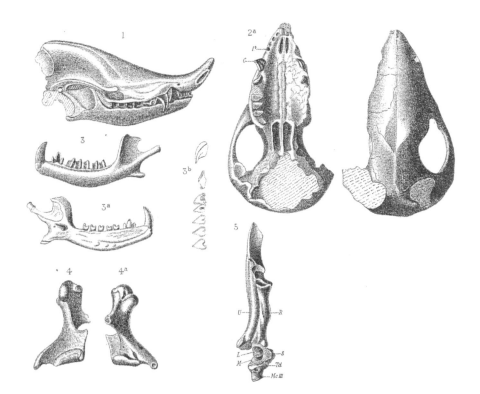

3.15. Skull and other remains of the putative marsupial *Necrolestes patagonensis* (from Scott, 1905), an enigmatic fossil mammal from the early Miocene of Argentina. It has eluded confident assignment to a higher-level clade since its discovery in the late nineteenth century (Ameghino, 1891b). Much of the skeleton of this fossil is known, but the mosaic of features of *Necrolestes,* many correlated with fossoriality, do not provide an unequivocal signal for the affinities of this genus (Ladevèze et al., 2006).

of La Venta, suggesting that isolating mechanisms between the low and middle latitudes were in place during the early and/or middle Miocene (Croft, 2007). The oldest marsupial sabertooth, *Anachlysictis gracilis*, is found here (Goin, 1997).

Slightly younger is the Mayoan Age (12 to 10 Mya), whose first paleontological outcrops were discovered in the cliffs of the Mayo River, Chubut Province, Argentina. Phorusrhacids continue to be the top predators, and small carnivorous marsupials shared those trophic habits. Xenarthrans and native ungulates are well represented, and hystricomorph rodents underwent a significant radiation.

The final age of this subcycle is the Chasicoan Age (10 to 9 Mya). Its sediments began to be explored about 1916 in the area of Chasicó Creek, in the south of Buenos Aires Province, Argentina. The first publications on its mammalian fauna were those of Cabrera (1928) and Cabrera and Kragliev-ich (1931). This fauna includes forms related to those of the Friasian, but with relicts like the notoungulates *Nesodon* and *Theosodon* that survived also into the Huayquerian. In the Chasicoan, *escorias* (glassy slabs) and *tierras cocidas* (red, bricklike fragments) provide evidence of an asteroid impact dated 9.23 Mya (Zárate et al., 2007).

Neocenozoic Megacycle

Mesocenozoic Supercycle
    Panaraucanian Cycle
        Araucanian Subcycle
With the Huayquerian Age begins a new subcycle, the Araucanian, which also marks the final part of the Panauracanian Cycle. This age is characterized by a shift observed in the contents of ^{13}C in the teeth of fossil mammals, revealing a change in the vegetation, which in this age starts to be

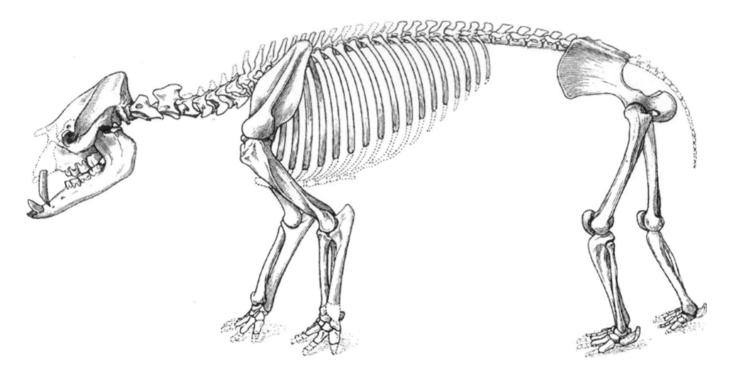

3.16. Reconstruction of the skeleton of *Astrapotherium magnum* (Riggs, 1935: fig. 36), based on remains housed in the Field Museum of Natural History, Chicago. Restored parts indicated as dotted lines.

Reproduced by kind permission of the Field Museum of Natural History.

dominated by plants with C_4 metabolism (MacFadden et al., 1996). This metabolic pathway of fixing carbon (see Chapter 8) gives the plants that bear it (herbaceous, tropical, arid-adapted grasses and warm/dry herbs) a competitive advantage over those possessing the more common C_3 carbon fixation pathway under conditions of drought, high temperatures, and nitrogen or CO_2 limitation. Therefore, the early appearance of hypsodont herbivores in the South American mammalian faunas must indicate the existence of hard, abrasive plants and not an early predominance of grasses.

The Huayquerian Age (9 to 7 Mya) is named after the Huayquerías (from Quechua, *huayco*, "dry stream bed") de San Carlos, in Mendoza Province, Argentina, but it is better represented in eastern La Pampa and western Buenos Aires provinces. It is during this age, 8 or 9 Mya, that the initial phases of a complete land connection between North and South America got underway, allowing the beginning of an interchange of mammalian species (and a few nonvolant birds) between these continents, called the Great American Biotic Interchange (GABI; Stehli and Webb, 1985; see Chapters 5 and 9). As a consequence, placental carnivorans, such as the procyonid *Cyonasua*, arrived in South America in this age, heralding a larger biogeographic process. Also, a gigantic caiman (*Dinosuchus terror*) is known from the Solimões Formation in Brazil.

This age is well represented by fossils found in the west bank of the Paraná River. They have been known for a long time, as that knowledge began with descriptions by the French naturalist Alcide Dessalines D'Orbigny (1842). Huayquerian Age deposits are found in Argentina, Uruguay, Bolivia, southern Peru, and the western Amazon basin. Among the mammals, glyptodonts in particular underwent a remarkable radiation. *Thylacosmilus atrox* is the better-known name for a 90-kg (i.e., jaguar-sized) marsupial sabertooth, one of the most specialized species in the whole clade, which must have been a hypercarnivorous top predator. However, Ameghino's (1891b) *Achlysictis*, though based on fragmentary remains, was the first-used

3.17. Reconstruction of *Argentavis magnificens* driving away the jaguar-sized, marsupial saber-tooth *Achlysictis* from its prey, a large ungulate carcass. This aggressive scavenging, termed kleptoparasitism, is not known in modern vultures, but *A. magnificens* might have used its large size for intimidating these marsupial predators and taking their food, as spotted hyaenas (*Crocuta crocuta*) often do with the prey of wild dogs (*Lycaon pictus*). As noted by Elissamburu and Vizcaíno (2005), the high energetic cost of hunting (25 times the basal metabolic rate for wild dogs; Gorman et al., 1998), loss of prey by kleptoparasitism would have greatly affected the amount of time that such sabertooth marsupials devoted to hunting in order to achieve their energy balance.

Image from Elissamburu and Vizcaíno (2005), by kind permission of Asociación Paleontológica Argentina. Drawing by Néstor Toledo.

name for this genus. Given its priority, it should be recognized as the valid name unless the International Committee of Zoological Nomenclature, the world's authority on those matters, is petitioned to preserve *Thylacosmilus*.

This fauna also includes remains, found in the plains of La Pampa Province, Argentina, as well as near the Andes, of the largest flying bird ever found, the giant teratornithid *Argentavis magnificens*, with a wingspan of 7 m and an estimated body mass of 60 kg (Vizcaíno and Fariña, 1999). It may have been a carrion eater that took advantage of the leftovers of the marsupial sabertooth (Fig. 3.17).

The next age is the Montehermosan (7 to 4 Mya, Flynn and Swisher, 1995; or between 6 and more than 4 Mya, Cione and Tonni, 1995c), which is recorded in the coastal cliffs of southwestern Buenos Aires Province, a few kilometers from Pehuén Có (see Chapter 5). The first person to collect fossils here was Charles Darwin in 1832. *Argyrolagus palmeri* was a marsupial that shared many features with kangaroo rats and jerboas, suggesting semiarid environments. The Holarctic-origin cricetids, the first among the many mammalian clades of the many groups that participated in the GABI, made their first appearance. Their arrival and ecological success in their new home will be further dealt with in the following chapters.

Immediately following the Montehermosan, the Chapadmalalan Age is between 4 and 3.2 Mya, according to Cione and Tonni (1995b), who subdivided it into early and late, with their boundary at 3.5 Mya. The name of the Chapadmalalan Age (originally spelled Chapalmalalan) was first applied by Florentino Ameghino (1908) to refer to the cliffs in Buenos Aires Province, Argentina, from south of Punta Mogotes near Mar del Plata north to Miramar, an area known during Ameghino's time as Chapalmalán. During this age at the end of the Pliocene, South America and North America were completely united by the Isthmus of Panama. Among its mammals are the tayassuids (peccaries), representing what is usually considered the first occurrence of the clade in South America (see Chapter 5), and *Chapalmalania*, a large, bearlike procyonid, another herald (in the words

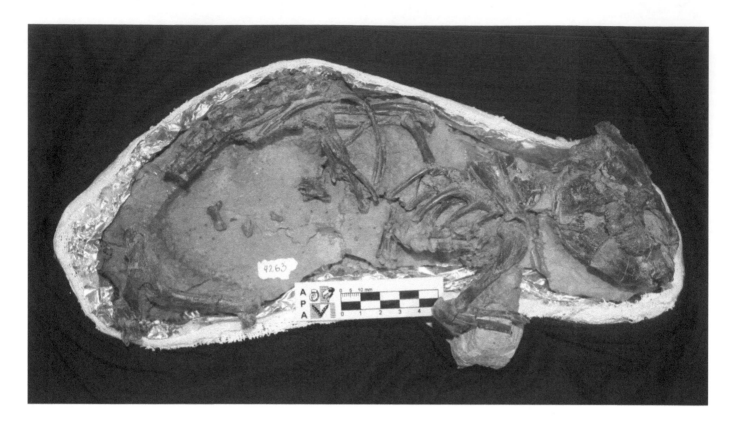

3.18. Remains of *Interatherium* from the Santa Cruz Formation of Argentine Patagonia. This is the smallest of the Santacrucian notoungulates, with an estimated body mass of 500 g. Interatheriids include species with complete dentition and without diastema or hypertrophied incisors. Thus, their incisors are all the same width. Cheek teeth are high crowned and ever-growing, and the masseteric muscle attachment areas are larger than those for the temporal muscles.

Specimen in the Museo Regional Provincial "Padre Manuel Jesús Molina," Rio Gallegos, Santa Cruz, Argentina.

of Simpson, 1953b) of what would turn into a massive invasion of North American mammals. Moreover, *Achlysictis* extends its biochron to this age. *Interatherium*, described by Ameghino (1887), was a small notoungulate (Fig. 3.18). *Actenomys*, a ctenomyid rodent, was probably the builder of the many small paleoburrows found in the Mar del Plata cliffs.

Neocenozoic Megacycle

Supracenozoic Supercycle
    Panpampian Cycle
       Pampian Subcycle

It is in the Supracenozoic Supercycle that a radical change in the composition of the South American mammals took place (Pascual and Ortiz Jaureguizar, 1990). It includes only the Panpampian Cycle and two subcycles, of which the Postpampian Subcycle, equivalent to the Holocene, is of little paleontological importance.

The earlier age is the Marplatan, proposed by Cione and Tonni (1995a) to replace the less clearly defined and also less accessible Uquian, and divided into the Sanandresan (2.5 to 1.9 Mya), Vorohuean (2.9 to 2.5 Mya), and Barrancaloban (3.2 to 2.9 Mya) subages. The beds deposited during the Marplatan Age are found in the cliffs south of Mar del Plata, such as Barranca de los Lobos, Punta Vorohué, and Punta San Andrés, from which the names of the subages are derived. There is an important faunistic replacement at 3.3 Mya: 37% of the genera and 53% of the species became extinct between the Chapadmalalan and the Barrancaloban (i.e., the earliest of the Marplatan; Vizcaíno et al., 2004b)—more dramatic, at a local scale, than at the K-T transition, although the latter was of course global. The Chapadmalalan fauna was diverse and balanced from an energetic point of view, while the Barrancaloban represents a poorer environment, and the proportion of herbivores to carnivores is apparently unbalanced as a result of the absence of medium- to large-sized predators, either birds or mammals (see Chapter 8). The traditional cause proposed for this faunistic replacement is climate (Cione and Tonni, 1995a, 1995b), perhaps related to tectonics (Cione and Tonni, 2001), but there is also evidence of an asteroid impact during this age (Schultz et al., 1998), which may have caused important turnover in the mammalian faunas (Vizcaíno et al., 2004b).

Another important event that occurred during the Marplatan Age is the arrival of some newcomers of northern origin, but we will defer their story until their history in the northern continents is detailed in Chapter 4. We will also interrupt our narrative to treat the events that occurred in South America after the Marplatan (we'll return to this thread in Chapter 5), both because what happened during Pleistocene times requires special consideration, and because we must first deal with the tectonic events, mainly the emergence of the Isthmus of Panama, which made possible the subsequent intermingling of North and South American forms in the GABI (also considered in Chapter 5).

4.1. North and South American Land Mammal Ages.

Drawn by Sebastián Tambusso from several sources, especially the Paleobiology Database (http://paleodb.org/).

My	Period	Epoch		Stage	Polarity Chron	South America	North America
0	Neogene		Holocene	Upper	C1	Lujanian	Rancholabrean
			Pleistocene	"Ionian"		Bonaerian	Irvingtonian
				Calabrian		Ensenadan	
				Gelasian	C2	Marplatan	Blancan
		Pliocene	L	Piacenzian	C2A	Chapadmalalan	
5			E	Zanclean	C3		
		Miocene		Messinian		Montehermosan	Hemphilian
					C3A		
					C3B	Huayquerian	
			L		C4		
				Tortorian	C4A	Chasicoan	
10					C5	Mayoan	

North American Late Cenozoic Faunas

4

North America also had a varied mammalian fauna during the late Tertiary and Quaternary, and its importance for understanding the Lujanian in South America has to be emphasized because the connections between the two continents are strong and relevant to our main subject. The South American faunas certainly made their mark in North America, but there were numerous other interesting creatures. The reasons for northern diversity were partly due to continued migration and evolution of forms of Eurasian origin and the arrival of the South American mammals, as well as indigenous evolution in North America. The megafauna was spectacular here as well, and like that of Pleistocene South America, it easily surpassed that of modern Africa. The period of interest to us, in North American land mammal ages, spans part of the Hemphillian (pronounced Hemp-hillian; from 10.3–4.9 Mya), plus the Blancan (4.9–1.8 Mya), Irvingtonian (1.8–0.3 Mya), and Rancholabrean (0.3–0.011 Mya), as seen in Fig. 4.1.

Many spectacular fossil localities represent these periods, spread out all over the territory but particularly abundant in the United States. We will look at some of these later. We have already mentioned the migrations between North and South America in Chapter 3, and this subject will be further treated below, so we should examine briefly the reasons for migration between North America and Eurasia.

The Pleistocene was the time of the great ice ages, and North America (as well as northern Eurasia) was subjected to repeated glacial advances, when much of its northern half was covered by continental ice masses. The advances were separated by interglacials or recessions, during which the land was essentially free of ice. There were four main, or classical, ice ages in North America, named the Nebraskan, Kansan, Illinoian, and Wisconsinan, in order of oldest to youngest; the intervening interglacials are the Aftonian, Yarmouthian, and Sangamonian. There were two main ice sheets. The larger Laurentide ice was centered around Hudson Bay and covered most of the current territory of Canada and a large portion of the northern United States, including what today are the cities of New York and Chicago. At times of glacial maxima, the ice sheet would even enlarge sufficiently to coalesce with the smaller Cordilleran ice of the northwest mountains (Dyke and Prest, 1987), as shown in Fig. 4.2.

An important event that happens during glacials is that water becomes locked in ice, so that global sea levels fall to expose vast areas of previously submerged land. This is what happened with Beringia, the shallow platform that links Siberia and Alaska, named after the discoverer of that region, the Danish explorer Vitus Jonassen Bering (1681–1741), hired by the Russian czar Peter the Great, an enlightened despot who wanted to map the most remote regions of his vast empire, as represented in Fig. 4.3.

Meanwhile, back in the north . . .

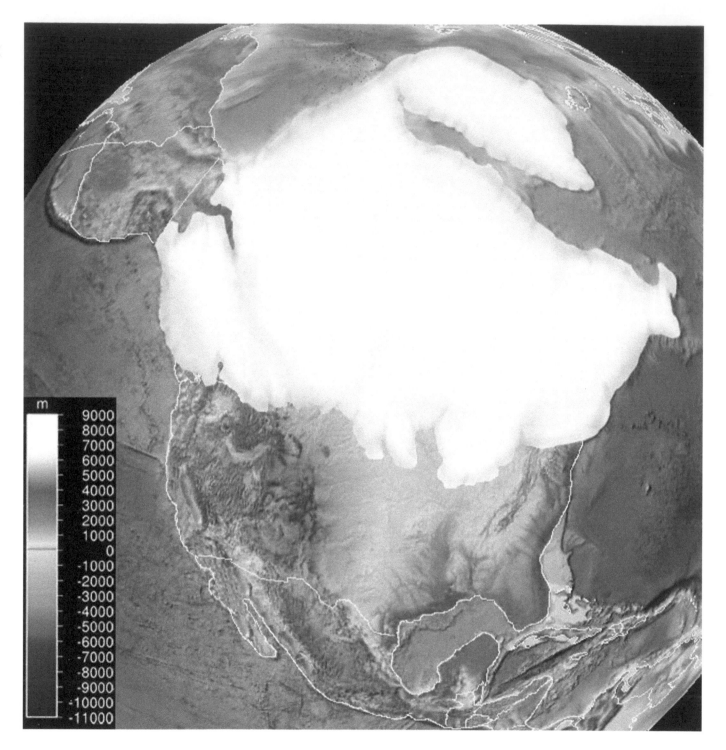

m	
	9000
	8000
	7000
	6000
	5000
	4000
	3000
	2000
	1000
	0
	-1000
	-2000
	-3000
	-4000
	-5000
	-6000
	-7000
	-8000
	-9000
	-10000
	-11000

4.2. Reconstruction of the ice sheet on the map of North America during the last glacial maximum, approximately 18 kya. The two large ice sheets that covered a great part of North America, the Cordilleran sheet in the western mountain regions, and the Laurentide sheet over the rest of the continent, coalesced to form a single sheet.

Illustrated by Sebastián Tambusso, based on data from Andrews (2006).

Although Siberia and Alaska seem a world apart on conventional maps, they are separated by the Bering Strait, which is only some 100 km (60 miles) wide. Thus, when the ice fields were at a maximum, Beringia was left high and dry, and although it seems illogical, it was mainly ice free. The reason is that this region was too dry, so there was not enough snow to produce great masses of ice (Edwards et al., 2000).

Animals that migrated to North America would have done so in two stages. They would have occupied Beringia during glacials, but their path to the main part of North America was blocked by ice. They had to await an interglacial and for sufficient thawing to occur to produce an ice-free

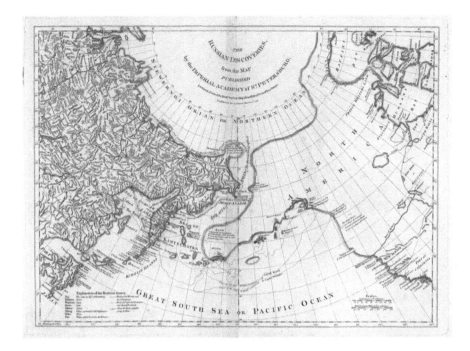

4.3. Map of Bering's land discoveries in eastern Siberia and Alaska (Jefferys, 1775). During Bering's travels, organized by Czar Peter the Great, it was discovered that Asia and North America were separate landmasses. The expedition claimed the lives of 21 of the 77 men aboard the ship, Bering included.

corridor between the Cordilleran and Laurentide ice sheets (which, incidentally, would cause sea levels to rise and reestablish the strait). Only then could animals have continued their journey from Alaska (Lacourse and Gajewski, 2000).

Proboscideans

North American megafauna

Proboscidea includes the elephants, the largest living land mammals, and their extinct relatives, such as mammoths, mastodons, gomphotheres, and barytheres, among others (Rose, 2006). They are generally characterized by their gigantic size and long, flexible nose, drawn out as a trunk, although several of the earliest proboscideans were smaller and trunkless. Today there are only two genera of proboscideans, *Loxodonta* from Africa and *Elephas* from southeast Asia, both belonging to Elephantidae, but the group has a rich fossil history. Proboscideans probably originated in Africa during the Paleocene (and possibly earlier) and became diversified and widespread by the Miocene.

A main trend characterizing proboscidean evolution is the development toward large and even gigantic size, accompanied by the presence of a trunk. The evolution of the latter has generally been explained as a consequence of the former, especially in the case in which large tusks add both weight and leverage (McGowan, 1994). The tusks are stereotypical features of elephants, but we should note that in some proboscideans, the tusks took on different forms and functions from those of living elephants; in some, for example, there were tusks in both the upper and lower jaws. With increased size, the weight and height of the body increased, and with a concomitant increase in the size of the head (although we should note that in the case of *Paraceratherium*, the largest of land mammals, this paradigm is not applicable), it was biomechanically advantageous in such circumstances to shorten the skull and neck. In addition, the forelimbs and hind limbs became permanently extended at the elbow and knee joints, forming columnar supports for the increased weight. This further increased the distance of the head from the ground. Elongating the nose as a trunk became a solution for the safe and efficient procurement of food and water. Another notable feature of proboscideans is the already mentioned enlargement of a pair of incisor teeth (those at the front of the oral cavity) into tusks, combined with the reduction or loss of the other incisors as well as the canine teeth (Appendix 1). Yet another feature of elephants, possibly also associated with the change in skull geometry, is the mode of tooth replacement. Rather than maintaining most of their teeth in the oral cavity at the same time, and replacing them vertically (that is, a tooth in one position in the jaw is replaced by a tooth at the same position), elephants have horizontal tooth replacement. In this mode, usually just one tooth (or parts of two teeth) is present in each jaw quadrant at any one time. As the tooth is worn, it is pushed forward and gradually replaced by the tooth behind it.

Among the several archaic proboscideans (leaving aside the Asian anthracobunids, possible but not certain members of this clade), we may mention the African *Phospatherium*, *Moeritherium*, and *Numidotherium*. These early forms (ranging from the Eocene to early Oligocene) were plesiomorphic proboscideans, and they lacked the exaggerated, telltale features of later members of the group, such as the presence of long tusks and, as far as can be inferred, trunk (though *Numidotherium* appears to have had a short, tapirlike proboscis), while retaining a nearly full complement of teeth. A distinct group of proboscideans includes the deinotheres. These Old World proboscideans, known from the early Miocene to about 1.5 Mya, were elephantine in proportions, but the skull remained relatively long; and though the nasal openings were large, the trunk of deinotheres was probably short. Perhaps their most distinctive feature was the form of their tusks. Only a lower pair, curving down and under the chin, was present (Benton, 2005).

Beyond the deinotheres and the early forms mentioned above, most other proboscideans are generally considered as Elephantiformes, including the basal *Phiomia* and *Palaeomastodon* from the African Oligocene (Gheerbrant and Tassy, 2009). More derived proboscideans comprise Elephantomorpha, a clade that includes several lineages. Major radiations in the Miocene gave rise to mammutids (or mastodons, not to be confused with the mammoths, which are elephantids), gomphotheres, and stegodontids. These became extinct in the Pliocene or Pleistocene, but a subsequent radiation produced the elephantids, which includes the modern forms, as well as the mammoths, *Mammuthus*, which are characterized by ridged molar teeth and tusks that lack enamel (Gheerbrant and Tassy, 2009). Of these groups, all but the stegodontids played a role in North America (and some as well in South America; see Chapter 5).

Mammutids, represented by *Zygolophodon*, are known from about 17 Mya in North America, but the remains of these early representatives are rare. *Mammut* is known since the early Miocene and survived until the end of the Pleistocene. Remains of this genus are much more abundant, particularly of its later representative, *Mammut americanum*, the American mastodon (Gheerbrant and Tassy, 2009). The latter is known for its long, curved upper tusks, although small tusks in the lower jaw were sometimes present. Pleistocene mammutids have generally been considered to be

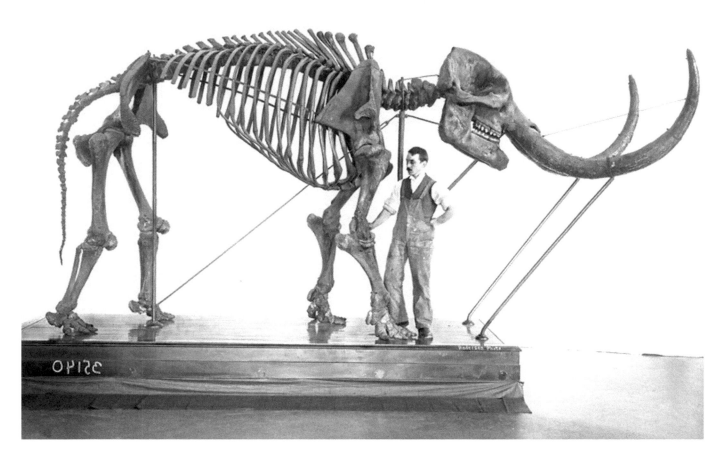

browsers that occupied forest rather than grassland savanna, but they may also have been habitat generalists, given their wide distribution across North America. Stomach contents recovered from *Mammut* include generally coarse vegetation, such as spruce needles, pinecones, and grass (Lambert and Shoshani, 1998). *Mammut americanum* was generally smaller than the woolly mammoth, but reached the height of the modern Asian elephant, though it was stockier and had a low, flattened skull (Fig. 4.4).

Gomphotheres, formally referred to as Gomphotheriidae (although this is apparently a paraphyletic grouping), comprise several lineages. *Gomphotherium*, the namesake of the group, is probably representative of the basal gomphothere stock. It entered North America in the middle Miocene and quickly became abundant and widespread. It became scarce in the latest Miocene but survived until about the end of the Pliocene (Lambert and Shoshani, 1998). Mammutids, and gomphotheres as well, are commonly referred to as mastodons (e.g., Mothé et al., 2010). This vernacular term, meaning "breast tooth," is derived from the odd form of their teeth, which bear two rows of blunt cusps resembling breasts—or at least they seemed so to Cuvier, who coined the term (Fig. 4.5).

Two other gomphothere lineages are important to our story. One of these is the group known as the shovel-tuskers, which were fairly common through the latter half of the Miocene and represented by such genera as *Amebelodon* and *Platybelodon*. These shovel-tusked gomphotheres were characterized by broad and flattened lower tusks. There has been some uncertainty over the use of these oddly shaped tusks. The traditional view of shovel-tusker feeding behavior was that the tusks were used to scoop up aquatic plants from shallow ponds and lakes. Also, these beasts have

4.4. Skeleton of *Mammut americanum* mounted in 1907 at the American Museum of Natural History in New York City. Fossils of this huge proboscidean were spectacular finds made during the early years of the United States. In the hands of Thomas Jefferson, its remains would go on to play a prominent role in the development of vertebrate paleontology in that country (see Box 4.1).

Image 35140, American Museum of Natural History Library.

4.5. Molar of *"Stegomastodon" platensis,* housed in the paleontological collection of the Facultad de Ciencias, Universidad de la República, Montevideo, Uruguay. Scale bar equals 5 cm.

Courtesy of Sebastián Tambusso.

generally been reconstructed as having a short, flaplike trunk. However, Lambert (1992) demonstrated that *Amebelodon* was a feeding generalist, feeding as portrayed in the traditional view, but also by scraping bark and stripping leaves and twigs from trees, and aided by a long, flexible trunk. *Platybelodon* was similarly viewed traditionally as using its shovellike tusks to scoop up aquatic plants, but Lambert (1992) found no evidence for such behavior and instead suggested that the tusks were used together with a long, flexible trunk to cut tough vegetation in sawlike fashion.

The other gomphothere lineage important here is generally referred to as the New World proboscideans; it includes *Rhynchotherium, Stegomastodon, Cuvieronius,* and, in South America at least, *Haplomastodon* (see Chapter 5). *Rhynchotherium,* like *Gomphotherium,* was a long-jawed gomphothere, with long tusks in the upper and lower jaws. Lambert and Shoshani (1998) noted that it probably evolved in Central America and then dispersed northward into North America during the late Miocene. It apparently did not survive into the Pleistocene. *Cuvieronius* was likewise a neotropical form, but it was a short-jawed gomphothere and abundant during the Pleistocene. Its entrance into South America was generally considered to have occurred near the beginning of the Pleistocene, but more recent analyses suggest a late middle to late Pleistocene dispersal (Ferretti, 2008). Further, it appears there was only a single dispersal event, with *Cuvieronius* diversifying in South America (see Chapter 6). *Stegomastodon,* another short-jawed gompothere genus present in North America, may also have entered South America, but some authorities consider that the South American form may not actually belong to this genus, as noted by Ferretti (2008; see Chapter 6), but retain it tentatively as *"Stegomastodon."*

Of all the great mammals, a proboscidean, the woolly mammoth *Mammuthus primigenius,* is probably the most characteristic of the ice ages—so much so, in fact, that it has become symbolic of this episode in earth's history. We are all familiar with artistic renditions, including popular cartoon movies, of these majestic beasts, with their high-domed heads, massive curving tusks, great sloping backs, and shaggy fur coats, struggling against the wind on barren fields of ice and snow. They were certainly well adapted to the cold, but the idea that they lived on the ice and snow is fanciful: what would there have been for them, committed herbivores that they were, to eat on the ice? Thus, in spite of the popular image of the mammoth's habitat, snow rarely covered much of its range. They lived south of the ice sheets, on grassy steppe with few trees but rich in herbaceous vegetation

(Stuart et al., 2002). They were probably also only around for the last ice age, the Wisconsinan, because they evolved only 250,000 years ago. Finally, although they were gigantic mammals, the woolly mammoth was in fact the smallest Pleistocene mammoth — some 3 m in height and 4–6 tons in mass. Mammoths are elephantids and thus are much more closely related to the living African and Indian elephants than to the American mastodon and the gomphotheres.

In addition to the woolly mammoth, whose range cut a great southeastern swath from Alaska through Canada to the Atlantic coast of the northeastern United States, there were two other (at least) species present in the North American Pleistocene. One was *Mammuthus meridionalis*, which made its appearance in the Old World 3 Mya and managed to get to North America by 1.5 Mya. It is known from a few localities in the central parts of the continent. After the subsequent separation of the Eurasian and North American landmasses, this form apparently continued along two lineages. In the Old World, the lineage continued into *M. trogontherii*, the steppe mammoth, 750,000 years ago, and then into the woolly mammoth by 250,000 years ago, which entered North America 100,000 years ago. In the New World, the lineage led to *M. columbi*, the Columbian mammoth, which extended as far south as southern Mexico.

Thus, the woolly and Columbian mammoths probably represent separate evolutionary paths; one did not give rise to the other. Some researchers have recognized the existence of additional species: the imperial mammoth (*M. imperator*) and Jefferson's mammoth (*M. jeffersonii*). Thus, the third United States president has yet another link with life of the past (Box 4.1). On the taxonomy, current evidence suggests that there is not much justification for separation of these species from *M. columbi*.

Thomas Jefferson, naturalist   Box 4.1

In many ways, Thomas Jefferson's (Box Fig. 4.1) interest in the life of the past was the result of his involvement in the exploration of and expansion into the North American West, although Jefferson's natural curiosity was certainly a factor — indeed, Thomas (2008) counted Jefferson as among the two American intellectual heavyweights of his day, Benjamin Franklin being the other. Jefferson believed that these great beasts might still have been roaming around in the huge uncharted territory west of the Mississippi River, but the main reasons for his passion was to counter the prejudice of European (mainly French) intellectuals against the New World, including its endemic organisms. Comte de Buffon's views on the inferiority of most of the American fauna (including its indigenous peoples, whose men were characterized by small and feeble organs of regeneration, and lack of hair, beard, and ardor for women) were leveled not only at South America, but at North America as well — in ironic contrast to the pejorative views of much later North American scientists with regard to the South American fauna. Buffon was among the most important and influential naturalists of his time. His opinions held great sway and were propagated by numerous researchers, most of whom (Buffon included) never set foot on American (meaning from the Americas) soil. The great naturalist did concede to the

Box Fig. 4.1. Thomas Jefferson, the third president of the United States and a devoted naturalist.

Portrait by Rembrandt Peale.

superiority of America's reptiles, amphibians, and insects — in other words, a nod to its status as being better at being "lowly."

Jefferson struggled to demonstrate that such views were inaccurate, and that American creatures were every bit as full of vim and vigor as their Old World counterparts. Indeed, the only book he ever wrote, *Notes on the State of Virginia*, was undertaken mainly as a response to Buffon, and contains "page after page of tables comparing the sizes and weights of European and American animals" (Thomas, 2008:39). The fossils of great beasts were among the more forceful examples in Jefferson's arsenal — thus his great interest in the American mastodon and Jefferson's ground sloth. Jefferson wanted the mastodon to be something entirely new to science (which it was) and different than the elephantine creatures that had already been discovered, so he tried to make it into a ferocious carnivorous predator. Likewise with the sloth, referred to as the "great-claw," which Jefferson dealt with in a later publication. Its claws were indeed large, larger than a lion's, which was the animal to which Jefferson compared it. He wanted very much for the sloth to be a proud, regal American king of beasts, and he suggested, on the basis of its claws, that it may have been three times as large as a lion (although he did have to concede in a last-minute appendix that *Megalonyx* was indeed a sloth; Thomas, 2008). Jefferson, of course, was incorrect on both counts in our modern view: according to current evidence, the mastodon and sloth both turned out to be herbivores. They were nonetheless spectacular new discoveries, and Jefferson's efforts did much to spur the development of North American paleontology.

All the other mammoths mentioned were quite a bit larger than the woolly mammoth. The steppe mammoth was the true giant of the lot, with maximum shoulder height of 4.3 m (or 14 ft) and mass of at least 10 tons. The Columbian was almost as large, with a height of 4 m and a mass of 10 tons. Although some gomphotheres managed to get into South America, the mammoths (and mastodons) did not, although a recent discovery (yet to be published) in Amazonia by Nascimento et al. (2010) may represent an exception. One possible reason is that the Isthmus of Panama was mainly a tropical to semitropical forested region. The browsing gomphotheres thus had food along the way, whereas the grazing mammoths, which needed huge amounts of grassy plain, were barred from entry. Both *Mammuthus columbi* and *Mammut americanum* were present in Central America. The former was more widespread, whereas *Mammut* was considerably less abundant south of central Mexico (Arroyo-Cabrales et al., 2007).

Artiodactyls

Artiodactyls include the suids (pigs, peccaries, and hippopotami) and ruminants (antelopes, giraffes, cows, and camels), the most derived and diverse ungulates (Fig. 4.6). In the past, these mammals were grouped together as Artiodactyla, but more recently, researchers have discovered that these animals are closely related to the whales, or cetaceans, so that the two groups are now classified together as Cetartiodactyla. It is unclear whether the artiodactyls all group together more closely with each other than with cetaceans or whether some may be more closely related to cetaceans, and so use of the formal name Artiodactyla (to refer exclusively to suids and ruminants) is questionable. Indeed, both molecular and morphological (e.g., Goloboff et al., 2009; Boisserie et al., 2005a,b,c) evidence generally indicates that cetaceans are more probably related to hippopotami, so that Artiodacyla as a formal name cannot be used to the exclusion of cetaceans (for it would be paraphyletic, a term explained in Chapter 1). For this reason, we will avoid using Artiodactyla as a formal name, but we will retain the informal term *artiodactyls* to refer to suids and ruminants. Perhaps the most distinctive feature of the highly cursorial and strictly herbivorous ruminants is, as their name suggests, their habit of ruminating or chewing cud, an activity for which they are well equipped by way of a multichambered stomach. One of these chambers, the rumen, acts as a fermentation center and is home to microorganisms that produce the enzymes to break down the tough walls of plant cells so that the animal can make use of the carbohydrates (the cell wall is formed largely of cellulose, a complex carbohydrate), and also gain access to the proteins, lipids, and carbohydrates within the cell (Vaughan et al., 2000). The cursorial nature of ruminants is reflected in elongation of the distal limb segments, fusion of the carpals and tarsals (the bones of the wrist and ankle), and a tendency, sometimes marked, toward reduction or loss of the side toes, leaving the third and fourth as the main supports of the body (Appendix 1).

There are essentially two types of ruminants, the camels and llamas on the one hand and the pecorans on the other, which include the ungulates, apart from the horse, that are most familiar to us. Horns or antlers are commonly present in pecorans and are used for functions such as defense or interspecific male dominance behaviors. Two characteristic features of ruminants are fusion of the third and fourth metapodials (the bones of the palm and sole) into a single structure, termed a cannon bone, and unguligrady, the practice of raising the sole off the ground and walking or running on the tips of the toes or hooves (the equivalent of claws or nails, for which the term has been appropriated from the Latin; Appendix 1). Among pecorans are Bovidae, a clade that includes sheep, goats, and cattle. The bison belongs to this group and is especially closely related to *Bos*, the domestic cattle, as well as *Syncerus* (the African or Cape buffalo) and *Bubalus* (the Asiatic buffalo). These bovids make up a compact group known as Bovini.

Bovini are originally an Old World group, represented as early as the late Miocene in Eurasia and Africa. The modern genera had evolved by the late Pliocene and were widespread on these continents, but only *Bison*, the most northern bovine representative, became widespread across Eurasia and North America. That it managed to cross Beringia is not particularly surprising, as it is better adapted to cooler or temperate climates; indeed, it is the only living bovine that does not have, at least partially, a tropical distribution. Its arrival in North America is fairly recent and an important indicator of time, for the beginning of the Rancholabrean land mammal age is marked by its presence in North America.

Although a general agreement on *Bison* systematics has yet to be reached, various species of *Bison* were present in the North American Pleistocene, of which three appear to be autochthonous (i.e., North America is the place where they evolved their particular characteristics), whereas the

3. Kisamthier. (Moschus.) 3 c. 3 b. 2 b. 2. Trampelthier.

5. Giraffe. (Camelu pardalis.) 5 b.

1. Lama. (Camelus.) 5 c.

4. Rennthier. (Cervus.)

6. Muflon. (Ovis musmon.) 6 b.

6 b.

7. Steinbock. (Capra ibex.)

8. Spiessgemse. (Antilope leucoryx.)

9. Urochs. (Bos urus.)

9 b.

4.7. *Bison latifrons.* This extinct species had horns that spanned over 2.5 m, like this specimen exhibited in the American Museum of Natural History.

Photograph by Denis Finnin, © American Museum of Natural History.

other two are Eurasian. Although there are some complications, there were essentially two lineages, one including the Eurasian species *B. priscus* (the steppe bison) and *B. alaskensis*, the other the North American authochthons *B. antiquus* (the ancient bison), *B. latifrons* (the long-horned bison, Fig. 4.7), and the modern *B. bison* (which includes the larger, more northern wood bison, *Bison bison athabascae*, and the plains bison, *B. b. bison*), whose illustration in Fig. 4.8 is reproduced from the pioneering work on animal locomotion by Eadweard Muybridge (1899). Researchers disagree on the evolutionary relationships of the species; some believe that the modern bison followed an evolutionary path through the *B. latifrons–B. antiquus* lineage, whereas others prefer a descent from *B. priscus*. Notwithstanding these disagreements, certain observations are beyond dispute.

These animals were widespread, ranging as far south as Mexico (indeed, they are among the most common ungulates in the fossil localities in which they are found), but they were particularly prevalent throughout the Great Plains, a region that includes much of central and western North America. *B. latifrons*, typically from the middle Pleistocene, is generally acknowledged as the oldest of the three species. *B. antiquus* certainly makes its appearance by the end of the middle Pleistocene, but both of these species occur together in some localities, notably Rancho La Brea. *B. bison* appears well into the Holocene and thus is a recently evolved species. Two main trends characterize bison evolution: reduction in body size and shortening of the horns.

The modern bison is a large mammal (Box 4.2). A large bull can have a mass of 1000 kg and stand nearly 2 m at the shoulder. This is about the size limit of *Bos taurus*, which includes the familiar domestic cattle used in dairy and meat production, as well as the zebu, the sacred cattle of India, which, incidentally, also has a prominent hump. The extinct bison were larger, with *Bos latifrons*, probably reaching 1250 kg, being the giant of the group. The long-horned bison is well named, for its horns were incredibly long. The longest recorded spread measures nearly 2.2 m from tip to tip, compared with approximately 0.6 m in modern bison. *Bos antiquus* was smaller both in body size and horn length than *Bos latifrons* (Fig. 4.7), with the longest recorded horn spread being just over 1 m.

4.6. *Opposite.* Representative artiodactyls. Center, giraffe. Left side, from top to bottom, musk deer, llama, mouflon, Arabian oryx. Right side, from top to bottom, camel, reindeer, Alpine ibex, auroch.

Image reproduced from Oken (1833).

Copyright, 1887, by Eadweard Muybridge.]

TRANSVERSE-GALLOP.
ONE STRIDE.
The Buffalo, or Bison.

SERIES 52.

Box 4.2   The extinction of the bison

We note here the fate of a group that very nearly disappeared, though more recently than all the other megamammals, perhaps implying that the great extinction to be dealt with in Chapter 9 has not completely ended. In this case, the scythe was definitely wielded by humans, but of the fully modern, supposedly more learned type.

The North American bison or buffalo, a majestic and dignified creature, symbolizes the romanticized era of the Wild West, when North America's frontiers were opened and peopled (that is, of course, by nonnatives). The rapid decline and near extinction of the bison in North America is a tragedy of monumental proportions, the result of greed, lack of foresight, and, surprisingly, even well-planned public policy. The relationship between humans and bison is an ancient one, probably as old as when humans first stepped on this hemisphere.

By the early 1800s, the bison had become a steady and available food supply (bison are, after all, a form of cattle). More importantly, the fur trade had begun to move in on these basically docile creatures, mainly to satisfy the eastern U.S. market for buffalo robes. For many of the middle years of the 1800s, the number of robes sent to market has been estimated

4.8. *Above.* A bison galloping, taken by the cameras of Eadweard Muybridge (1899), in one of his examples of motion capture through 24 successive cameras that were activated by the photographed subject.

at 100,000 per year. This is clearly too low a figure, given that some trading companies each reported handling about 100,000 robes during peak years (Barsness, 1985).

But bison had other concerns. With the opening of the West, taking out a beast became a sort of emblem of arrival. You could not become a real frontiersman until you'd bagged your own buffalo; and why stop there? Buffalo hunting became sport among Westerners and tourists (there often mainly for the chase) alike. There was no particular concern for their plight, and although a few enlightened souls did try to introduce means of protection, their efforts were usually killed by Congress or presidential veto. As told by Barsness (1985), the few feeble laws in existence in some states were ignored or unenforced. Indeed, the slaughter of bison was tolerated essentially as a matter of policy. Before the Civil War, the U.S. government saw bison as a means of keeping Indians fed once they had been moved to reservations. Such civility degenerated once the war ended and the army found that they were having a hard time of managing the Indians—meaning that they were losing, and at great price, the Indians wars. It became common wisdom then that the easiest way to subjugate the Indians was to exterminate their food source. The story goes, perhaps apocryphally, that

Box Fig. 4.2. Photograph from the mid-1870s of a pile of American bison skulls waiting to be ground for fertilizer.

General Sheridan, in rebuking a joint session of Texas legislators for even considering a bill to ban buffalo hunting, noted that they instead should give thanks to the hunters and award them medals for producing a dead buffalo on the one hand and a discouraged Indian on the other. Lest we jump into being too judgmental about the white man—though by no means are we trying to defend his actions—much the same strategy was used by one tribe against another. The Crow, for example, would rather have killed bison than have left them to feed the Sioux.

In any case, almost overnight, the bison were gone. Their numbers before the arrival of the Spaniards have been pegged at 50 to 60 million. By the time the hide hunters got into the act, six or seven million were left; by the turn of the twentieth century, some 75 animals in the United States and a few hundred in Canada remained. All that was left after the hunting stopped were their bones, which were left strewn about the prairies once their hides and flesh had been stripped away. Yet not even their skeletal remains were left in peace, as they too quickly found a market: bone china manufacturers, sugar refiners, glue makers, and fertilizer producers bought all they could. The hunt was on again, this time by men scouring the countryside for bison bones. It has been estimated that nearly 1.5 million tons of bone were sent east to be processed (Barsness, 1985).

Today, the bison is considered endangered by the International Union for Conservation of Nature, and it has been reintroduced into a number of places, mainly national parks and wildlife refuges, around North America, including Alaska, Alberta, and Oklahoma. They are also farmed for meat, which opens the possibility for a brighter future for the species.

Let's look next at the North American success story of a group that has ranged far and wide from their beginnings on this continent. We are speaking here of the camels, not the horses, which also spread widely from their North American base and, like the horses, could not manage to hold on in their native lands. It surprises most people to learn that camels are indeed North American, partly because there have not been any camels there for 10,000 years, but also because of the familiar image of camels as desert beasts, able to go without drinking for weeks during marathon crossings of the Sahara. This image is accurate—in recent history camels have been used as dependable transport during long desert treks—but there is much more to their story.

There are two types of camel, the camelines, including *Camelus dromedarius* and *C. bactrianus* (the one-humped and two-humped camels), which before domestication occupied the Arabian region and Central Asia, respectively, and the lamines, the llama, guanaco, alpaca, and vicuña that roam the mountainous Andes. These two groups differ considerably in size today, but both have small heads on long, thin necks, a slender rostrum (so much so that from the top or bottom the skull looks rather like a bowling pin), and thin, slitlike nostrils and a cleft upper lip that gives them a permanent and rather derisive sneer. Camels are ruminants, with three-chambered stomachs, and are unique among mammals in having crescentic or oval rather than circular red blood cells.

Particularly characteristic of camels are their feet. Being artiodactyls, camels originally tended toward a didactyl, unguligrade stance, but early on began to revert to a digitigrade stance with splayed toes on the ends of a distally divergent cannon bone and a padlike rather than hoofed foot (Fig. 4.9). In fact, camels are classified as Tylopoda, distinct from the pecorans, a name derived from the Greek τυλος and πούς (πόδε in the plural), the latter meaning "foot." Τυλος referred to a variety of things, such as a knot, a callus, a swelling, a knob on a club, or a cushion, all of which, but particularly the latter, describe camel feet. Τυλος apparently also referred to the phallus, but we think it is unlikely that this structure, even in the imaginative mind of the taxonomist who coined Tylopoda, could be confused with camel feet.

Such feet are remarkably useful for walking over loose, sandy soils and are no doubt an important factor in the camelines' legendary ability in desert extremes, as well as in lamines' unhindered movement over mountainous terrain. However, verified reports of Saharan crossings of 600 km and lasting 3 weeks with little vegetation and no water along the way clearly point to more than just feet as important camel characters. Camels can routinely go without water for a week in the desert heat, enduring as much as a one-third loss of body fluids. To put this in perspective, consider that humans can barely tolerate a 5% water loss; a 10% loss usually results in delirium and 12% in death. The main reason is that the blood becomes more viscous with loss of water, which stresses the heart and slows circulation, thereby impeding heat dispersal and leading to a rise in internal temperature. Camels do not lose water as readily, mainly as a result of the unique nature of their blood cells and concentrated urine, but also because the hump stores fat that can be oxidized to release water. Further, water can be stored in the rumen, one of the stomach chambers.

Although camels no longer grace the North American landscape, in the Pleistocene and earlier times, both camelines and lamines were integral parts of the fauna. Their remains are among the most abundant elements of the Plio-Pleistocene deposits of the New World. Many of these camels were smaller than modern camels (the lamines only produced two really large forms, for example), but the trend toward large body size in the later part of the Cenozoic resulted in forms of gigantic proportions. Consider that the modern *Camelus*, which reaches a mass of nearly 700 kg and a shoulder height of 2 m, is the smallest of the giant camels! During the Blancan and Irvingtonian, at least three genera were considerably larger. *Megatylopus* is geologically the oldest of the giant camels, followed by *Gigantotylopus* and *Titanotylopus* (one might quibble here with the less than inspired choice of names, but we're stuck with them). All are relatively common throughout the central United States, but less so in the West Coast and, particularly, the East Coast.

Cervids, the antlered artiodactyls, are typically browsing, forest-dwelling mammals. Cervids are rarely included among the megamammals, although it should be noted that one European species, the extinct giant Irish elk (see Plate 5, a beautiful painting in Lascaux cave), has received some press for the incredible size of its antlers. However, this cervid was not an elk (Box 4.3), not entirely Irish, and certainly not American.

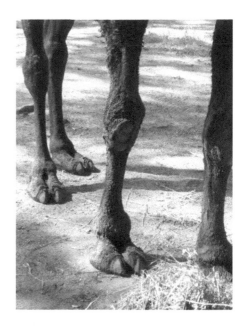

4.9. Camel feet, showing the feature that gives name to the whole group of artiodactyls. Artios (αρτιος) is ancient Greek for "even numbered," and dactylos (δάκτυλος) means "finger" or "toe," which reflects the usual pattern of having an even number of toes (2 or 4) on the front and back feet. In addition to being even in number, the toes are more or less arranged symmetrically around an axis passing between digits 3 and 4 (those visible in the image), producing a cloven hoof. This pattern of symmetry is termed *paraxonic*. In contrast, the perissodactyl (horses, tapirs, and rhinos) pattern is termed *mesaxonic*, in which the toes are usually odd numbered (i.e., 1, 3, or 5) and the axis of symmetry passes through the middle of digit 3.

Photograph by Eva Fariña.

Box 4.3

The formal name of the moose is *Alces alces*, but its common name is cause for considerable confusion. In North America, *A. alces* (though technically referred to now as *A. americanus*) is termed the moose but it is known as the elk in Europe. This elk, as already noted, belongs to Odocoileinae, but its name has been applied to a member of a different clade that North Americans also refer to as the wapiti, which is closely related to the North American red deer. It is so closely related, in fact, that it is now usually considered to belong to the same species (*Cervus elaphus*), and this cervid may thus be termed the red deer, wapiti, or elk; but this elk is often termed a moose in Europe! To make things even more confusing, *Cervalces scotti* is commonly referred to as the elk-moose or stag-moose. Confusion over cervid names doesn't end here. Consider the reindeer, *Rangifer tarandus*, which in North America is referred to as the caribou. And we have already mentioned the giant Irish elk . . .

Antlers are the conspicuous ornamental cranial appendages of male cervids (but the females of one species also has antlers), which serve in social interactions during the fall breeding season. They are handy, however, for more than just their good looks; they come into play during clashes between males competing for females. Antlers differ from horns, the cranial appendages of various other artiodactyls. True horns consist of a bony core covered by a hardened epidermal sheath and are neither shed nor branched (for the latter feature, the pronghorn is an exception). The vast majority of cervids bear antlers, but there are two that do not. Apparently to compensate for this absence, the canines are enlarged saberlike structures.

Antlers are formed of bone, and when mature, they consist only of bone. Also, they are branched and shed annually. They are interesting for many things, but their annual shedding and rapid rate of regrowth are truly remarkable. This annual cycle is governed by an intricate interplay of hormones and is spurred by increased spring day length. Growth rates in the antlers of the white-tailed deer (*Odocoileus virginianus*) may reach nearly 2 cm per day. During growth, the bone is protected by fur-covered skin, or velvet, which carries nerves and blood vessels that provide nourishment. In the autumn, the velvet dries up and is shed, normally through rubbing the antlers against vegetation, so that only the bony part, by now polished and stained brown by resins in the vegetation, remains. During late winter, the antlers are shed and the males remain antlerless for a few months.

There are several cervids that we may consider, all belonging to Capriolinae, a group that also includes the familiar Virginia or white-tailed deer, an unfortunate favorite target of North American hunters (and automobiles). The largest of the deer is the moose, a magnificent mammal, easily recognizable by its size, massive palmate antlers, broad overhanging muzzle, and the flap of skin, or bell, beneath its throat. Although its range has been restricted in modern times, its natural range nearly circumscribed the Arctic region and extended from northern Europe to the northeastern United States. Its shoulder height may reach 2.3 m and its mass 830 kg;

record antler spread is nearly 2 m. Specialists have generally recognized a single species for the moose, but recent evidence indicates that there may actually be two, *Alces alces* from Europe and western Siberia, and *Alces americanus* from eastern Siberia and North America (Groves, 2007). The extinct *Cervalces* is a late Pleistocene form that occurred as two distinct populations in North America. One population lived south of the Laurentide ice sheet and the other in Beringia, although it is unclear whether these two populations represent different species (Schubert et al., 2004). Mooselike, but slightly larger in body proportions, it had the muzzle of deer. While authors like Lister (1993) contend that *Cervalces* is part of the *Alces* lineage and include it in the latter genus, several publications, such as that by Breda and Marchetti (2005), maintain a distinction between *Alces* and *Cervalces*. The North American *Cervalces* is generally considered a distinct species, *Cervalces scotti*.

Perissodactyls

The other main group of ungulates is Perissodactyla, or the odd-toed ungulates, including horses (like the one studied by Leonardo in Fig. 4.10), tapirs, and rhinoceroses among surviving members, and the extinct chalicotheres and brontotheres. Perissodactyls, like ruminant artiodactyls, use microbial fermentation to process the tough cell walls of plants, but whereas ruminants possess complex stomachs for this purpose, perissodactyls use a cecum, a specialized outpocketing or expansion of the large intestine.

Though less diversified than the artiodactyls, the perissodactyls, particularly the horses, have a rich fossil history and were important members of the North American fauna. The horses, more properly equids, including the familiar modern horse as well as donkeys, asses, zebras, and a host of fossil kin, are a North American success story, though their phylogenetic longevity was achieved away from home. Horses originated in North America and much of their initial diversity occurred in this continent, but they became extinct here (as well as in South America) at the end of the Pleistocene. Of course, horses were also widespread and diverse throughout Eurasia, and the Americas were repopulated with horses of Eurasian descent during the 1400s.

The evolutionary history of equids is complex, although for nearly a century, horse evolution was used as the textbook example of simple and gradual evolutionary change. Several trends prominent in the eventual evolution of the modern horses were identified and presented as an evolutionary progression leading inexorably to modern horses. Thus, we had a sequence beginning with the small fox-sized Eocene ancestor, *Hyracotherium*, which possessed relatively simple, low-crowned teeth, and four toes in the forelimb and three in the hind limb. The next stage was represented by the Oligocene *Mesohippus*, which had similar teeth but was larger and had three toes in both the fore- and hind limb. The sequence marched on through the remaining epochs of the Cenozoic, represented by forms that seemed progressively to anticipate the evolution of the final form of the modern horse. In the Miocene, *Parahippus* followed by *Merychippus* were progressively larger and had three-toed feet with a central toe in each foot flanked by reduced side toes; the teeth were progressively more complex

4.10. Study of horse by Leonardo da Vinci. The brilliant Italian Renaissance painter, sculptor, and inventor's interest in anatomy is well known; he pursued its study in part to help him achieve a better understanding of form and function as applied to art. This study of a horse was made for use in the preparation of a statue.

Original study in the Royal Library, Windsor, UK.

and higher crowned, particularly in *Merychippus*. In the Pliocene *Pliohippus* appeared, larger still and with higher-crowned teeth and single-toed feet (the side toes reduced to nonfunctional splints), and finally, at the pinnacle of the sequence, the Pleistocene (and still living) *Equus* strode onto the scene, larger and with higher-crowned teeth and single-toed feet (Appendix 1). The earlier forms, *Hyracotherium*, *Mesohippus*, and *Parahippus*, were forest-dwelling creatures that used their low-crowned teeth to browse on leaves. The appearance of the later forms, first *Merychippus* and then *Pliohippus* and *Equus*, coincided with a major environmental shift that produced widespread plains or grasslands during the Miocene. The higher-crowned teeth were thus put to use in grinding tough grasses (and the accompanying grit), and the single toed-feet were used for increased running ability, or cursoriality. Although early perissodactyls generally had running abilities, in these later equids, cursoriality was emphasized by an increase in the length of the distal limb segments (as well as overall size increase) to produce longer strides (Appendix 1).

We began to realize several decades ago that although this scenario summarizes the trends among some lineages of equids, the overall evolutionary picture was much more complex than a simple, linear progression from one form to the next. For example, the acquisition of toe reduction to one central toe in each foot was accomplished independently in *Pliohippus* and *Equus*. Equid evolution was complex, with many side branches, so that the appropriate analogy would be that of a dense bush, as described by MacFadden (1992). Indeed, the idea of a single evolving main lineage is achieved only in hindsight by pruning off and ignoring all the other branches.

An early group of large-sized, perhaps basal perissodactyls (Hooker and Dashzeveg, 2004) or related to equids (Benton, 2005) were the brontotheres (titanotheres, or, more evocatively, thunder beasts). This was a short-lived group, appearing in the Eocene but dying out by the end of this epoch in North America, though they survived slightly longer in Eurasia. These relatively short-limbed and noncursorial beast with low-crowned teeth were browsers, but they might also have fed on fruit (Benton, 2005; Rose, 2006). They exhibited a clear trend toward increased size, with the largest achieving elephantine proportions (Rose, 2006). In addition to increasing size, brontotheres were notable in the development of blunt, bony horns, probably hide-covered rather than cornified as in cows, over the nasal region. Some grew to enormous size and were distinctly bifurcated, as in the gigantic *Brontops* (Fig. 4.11) and *Megacerops*, which stood approximately 2.5 m high at the shoulder (Benton, 2005; Rose, 2006). Much of the variation in horn size and form is apparently related to intraspecific sexual dimorphism, used perhaps for display, species recognition, and intraspecific combat, perhaps in the head-ramming manner of sheep (Stanley, 1974; Mader, 1998).

The tapirs comprise another group of perissodactyls important to our story, though they were less abundant and diverse than horses. They are represented today by four species, three in the Americas and one in Southeast Asia. They are large, browsing, forest-dwelling mammals, without the cursorial or dental specializations present in horses. The earliest tapirs, represented by *Heptodon* from the Eocene of North America, were much smaller than their surviving relatives. As a whole, the group was rather conservative, retaining a generalized, somewhat piglike body shape, with low-crowned teeth and four toes in the forelimb. Their characteristic feature is the extension of the nasal region into a short, prehensile proboscis or trunk. Tapir evolution was centered largely in North America, but the group spread to Asia during the Miocene and to South America during the Great American Biotic Interchange. Like the horses, they became extinct in North America (and the rest of the Northern Hemisphere). Unlike the horses, however, they managed to hold on in South and Central America (as well as Southeast Asia).

The rhinos and their kin, or rhinoceratoids, also appeared in the Eocene of North America and Europe; they diversified into at least three lineages, including amynodonts, hyracodonts, and rhinocerotids. During the Miocene, the latter group replaced the others and moved into Africa as well. Rhinocerotoids were generally medium- to large-sized mammals, although a few giants developed among them. One such group produced

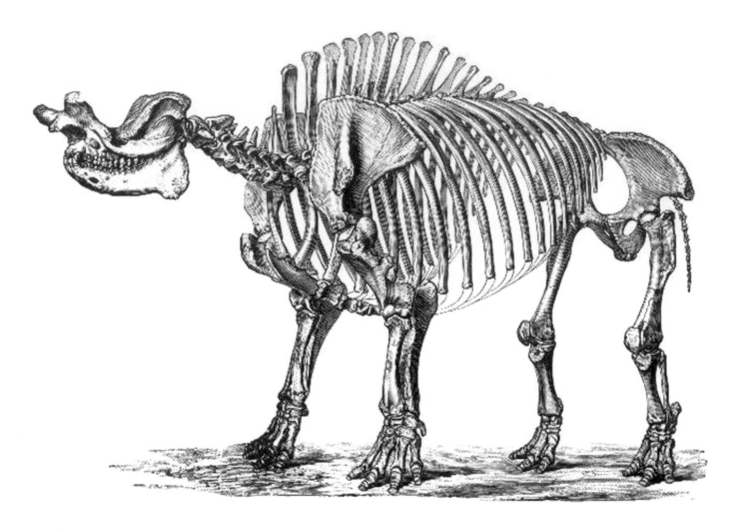

4.11. The skeleton of *Brontops,* a large brontothere (a group belonging to Perissodactyla). These beasts were enormous, reaching some 2 m at the shoulder, and characterized by a bifurcated hornlike structure on the snout. The horns were probably used for display, species recognition, and intraspecific fighting. They may have been used for head-on ramming, much as sheep do.

Image from Lankester (1905: fig. 101).

the largest mammal ever known, the immense *Paraceratherium* (also probably including *Baluchitherium* and *Indricotherium*, which may all belong to the same genus; specialists are still not quite decided on this point). *Paraceratherium* has been estimated at 5–6 m high at the shoulder, although this animal was not North American (Rose, 2006).

Another interesting group are the chalicotheres (or ancylopods), Eocene to Pleistocene perissodactyls that included medium to large forms, such as the Miocene *Chalicotherium* and *Moropus*. These beasts were rather bizarre, particularly in the presence of claws and proportions of the limbs. The forefeet of the latter ended in small hooflike structures, but its hind feet bore claws, as did all the feet of *Moropus*. Also, while their limbs were generally elongated, the front limbs, particularly in *Chalicotherium*, were markedly longer than the hind limbs, so that these beasts probably spent a portion of their time in a bipedal posture, much like gorillas do (Coombs, 1983), and when on all fours, they probably walked with their forefeet curled up, with the weight resting on the sides of their hands. Such features must have made them resemble gorillas, at least in their knuckle-walking mode (Benton, 2005), but all this was topped with a horselike skull!

a)

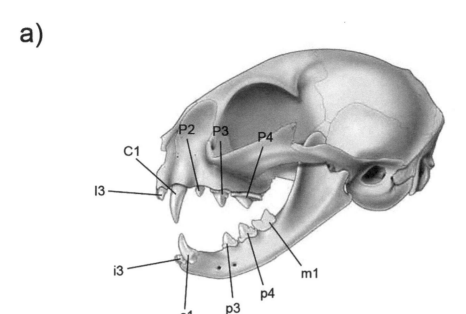

b)

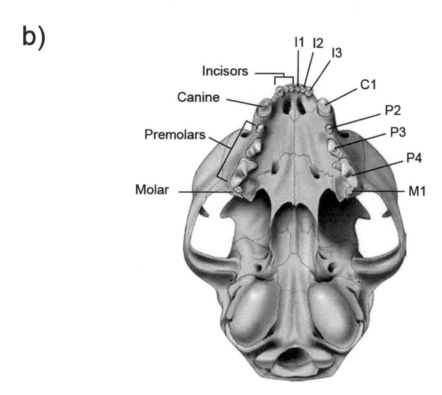

4.12. Illustrations of the skull and mandible of the domestic cat, *Felis domestica*. The top figure shows the skull and mandible in lateral (side) view, and the bottom figure shows the skull only in ventral view (i.e., the underside of the skull). The teeth are labeled with uppercase or lowercase letters, indicating their position in, respectively, either the upper or lower jaw. C or c, canine tooth; I or i, incisor tooth; P or p, premolar tooth; M or m, molar tooth. The elongated, pointed, and curved form of the canine teeth, particularly C, is typical of carnivorous mammals. A pair of teeth usually become modified as carnassials, which are elongated, bladelike teeth that have a scissorlike action as they pass each other during closing of the jaw. Such teeth are efficient in shearing and cutting through flesh and tendons. In carnivorans, the carnassial pair is formed by P4/m1. The teeth behind the carnassial pair are strongly reduced, and some are lost in the cat (M1 is barely visible, for example, in lateral view; the molars other than M1 and m1 are absent), but this is not true of all carnivorans. In bears, such teeth are better developed.

Image © Dino Pulerà, labeled by Sebastián Tambusso. Reproduced by courtesy of Dino Pulerà and Elsevier Inc.

Carnivores (or, more accurately, carnivorans)

The term *carnivore* is applied by biologists to any organism that relies largely on animal matter for its nutritional requirements. Of course, many mammalian and nonmammalian organisms are carnivores, but among placental mammals, there is a group named Carnivora — which, of course, closely resembles the word *carnivore*. For many years, carnivore, when applied to mammals (and especially placental mammals), was used interchangeably with Carnivora, and the potential for confusion is obvious. Thus, over the past several decades, biologists, though not the media or

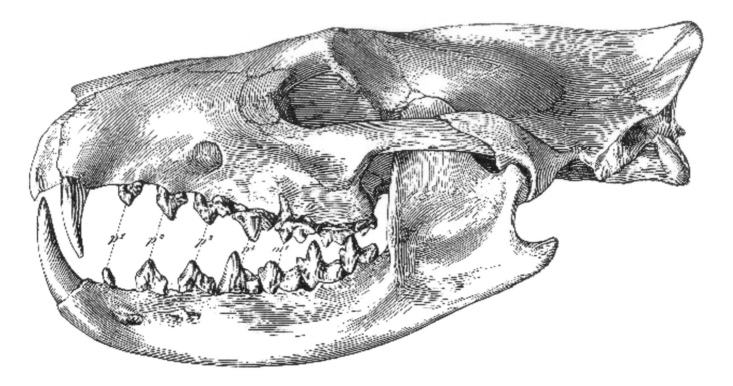

4.13. Skull and mandible of the creodont *Sinopa grangeri* in lateral (side) view. This species belonged to a group of ancient carnivorous placental mammals, Creodonta. With its long snout, the skull is rather doglike in appearance. A main difference between these groups is that in creodonts, the carnassial pair was formed by teeth farther back in the jaws (compare with Fig. 4.12). The teeth are designated by p for premolar and m for molar.

Image from Matthew (1906: fig. 3). Reproduced by kind permission of the Smithsonian Institution, National Museum of Natural History.

general public, have used the term *carnivoran* to refer to members of Carnivora (and Carnivoramorpha), whereas *carnivore* may be used to refer to any carnivorous organism. Among placental mammals, which are of interest to us here, there are two groups of mainly carnivorously adapted creatures. These are Carnivora (more broadly, Carnivoramorpha), already mentioned, and Creodonta. These clades have been traditionally considered to be sister groups (see Chapter 1), usually referred to as Ferae, but a sister group relationship is not particularly well supported (Flynn and Wesley-Hunt, 2005; Rose, 2006).

Creodonts comprise an archaic group that was successful and diverse in the early part of the Cenozoic but that had become extinct by the end of the Miocene. Creodonts resembled to a large degree some of the living carnivorans, but they differed especially in which teeth were specialized as carnassials. The carnassials are a pair of cheek teeth (one from the upper jaw and the other from the lower jaw) that become enlarged and specialized as the main teeth that slice or shear flesh and tendons; the carnassials thus often assume an elongated, bladelike form (Fig. 4.12). In creodonts, the carnassials were formed by either the first upper molar (M1) and the second lower molar (m2) or M2 and m3 (Appendix 1). Among them were such forms as *Sinopa* (Fig. 4.13) and *Hyaenadon*.

Carnivorans include the animals most familiarly perceived as mammalian carnivores, such as cats and dogs, although not all carnivorans (recall this is a genealogical term) are strictly or even largely carnivorous. Raccoons and bears, for example, are omnivores, and one bear—the giant panda—has an almost entirely vegetarian diet. Carnivorans were (and remain) the main predaceous mammals for much of the Cenozoic (Rose, 2006), and they are considerably more diverse than usually realized. Among them are several extinct fossils lineages, the viverravids and miacids (the two are usually referred to paraphyletically as miacoids; see Chapter 1), and a clade that includes several groups of more modern forms. The tendency these days is to restrict the name Carnivora to the latter clade (that is, the one that has still living representatives), whereas the more inclusive clade, with viverravids and miacids, is referred to as Carnivoramorpha. Following this definition, Carnivora comprises two broad groups, Feliformia and Caniformia.

Feliformia includes the more catlike carnivorans: nimravids and barbourofelids (the extinct, so-called false sabertooths, though "false" alludes to their relationship to sabertooth felids, because their upper canine teeth were quite large in some forms), felids (true cats), herpestids (mongooses), viverrids (civets), and hyaenids (hyenas), among others. Caniformia is subdivided into Canoidea, the dogs and relatives (wolves, foxes, and jackals), and Arctoidea, which includes several groups, such as ursids (bears), amphicyonids (bear-dogs), pinnipeds (seals and walruses), mustelids (weasels), and procyonids (raccoons). In carnivorans the carnassials are formed by the fourth upper premolar (P4) and m1; in other words, they are farther forward compared to creodonts (Fig. 4.13). Several carnivorans, as noted above, are not particularly carnivorous, and their teeth are modified in other ways. In most bears, for example, the carnassials are not bladelike and so are not particularly specialized for shearing. Rather, the molar teeth behind the carnassials are enlarged with a broad crown, a form better suited to crushing, although the molars are narrower in polar bears, which are strict carnivores specialized for eating seals. In most pinnipeds, the cheek teeth tend to be essentially conical, with none being modified as carnassials.

Several carnivorans, mainly among nimravids, barbourofelids, felids, and ursids, attained large size. The nimravids survived into the Miocene (Tseng et al., 2010), and some of their members early developed a large size and large canine saber teeth. Somewhat later, the barbourofelids also developed saber teeth and large size, with *Barbourofelis fricki* reaching the size of a lion. A digression on current taxonomy with these last two groups is warranted. The nimravids and barbourofelids have generally been grouped together as nimravids (in which case the barbourofelids are referred to as barobourofelines), as noted by Tseng et al. (2010). Therrien (2005) noted, however, that recent discoveries indicate that the two represent independent phylogenetic lineages, with barbourofelids more closely related to felids. As noted by Turner and Antón (1997), *Barbourofelis fricki* was probably the most sabertooth carnivoran ever, with its large canine teeth exceeding in size those of *Smilodon*, which is probably the most famous sabertooth and whose name translates as "knife tooth" (from the ancient Greek σμίλη, "carving or cutting knife"). In addition to the long canines, *Barbourofelis* also had a long bony flange extending down from the front of its lower jaw to help protect its massive teeth. The nimravid *Dinictis* (Fig. 4.14) is another example of a sabertooth. Though smaller in overall size, its well-developed saber teeth were nonetheless formidable weapons.

Barbourofelis and *Dinictis* represent two of the common ecomorphs among sabertooth carnivorans. The former is an example of "the dirk-toothed ecomorph, characterized by long, narrow canines with fine or no serrations and postcranial adaptations suggestive of an ambush predator," whereas *Dinictis* represents the "scimitar-toothed ecomorph, characterized by relatively short, broad canines with coarse serrations and a body form

4.14. *Dinictis* is a member of Nimravidae, a group of feliform carnivorans. The teeth are indicated as follows: i, incisor; c, canine; premolar; m, molar. Also indicated are ms. p., mastoid process, and pa. p., paraoccipital process, which are prominent protuberances at the back of the skull. As is usual in Carnivora, the carnassial pair is formed by p4 in the upper jaw and m1 in the lower jaw. Although not true cats (i.e., Felidae), some nimravids evolved a sabertooth morphology, including saber teeth as well as associated features of the skull and mandible, independently of the sabertooth cats (Machairodontinae, which are members of Felidae). Such independent acquisition of characters between relatively distantly related organisms is termed *convergence*.

Image from Matthew (1910: fig. 10), courtesy of the American Museum of Natural History.

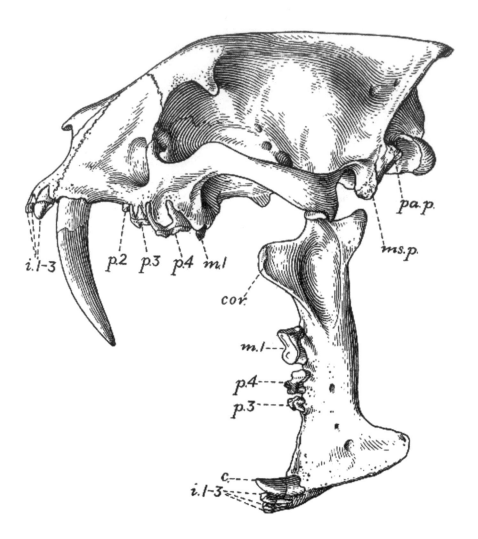

with cursorial adaptations" (Therrien, 2005:394). *Nimravus* (Box. 4.4) is another example of a scimitar-toothed nimravid (though not all were scimitar-toothed; *Hoplophoneus* is a dirk-toothed nimravid, for example). The living felids, on the other hand, are characterized by conical-shaped canines. Therrien (2005) noted that dirk-toothed and scimitar-toothed ecomorphs killed their prey differently. The dirk-tooths tended to emphasize a powerful canine bite to subdue their prey. With long sabers and an inferred ambush predatory style, they probably drove their sabers deep into their prey to kill it quickly. Scimitar-tooths, on the other hand, probably did not rely on a single, powerful canine bite. Instead, their cursorial adaptations likely allowed them to pursue prey, delivering slashing canine bites to weaken it, a strategy for which their short, coarsely serrated sabres were ideal.

Box 4.4

Mortal combat

Locked forever in mortal combat? The magnificent specimen depicted in Box Fig. 4.4 represents the skull of *Nimravus brachyops* (noted in the text as a scimitar-toothed nimravid), with one of its sabers piercing a humerus (upper arm bone), also belonging to a *Nimravus* individual. It was uncovered in 1932 from Oligocene deposits (about 25 Mya) in Black Hank Canyon, Morrill County, Nebraska, USA, by a paleontology crew from the

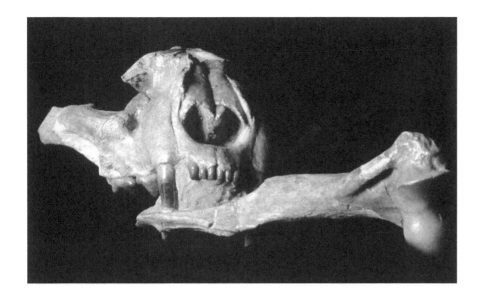

Box Fig. 4.4. *Nimravus brachyops* (noted in the text as a scimitar-toothed nimravid), with one of its sabers piercing a humerus (upper arm bone), also belonging to a *Nimravus* individual.

Courtesy of the Nebraska Land Magazine/ Nebraska Game and Parks Commission.

University of Nebraska. Upon discovery, its excavators were awestruck by the thought that the outcome of the final battle between these two predators had been preserved for so many millions of years, with the participants likely having been involved in male–male combat. The fighting had been so fierce and the blow so devastating that neither animal survived. The scene of discovery moved Loren Eisley, then a student on the dig and later a noted academic and writer, to pen the poem "The Innocent Assassins." Of course, the combat scenario depends on the skull and humerus belonging to different individuals, but there is also the possibility that they belong to the same individual (Toohey, 1959). In this case, the specimen is more likely the result of postmortem depositional compaction. This makes more sense because the long sabers would have been unlikely to withstand such a powerful blow to a stout, bony region and the wild thrashing that would necessarily have followed; close inspection of the fossil reveals that only the tip of the saber is broken and displaced—anticlimatic perhaps, but the fossil is still a wonderful specimen.

Smilodon, a dirk-toothed Felidae, was the dominant carnivore during the Pleistocene of North America; it also spread into South America (see Chapter 6). It is the most recent of the sabertooth forms, with *Smilodon fatalis* becoming extinct only 10,000 years ago (Fig. 4.15). Its numerous remains, particularly from the asphalt deposits of Rancho La Brea in Los Angeles, California, have been well documented and make it the best-known fossil mammalian carnivore (Turner and Antón, 1997).

The great similarity between *Barbourofelis* and *Smilodon* is a striking example of the processes termed convergent evolution and iteration (i.e., repeated process in different lineages), in which relatively unrelated taxa acquire similar characteristics. In other words, *Barbourofelis* and *Smilodon* did not possess saberlike teeth because they inherited them from a common ancestor; rather, the teeth (as well as several other associated features of the skull) evolved independently in the lineages leading to each of them. In addition to these two, there is another example of convergence of the sabertooth type: the South American marsupial formerly

4.15. The famous sabertooth cat *Smilodon fatalis* from Rancho La Brea, Los Angeles, California. This carnivoran belongs to Felidae and is thus more closely related to living cats than to *Dinictis* (Fig. 4.14). As in living cats and *Dinictis*, though, the carnassial pair is formed by P4 above and m1 below (abbreviations as in Fig. 4.14). The longer canines, which form the saber teeth, meant that *Smilodon* needed a wider gape in opening its jaws (compare with *Dinictis*), a feature made possible by modifications of the jaw joint.

Image from Matthew (1910: fig. 15), courtesy of the American Museum of Natural History.

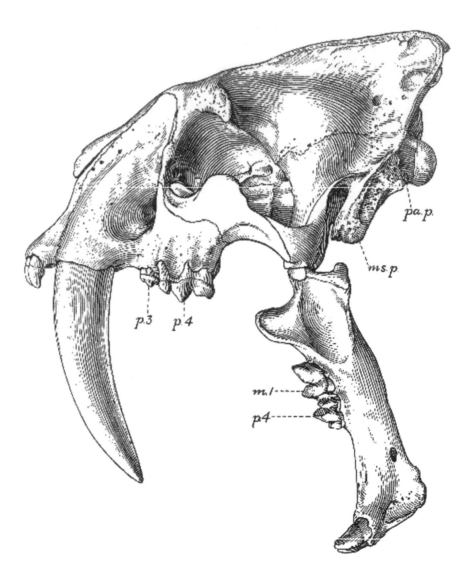

called *Thylacosmilus atrox,* but now assigned to the genus *Achlysictis* (see Chapter 3). A recent morphometric and phylogenetic study by Prevosti et al. (2010) revealed that the form of the skull of sabertooth forms is closely correlated with the development of a sabertooth morphology of the upper canine teeth. According to this finding, less derived members of a clade maintained the general morphology of the groups to which they belong, whereas derived members of distantly related groups independently evolved the shared suite of features arising from the development of sabertooth morphology.

Among felids, the sabertooth cats, with elongated and flattened canines, are generally included in Machairodontinae, whereas the cats with shorter, conical-shaped canines are included in Felinae. In addition to *Smilodon,* there was quite a diversity of large felids in the Pliocene and Pleistocene of North America (as well as the Old World, but not *Smilodon,* which is a New World taxon). Among these is *Dinofelis,* which was perhaps the size of a jaguar and is often also termed a false sabertooth, in this case because its characteristics are somewhat intermediate between those of sabertooths with flattened canines and those of felids, with shorter canines (Turner and Antón, 1997). *Homotherium,* species of which were of lion size, is another example of a large machairodontine.

Among felids, the jaguar, *Panthera onca,* and the puma (also known as the cougar or mountain lion), *Felis concolor* (considered by some authorities as *Puma concolor*), are the two large carnivorans currently inhabiting the New World, and both were important faunal elements during much of the Pleistocene. The puma is widely distributed, ranging from British Columbia in Canada to the southern parts of South America, whereas the jaguar is restricted to Central and South America, although it ranged farther north into the United States and also in Patagonia during the Pleistocene and even in prehistoric and historic times. The jaguar is larger, with large males reaching 120 kg, whereas puma males may reach about 100 kg (Turner and Antón, 1997).

Lions, also members of Felinae, were the most widespread carnivorans during the Pleistocene (Barnett et al., 2009) and survived in North America until 11,000 years ago. The modern species, currently present in Africa and India (as a relict population) but more widely distributed in historical times (in, for example, Greece and Iran), is known as *Panthera leo* (Turner and Antón, 1997). Remains of fossil lions have been found in Africa, Europe, northern Asia, and North America, but their taxonomy is not entirely decided. Among the usual names applied to fossil remains outside of Africa are *Panthera spelaea,* the Eurasian cave lion, *P. atrox,* the American lion, and *P. vereshchagini,* the Beringian lion (from northwestern Asia, Alaska, and the Yukon). These are sometimes considered as distinct species or as subspecies of *Panthera leo,* and at least one author (Christiansen, 2008) did not consider *P. spelaea* and *P. atrox* to be lions in the traditional sense, as the closest relative of *P. leo* was found to be the leopard, *P. pardus.* Although this study, which was based on morphological features, has not cleared up all aspects of lion taxonomy, it does suggest that *P. spelaea* and *P. atrox* were indeed distinct species. The work of Barnett et al. (2009), based on molecular data, suggests that these taxa were indeed lions and recognized three distinct lion groups (whether these be species or subspecies is still open to question) as present during the late Pleistocene, as follows: (1) the ancestors of the modern species, *P. leo;* (2) a group (i.e., *P. spelaea*) that ranged from Eurasia and extended into Alaska and the Yukon (including the Beringian lion); and (3) the American lion (i.e., *P. atrox*). Barnett et al. (2009) suggested that the latter became genetically distinct 340,000 years ago.

The other carnivorans that attained large size were members of Ursidae, or bears. Modern bears are generally quite large and omnivorous mammals, but there are exceptions. For example, the Malayan sun bear (*Helarctos malayanus*) reaches a maximum of 65 kg (Christiansen, 1999)—still large for a mammal, but not for a bear. The giant panda (*Ailuropoda melanoleuca*), though large, feeds almost entirely on bamboo shoots; the polar bear (*Ursus maritimus*) is a hypercarnivore with a fairly restricted diet consisting mainly of seals; and the sloth bear (*Ursus melursus*) is an insectivore, feeding mainly on ants and termites (Sacco and Van Valkenburgh, 2004). Bears are currently widespread, as they have been since the Miocene. Today, they are represented in North America by species of the genus *Ursus,* including *U. americanus,* the black bear, *U. maritimus,* the polar bear, and *U. arctos,* the brown bear. *Ursus,* incidentally, is Latin for "bear," as is *arctos* in Greek; so *Ursus arctos* therefore means, rather emphatically, "bear bear." The Arctic was named after the bear, rather than the other way around—which, we suppose, makes the Antarctic the antibear (our apologies—we couldn't resist).

The polar bear and brown bear (represented by several subspecies, among which in North America are the grizzlies and Kodiaks) are similar in size and usually considered the largest living carnivorans. Although the upper size range for male polar and brown bears may reach 800 kg and possibly more, masses of 500 kg are not unusual, whereas the smaller black bear has a mass of about 105 kg (Christiansen, 1999). *Ursus* includes the last Eurasian ursids that ranged into North America, with the brown and black bears apparently representing separate dispersal events (Hulbert et al., 2001). The brown bear and polar bears are regarded as sister species, although recent research (Bon et al., 2008) suggests that *Ursus* may be paraphyletic; that is, some brown bear populations are more closely related to polar bears that to other brown bears.

Another group of bears, Tremarctinae, was also present in North America (and South America; see Chapter 6) during the Pleistocene (Schubert et al., 2010). Tremarctine bears are represented today only by the South American spectacled bear (*Tremarctos ornatus*), but they were considerably more diverse in the past. They are characterized by a shorter snout than other bears and are thus termed short-faced bears. Among extinct Tremarctinae was *Arctodus simus*, the giant short-faced bear, whose habits will be discussed below. *Arctodus pristinus* (the lesser short-faced bear) and *Tremarctos floridanus* (the Florida cave bear) were also present during the North American Pleistocene, although the former apparently survived only until the mid-Pleistocene (Schubert et al., 2010).

Arctodus simus was widespread, ranging from Alaska and the Yukon to central Mexico. Members of this species came in large and extremely large sizes. In the past, these groups were often recognized as distinct subspecies (as by, for example, Churcher et al., 1993; Richards et al., 1996; and Gobetz and Martin, 2001), but Schubert et al. (2010) indicated that they represent a single species with marked sexual dimorphism (a difference between males and females of a species), as occurs in all extant bears, with the larger individuals probably representing males and the smaller representing females. Christiansen (1999) estimated that the mass of some of the largest males may have exceeded 800 kg and noted that *A. simus* was larger than another very large bear, the European cave bear (*Ursus spelaeus*).

We have already noted the shorter snout of *Arctodus simus*. It also had a wider skull than bears generally do, and these features imparted a somewhat catlike appearance to the skull (Christiansen, 1999). Among other features that distinguish *A. simus* from other ursids are that it was distinctly longer limbed, but with a shorter body, more well-developed carnassials, and a possibly more digitigrade stance (the tendency to raise the posterior parts of its manus and pes from the substrate during locomotion, rather than also using its palms and soles, a condition termed plantigrade; Appendix 1), all features suggesting that the giant short-faced bear was relatively fleet of foot. In turn, this may indicate that it led a more active and predaceous—in other words, carnivorous—lifestyle than the stereotypically more omnivorous habits of bears (Christiansen, 1999). Schubert et al. (2010) presented isotope analysis data that also suggested that *A. simus* (at least from Beringia) consumed a high proportion of meat, although these authors cautioned that it is not known whether such a high-protein diet was obtained primarily from predation or scavenging. With regard to these possibilities, the authors noted that the longer limbs of *A. simus* may indeed indicate that it was a fast runner that chased down its prey, but they raised the possibility that the limbs may have been adaptively important for efficient long-distance travel to locate carcasses. An ecomorphological study by Figueirido et al. (2010) concluded that this species was an opportunistic omnivore that took advantage of whatever was available, a finding based on cranial and limb proportions as compared to those of other Carnivora.

An interesting scenario occurred during the late Pleistocene of Florida (at least), where *Arctodus simus* lived contemporaneously with two other bears, *Tremarctos floridanus* and *Ursus americanus* (the black bear, noted above as still extant in North America, including Florida). Apparently these three coexisted happily, at least in terms of dietary preferences, as each occupied a different niche. *Arctodus simus*, as already noted, was more carnivorous, while *U. americanus* tended to be more of a generalist omnivore and *T. floridanus* was primarily herbivorous (Schubert et al., 2010).

This impression of *Arctodus simus* as an active predaceous carnivore has been absorbed by the scientific community (and conveyed to the general public) over the last decade or so, but it may need revision, as several recent efforts (Figueirido et al., 2008; Figueirido and Soibelzon, 2009; Figueirido et al., 2010), which used a combination of more sophisticated analytical tools, are beginning to paint a different picture of this ursid. This is the way science works: from initially understanding little about extinct bear paleoecology (other than what we could deduce from living bears in general), we came to accept one interpretation (not, as we'll soon see, that it was entirely wrong), and we may now have to change our minds as we have acquired a more profound understanding of the bits and pieces that have been left for us to cobble together a coherent picture of their morphology and paleoecology. Together, these recent efforts analyzed a variety of living and extinct bears, as well as other carnivorans, in attempts to tease out the functional versus phylogenetic signals of the skeletal morphology. As an example of how this might work, let's say the morphological features of relatively distantly related bear species group together, and that we can correlate diet on the basis of the morphology of living ursids. We can then use this information to infer that morphological similarities among living and extinct bears indicate similarities in function and hence dietary preferences in fossil bears.

It turns out that the broader comparisons to which fossil bears have been subjected do not support specializations toward hypercarnivory or bone-cracking behavior. Indeed, in their delightfully titled "Demythologizing *Arctodus simus*, the 'Short-faced' Long-Legged and Predaceous Bear that Never Was," Figueirido et al. (2010) noted that *A. simus* was neither particularly short-faced nor possessed of especially elongated limbs, and that it was neither a fast-running predator nor a specialized scavenger. These authors also found that a mass of 1000 kg was more common than previously believed. The picture emerging from these studies is that *A. simus* was a colossal omnivorous bear whose diet likely varied according to resource availability. This is not really saying anything new about the general ursid pattern: "Bears are one of the most ecologically and morphologically

4.16. Although it is not from North America, the massive skull of *Andrewsarchus*, about 83 cm in length, is the largest of any mammalian carnivore. The top figure shows its skull in lateral (side) view. The lower figures show it in ventral (underside) view on the left and dorsal (top) view on the right. The skull of a grizzly bear (*Ursus arctos*) is set between them for comparison.

Image modified from Osborn (1924: figs. 1 and 2), courtesy of the American Museum of Natural History.

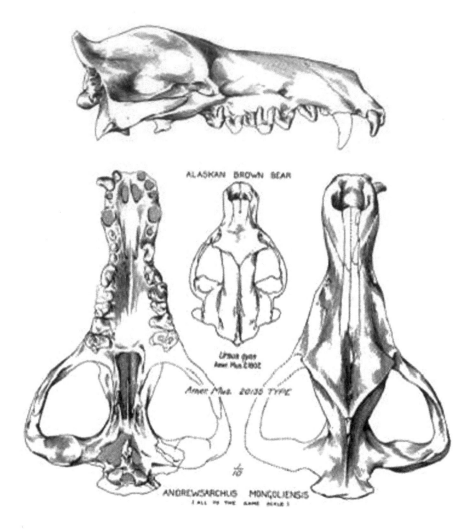

ALASKAN BROWN BEAR

Ursus feos
Amer Mus. E180E

Amer. Mus. 20135 TYPE

ANDREWSARCHUS MONGOLIENSIS
(ALL ¾ THE SAME SCALE)

adaptable groups among the large carnivore guild. Omnivorous bears retain the ability to behave as carnivores or herbivores according to resource" (Figueirido and Soibelzon, 2009:12). We should emphasize that this new outlook on *A. simus* is a subtle change. We are not precluding the possibility that it may have been more carnivorous in parts of its range, as suggested by Schubert et al. (2010). It may have been, given its dietary adaptability and the available food resources, but we are no longer pigeonholing *A. simus* into its previous narrow role.

We noted earlier that *Arctodus simus* may have been the largest terrestrial mammalian carnivore ever (but see Chapters 5 and 8). However, several others are also usually cited as candidates for this role—though the difference among them does not seem to have been particularly great. Although these other carnivores are not directly relevant to our story, it is worth mentioning them because they are as spectacular in their own ways as *A. simus*. Several authors (e.g., Farlow, 1993; Christiansen, 1999; Benton, 2005; Rose, 2006) have noted that the mesonychid (a group apparently closely related to ungulate or hoofed and generally herbivorous mammals) *Andrewsarchus* and the creodont *Megistotherium* are to be counted among the largest terrestrial mammalian carnivores. *Megistotherium* (from the Miocene of Egypt) had a skull that, at 66 cm in length, was nearly twice that of a bear or lion (Rose, 2006) and may have had a mass of as much as 880 kg (Farlow, 1993). The skull of *Andrewsarchus* (from the Eocene of

Mongolia and China) was nearly 83 cm in length, larger than that of any other terrestrial mammalian carnivore (Benton, 2005). Even a glance at its skull (Fig. 4.16) is enough to convey its terrifying power.

So far, we have briefly considered the megafauna present in North America. In the previous chapter, as part of our coverage of the South American land mammal ages, we introduced many of the mammals (as well as other organisms) of the South American fauna. For most of the latter half of the Cenozoic (the Age of Mammals), South America was largely separated from the other continents, and so the South American forms were essentially isolated from other life on earth. But this began to change about 12 Mya. It was at this time that geologic forces began the processes that led to the formation of a complete land connection (i.e., essentially Central America) between North and South America. The progressively emerging land connection and its final emplacement (about 3 Mya) ended the long isolation of South America and produced one of the most profound episodes in the history of mammals, as mammals from North and South America, long isolated and evolving independently of each other, began to migrate and intermingle. The consequences of this exchange, dubbed the Great American Biotic Interchange, are generally considered as one of Nature's great experiments. The next part of our story traces the events and causes of this exchange, and discusses aspects of life in South America until the end of the Pleistocene, 10,000 years ago, when a great extinction event changed the characteristics and distribution of life yet again.

5.1. Maps of Central America illustrating the progression of geologic events during the late Cenozoic that led to closure of the Central American Seaway. Current coastline of the region shown in white outline. Emergent land represented by dark regions, shelf (shallow ocean) sediments by light gray, and abyssal (deep ocean) sediments by dark gray. Central American region during the (A) middle Miocene (approximately 16–15 Mya), showing relatively small areas of emergent land; (B) late Miocene (7–6 Mya); and (C) late Pliocene (about 3 Mya). As the formation of the isthmus neared completion, a few relatively narrow marine corridors remained.

Illustration by Sebastián Tambusso based on Coates and Obando (1996) and Woodburne (2010).

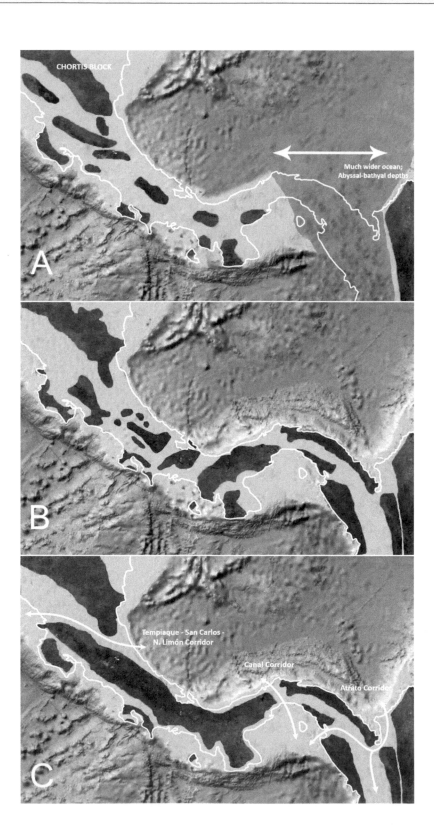

The Great American Biotic Interchange and Pleistocene Habitats in South America

5

Most people, including those reasonably familiar with the modern South American fauna, would be surprised to learn that most of the creatures currently inhabiting South America are relatively recent immigrants. Deer, pumas, jaguars, llamas, foxes, field mice, otters, and possibly peccaries and tapirs have ancestors that reached this continent less than about 3 Mya. The situation was much different in the Lujanian, which will be discussed once we examine how the modern assemblage of South American mammals was shaped.

The extinction of the dinosaurs at the end of the Cretaceous period, which was likely triggered by an extraterrestrial impact 65 Mya left the field open for the surviving mammals. What followed was an evolutionary phenomenon referred to as a radiation, whereby a multitude of ecological niches, which had previously been occupied by the dinosaurs, were filled by newly evolving mammals.

At this time, and continuing up until nearly 3 Mya, the southern supercontinent of Gondwana had split apart, and the Isthmus of Panama did not exist. For a while, South America maintained a tenuous connection to the rest of the world through a transantarctic connection with Australia that ended in the early Tertiary (see Chapter 3). From this point and for a long time afterward, a peculiar group of South American mammals evolved that had little interaction with those in the rest of the world. This began to change with the gradual closure of the Central American Seaway, the ancient body of water separating North and South America, by the landmass known as Central America (which is usually considered part of North America). Its southern ribbon, known as the Isthmus of Panama (or the Isthmus of Darien), was among the last emergent bits that finally allowed the separation of the Pacific Ocean from the Caribbean Sea.

The formation of the Isthmus of Panama was an extended geologic process that produced fundamental changes to the biogeography, ecology, and evolution of terrestrial and marine organisms and had (and continues to have) an enormous effect on ocean circulation and global climatic patterns. The final closure of the isthmus, timed at about 3 Mya (Coates et al., 2004; Kirby et al., 2008), was due mainly to the growth and migration of the Central American volcanic arc and its collision with South America (Coates and Obando, 1996; Coates et al., 2004). The arc is a chain of (still active) volcanoes extending along the western margin of Central America (technically on the southwest part of the Caribbean Plate), formed at the subduction zone where the Cocos, Nazca, and Caribbean plates meet (see Chapter 3). The Caribbean Plate moved generally eastward during the Cenozoic and began colliding with South America about 12 (and perhaps as late as 9.5) Mya (Coates et al., 2004), as illustrated in Fig. 5.1a.

There are two hypotheses on the earlier history of Central America. Coates and Obando (1996) suggested that the closure of the Central American Seaway occurred primarily through expansion of the arc, progressive shallowing of marine depths, the development of several interconnected forearc and back-arc sedimentary basins (see Chapter 3) adjacent to the extended archipelago, and uplift. These authors suggested that widespread uplift had apparently produced substantial stretches of emergent land between about 8.5 to 7 Mya (Fig. 5.1a). By about 6 Mya, although there was a brief increase in marine flooding, a series of islands (essentially a subcontinuous isthmus with a few marine corridors) was strung across the region between Nicaragua and the northwestern coast of South America (Fig. 5.1b). This was followed by further strong uplift (Fig. 5.1c), and the gap between the Americas was finally sealed at about 3 to 2.8 Mya by the emergence of the Isthmus of Panama, which has since remained essentially unchanged—as a barrier, at any rate (Coates et al., 2004; Woodburne, 2010).

Although the timing of the last stage has been constrained by several lines of evidence, Kirby et al. (2008) suggested a different scenario for the earlier stages. Their hypothesis rests on extensive analyses leading to a reinterpretation of the formations exposed along the Gaillard Cut (the artificial valley created as part of the Panama Canal). In contrast to the gradual emergence of Central America and several significant seaways over 15 Mya described above, Kirby et al. (2008) proposed that the main axis of the volcanic arc had coalesced into an extensive and continuous Central American peninsula, connected to North America, as early as 19 Mya. Panama was the extreme southern end of the peninsula. This suggests a long and near-continuous biotic connection (short-lived straits did exist) between North and Central America.

In any event, the closure of the seaway had several profound consequences. The effect of changes in oceanic currents (the conduction of warmer water from the gulf coast up the eastern North American shores) and possibly the decrease and eventual cessation of the transfer of low surface salinity concentrations from the Pacific to Atlantic oceans apparently produced an intensified moisture supply to northern high latitudes. This in turn may have been an important causal factor in initiating, about 3.15 Mya, the production of Northern Hemisphere glaciation, or the ice ages (Bartoli et al., 2005). The rise of the Isthmus of Panama also produced profound changes in the biogeography of organisms in an episode termed the Great American Biotic Interchange (GABI; also called the Great American Faunal Interchange, GAFI, by authors such as Campbell et al., 2010). We will focus here on the resulting extensive intermingling of terrestrial mammalian faunas that had been separated for many millions of years. The interchange began and set the stage for the eventual production of the modern faunal composition just after the end of the Pleistocene, 10,000 years ago.

Before we begin to explore the dynamics of the GABI, it is useful to consider the broader context of South American mammal distribution during the Cenozoic. In doing so, we will make use of Simpson's (1950) categorization of the faunal structure of terrestrial mammals (even though it has been criticized as inadequate to describe this history; Reig, 1981), because Simpson's view helps us visualize the timing of appearances of ancient inhabitants and newcomers. Simpson (1950, and thereafter) recognized three faunal strata, with stratum 1 including the ancient inhabitants, the mammals that had been in South America essentially from the start of the Cenozoic and evolved in almost complete isolation from the world's remaining mammalian fauna—in other words, we can consider it a closed system. Stratum 2 includes an episode of restricted input, the ancient immigrants, into this closed system that occurred more or less during the late Eocene and Oligocene. Stratum 3 includes the arrival into South America of the vast majority of mammals from North America during the GABI. This episode occurred over an extended period, lasting from the late Miocene to the Pleistocene. It also was generally regarded as comprising two broad dispersal patterns: an older, restricted migration accomplished by island hoppers and a later, intensive fast lane invasion by recent immigrants over a continuous land route.

The traditional idea of exchange between the two continents has thus been that there were some isolated episodes of faunal interchange, first when the ancient immigrants arrived and later when the island hoppers made their entrance. Such events had little effect on the overall faunal composition of South America, which remained relatively uniform until the recent immigrants began to arrive some three million years ago. However, the relative importance and disjunct nature of the last two migration episodes is not as clear today as was thought in Simpson's time (and hence there is some justification in Reig's comments), but we will leave this for now and return to it later in our discussion. Another point we should note is that there is a difference among authors on how the GABI is interpreted. The traditional view is that described here, essentially following Simpson (as by Campbell et al., 2010). Woodburne et al. (2010) restricted the GABI to the episode essentially equivalent to the exchange that swept in the recent immigrants to South America—that is, beginning near 3 Mya. We must also note that many mammals of South American origin made their way north during the period beginning with the initial stages of the isthmus's formation and continuing on until the late Pleistocene. The discussion below emphasizes and is set within the framework of the mammals present in South America over this time interval, but it includes mention of the mammals that went the other way.

Ancient inhabitants

What mammals were present in South America when it was an island continent, or nearly so? As we saw in Chapter 3 in the account of faunal succession throughout most of the South American Cenozoic, there were just three types, or stocks: xenarthrans (Fig. 5.2), marsupials, and various kinds of ungulates.

Let's consider the marsupials first. This group is characterized by the presence of a particular type of placenta (choriovitelline placenta) and a pouch (though not ubiquitously present), where the young complete their development after birth. Fortunately for paleontologists, marsupials also have several skeletal features, such as teeth and bones that fossilize. Although the present home base of marsupials appears to be Australia, a land that conjures up images of kangaroos and the like, evidence from as early as the second half of the Cretaceous points to a North or South American

Origin of the South American mammals

5.2. Skull and mandible of the Miocene specialized armadillo *Peltephilus*. Scale bar equals 1 cm.

From Vizcaíno et al. (2004a), by kind permission of Asociación Paleontológica Argentina.

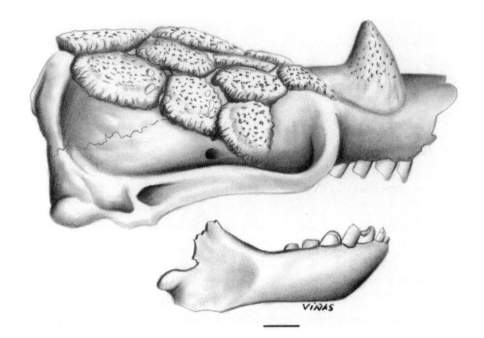

origin for marsupials. Also, molecular evidence is congruent with a time of divergence between marsupials and placentals near the boundary between the Jurassic and the Cretaceous, 150 Mya (Bininda-Emonds et al., 2007).

From this beginning, they dispersed on the one hand to Eurasia, and on the other to Antarctica and Australia, which were then still joined. As noted in Chapter 3, Antarctica was not yet covered in ice because a thermically isolating circumpolar current formed only after the Drake Passage opened, and the Antarctic peninsula ceased to be contiguous with Patagonia, in the early Miocene (about 23 Mya) and hence much later than the events now being discussed.

During a good part of the Cenozoic, South America hosted a marsupial fauna nearly equal in number and variety to that of Australia. Among them is the opossum, which has survived to the present with such vigor that it has made its way farther and farther north in the United States, and several living species of small insect eaters (including a couple of clades with rodentlike dentition) and herbivores, as well as the spectacular bearlike *Callistoe* and sabertooth *Achlysictis* (Fig. 5.3).

In South America a rather odd assemblage of native ungulates, grouped under the name Meridiungulata, evolved (and some became extinct) during the Tertiary, including those mammals mentioned in Chapter 3 with mythological sounding names: pyrotheres, astrapotheres, and xenungulates, some of very large size and great interest that did not survive into the Lujanian (see Box 3.1). As noted in Chapter 3, notoungulates and litopterns continued their existence to Pleistocene times, although at considerably reduced diversity, contributing by their peculiar genealogical history to the odd mix of the fauna we are studying. Chapter 6 provides further details of some of the last representatives of these groups, the impressive *Toxodon* and *Macrauchenia*.

Finally, there are the xenarthrans, such as the living armadillos, tree sloths, and anteaters, and the fossil glyptodonts and ground sloths. During the long stretch of Tertiary time, xenarthrans had their moments of rise

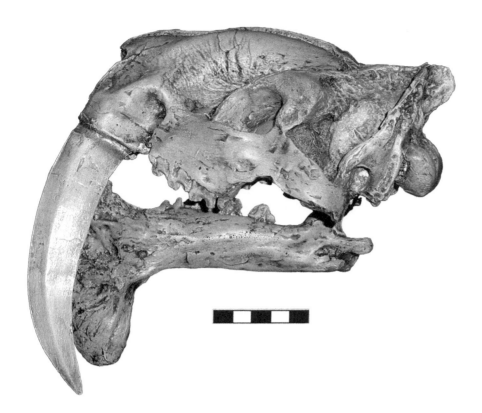

5.3. The marsupial sabertooth formerly known as *Thylacosmilus*, now named *Achlysictis*. Scale bar equals 5 cm.

Photograph of a cast kept in the Museo de La Plata, courtesy of that institution.

and fall; they were so diverse that they merit a book all to themselves (and at least two books deal with them: Montgomery, 1985; and Vizcaíno and Loughry, 2008). With our readers in mind, we will not tell the story of their epic tale here, although, of course, this is certainly not the last we'll hear of them.

This was the basic makeup of the South American fauna when the first invasion occurred—not particularly large, but interesting.

Ancient immigrants

During the late Eocene–early Oligocene, two quite new (for South America, anyway) groups appeared rather suddenly, meaning that they did not have any previous record in this continent. These are the old native South American rodents, known as caviomorphs (or more broadly hystricomorphs), and the primates generally referred to as platyrrhines. The latter are distinguished from the Old World monkeys (or catarrhines) by widely separated nostrils that generally open to the side. They retain the ancestral primate dentition and are characterized by being exclusively arboreal, with quadrupedal locomotion; some species have grasping tails, a unique arboreal specialization.

Zoologists and paleontologists have done a lot of head-scratching in trying to defend a North American origin for these mammals. Several researchers have postulated the mind-boggling solution that the origins of nutria (or coypo), capybara, marmosets, and howler monkeys and their kin are to be sought elsewhere, in the forests and plains of a continent that is far away now, but that during the Oligocene, 35 million years ago and containing the first records of these mammals, was a lot closer: Africa.

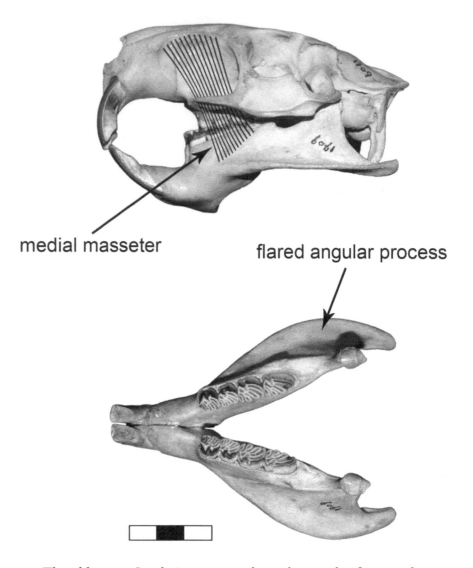

5.4. Skull and mandible in lateral view (top image) and mandible in occlusal view (bottom image) of *Myocastor coypus* demonstrating the hystricognathous lower jaw and hystricomorphous zygomasseteric system. Scale bar equals 3 cm.

Modified from a photograph of a specimen kept in the Museo de La Plata, courtesy of that institution.

medial masseter

flared angular process

The old native South American rodents share with African rodents a condition known as hystricognathy, which means that they have a mandible similar to that of *Hystrix*, the Old World porcupine (Fig. 5.4). Such a mandible has a posterior part that is deflected laterally to the tooth row, which indicates that the masseter muscles (important for the peculiar chewing and gnawing styles of rodents) are very large.

The situation is more complex, however, and requires the introduction of another term. Caviomorphs (with one possible fossil exception, in which it is difficult to tell) and a few other rodents share a condition known as hystricomorphy. This denotes that the masseter muscle (the jaw-closing muscle between the cheekbone and lower jaw) has become so enlarged and complex that one of its subdivisions attaches far forward on the side of the rostrum (or snout) in front of the eye socket, rather than only to the cheekbone as in most mammals, including many, but not all, other rodents. In extending to the lower jaw, this subdivision of the masseter passes through a large passageway in the bony element that forms the front of the cheek. This passageway is termed the infraorbital foramen. Other mammals have it too, although it is usually relatively small and restricted to the passage of blood vessels and nerves, rather than muscles. Most rodents possess an arrangement similar to that of the hystricomorphs (Appendix

1). This is termed the myomorphous condition and is present in rodents such as murids (a group that includes the mouselike rodents and comprises approximately 65% of living rodent species). In the myomorphous condition, the infraorbital foramen is only moderately enlarged, and a different subdivision of the masseter muscle passes through it (Vaughan et al., 2000).

The idea is that the hystricognathous and hystricomorphous conditions are so odd that their combined presence indicates that they evolved in caviomorph and *Hystrix*-like rodents through common ancestry, and thus that caviomorphs have an African origin (Lavocat, 1969; Huchon and Douzery, 2001). However, several researchers disagree (e.g., Wood, 1985). Instead, they suggest that caviomorphs originated from North American ancestors and that the caviomorphs then developed the odd conditions in parallel with African rodents, although nothing similar is present in rodents originating from North America. In the case of monkeys, a biogeographical juggling act is required to support a North American origin for their dispersal.

According to the African-origin hypothesis, what might have happened, at least once, is that floating islands, small portions of the African continent cast adrift into the ocean by the formation of large rivers, managed to reach the South American coast. Biological evidence supports such a scenario because African and South American rodents resemble each other in features that are hard to confuse. It is possible that a similar dispersal occurred with primates. This dispersion scenario is improbable, but it is not impossible.

In any event, if the origin were North American, we would also have to postulate a crossing over of a large body of water because the Isthmus of Panama had yet to be formed. One interesting hypothesis facilitating this leap is a variant of the rafting idea and suggests that North American mammals may have been carried across by smaller chunks of the earth's crust that shifted their attachment from the Northern Hemisphere to the Southern Hemisphere. Although tectonic studies seem to show that this was improbable, the idea is known by the charmingly named Noah's Ark hypothesis.

The absence of the isthmus also implies a totally different pattern of ocean currents. There seems to have been a strong current flowing westward between the two Americas, making a southern and mainly eastern (remember that South America lies primarily east of North America) journey trying at the best of times. On the other hand, the modern Brazilian current probably did not exist, and a southern current must have flowed on between the two great southern continents, so our hairy Ulysses could have navigated from Africa with the wind at their sails. The fact that the Atlantic was probably only half as wide as it is today, and possibly even less in some areas, may have made for a less harrowing journey.

Finally, today's flooding of the several-thousand-mile-long Uruguay and Paraná rivers produces islands of floating camalote river plants that make their way as far as the Rio de la Plata. Anyone who has seen this phenomenon may be easily convinced that the Tertiary flooding in Africa could have cast into the ocean an island large enough to have found its way, though perhaps not in perfect shape, to South American shores.

It is possible that a new discovery tomorrow might lead to different interpretations and show that this idea is a royal blunder. But this is the way science works: it is a string of temporary answers and errors that lead us along an irregular path in our attempts to pry away at Nature's secrets. What matters is that a hypothesis accounts for the known facts, and that it be in agreement with other hypotheses—or, alternatively, that they disagree so much that we refute the old paradigm and create a new one. Anything is better than blissful ignorance, even a bad hypothesis, because at the very least it makes us think.

Naturally the situation is more complex than we have made it out to be, but North American and African origins for platyrrhine primates and caviomorph rodents were the two traditional competing hypotheses. More recent studies (e.g., Poux et al., 2006; Honeycutt et al., 2007) have helped clarify some aspects of the possible origins, timing, and migration routes of both platyrrhine primates and caviomorph rodents. Such studies indicate that a North American origin is unlikely, but they raise the possibility that the lineages may have reached South America either from Africa or Asia, from the latter by way of the connections through Australia and Antarctica. Paleontological evidence suggests that hystricognath rodents have an Asiatic origin, so caviomorphs may have been derived from an Asiatic or African ancestral stock. Similarly for the primates, the Old World monkeys and apes, or catarrhines, are from Asia and Africa, and form the sister group to the New World monkeys, together comprising Anthropoidea. Fossil anthopoids that share most similarities with South American primates are from the Eocene–Oligocene of Egypt (Fleagle, 1999), providing some support for an African rather than Asiatic origin for the South American primates.

The current weight of evidence seems to tip the scales in favor of an African origin for both the South American primates and caviomorph rodents by way of a transatlantic migration (Huchon and Douzery, 2001; Honeycutt et al., 2007). For the primates, a transatlantic crossing is favored for two reasons: first, fossils considered as representing early platyrrhines and early catarrhines have so far only been discovered in Africa; and second, a migration through Antarctica is unlikely because at the time of divergence of these two primate groups, at most 37 Mya, the southern landmasses were no longer so closely connected and Antarctica was covered by ice (Poux et al., 2006). An Atlantic crossing, though a long distance at approximately 1700 km (Honeycutt et al., 2007), was much less than the current gap. Houle (1999) suggested that a small 1-kg primate could have survived 13 days on a raft of vegetation. Given the shorter distance and possibly aided by marine currents, paleowinds, and stepping-stone islands, such a voyage seems less harrowing than might at first appear. For the rodents, the situation is more ambiguous; African, Asian (by way of Antarctica), and even North American routes remain possibilities. The rodents probably dispersed to South America earlier than the primates, and the transantarctic route remains possible because Antarctica was then still temperate. On the other hand, the sea gap between Australia and Asia was at the time wider than that between South America and Africa. The other weaknesses in the Asia or North America to South America routes is that no fossils of what might be considered a protocaviomorph has been found in Australia or Antarctica (although fossils from this continent are hard to come by), and those reported from North America were apparently misidentified (Poux et al., 2006).

The caviomorphs radiated soon after their arrival, before the Oligocene glaciations, and these early caviomorph lineages persisted until the present. In contrast, few platyrrhine fossils are known in the Oligocene. According to Poux et al. (2006), the late Eocene transatlantic migration was followed by extinction of all but one of the few earlier diverging lineages, and thus modern platyrrhines stem from an early Miocene diversification.

For our purposes here, it does not matter too much from where and how the caviomorphs arrived in South America, because by the time we get to the late Tertiary, caviomorphs were thoroughly South American. The same goes for the primates, represented so abundantly today in neotropical forests. The fossil primate record from South America is poor, but they clearly must have been there, and indeed were rather successful. Their depauperate record is not odd; they were forest dwellers, and we noted in Chapter 1 the reason why forest environments are not the most favorable for preservation.

GABI: island hoppers

The presence of a continuous or near-continuous land connection between North and South America should have fostered (and did) the movement of terrestrial mammals from one continent to the other. Although it might seem that such transfers would await a near-final closure of the land bridge, it is evident that this was not the case. It is certainly true that the main episodes of exchange, as indicated by the timing of the presence of fossils in either one or the other continent (and for North America, this includes Central America), occurred near the proposed emergence of a continuous land bridge, 3 Mya. However, there is also indisputable evidence that several groups dispersed across the isthmus during its embryonic stages, when it was still mainly a collection of volcanic islands (an analogous situation today is the constellation of the Caribbean Islands, formed by similar processes). Indeed, such dispersals began to occur relatively soon after the initial stages of closure about 12 Mya (Woodburne et al., 2010). We might regard these as the first few tentative steps. Although hopping across islands might seem improbable (much as we noted for the ancient immigrants), time is powerful. It is clear that the arrangement of islands between North and South America permitted several terrestrial animals to cross from one to the other, and thus gradually breach the distance between the continents, much as we would cross a river over stepping-stones.

Let's consider first the island hoppers into South America and the traditional interpretation of the sequential arrival of immigrants. For the purposes of the following discussion, we direct your attention to Fig. 5.5, as a visual presentation facilitates an appreciation of the timing and groups involved. Among the first were raccoons or procyonids (Carnivora; see Chapters 3 and 4). The earliest South American record goes back to the Miocene, 7.3 Mya, with the presence of *Cyonasua*. This raccoon lineage persisted into the Pliocene before apparently becoming extinct, and thus, this dispersal did not have a notable effect on the South American fauna, in contrast to the Oligocene arrival of primates and caviomorph rodents. As an aside for the moment, procyonids are represented today in South America by various types, such as *Procyon cancrivorus*, but this lineage of

5.5. Taxa that participated in the GABI, distributed among higher-level taxonomic groupings and direction of movement. Mammals of South American origin that entered North America are listed as North American immigrants and vice versa for mammals of North American origin. Often, these sorts of tabulations present taxa as distributed by the rank family (traditionally designated by the -dae ending, as in Nothrotheriidae). We use this figure to convey several points. One is its obvious use in providing a straightforward visual reference to the comings and goings of the mammals involved in the GABI. But it also illustrates how the organization of data can influence interpretations we make about the mammalian groups (South American versus North American immigrants) and their success relative to one another. There are several aspects to teasing apart the reasons about how we make such interpretations. Among them is the resolution of the data as determined by the way we tabulate taxa. Traditionally, as is also done here (in part to illustrate just this point), the taxa are grouped as families. Chapter 1 provides reasons why such ranking is falling into disuse, but we retain it here to emphasize why ranking can lead to bias: the use and recognition of ranks is subject to the vagaries of taxonomic opinion (which is subjective). Thus, whether a particular clade is considered a family or another rank such as subfamily can change. Currently, most authorities (if they use ranks) would recognize the sigmodontine rodents as a subfamily of the Family Cricetidae, which also includes Cricetinae (subfamilies are designated by the -nae ending). Clearly, whether we choose to list these two groups of rodents (both ultimately South American immigrants) as a single family, Cricetidae, or independently as two subfamilies, Cricetinae and Sigmodontinae, influences the number of groups we would recognize as being South American immigrants. Felids are another example. Usually, the two main groups of Felidae are Felinae and Machairodontinae (see Chapter 4). However, Woodburne et al. (2010) raised both to family status: Felidae and Machairodontidae. We have done a similar thing ourselves: Nothrotheriinae was until recently considered a subfamily of either Megatheriidae or Megalonychidae, and thus never appeared independently in the tabulations of past authors. It is clear, then, that how we choose to represent groups can influence our interpretations of how many groups went one way as opposed to the other. A second reason why the use of higher-level groupings can obscure the true underlying biological patterns is that higher-level groups contain different numbers of taxa. Sigmodontines are a huge group, including (depending on the authority followed) some 100 genera, and either 300-odd or 500-odd extant species. On the other hand, Myrmecophagidae (anteaters) includes two genera and three extant species (and here only one species actually represents the family). Although there are biological reasons for such differences in groups (which are noted in relation to sigmodontines; see Table 9.1), any measure of success that depends on "the numbers of groups" (be they species, genera, or some other higher-level group) will also be influenced by tabulations of higher-level groups. A simple way to think about this is to ask: are we considering the same thing when we count myrmecophagids (one species) as equivalent to sigmodontines

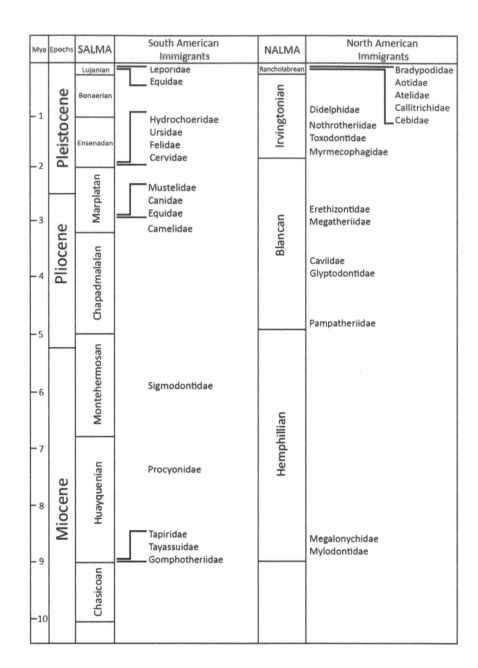

modern South American raccoons seems to have entered the continent only recently, according to molecular data (Koepfli et al., 2007).

Another group traditionally regarded as among the early arrivals to South America are the sigmodontine rodents or field mice. We emphasize that the existing South American rodents represent two separate invasions: an ancient, already mentioned African invasion of hystricognath ancestors (that is, the caviomorphs noted as ancient immigrants) and a more recent Nearctic invasion of murid ancestors of Sigmodontinae. The sigmodontines represent a New World group of murids (which we noted as among those rodents that represent the myomorphous condition) that some researchers divide into North and South American clades. The major controversy surrounding South American sigmodontines relates to their diversification and the timing of their invasion. Some authors (e.g., Reig, 1981) have suggested an old invasion (early Miocene) by a sweepstakes route before the complete emergence of the Panama land bridge. Others (e.g., Flynn et al., 2005) argued for a more recent invasion across the land bridge, thus suggesting rapid speciation events after crossing a complete land bridge.

Among the evidence that can help decide the issue is that fossil remains have been reported from Miocene strata aged at about 6 Mya (Pardiñas et al., 2002; Woodburne et al., 2010). In dealing with small organisms, we have to be doubly cautious in our interpretation of absence in the fossil record. Smaller animals are less likely to survive postmortem destructive forces to become fossils at all, so it may be that these rodents crossed over a little earlier than this date. Also, molecular studies indicate that South American sigmodontines are a monophyletic group, and thus there was only a single ancestral stock that crossed into South America. In addition, molecular clock time estimates essentially fall between the earlier and more recent invasion scenarios (Honeycutt et al., 2007). Such diverse lines of evidence seem to favor an earlier invasion by ancestors that entered South America well before the formation of a complete land bridge, and thus the position of sigmodontines in Fig. 5.5.

The recent work of Campbell et al. (2010) may be cause for revising the earliest appearance of South American immigrants. In their paper that delved into dating the first "pulse" of the GABI, these authors argued for a much earlier appearance of proboscideans in South America than has usually been recognized: they dated the Amazonian remains of the gomphothere *Amahuacatherium peruvium* as about 9.5 Mya, or late Miocene. There has been some debate on the age of this proboscidean, as well as other Amazonian mammals, because it is from Amazonian lowland deposits, which had proven notoriously difficult to date. In contrast, A. *peruvium* has also been considered as a variant of the Pleistocene *Haplomastodon*, another gomphotheriid proboscidean (e.g., Ferretti, 2008; see Chapter 6), based almost entirely on its (i.e., of A. *peruvium*) supposedly Pleistocene age. However, the methods of Campbell et al. (2010) seem fairly secure because they are based on a combination of radioactive dating and magnetostratigraphic correlation. In addition to proboscideans, these authors also reported a late Miocene age for remains of peccaries (Tayassuidae) and tapirs (Tapiridae). Such records are much earlier than usually recognized for these groups (as, for example, by Woodburne et al., 2010), and the opinions of Campbell et al. (2010) are reflected in Fig. 5.5. Perhaps the early existence of the Central America peninsula and the long-standing biotic continuity with North America proposed by Kirby et al. (2008) render the early presence of these taxa less surprising.

Given these new data on the appearance of South American immigrants, those tentative steps take on a much bolder flavor and may require that we alter our perspective. Not only did they begin earlier than we thought, but there were also more of them, and the immigrants themselves—at least the proboscideans, tayassuids, tapirids, and sigmodontines—played large roles in subsequent South American faunas. We can no longer easily argue that the island hoppers had little effect on the composition of the South American fauna.

However, island hopping was not a one-way affair, because South American forms also hopscotched their way north. The records of such journeys also go back to the Miocene. Sloths have generally been regarded as the first mammalian migrants (at least before the Campbell et al., 2010,

(several hundred species) in evaluating how well South American immigrants fared with respect to North American immigrants? Yet another aspect that can bias our interpretations has to do with just which groups we decide to include. We noted that primates and living tree sloths are not usually considered, and we have wondered why this should be so. In this respect, we note McDonald's (2005) prudent words. Central America is part of North America, so why exclude them? Their fossil records might be poor (and these sorts of analyses are usually done by paleontologists), but they are certainly present today in North America, they clearly have a South American ancestry, and they must have got there by crossing the isthmus. If we choose to represent things as we have done, then a different picture emerges: there are now more groups of South American mammals invading North America than the other way around. This is the reverse of the usual view, and one part of the reason on which a case of North America superiority is based. The other is the eventual distributions of mammals following the end-Pleistocene extinctions. A final word: of course, we are not attempting to stake out a claim that South American mammals were superior to North American mammals. We have presented this argument only to illustrate that how we choose to arrange our data can have significant consequences on the interpretations we derive from them.

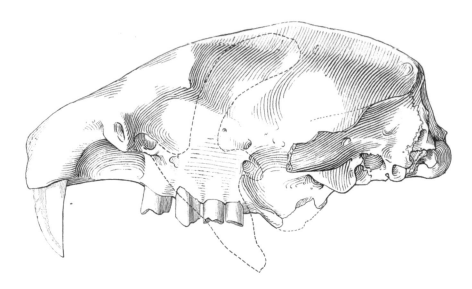

5.6. Skull of *Acratocnus,* a Pleistocene–early Holocene ground sloth found in Cuba, Hispaniola, and Puerto Rico. Total length of skull: about 20 cm.

From Anthony (1926).

report of a Miocene appearance of South American proboscideans). They appear in North America about 8.5–9 Mya (Morgan, 2008), but they may have been present even earlier in Central America, from which our fossil record is less complete. Woodburne et al. (2010) suggested that Central America may have functioned as a holding pen, where mammals may have paused on their way north or south until conditions favorable to their continued journeys arose, although this hypothesis remains open to further testing. In this respect, the appearance of South American gomphotheriids 9.5 Mya may be more than an interesting coincidence.

The early sojourns into North America involved two groups of sloths, the mylodontids and megalonychids. The latter were represented by *Pliometanastes,* which founded a lineage that was to become successful in the north, eventually culminating in *Megalonyx,* which wound up in Thomas Jefferson's hands. The mylodontids were represented early by *Thinobadistes* and later by *Glossotherium* and then *Paramylodon,* apparently derived from *Glossotherium* (McDonald and Naples, 2008). The megalonychids, incidentally, also colonized the Caribbean Islands. How they got there is, like the story for platyrrhines and caviomorphs, the subject of debate. One robust hypothesis is that the ancestors of the Caribbean sloths made their way over a partial and possibly at times continuous land bridge postulated as having existed for a short period (about one million years) 35 Mya (Iturralde-Vinent and MacPhee, 1999; MacPhee and Iturralde-Vinent, 2005; Taboada et al., 2007). This land span, postulated as having linked northern South America and the proto–Greater Antilles, occurred over the Greater Antilles and the Aves Ridge (now almost entirely submerged), just west of the present-day Lesser Antilles. The name coined for this route is the captivating partial acronym GAARlandia (Greater Antilles–Aves Ridge–landia). Among the Caribbean sloths we may mention such genera as *Acratocnus* (Fig. 5.6), *Megalocnus,* and *Neocnus* (Taboada et al., 2007).

GABI: recent immigrants

During the later Pliocene, 3 Mya, the final emergence of the Isthmus of Panama finally ended the long isolation of South America. The flora and fauna of North and South America, long separated and evolving independently (other than the limited exchanges noted above), begin to mix extensively. What began as a trickle became a flood as hordes of mammals trekked north or south from their lands of origin to their new homes. However, the proportions going in opposite directions were unequal.

Even so, new clades (traditionally designated as families; Fig. 5.5) of South American mammals trekked north across the bridge. Among the earliest to make the trip were the caviomorph rodents, glyptodonts, and another sloth, *Glossotherium*. The sloths kept coming, with *Nothrotheriops* and *Eremotherium*—about as large as *Megatherium americanum*—being successful in the north. Cingulates (including armadillos, pampatheres, and glyptodonts) and anteaters, also xenarthrans, as well as opossums and toxodonts made the trip too, although not all were equally successful or ventured equally far in their new land. In all, 10 clades (or families) are represented; here, however, we recognize 11, because one of the lineages, Nothrotheriidae, is now regarded as representing a clade separate from Megalonychidae, in which nothrotheres were traditionally included, and Megatheriidae, in which they have been more recently included. For the sake of completeness, we should note that this roster was increased by various representatives of originally North American stock that returned to the north after substantial evolution in South America.

Traffic in the other direction was a bit more congested, with some 15 clades (designated as families) of North American land mammals appearing in South America over the Montehermosan, Chapadmalalan, and Marplatan. Among them were the procyonids, which had already arrived much earlier by way of island hopping. However, these ancient representatives had gone extinct, so the bridge crossers represent a new bunch. Besides this group, all the others were new—quite a parade! They include other carnivorans (cats, dogs, ferrets, skunks, and bears, among others) and shrews, rabbits, horses, camels, deer, mastodons, tapirs, and peccaries (although these last three apparently first entered much earlier). Representatives of some of these groups will be considered further in the Bestiary of Chapter 6. The appearance of these mammalian groups, in both North and South America, were not simultaneous events. Indeed, there appears to have been several main bursts of migration since 3 Mya. Woodburne et al. (2010) designated these phases as GABI 1 to GABI 4. Webb (1978, 1985a) noted that the corridor along the isthmus connecting North and South America acted as a filter that prevented a wholesale exchange of taxa. The timing of the phases of periodic migrations may reflect the progressive environmental changes that occurred in the Central American region over the course of the last three million years.

As noted, the GABI had a profound effect on the existing faunas. This is clearly true for South America because nearly half of the modern fauna is composed of descendants of North American immigrants. It was equally true of North America. The mammals of South American origin diversified and spread far and wide throughout the Blancan, Irvingtonian, and Rancholabrean, and established various independent evolutionary lineages. The range of the sloth *Nothrotheriops*, for example, stretched from Florida to the tar pits of Los Angeles, and extended into Canada. Another sloth, Jefferson's *Megalonyx* (Fig. 5.7), reached Alaska (McDonald et al., 2000).

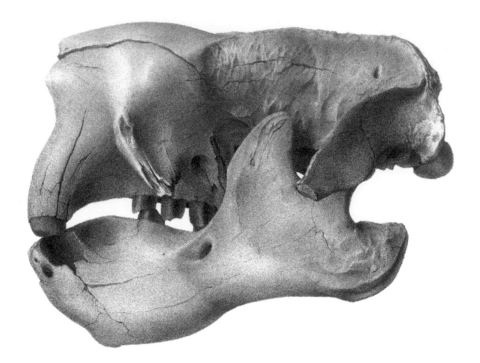

5.7. Skull of *Megalonyx*. This North American ground sloth was one of the many intellectual interests of President Thomas Jefferson. Length of skull approximately 36 cm.

From Leidy (1855).

One way of measuring success of the South and North American clades is by tallying up the numbers of clades that went one way as opposed to the other, as has been traditionally done. Fig. 5.5 is useful in this regard, although the validity of tallying up groups is questionable. The mammals of South American origin were successful for some three million years, at least, and became widespread. This did not last; the modern fauna of North America has few representatives of originally South American ancestry. According to the traditional way of viewing things, there remain only three, the armadillo, opossum, and porcupine, although we may note that they (particularly the last two) are quite widespread. There is a second opossum, the mouse opossum, that is abundant in some areas of Mexico.

What are we to make of the resulting imbalance in the final (or at least modern) faunal compositions of these two continents with respect to the relative contributions that original members of one continent make to the fauna of the other? What were the factors contributing to these patterns? Should we attempt to evaluate the relative fitness of one over the other group of migrants as part of our explanation? In anticipation of a detailed consideration of such questions in Chapter 9, we can note here that several authors have attempted to provide answers, and the most prominent opinions (until fairly recently) involved some idea of inherent inferiority of mammals of South American origin (i.e., when crunch time came, they were not able to compete with their North American counterparts). However, other factors enter into the equation. First, consider the fact that North America had—and still has—considerably more landmass: 24 million km^2, as opposed to South America's 18 million km^2. Further, the northern fauna was closely tied to the Eurasian fauna, with which it has had frequent contact over millions of years, primarily through the periodic emergence of the Beringian land bridge linking Alaska and Siberia.

If we want to get picky—and consistent, as we will see in minute— we can count two more South American groups. The modern sloths and anteaters range into the forests of southern Central America, with the latter reaching Mexico. It is true that they do not get into temperate North America, but this is another matter and serves to illustrate one of the problems in trying to determine possible reasons for the apparently greater success of the North American immigrants. In such cases (i.e., where we consider sloths and anteaters as being excluded from North America), we are revealing ourselves as being fairly flexible in what we consider North America to be. Does it include Central America, or are the latter and parts of Mexico more logically a continuation of South America? (There is clearly a bias based on our preconceived notions of political boundaries.) Perhaps if we juggle around these definitions, we can come up with more favorably equable proportions—we are being facetious, of course! Trying to explain the patterns as largely due to inferiority of one or the other group is merely a continuation of the sorts of European attitudes toward America in general that so infuriated Thomas Jefferson and that Charles Darwin initially debunked, or akin to viewing marsupials as representing some sort of inferior mammal because of current marsupial biogeography. It is worth keeping in mind that mammals of North and South American ancestry lived together and thrived in their own ways for nearly three million years.

To get back to the question we raised above, is it unfair to count sloths and anteaters? The best answer may be that it depends on how we choose to look at it. Shrews, for example, are still counted as one of the North American groups that managed to conquer South America. However, this group (although no longer considered monophyletic, i.e., descending from a common ancestor) is represented by a single species of shrew scattered in parts of Venezuela, Colombia, and Ecuador. In other words, they barely have a foothold in the northern margin of South America, and their presence here is clearly attributable to their close relationship to Central American shrews.

Many of the South American taxa that passed to the north are representatives of Xenarthra, which are generally restricted by either food availability (anteaters) or aspects of their physiology to warmer, largely tropical climes; that is, given their generally lower metabolic rates, temperate forests as are present in much of North America north of Mexico are not their preferred environments—think of sloths, for example. Although it is true that there were many potential candidates to choose from as dispersers, generally only one taxon from each major group of xenarthrans passed through the corridor and reached temperate North America, as noted by McDonald (2005:325), who advanced the useful concept of the isthmus acting as a nested sieve: the different sized-openings "in each layer of the sieve represent different ecological parameters that allowed different taxa to disperse varying distances northward but at the same time held back other taxa." On the other hand, not all the potential xenarthrans (yes, there were many in South America as a whole) may have initially been distributed near the region where they might have crossed the isthmus. *Eremotherium* was a largely tropically to semitropically adapted giant ground sloth. Its range extended to the northern limits of South America, and it did cross the isthmus to the north (probably initially as the species *E. eomigrans*). The similarly sized *Megatherium americanum*, however, was adapted to a more temperate environment. Perhaps the reason that it did not manage to cross to the north has something to do with its initial distribution in South America rather than having been filtered out or held back by ecological parameters along the corridor.

McDonald (2005) also noted that unlike some of the earlier immigrants to North America (the island hoppers and earlier recent immigrants), which crossed northward over a mosaic of habitats, most of the more recent newcomers probably crossed when the isthmus had become established as a tropical rain forest, and the rain forest had a relatively restricted distribution. Although the Central American fossil record is relatively poorly documented, McDonald (2005:335) noted that the modern presence of xenarthrans suggests "that the dispersal of xenarthrans out of South America is richer than the traditional view of only looking at those forms that made it into the North American temperate zones." McDonald cited the presence of the large megalonychid sloth *Meizonyx*, known only from a single specimen from El Salvador, as a good example of this potential richness. Continued fieldwork in Central America will likely result in the discovery of other such examples.

The extraordinary episode of extinction of the megafauna at the end of the Pleistocene, and the resulting distribution of faunas, has traditionally been linked to this great faunal interchange, and this topic will be discussed in Chapter 9. It is enough for now to note that this extinction affected those that remained, those that left, and those that arrived, although perhaps to different degrees.

Stratigraphy and sequence of Pleistocene vertebrate faunas

We may now resume our description of the ages, interrupted in Chapter 3, and consider those of the Pleistocene, the final epoch of the Cenozoic. The ages mentioned below correspond to the same megacycle (the Neocenozoic), supercycle (Supracenozoic), cycle (Panpampian), and subcycle (Pampian), all of which were noted in Chapter 3 during the discussion on the Marplatan Age.

An overview of their paleoecological settings is presented later in this chapter, but this subject is further considered in Chapter 8. By the Chapadmalalan Age, as noted in Chapter 4, the ancient South American fauna began to incorporate such Northern Hemisphere newcomers as canids, mustelids, mephitids, camelids, proboscideans, and equids. However, it was during the Pleistocene that significant faunal changes in the midlatitude region of South America took place. These began in the Ensenadan Age (to early to middle Pleistocene, from ~2.6 Mya to 0.78 Mya) and involved the appearance of bears, cats, and deer (among others), as well as a remarkable diversification of the native Cingulata and Tardigrada among Xenarthra (Cione and Tonni, 1995b). Moreover, during the Ensenadan, this diversification was particularly marked among mammals with a body mass of more than 1 metric ton, or megamammals (Cione and Tonni, 1995b). The mammals of the late Ensenadan Age and of the early Lujanian Age indicate arid and cold conditions. For most of the early Lujanian, taxa adapted to arid conditions are recorded, but during the end of the late Lujanian, two different associations of mammals are recorded: one with a decrease in the frequency and diversity of the browsers and the other with an increase of grazing megamammals (Cione and Tonni, 1995a; Tonni et al., 1999), although the autecology of many of the species has yet to be more accurately assessed (Vizcaíno et al., 2008).

Cione and Tonni (1995a) divided the Lujanian into two subages, the Bonaerian (300 to 30 thousands of years ago, kya) and the Lujanian sensu stricto (30 to 10 kya), i.e., as a subage, so the Lujanian Age included the Bonaerian and Lujanian subages. Later, these authors (1999, 2001) considered the Bonaerian as a stage/age independent from the Lujanian (thus both are now considered ages, as depicted in Figs. 3.5 and 4.1), comprising the middle Pleistocene, between approximately 0.78 and 0.13 Mya (or 780 and 130 kya). The beginning of the Lujanian Age is placed at the base of the late Pleistocene (0.13 Mya or 13 kya) and its end in the early Holocene (approximately 0.085 Mya or 8.5 kya). More recently, Cione and Tonni (2005), fundamentally on the basis of the biozone of micromammals (the biozone of *Ctenomys kraglievichi*) at the base of the Bonaerian (Verzi et al., 2004), restricted the beginning of the Bonaerian to approximately 0.5 Mya or 500 kya.

We have thus far dealt with the broad compositions of the North and South American faunas throughout most of the Cenozoic. We have also covered the historical geologic processes that resulted in the formation of the Isthmus of Panama, which linked the two continents and triggered the GABI, bringing together the faunas of the two continents. We next turn our attention to some of the physiographic and climatic features of the Pleistocene of South America to place its amazing fauna in the proper context and to highlight some of the more spectacular Pleistocene South American fossil sites.

Useful approaches: astronomical forcing

The Quaternary has traditionally been defined as the time interval during which glaciation cycles occurred (or occur, as we are currently in an interglacial), most notably in the Northern Hemisphere, and in which human fossils were found. Despite the new light shed on our ancestry by the many discoveries made over the last few decades of truly ancient human remains (the oldest hominid fossils date from about 7 Mya) in Africa and other parts of the world, the Quaternary has retained its usual connotation of the time of glaciation. Currently, however, its start is set to coincide with the Gelasian, the last age of the Pliocene, at 2.6 Mya (Gradstein et al., 2004), thus breaking its 150-year-old definition of including the Pleistocene plus the Holocene (Hörnes, 1853; Lyell, 1830–1833).

The Quaternary witnessed several glaciations, during which, for one or several reasons, temperature fluctuations produced the accumulation of ice, particularly at high latitudes; major climatic, environmental, and even geographical changes also occurred. Temperature variations in the amount of solar radiation reaching the earth are due to interactions among three astronomical parameters that modulate the pattern of insolation. The relationships among and the effects of these parameters were first worked out by Milutin Milankovitch (1941; Fig. 5.8), who considered that the variation in radiation from the sun would cause the glacial–interglacial cycles, according to the changes in the eccentricity of the earth's orbit (with a

5.8. Milutin Milankovitch (or Milanković, 1879–1958), Serbian geophysicist and engineer, whose work linked cyclical climatic change with the earth's orbit.

Portrait by the Serbian painter Paja Jovanović (1943).

Geography and climate in the Pleistocene of South America

period of about 110,000 years), the obliquity of the rotation axis (with a period of 41,000 years), and the precession of the rotation axis (with a period of 26,000 years) (Box 5.1).

Box 5.1

Milankovitch's three astronomical parameters

Eccentricity of the earth's orbit. Every 110,000 years or so, the earth's orbit changes its shape from nearly circular to elliptical to circular again (Box Fig. 5.1a). In the most extreme cases, the eccentricity varies from only 0.5% (virtually circular) to 6%. As the sun occupies one of the foci of the ellipse, when the highest eccentricity is reached, the seasons are more marked in one hemisphere and more moderate in the other. Today the eccentricity is about 1.7%. When the earth, during its annual orbit, is closest to the sun, it is said to be at perihelion, and when farthest, at aphelion. This in itself affects the intensity of solar radiation reaching the earth. At present, perihelion occurs in January (northern winter) and aphelion in July (austral winter). This would be expected to lead to less seasonal contrast in the Northern Hemisphere, but in fact, seasonality is generally more marked in the north because other factors are important, such as the proportion of land to water (in which the Northern and Southern Hemispheres differ significantly), and hence the amount of radiation absorbed by land surfaces. In the long term, not only does eccentricity vary, but so does which of the earth's poles is tilted toward the sun at perihelion and aphelion.

Obliquity of the rotation axis. Every 41,000 years, the earth's rotational axis varies in its obliquity with regard to the perpendicular of the plane of its orbit between 21.5° and 24.5° and back (Box Fig. 5.1b). As the angle is increased, seasons become more extreme in both hemispheres, with warmer summers and cooler winters. The effects at higher latitudes are apparently the most influential: the amount of solar flux during the summer may determine whether winter snows melt or remain, and thereby accumulate. Currently, the rotational axis is deviated 23.44°.

Precession of the rotation axis. Every 26,000 years, the precession of the earth's rotational axis describes a complete circle (Box Fig. 5.1c). The precession is the gradual change in the orientation of the earth's spin axis. The spin axis traces a cone, completing a full circuit in 26,000 years, like the wobbling rotation of a top as it spins. The precession of the axis means that the equinoxes and solstices (marking the start and midpoint of the seasons) change over time. The effect would not be so important if the earth's orbit were circular, but the fact that it is elliptical means that the direction of a particular pole with respect to the sun will change depending on whether the earth is closest to (perihelion) or farthest from (aphelion) the sun. Such phenomena determine whether the summer (for example) in a certain hemisphere starts at a closer or more distant point of the orbit with regard to the sun. The result of this is a more marked seasonality when the maximal distance coincides with the maximal inclination. When these two factors have the same effect in one of the hemispheres, there are opposite effects in the other. Although this is relatively straightforward, the situation is not so simple because the precession of the earth's axis of rotation is not the only factor determining the periodicity of the equinoxes, and hence the

A

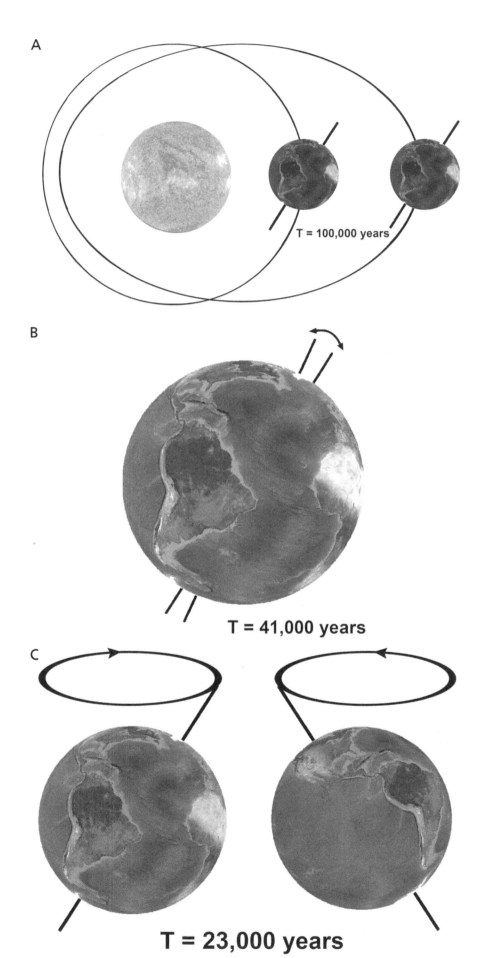

T = 100,000 years

B

T = 41,000 years

C

T = 23,000 years

Box Fig. 5.1. Astronomical parameters that influence the Earth's orbit and hence its climate (Milankovitch, 1941). A) Eccentricity of the orbit. B) Obliquity of the rotation axis. C) Precession of the rotation axis.

Drawn by Sebastián Tambusso after various sources.

effect on climate. The plane of the earth's orbit also undergoes a precession, with a period of about 70,000 years. The direction of this precession is opposite to the precession of the axis of rotation. The result is a shortening of the precession of the equinoxes relative to perihelion and aphelion to about 21,000 years.

Useful approaches: isotopes

As with Wegener, and for similar reasons, Milankovitch's ideas had to await recent advancements before they were generally accepted. The 1960s and 1970s saw improved methods and technology for calculating planetary motions and insolation. In addition, the development and refinement of techniques for radioisotope and other isotopic analyses made it possible to date rocks and decipher ancient temperatures (Kandel, 2003). The ability to obtain and analyze gas trapped in air bubbles (later to be smashed under the pressure of the accumulating ice) in glacial ice, for example, yields data on age and concentration of chemical elements and compounds that allow an estimate of past climates. Stable isotopes are useful as paleobiological and paleoclimatological tracers because, as a result of mass differences, different isotope ratios of an element (e.g., ^{12}C versus ^{13}C, ^{14}N versus ^{15}N, ^{16}O versus ^{18}O, ^{1}H versus ^{2}H) have different thermodynamic and kinetic properties. The changes between the gaseous, liquid, and solid states of water are affected by the temperature and the mass of the H_2O molecule. By using the isotope ratios from a sample of ancient ice, the temperature of formation of the snow that formed the ice may be estimated (Kandel, 2003). Box 5.2 provides an example of how this may be achieved.

Box 5.2

Using isotope ratios to estimate temperature

For elements with an atomic mass of less than 40, differences in atomic mass among isotopes of an element can lead to measurable isotopic partitioning between substances during physical and chemical processes that labels the substances with distinct isotopic ratios. Natural fractionations are small, so isotopic ratios are reported as parts per thousand (‰) deviation in isotope ratio from a standard, using the δ notation, where $\delta X = [(R_{sample} / R_{standard}) - 1] \times 1000$, with X = C, N or O, and R = $^{13}C/^{12}C$, $^{15}N/^{14}N$, or $^{18}O/^{16}O$. R_{sample} and $R_{standard}$ are the high-mass to low-mass isotope ratios of the sample and the standard, respectively: commonly used standards for $\delta^{13}C$, $\delta^{15}N$, and $\delta^{18}O$ are Pee Dee belemnite (PDB), atmospheric N_2, and standard mean ocean water (SMOW), respectively. Positive δ values indicate enrichment in the high-mass isotope relative to the standard, whereas negative values indicate depletion.

Thus, the temperature at which the biological molecules were formed can be calculated as

$$\delta\,^{18}O = \frac{(^{18}O\!:\,^{16}O)\ sample - (^{18}O\!:\,^{16}O)\ standard}{(^{18}O\!:\,^{16}O)\ standard} \times 10^3$$

The standard for carbonates is PDB, a belemnite of the Pee Dee Formation from South Carolina (USA), and for the water is SMOW (standard mean ocean water). The paleotemperature of calcite can be estimated, with an error of about 0.5°C, as

$$T = 16.0 - 4.14\Delta + 0.13\Delta^2,$$

where T is temperature (°C), and Δ is $\delta^{18}O$ calcite (versus PDB) minus the $\delta^{18}O$ of water (versus SMOW).

Equations are from Anderson and Arthur (1983).

Another approach used is that of the hydrogen isotope deuterium (^2H or D) as a source of climatic inferences. Indeed, heavy water (D_2O) has a slower evaporation rate than normal water. Therefore, water vapor is enriched in normal water while glaciers have a larger share of deuterium-rich water. This imbalance is more marked for colder climates than for warmer climates.

The observed changes in the earth's temperature and climate, which are associated with sea level, are influenced by the astronomical variables referred to above in a process known as astronomical forcing. However, greenhouse gases (for example, CO_2) and particles shed by volcanoes have had a complex pattern across the last few hundred thousand years and could have played a major role in the glaciation–deglaciation process.

One important consequence of these past climatic changes is that the amount of free water, i.e., not trapped in high latitude or high altitude glacial masses, has varied accordingly. Indeed, over the past three million years, huge amounts of water have moved between two of its largest reservoirs on earth, the ice sheets and the oceans. Since the last glacial maximum (LGM, less than 20,000 years ago), the incredible amount of about 50 million cubic kilometers of ice has melted from land-based ice sheets, raising global sea level by about 130 m over a few millennia (Lambeck et al., 2002).

The territorial effects of these changes are tremendous, creating or obstructing the connections between land or water masses. For instance, North America had an extensive land connection with Asia during the LGM, when the now-submerged territory of Beringia formed a bridge over which many species, possibly including ours, traveled from one continent to the other. In South America, the Atlantic continental shelf is relatively steep to the north but becomes more gently sloped as the pole is approached, and hence a smaller area of emergent land occurred along the northern coast of Brazil (Plate 6). Farther south, the Falkland (or Malvinas) Islands could have been united to the mainland or nearly so. The Pacific coast, on the other hand showed little variation, as this is the subducting, tectonically active margin and rapidly falls off to great depths.

In the midlatitudes, where the fauna studied here is better known, a dramatic physiographic change took place between the LGM and the present. The present Río de la Plata, actually a huge estuary, did not exist. Instead, the Paraná River followed through what today is the *thalweg* (a German word that means the line of lowest points in a streambed) to end in an immense delta, whose biogeographical and paleoccological consequences will be discussed in the next chapters. Between latitudes 33 and 37°S, and east from the dune fields depicted in Chapter 8 (Fig. 8.2), exposed land that is today underwater increased in area by 60% over the present-day area.

Pleistocene habitats in South America

As can be seen in Fig. 5.9, the northern part of South America is mainly covered by tropical rain forest (minimum normal annual rainfall is between 1750 and 2000 mm), which contains a great proportion of the world's terrestrial biodiversity. Although the general image has long been that this region has been stable and nearly unchanged, more recent studies indicate that it has undergone great variations in the last tens or hundreds of thousands of years (Hooghiemstra, 2002). In the midlatitude regions, on the other hand, a much more open habitat is present, in which precipitation seldom goes much above 1500 mm and in which human population is rather dense.

As might be expected, the periodic climatic alternation of glacial and interglacial epochs during the middle–late Pleistocene dramatically modified the distribution, composition, and biomass of plant and animal communities in South America (Cione et al., 2003). In northern South America, palynological (i.e., that derived from the study of fossil pollen and spores) records from Colombia (Van der Hammen, 1974; Hooghiemstra, 1984) indicate that since Pliocene times, the vegetation of the lowlands was being replaced by montane vegetation, with species from more temperate (both southern and northern) zones as a consequence of the lifting of the Andes, and gradually developed the typical high-montane vegetation found there today, i.e., the montane cloud forest and the peculiar high-latitude tropical ecosystem called *páramos* along the glacier valleys. Moreover, the records also indicate that communities that are typical of the Caribbean coast, such as mangroves, are dominant throughout the Neogene.

Because of the fluctuations in Pleistocene climates, in which glacial, colder, and more arid periods were cyclically followed by wetter and warmer interglacials, reconstruction of the distribution of Pleistocene biomes is a difficult task. However, it may be said that during the LGM, about 20 to 15 kya, most of what today is the temperate region of South America was more arid, perhaps with precipitation amounting to one-third that of modern levels (Iriondo and García, 1993) and temperature only 1°C lower than today, although with more marked seasonality. The tropical region, on the other hand, must have been more humid. This was due to the proposed displacement of the tropical anticyclone, an atmospheric phenomenon in which there is a descending movement of air toward the continent, bringing moisture from the Atlantic to an otherwise dry region, as suggested by oxygen isotopic ratios found in Bolivian ice cores (Thompson et al., 1998).

As stated by Ortiz Jaureguizar and Cladera (2006), the most evident differences between Tertiary and Quaternary climatic–environmental conditions are related to the amplitude and frequency of environmental changes, although the disparity of resolution should not be ruled out as a source of bias, with their consequent effects on the biota. In the Pleistocene, South American faunas were affected by two main factors: glaciations, which changed temperatures and arrangement of arid–humid biomes, as well as sea-level changes; and the arrival and diversification of North American

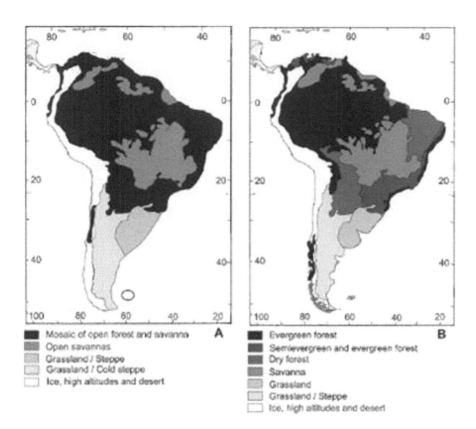

Mosaic of open forest and savanna **A**
Open savannas
Grassland / Steppe
Grassland / Cold steppe
Ice, high altitudes and desert

Evergreen forest **B**
Semievergreen and evergreen forest
Dry forest
Savanna
Grassland
Grassland / Steppe
Ice, high altitudes and desert

5.9. Maps showing changes in South American biomes in the (A) Pleistocene and (B) today.

Modified from Cione et al. (2009).

mammalian contingents (Lessa and Fariña, 1996; Lessa et al., 1997). During time intervals of cold and dry climates, forest areas decreased; consequently, open areas had wider connections, and east of the Andes, a north–south savanna corridor was established that reached even beyond the Isthmus of Panama, favoring dispersal of steppe and grassland organisms. During interglacials, forests expanded, but a corridor across the eastern South America and Atlantic coastal areas remained, permitting dispersion of mammals more adapted to relatively closed and mesic environments, such as the proboscidean "*Stegomastodon*" (Sánchez et al., 2004). During the early Pleistocene, pollen of palms and grasses of the Río Paraná Basin suggest the presence of warm and wet climates, but with a greater aridity than in the late Miocene for that area. In the Río Uruguay Basin during the middle and late Pleistocene, wet and warm conditions seem to have been prevalent, according to the plant fossils.

In the Pleistocene, glacial periods yielded a steppe in southern Argentina and Chile, with forest restricted to the western unglaciated areas, whereas during interglacials, mesic climates allowed the expansion of grasslands and the dominance of *Nothofagus* in the Andean forest, whose changes in conditions were less disruptive than on northern continents. After the early Pleistocene, the composition of South American faunas varied more regionally with the environments than in preceding times, influenced by pronounced habitat and altitude heterogeneity, with the consequence that the remains of some taxa across South America do not imply synchronicity, making correlation more difficult. The early Pleistocene land-mammal fauna of southern South America was dominated by grazers, followed by mixed feeders and carnivores, with browsers being much less diverse. This suggests that grasslands and steppes were widespread, with trees probably restricted to gallery forests.

Prevalent conditions in the Pampean region were those of an arid to semiarid environment with lower temperatures, according to biogeographical evidence and to sedimentary facies of the mammal-bearing sediments. Similar climates existed in the central regions of South America, although with higher rainfall. In Bolivia, the classic Ensenadan Tarija fauna is from moderate elevations (just below 2000 m) and indicates a mixed savanna–grassland–woodland mosaic, with a mixture of grasses as well as a few trees and shrubs (MacFadden and Shockey, 1997).

In the late Pleistocene, the land-mammal fauna of southern South America was again dominated by grazers, followed by carnivores and mixed feeders, with browsers having a lower diversity, which once more suggests open habitats were widespread and arid, cold conditions prevalent, but interspersed with short, warmer, and wet climatic periods. Conditions in the central region of South America were similar to those in the early Pleistocene, while in southern Patagonia, cold and arid climates were predominant most of the time, but with wetter and warmer periods that help explain the presence of the ground sloth *Mylodon* and the jaguar *Felis palustris* in southernmost Patagonia during the late Pleistocene.

As seen in Fig. 5.9, during the Quaternary fluctuations described above, current South American biomes shifted some degrees of latitude north and south, or expanded and contracted (Cione et al., 2009). In many places along the continent, these changes shaped the biotas that inhabited its regions. Although in this book we focus on the large mammals of the midlatitude plains and low hills of Argentina, Uruguay, and the southernmost part of Brazil, fossils of large mammals have been found in other places as well. Sometimes those other faunas have a slightly different taxonomic composition, albeit with allied genera. The main distinction among the faunas may be neatly summarized by the differential presence of two closely related taxa of giant ground sloths: the megatheriines *Megatherium americanum* and *Eremotherium laurillardi* (sensu Cartelle and De Iuliis, 1995, 2006, *nec* Guérin and Faure, 2000). The former, already mentioned in Chapter 1, is known primarily from the Pleistocene of Argentina, but also from several localities in Uruguay and Tarija, Bolivia (De Iuliis, 1996). The latter has a more tropical distribution and is known mainly from the Pleistocene–Holocene of Brazil and the southeastern United States, as well as several Central American and northern South American countries, such as Panama, El Salvador, Mexico, Venezuela, Ecuador (Cartelle, 1992; Cartelle and De Iuliis, 1995; De Iuliis, 1996), and Peru (Pujos, 2002, 2006), and is the most representative member of Hoffstetter's (1952) fauna of *Eremotherium*.

Many of the animals this book deals with may be found in the tropical part of South America, but different macraucheniid (*Xenorhinotherium*) and toxodontid (*Trigodonops, Mixotoxodon*) genera are present. Excellent records of these faunas are found in Brazil (most particularly in the wonderful finds of Lagoa Santa, Minas Gerais; Cartelle, 1994a), Venezuela (Rincón, 2003), Ecuador (Hoffstetter, 1952), and Bolivia (Hoffstetter, 1963). However, some of the ground sloths and glyptodonts seem to have felt more at home in the midlatitude regions, and therefore, the fauna reached its highest diversity there.

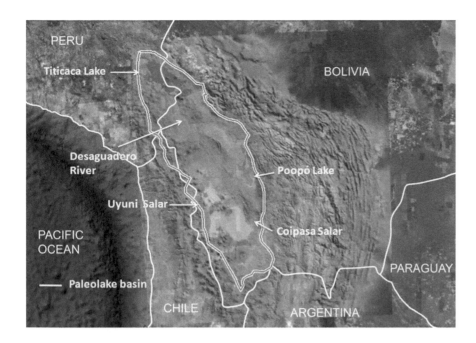

5.10. Giant lakes during the LGM. Basin in which the giant Pleistocene lakes formed.

Based on Blodgett et al. (1997), drawn by Andrea Sánchez.

The LGM was probably the moment in the last hundred thousand years in which the earth had its minimum temperature. As already noted, dramatic changes took place in the climate and physiography of the continents. Although the overall climate of the world must have been colder and hence drier, thanks to the great amount of water locked in the larger glaciers, there is evidence in the Bolivian altiplano that precipitation was higher in the tropical part of South America (Thompson et al., 1998). Indeed, the *salares* were giant lakes during the LGM, as illustrated in Fig. 5.10 (based on Blodgett et al., 1997). It could be therefore concluded that the habitats in the Amazonian basin and other intertropical regions must not have been very different from what they are today, while middle and high latitudes must have supported a more arid landscape, which must have been barren in the southernmost regions of the continent.

Pleistocene communities

Vegetation history is of the same importance as that of animals to reconstruct life in the past. However, its study is more difficult and more dependent on technology. Apart from that, pollen, the most used plant remains thanks to the abundance of their production, is composed of a material that proves resistant to many destructive agents, but that is prone to oxidization (Allison and Bottjer, 2010). This is the reason why pollen remains from arid environments are so scarce. Most studies on Pleistocene pollen have been carried out in high-altitude or in high-latitude lake sediments, thanks to the pioneering work of Markgraf (2001). The extra-Andean Patagonian records of *Ephedra*, a shrub of medicinal importance, and chenopodiines, the group that comprises such modern plants as quinoa and spinach, for pleniglacial (or full glacial) age suggest extremely cold conditions with marked continentality. Aridity and cold are also the hallmarks of the studies of pollen in the subtropical and tropical Andes, as well as in the altiplano (Villagrán, 1993).

On the other hand, much less is known about the vegetation history in the rest of the continent. One of the difficulties has to do with the paucity of pollen records extending to the LGM in tropical or subtropical areas. In spite of these limitations, pollen data from marine core GeoB 3104-1 (3°40'S, 37°43'W, Behling et al., 2000) suggest that forest or woodland occupied the eastern *caatinga*, a cactus scrub implying an arid area, since the LGM, which is consistent with the hypothesis that precipitation was even higher at the LGM than today (Mayle, 2004, 2006).

Other authors (Prado and Gibbs, 1993; Pennington et al., 2000) compared the current distributions of dry forest species across the South American tropics and showed that more than 100 phylogenetically unrelated species have similar geographic patterns, forming four disjunct (i.e., separated) dry forest blocks or nuclei (Caatinga, Misiones, Chiquitano, and Andean Piedmont nuclei) arranged diagonally south of Amazonia. Small, isolated patches of dry forest occupy dry valleys in the tropical Andes between Colombia and Bolivia, forming what the authors proposed to be havens during drier times and hence called the Pleistocene Arc. However, the key taxon *Anadenanthera colubrina* (a tree related to the mimosa; Fig. 5.11), a species that is either dominant or frequent in all the major dry forest nuclei of South America, does not occur in the admittedly scarce pollen records until after the Pleistocene, 9000 years ago (Maley, 2004).

In the region where the majority of megafaunal fossils have been reported, the midlatitude eastern plains, the markedly abundant subaerial deposits of loess (sediments formed by wind-borne particles) reflect sedimentary environments that are most inappropriate for pollen fossilization as a result of the pollen's susceptibility to oxidation. However, it has been proposed on abundant, multisource evidence that during the last part of the Pleistocene, the plains of Argentina, Uruguay, and the southern part of Rio Grande do Sul, Brazil, experienced a cool, dry climate (Clapperton, 1983; Tonni, 1985a; Prado et al., 1987; Alberdi et al., 1989; Iriondo and García, 1993) similar to that of modern northern Patagonia at about 40°S. Moreover, on the basis of an analysis of the isotopic composition of the enamel of high-crowned ungulate teeth, MacFadden et al. (1996) proposed a mixed ecosystem, such as wooded grassland. Finally, palynological evidence, although scarce, suggests mainly herbaceous psammophytic steppes (i.e., those in which plants grow in sand or sandy soil) for the southern part of Buenos Aires Province, dominated mainly by hard grasses, associated with xerophytic (i.e., related to a dry, arid habitat) woodland in the southwestern part (Prieto, 1996).

Another interesting consequence of past vegetation can be observed today in Janzen and Martin's (1982) "neotropical anachronisms." Some plants have been proposed to have exhibited what is referred to as a syndrome of dispersion by megaherbivores — that is, the fruit is eaten by these mammals, but the contained seeds pass through their digestive tract to be deposited at some distance from where the fruit was eaten. Such trees and shrubs display a set of shared fruit and seed traits that is puzzling if examined only in the context of the extant potential native dispersal agents (as in the example of the palm fruits presented by Janzen and Martin, 1982). Hence, these authors proposed the megafaunal dispersal syndrome, according to which the plants have specific features, some of which are detailed in Box 5.3 and elaborated upon by Guimarães et al. (2008).

5.11. *Anadenanthera colubrina.*

Photograph courtesy of Ludmila Profumo and Andrés Berruti.

Janzen and Martin's (1982) features of plants with the megafaunal dispersal syndrome

Box 5.3

1. The fruits are large and indehiscent, i.e., they do not open on their own to disperse seeds, and they contain a nutritive sugar-, oil-, or nitrogen-rich pulp.
2. The fruits look, feel, and taste like those eaten by large African mammals. Their seeds or nuts also resemble those eaten by African large mammals.
3. The large nuts or seeds are usually protected by a thick, tough, or hard endocarp or seed coat that usually allows them to remain intact as they pass through the molars and digestive tract when eaten by introduced large mammals, such as horses or cattle.
4. If the seeds are soft or weak, they are very small or imbedded in a hard core or nut. Fruits with soft seeds may also contain seed-free hard sections in the pulp or core that block occlusion of the molar mill—that is, the battery of grinding teeth of the animal eating them.
5. Many of the fruits fall off the tree upon ripening (even well before they ripen); this is best described as behavioral presentation of fruits to earthbound dispersal agents.
6. The fruits usually attract few or no arboreal or winged dispersal agents such as bats, guans, or spider monkeys. If these animals happen to be attracted, as they are to figs or the Asian *Spondias* fruits, they cannot eat them because of the large size of the fruit.
7. In present-day forests, a high proportion of a tree's fruit crop rots in the tree or on the ground beneath it without being taken by any potential dispersal agent. This is true even in those national parks where sizable wild vertebrate populations may equal or exceed their pre-Columbian densities.

Box Fig. 5.3. Tree and fallen fruits of *Butia yatay*, showing some of the features described.

Photograph courtesy of Eduardo Marchesi.

8. Peccaries, tapirs, agoutis, and small rodents usually act as seed predators and dispersers of these trees.

9. The fallen fruits are avidly eaten by introduced large mammals. Free-ranging populations of these animals at carrying capacity normally consume all of the fallen fruit in most trees' crops. At least some of the seeds pass through the digestive tract of these animals and eventually germinate. The introduced large herbivores may reenact many portions of the interaction the trees had with the extinct megafauna.

10. The natural habitats (such as alluvial bottoms or gentle slopes) of these trees are on the edges of grassland and in adjacent forest that are likely to be attractive to herbivorous megafauna and usually not on steep rocky outcrops and precipitous slopes.

Several plants in South and Central America show such a syndrome, and some of this flora, yet to be fully studied, may face extinction in the future as a result of the lack of dispersers.

The idea of the megafaunal dispersal syndrome (Box Fig. 5.3) is useful in explaining many of the aspects of the seed and fruit characteristics we mentioned, but some readers might wonder about several aspects of the hypothesis, particularly the comment that some modern South American flora may face extinction in the future as a result of the lack of dispersers. The question that immediately comes to mind is, how did such flora survive over such a large geographic area after the extinction of the megafauna (the tapir is one exception) 10,000 years ago? As possible solutions, Guimarães et al. (2008) suggested that secondary seed dispersal by small- and medium-sized scatter-hoarding rodents and interaction with humans may have been central to the flora's continued persistence.

Guimarães et al. (2008) characterized the plants that have the megaherbivore dispersal syndrome as belonging to two types. Type I includes fleshy fruits 4–10 cm in diameter with up to five large seeds (generally >2.0 cm diameter), and type II includes fleshy fruits >10 cm diameter with numerous (>100) small seeds. Most of those fruits are drupes or drupaceous (40.1% of the species), are berrylike (29.9%), or are legumes (18.6%). Contrary to fruit assemblages from different communities, the range of fruit colors of megafaunal species is restricted; they are predominantly brown, brown-red, or brown-greenish (24.8%), green or green-gray (34.5%), green-yellow (12.9%), or different tones of yellow or yellow-green (21.5%)—a marked contrast with the distribution of fruit color frequency in different communities worldwide, which are predominantly black-purple or red.

Other vertebrates

Apart from the spectacular megamammals, other vertebrates are important in the South American fossil record, contributing to our understanding of past biotas and their environments. Freshwater fishes are scarce as fossils in the localities where the late Pleistocene megafauna is found, although some catfish are found in Lujanian sediments in Buenos Aires Province, as well as one perciform (i.e., belonging to the perch group; Cione and López Arbarello, 1995). As these authors pointed out, there is no evidence of the biogeographic pattern of fish having been influenced by the GABI.

Reptile and amphibian remains are scarce in late Pleistocene deposits, but some remarkable finds have come to light (de la Fuente, 1999). Amphisbaenids and colubrids (represented by undetermined genera and species), some lizards (the large teiid *Tupinambis*, the polychrotid *Leiosaurus*), freshwater turtles (Chelidae; *Hydromedusa*, *Phrynops*, *Trachemys*), and giant tortoises (Testudinidae: *Chelonoidis*) are known from Argentina, Uruguay, and southern Brazil. Giant tortoises are particularly remarkable in that they are now completely extinct on the South American mainland, although one species is fortunately still thriving in the Galápagos (Fig. 5.12). Their presence suggests that the climate cannot have been particularly cool, as these ectotherms would have had trouble surviving cold nights.

Some lizard remains have also been described from the late Quaternary Camping Americano locality, near Monte Hermoso, Buenos Aires Province, as belonging to two species of tropidurid *Liolaemus*, one gekkonid (*Homonota* sp.), and one teiid (*Cnemidophorus* sp.). Only the first two species live in the same area at present, which is interpreted as evidence of the climate having been drier and cooler then (Albino, 2005).

Even though the spectacular "terror birds" (a Hollywood-style name that includes the already mentioned running giant birds of diverse phylogenetic origins) had been long extinct, the South American late Pleistocene avifauna also shows interesting features. Indeed, despite the common (and often wrong) opinion that birds are seldom fossilized, they provide a noteworthy context for the megamammals (Tambussi et al., 1993). In general,

5.12. Galápagos tortoise. Tortoises are known for their longevity, and it is possible that among those living today are individuals that may have witnessed the presence of the most famous finch collector in the islands, Charles Darwin. However, the charming story (spread in both the popular press and scientific literature) about Harriet, Darwin's pet tortoise, seems, unfortunately, to be inaccurate. Harriet, supposedly one of the tortoises that Darwin collected and brought back to England on board the HMS *Beagle*, was reported to have wound up in an Australian zoo and lived to the ripe old age of 176, dying in 2006. Bauer and McCarthy (2010) dispelled this tale. These authors verified the existence of records that indicated that Darwin had presented the two tortoises in his possession to the British Museum of Natural History (now the Museum of Natural History) in 1837. One of the specimens, that from James Island (this was Darwin's pet—the other was from Charles Island and was said to be Darwin's assistant Syms Covington's pet) was still in the museum collections, but it had been "hiding in plain sight for over 170 years," having been overlooked and reregistered under a different catalog number (Bauer and McCarthy, 2010:274). After a close inspection of the stuffed specimen, these authors discovered the original number plainly written on the inner surface of the plastron (lower shell).

Photograph taken in October 2005 in Galápagos National Park, Puerto Ayora, Isla Santa Cruz, Galápagos, Republic of Ecuador. Courtesy of Eugenia Del Pino.

as is the case for other groups, South American birds are highly endemic to this continent, having evolved in relative isolation. A few groups have been cited for the late Pleistocene (Tambussi et al., 1993; Alberdi et al., 1995a; Tonni and Cione, 1999; Rinderknecht and Claramunt, 2000; Claramunt and Rinderknecht, 2005): Tinamidae (the distinctly South American, primitive tinamous, Fig. 5.13), Rheiidae (South American ratites), Anatidae (ducks), Cathartidae (New World vultures), Rallidae (moorhens), Strigidae (owls), Charadriidae (plovers), Ciconiidae (storks), Furnariidae (ovenbirds), Phalacrocoracidae (cormorants), Psittacidae (parrots), and Picidae (woodpeckers). Some of these forms are found today in open and relatively dry environments, while others (ducks, charadriforms) indicate aquatic conditions; and yet others suggest the presence of forested areas. It is likely that the record of the avifauna represents regional differences and varied time intervals during the Pleistocene, and thus indicates various habitats.

Small mammals

According to Pardiñas et al. (2002), the Pampean region includes the most complete fossil record of sigmodontine rodents, the main source of micromammal remains. The oldest, fragmentary specimens come from the Monte Hermoso Formation, dated as 5–4 Mya, southwest of Buenos Aires Province, although some possibly Miocene species were found in the Cerro Azul Formation in La Pampa Province.

During the Pliocene, the diversity of the group increased remarkably, with the appearance of both now-extinct and extant species. Some genera of rather large size for the group (i.e., >60 g) suggest the prevalence of arid or semiarid conditions, although with short periods of warm and humid conditions during the late Pliocene and early Pleistocene.

Some early Pleistocene assemblages, like that from the Ramallo Formation, support the hypothesis that the species structure of the modern sigmodontine fauna in Buenos Aires Province was established during the Ensenadan. These faunal changes, accompanied by those in other groups, were proposed to have been climatically controlled (Cione et al., 2007). In the late Pleistocene, the information is limited but suffices to provide evidence for the dominance of desert and semidesert environments in the last 0.5 Mya.

The other group of rodents is Caviomorpha (see Chapter 4). The smaller species (such as tucu-tucos and guinea pigs) are well represented in late Pleistocene sediments, with a taxonomic composition different from that of Pliocene times (Vucetich et al., 2005). They again indicate dry environments, but further refinement of their chronology is needed to make confident use of their presence in reconstructing the environment.

Other small mammals of note in the fossil record of South American Cenozoic sediments are Marsupialia. During part of the biochron of the clade, there were several species that cannot be called small, such as the early Tertiary bear-sized borhyaenids and the impressive marsupial sabertooth in the Mio-Pliocene. However, in Pleistocene sediments, marsupials are represented only by groups of small body size. In the late Pleistocene, the didelphids, i.e., the group including opossums, are predominant among them. According to Goin (1995), marsupials are not good environmental indicators because most of them are generalized.

5.13. Great tinamou, *Tinamus major.*

Engraving by George Edward Lodge in Evans (1899).

The smaller forms of Xenarthra, especially the armadillos, are ubiquitous in Pleistocene sediments, the result of the presence of armor, a feature that increased their chances of fossilization. In the Lujanian in particular, large forms are found, such as the 50-kg *Propraopus grandis* and *Eutatus seguini*, together with more modern species.

Apart from the remarkable sabertooth and the bears, which will be treated in greater detail in Chapter 6 (the Bestiary), Carnivora are also well represented among the smaller species. Felidae, Canidae, Ursidae, Mustelidae, Mephitidae, and Procyonidae are known from the start of the GABI, even as heralds, following Simpson's (1980) terminology. According to evidence analyzed by Soibelzon and Prevosti (2007), the diversity of fossil and Recent carnivorans in South America may be a consequence of several independent immigrations at different taxonomic levels, and the corresponding radiation once those taxa arrived in the continent.

The tracksites at Pehuén-Có

Seriously threatened by human activity, the paleoichnological site of Pehuén-Có (Fig. 5.14) is one of the world's largest deposits of mammalian fossil tracks. It is located near Bahía Blanca, a place visited by Darwin in 1831, on the Atlantic coast of Argentina (Fig. 5.15); it includes various ichnites of birds and mammals (Aramayo and Manera de Bianco, 1987). Along several hundred meters, the clay, at times now covered by the sand of the beach, was used as a passageway by many of the animals we discuss in this book. Although well known to scientists, the site became prominent when Teresa Manera won the Rolex prize for her work in studying and conserving it. Moreover, a recent episode of a popular local TV series (*Serie Documental Pueblo Mio*, Channel 9, Buenos Aires) dealt with this remarkable site.

When the tracks were formed, probably in late Pleistocene times, the place was likely a lake. Aramayo and Manera de Bianco (1987) suggested that the megafauna stepped on the silty clay, forming prints that dried and were covered over by new sediments. There are hundreds of Pleistocene tracks of extinct and extant animals. The most impressive record is a 35-m sequence of 35 footprints made by an individual of the giant ground sloth *Megatherium*, but other ground sloths and large mammals such as camelids, macrauchenias, bears, and mastodons are represented; there is also a single print of a glyptodont. Other species have also been found, such as deer, guanacos, pumas, canids, and a varied assemblage of birds: rheas, flamingoes, grouse, and waterbirds.

Of particular interest, the series of 35 footprints of *Megatherium* found at this location allowed calculation of the speed at which that animal moved, about 2 m per second, the equivalent of a quick pace for a person (and for the animal as well; Alexander, 1976). The locomotion of this peculiar mammal has long been the subject of speculation. In 1912, Abel was among the first researchers to imagine *Megatherium* as a biped. Cabrera (1929a:439–440) thought that *Megatherium* was not usually bipedal, although he was convinced that "Debía alzarse con frecuencia sobre las extremidades posteriores, como suelen hacerlo todos los mamíferos unguiculados de marcha plantígrada o semi plantígrada y, especialmente todos los xenartros" ["It must have raised itself frequently on its hind limbs,

Taphonomically interesting sites and specimens

5.14. Professor Neill Alexander analyzing the *Megatherium* ichnites in Pehuén-Có.

5.15. Map of Pehuén-Có, Buenos Aires Province, Argentina, where the footsteps assigned to *Megatherium americanum* were found.

Drawn by Sebastián Tambusso,

much the way all plantigrade or semiplantigrade ungulates, especially xenarthrans, occasionally do"].

The reconstruction of the track maker's locomotion, as explained by Aramayo and Manera de Bianco (1996), is that the animal was walking bipedally and briefly adopted a quadrupedal posture as a result of the difficulty of moving in soft mud (Blanco and Czerwonogora, 2003), as discussed in Chapter 7.

Unusually well-preserved specimens and coprolites at Última Esperanza

Since its discovery during the final years of the nineteenth century, the cave at the ends of Última Esperanza Sound in Chilean Patagonia (51°36'S; 72°36'W; Plate 7) has attracted the attention of many scientists and the general public alike as a result of its exceptional fossil contents: the dung and hide of a ground sloth belonging to the genus *Mylodon* and perhaps to the species *M. listai*.

In 1895, the German merchant captain Hermann Eberhardt (1852–1908, Fig. 5.16), already based in Puerto Consuelo, in the Magallanes Region of southern Chile, since 1892, and after considerable exploration along the coast of that impressive region, found this material when he was exploring the neighboring area. Some years later, enough material had been dug out, including the famous mummified hide with long, coarse hair (Fig. 1.9), to foster the hope of finding the animal alive in those isolated regions

(Ameghino, 1898). Unfortunately, the animal had been long dead, as shown by radiocarbon dates of between 13,560 ± 180 years ago and 10,200 ± 400 years ago (Tonni et al., 2003). However, this promoted new speculations, with some even suggesting domestication of this ground sloth, dubbed "the mysterious mammal of Patagonia, *Grypotherium Domesticum*" (Hatcher, 1899). The site has since been studied as one of the most interesting sources of paleobiological information on giant sloths. The dung of such beasts has provided information on their parasites and—based on included pollen— diet and habitat. Samples of hide have allowed studies of their molecular phylogeny.

As a final note on this site, there is a literary mention of its discovery in Bruce Chatwin's *In Patagonia* (1977). Chatwin was aware of the site because of an ancient seafaring relative's supposed procurement of a piece of *Mylodon* hide from the cave. Indeed, the cave was among Chatwin's inspirations in undertaking his quest to Patagonia. The result, as anyone who has read his legendary book knows, is among the most evocative travel stories ever written.

5.16. Captain Hermann Eberhardt in 1899. Seaman, explorer and diplomat.

Courtesy of the Eberhardt family.

Interactions between humans and megafauna at Estancia La Moderna

Unlike North America (Dillehay, 1999), only a few South American sites have evidence of people having exploited the extinct megafauna: Taima-Taima, a 13,000-year-old mastodont kill site in Venezuela; Fell's Cave and the surrounding Pali Aike and Cerro Sota in southern Chile, where sloths and fossil horses are found along with human remains; Quereo, near Los Vilos in central Chile, whose two cultural levels are accompanied by extinct fauna (sloths, mastodonts, *Palaeolama* sp., *Antifer* sp., and *Equus* [*Amerhippus*], among others); Tagua-Tagua, also in central Chile; and the promising Monte Verde, with levels dated as at least 13,000 years ago in south-central Chile.

Despite the evidence of direct interactions between Paleoindians and megafauna being elusive (Borrero, 2009), an interesting site in Buenos Aires Province can be added, namely the excavation in Estancia La Moderna, which showed that the place was once occupied by humans. Evidence from the upper units suggests the presence of historically recent native groups, while the middle unit, likely formed in mid-Holocene times, preserves human-made tools. However, the most interesting level (at least for our purposes) was extensively described by Politis et al. (2003). In the lower part of unit Y and in the following unit S, scrapers and other large archaeological items were found associated with some of the extinct mammals that are the subject of this book: the fossil camelid *Palaeolama*, the fossil horses *Equus* (*Amerhippus*) and *Onohippidium*, an armadillo, a fossil deer, the notoungulate *Toxodon*, two ground sloths (*Megatherium americanum* and *Glossotherium robustum*) and *Macrauchenia*, together with the guanaco, a camelid that is still living and was a preferred prey of the hunters that apparently inhabited the area. Also, the lower component of La Moderna has been interpreted as a procurement site, where during a single event, the glyptodont *Doedicurus clavicaudatus* was butchered at the edge of an ancient swamp using tools of crystalline quartz.

Despite the preference for guanaco, and rather important as a result of the scarcity of similar sites, *Megatherium* and the fossil horses were also consumed by those ancient inhabitants of the pampas. Those animals are represented at the site only by the more palatable parts, the forelimbs and hind limbs. This is interpreted as indicating that the humans, after having hunted the animals or found their carcasses, transported the most profitable anatomical parts, leaving the other, less attractive bits behind. Moreover, the fracture pattern of the bones found there is consistent with those made by humans breaking large bones; cut marks on muscle attachment surfaces were also present.

The chronology of this site, which was also used as a funeral mound, is still unresolved. The radiocarbon date of 8390 ± 140 years ago that was obtained from one of the bones of *Megatherium* was then corrected by using new techniques and found to be of Pleistocene (i.e., older than 10,000 years), rather than Holocene, age.

Interactions between humans and megafauna at Piedra Museo

Miotti (1992) studied another interesting site preserving evidence of interactions between humans and megafauna, that of Piedra Museo, a Patagonian rock shelter in Santa Cruz Province, Argentina. This locality is composed of six units, with the lowest of them, U6, dated as between 12,890 ± 90 years ago and 10,925 ± 65 years ago. This unit contains a mixture of extinct and extant species, such as the two species of living South American rheas, volant birds, guanaco, rodents, and a canid, as well as a few ribs and a portion of the skull of the ground sloth *Mylodon*, and some remains of a fossil horse and of *Lama* (*Vicugna*) *gracilis*, a small extinct camelid. Cut marks were observed on the bones belonging to guanaco and horse.

According to Miotti, the landscape surrounding the Piedra Museo locality was a place for hunting activities: the ambush, killing, and butchering of game. Hunter-gatherers would have found resources year round, thanks to a paleolagoon approximately 50 m away from the rock shelter, which was situated in the center of a smooth hilly area that concentrated abundant game and subsistence resources.

Interactions between humans and megafauna at Arroyo del Vizcaíno

During a severe drought in the austral summer of 1997, local enthusiastic students of the secondary school of Sauce, a small town 40 km northeast of Montevideo, Uruguay (and thus even less distant from where the first described glyptodont was found in 1838 by Vilardebó and Berro—see Chapter 6), recovered a number of fossil bones of Pleistocene megamammals from the riverbed of the nearby Arroyo del Vizcaíno. This stream is located at latitude of 34° 35'S and longitude of 56°03'W. A few weeks later, when normal rainfall levels resumed, this fossiliferous locality was covered by water, and hence the sediments and the rest of their contents became inaccessible again. This site is typically composed of reddish brown deposits of reworked loess with tectosilicate grains. From a geomorphological perspective, the site is a place where the stream becomes deeper, forming a natural pond on a substrate of Cretaceous sediments (Mercedes Formation).

The collected material includes skeletal remains of the ground sloth *Lestodon* (minimum number of individuals = 5) and of three genera of glyptodonts (*Glyptodon, Doedicurus,* and *Panochthus*), identified from carapace remains and an isolated tooth. As described by Arribas et al. (2001), it consists, apart from some teeth, of 979 specimens (36% identifiable), 36 (3.7%) carapace fragments of glyptodonts, and 587 (60%) fragmentary, unidentifiable bone fragments. Among the 356 identifiable specimens, most are rib portions (42%), followed by limb bones (23%) and vertebrae (also 23%), while cranial remains are about 10% and phalanges constitute a modest 0.6%. Except for phalanges, the proportion of anatomical regions represented in the assemblage are similar to those of large mammal bones exposed on the surface of Amboseli Park, Kenya, and studied by Behrensmeyer and Dechant Boaz (1980).

The bones seem to have undergone little, if any, transport before they were buried, according to current taphonomic criteria. Indeed, the ratio of isolated teeth to vertebrae (Shipman, 1981) is only 1:6, and the grouping of bones according to their transportability excludes the possibility of fluvial transport. As studied in the classical work by Voorhies (1969), when the frequency of heavier, difficult-to-move elements is high and that of the lighter, more transportable elements is low, it can be concluded that the watercourse took away the latter and left the former relatively untouched. This is the case at this site, as the frequencies are 67% for the least transportable elements belonging to group I, 23% for the intermediate group II, and 10% for most transportable group III.

The material also seems to have undergone little exposure to the elements before burial, as the outer bone surface shows a low degree of wind abrasion, and this is restricted to only one of the surfaces. Moreover, in some cases, the limbs of the giant ground sloth were in anatomical connection. This context became particularly important when Alfonso Arribas, from the Museo Geo Minero of Madrid, Spain, visited the site and found a clavicle with interesting marks, a story that will be discussed in the context of the megafaunal extinctions in Chapter 9.

lnud anımal eſt mılo fuuıo q̄d
dīcıtur̄ hydrus. phiſiologuſ dīcıt
de eo quod ſerıꝰ eſt hoc anımal ınımıcī
cocodrıllo. et hanc habet natūaṁ ꝛ con
ſuetudıneṁ. ōtū under cocodrıllūm
ın lıttorıbus fluṁınıſ dornıentem aꝑꝛo
oꝛe nadıt ꝛ ınuoluıt ſe ın luṁun luꝛ q̄

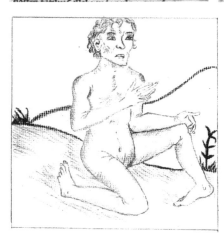

Bestiary

6

The creatures—some based on real beasts, others thoroughly imagined and mythological—adorning medieval bestiaries were meant to inspire solace, astonishment, and awe in readers; but bestiaries were ultimately allegorical and thus spiritual texts, rather than attempts to accurately portray the natural history of the included creatures (elephants, for example, were said to have no desire to copulate and to represent Adam and Eve; White, 1960). The animals were created, according to medieval thought, by God to provide people with food for the body and nourishment for the soul. The books on these animals were written by people for their own edification: the texts and images of bestiary animals embodied characteristics of Christ or the devil, and they usually had a moralizing purpose (Hassig, 1995). Bestiaries commonly organized animals into groups—mammals, birds, serpents, worms, and fishes, with priority often given to wild animals and emphasis placed on the power of their symbolism. The lion, for example, leads off (Clark, 2006), because of its strength, its status as kingly, and its other purported parallels with Christ: its ability to conceal its scent (as Christ concealed his love of mankind until summoned by the Father), its ability to sleep with its eyes open (as did Christ, who was crucified and buried, yet never died), and the ability of the female lion to give birth to dead cubs that were brought to life after three days by their father's breath (this one is too obvious to require explanation) (White, 1960).

Our modern bestiary on Lujanian mammals contains facts and figures of Lujanian beasts that are sure to produce similar emotions in modern readers; indeed, some of them are even weirder than those depicted in the aged manuscripts by imaginative monks; and whereas the griffon, unicorn, and bonnacon (its incurled horns being inadequate for protection, it defended itself by flatulating and expelling excrement that covered three acres, driving away pursuers and setting alight unfortunate trees within range) were imaginary, the Lujanian mammals were once real flesh-and-blood beasts. Although we cannot promise that their stories are balms for the soul, the following snapshots of past life are fodder for the imagination of interested readers: these portentous beasts ambled over what today are large cities in South America adorned with obelisks, cars, shopping malls, and fast food joints—as well as the museums that preserve and display their remains. Perhaps, on second thought, they might serve an allegorical purpose after all.

In this chapter, we will explore the Lujanian megafauna in all its fullness and diversity. Each section presents an account of the more representative genera, including anatomy, general life habits, body size, diet, and topics concerned with their inferred habits. (A more rigorous consideration of the biomechanics of some of these beasts is presented in Chapter 7.) The diagrams provide a reconstruction of the life form of these mammals or illustrations of their skeletal elements, and each is accompanied by generic and common names, and a classification. Whereas the medieval bestiaries were organized largely on the basis of strength of symbolism, our Lujanian bestiary is arranged according to the mammals' phylogenetic relationships as based on modern methods of scientific analyses. Thus, because we are dealing with a modern understanding of mammals, it is important that before we present the mammals themselves, we first consider the results of the analyses that inform us about the relationships of these mammals, as well as the characteristic features of the groups to which they belong.

Relationships among eutherian mammals

Recent studies of eutherian phylogeny based on molecular data (Madsen et al., 2001; Murphy et al., 2001a,b; Delsuc et al., 2002; Springer et al., 2004; Delsuc and Douzery, 2008) have clarified the main relationships among living placental mammals, resolving them into four main clades grouped in twos (although it should be noted that other results, such as those of Goloboff et al., 2009, include xenarthrans among Euarchontoglires). The four main clades are as follows:

Laurasiatheria. Carnivorans, pangolins, cetartiodactyls (including artiodactyls and cetaceans), perissodactyls, bats, hedgehogs, moles and shrews.

Euarchontoglires. Rodents, rabbits, tree shrews, flying lemurs, and primates.

Afrotheria. Elephants, hyraxes, manatees and dugongs, aardvarks, golden moles, tenrecs, and elephant shrews.

Xenarthra. Armadillos, anteaters, and tree sloths.

Laurasiatheria and Euarchontoglires are grouped as Boreoeutheria, while Xenarthra and Afrotheria are branches of Atlantogenata. Although these clades are well supported, some issues have yet to be resolved for both living and fossil groups. For example, it has become clear that Xenarthra is a natural or monophyletic group, an assertion widely corroborated by both molecular and morphological studies (Patterson et al., 1989, 1992; Gaudin, 1993, 1995, 1999b; Szalay and Schrenk, 1998; Gaudin and Wible, 1999; Czelusniak et al., 1990; DeJong et al., 1993; Cao et al., 1998; Stanhope et al., 1998; Van Dijk et al., 1999; Delsuc et al., 2001; Eizirik et al., 2001; Madsen et al., 2001; Murphy et al., 2001a). It has also become clear that it represents one of four major placental clades with origins in South America. However, a consensus has not emerged on the affinity of Xenarthra to other (both living and extinct) mammalian groups. Neither is it entirely clear where litopterns and notoungulates fit within this scheme, because they have no living representatives, essentially precluding molecular analyses, though hope remains that DNA may be recovered from their more recently extinct members. Our understanding of the relationships of litopterns and notoungulates must thus still be based on morphological evidence, but the phylogenetic implications of morphology are clouded by the long isolation of these groups—recall that they are among the native South American ungulatelike mammals that evolved independently from other, true ungulate groups, comprising Euungulata and classified among Laurasiatheria.

The South American ungulates (Litopterna, Notoungulata, Astrapotheria, Pyrotheria, and Xenungulata; see Box 3.1) have for several decades
been commonly grouped together as Meridiungulata (McKenna, 1975;
McKenna and Bell, 1997), implying a common origin for these forms, a
not illogical supposition given that they, like other South American mammals, evolved largely in isolation. However, support for such a group does
not appear to be particularly strong, with several authors having considered
them related to different groups of "condylarthrans." The latter is an assemblage comprising several lineages of generalized archaic ungulatelike
placental mammals (completely extinct groups such as phenacodonts,
periptychids, mioclaenids, and didolodontids) that include the earliest and
most generalized representatives of more recent ungulate groups. In the
past, these "condylarth" lineages were included in a formal group Condylarthra, implying that all placental ungulates ultimately had a common
origin. More recently, we have realized that although some condylarths
may indeed represent the earliest members of some modern ungulate
lineages, the condylarth lineages do not form a natural group, and thus
some extinct placental ungulates do not have a common origin with the
modern ungulates.

As these considerations apply to our discussion of South American
ungulates, Horovitz (2004) recognized a monophyletic Litopterna (as has
long been considered) that comprises the sister group of notoungulates and
that together, these two South America clades are related to phenacodontid
condylarths. On the other hand, Astrapotheria, a Paleogene group mentioned in Chapter 3, is more closely allied to periptychid condylarths, and
hence is unrelated to Notoungulata and Litopterna. Conversely, Muizon
and Cifelli (2000) defended a relationship between Litopterna, the South
American didolodontid and the North and South American mioclaenid
condylarths, a group they named Panameriungulata. These authors did
not preclude inclusion of notoungulates in this group on the basis of postcranial features. Also, among these South American ungulates, a strong
link between pyrotheres (see Chapter 3) and notoungulates has long been
postulated (e.g., Osborn, 1910; Loomis, 1913; Patterson, 1977), and although
at times questioned, Billet's (2010) recent analysis not only supported a
strong relationship, but also suggested that pyrotheres might actually be
notoungulates as traditionally defined. This author also found support for
a close relationship between these groups and Astrapotheria, but suggested
that Litopterna is not monophyletic.

Atlantogenata is an ancient Southern Hemisphere radiation that includes the South American Xenarthra and the mainly African Afrotheria.
This grouping reflects the hypothesis that the opening of the Atlantic
Ocean, which caused the breakup of Gondwana, isolated one branch of
this lineage in South America and the other in Africa during the early
Cretaceous. The fossil remains and recent molecular phylogenies bolster
suggestions that placental origins must be sought in the former Gondwanan
landmasses (Hunter and Janis, 2006), although the path of the successive
origins of higher-level clades is yet to be completely understood. This biogeographical scenario leaves open the possibility that Meridiungulata (if
they indeed do form a clade, possibly with at least some condylarthrans) are
not part of the lineage of other ungulates and should instead be considered

a clade within Atlantogenata that retained the ancestral ungulatelike form, in the same way that perissodactyls and artiodactyls conserved generalized ecomorphology in Eurasia. Thus, although the monophyly of Meridiungulata remains to be confirmed (and neither can we preclude the possibility that Meridiungulata is related or belongs to Boreoeutheria), there are biogeographic reasons to view it as the sister group of Xenarthra.

Chimento and Agnolin (2010) and Agnolin and Chimento (2011) recently suggested yet another possibility: litopterns and didolodontids are related to mioclaenid condylarths (and thus may ultimately have a northern origin), whereas the remaining South American ungulates are related to afrotheres (and thus ultimately have a southern origin); on the other hand, these authors favored a relationship of Xenarthra to Boreoeutheria, whereas xenarthans are more generally considered as related to afrotheres (in the group termed Atlantogenata). Among the characters Chimento and Agnolin (2010) and Agnolin and Chimento (2011) cited as uniting the South America ungulates and afrotheres is the presence of more than 19 thoracolumbar vertebrae. We may note, however, that some xenarthans (such as the Miocene sloth *Hapalops*) also have more than 19 thoracolumbar vertebrae.

Clearly, there is considerable confusion, and much remains to be resolved. A better understanding of the positions and origins of the South American ungulates depends on such resolution. In light of the new molecular hypotheses, more rigorous morphological investigation is required on the higher-level relationships between xenarthrans and afrotheres, as well as between meridiungulates and afrotheres and xenarthrans. In addition, these possibilities need corroboration by molecular studies. The potential of finding ancient DNA in the latest representatives of Litopterna and Notoungulata, which became extinct as recently as about 10 kya, remains an exciting possible source of such corroboration.

For our purposes in this book, we will follow the biogeographical scheme that views litopterns and notoungulates, whose relationships are far from being satisfactorily resolved, as probably part of the great southern radiation of mammals, and include them in Atlantogenata related to the afrotheres or, more interestingly, xenarthrans. Although this is provisional, so are the other possible scenarios outlined above. In any event, there are only two endemic South American ungulate groups—the litopterns and notoungulates—with which we need to concern ourselves, because only these two survived to the late Pleistocene.

Litopterns: camel and horse lookalikes

Litopterns are a diverse and abundant clade, already present in early Paleocene sediments (Bonaparte and Morales, 1997), that achieved its highest taxonomic richness in the late Miocene (Bondesio, 1986; Bond, 1999). Litopterns gradually diminished in taxonomic diversity throughout the Pliocene, with forms that became progressively more specialized, until their extinction in the late Pleistocene–early Holocene (Bond et al., 1995). Most authors consider them to be monophyletic (Cifelli, 1993; Bergqvist, 2010), in contrast to Billett's (2010) views. Bergqvist (2010) included Protolipternidae, along with the more commonly represented Prototheriidae and Macraucheniidae (as well as other clades like Adianthidae and Notonychopidae) as litopterns.

Macraucheniidae were large to very large animals, with an inferred general appearance resembling living camels or llamas (Scott, 1913; Bond, 1999). Among them, we may note the general evolutionary trends beginning with slightly retracted narial openings (i.e., nostrils in about the typical position) and a masticatory apparatus that includes a complete dentition, no diastema, and brachydont cheek teeth to markedly retracted narial openings (suggesting the presence of a proboscis or trunk) and slightly high-crowned cheek teeth of finite growth (protohypsodont). Proterotheriidae were medium- to large-sized mammals generally considered similar to primitive holarctic (i.e., Northern Hemisphere) horses as a result of the convergences of reduced digits and mesaxonic toes (see Chapter 4; Appendix 1), but without ever evolving hypsodont teeth (Bond, 1999).

Notoungulates constitute the most abundant and diverse clade of endemic South American ungulates, both taxonomically and morphologically (Simpson, 1936; Patterson and Pascual, 1972; Cifelli, 1993; Croft, 1999). Following Billet (2010), the clade comprises two main monophyletic groups, Toxodontia, large to very large beasts known from the Paleocene to the late Pleistocene (Ameghino, 1907; Scott, 1912; Bond, 1999), and Typotheria, small- to medium-sized, mostly rodentlike forms, known from the late Paleocene to early Pleistocene (Fig. 6.1). Toxodonts are often compared to hippos or rhinos because of their inferred general appearance and the grinding pattern of their molar crowns. Among them are such clades as Homalodotheriidae (rather large forms), Isotemnidae (also rather large, with a complete dental formula), Leontiniidae (such as the cow-sized *Leontinia*; perhaps this name was not the wisest of choices—at least not for domestic harmony, as Ameghino named it after his wife), Notohippidae (which resembled horses, as the name, meaning "southern horses," suggests, and among which the evolution of dental adaptations for eating coarse food is apparent) and Toxodontidae (with large-sized species, among which is the mighty *Toxodon*). The homalodotheriids may be singled out here for their odd combination of features, serving as another example of why paleontologists cannot reconstruct entire animals from a few bits of their remains. *Homalodotherium* (Fig. 6.2), a Santacrucian genus, had perfectly typical ungulatelike teeth (its name means "even-toothed beast" on account of its continuous and evenly graded teeth), but it bore claws at least on its front feet (Simpson, 1980). As well, it walked apparently on the balls of its front feet (digitigrade), but on the entire sole of its hind feet (plantigrade). This combination of features is our second example of a clawed mammal with teeth clearly used on vegetation, the first being the chalicotheres, members of Perissodactyla discussed in Chapter 4. In addition, these different mammals furnish us with another example of the presence of independently evolved characters, or convergence (see Chapter 4). Simpson (1980), following Riggs's (1937) opinion, noted that *Homalodotherium* was capable of digging in search of food and pulling down branches to feed on fruit or foliage. More recently, Elissamburu's (2010) morphofunctional analyses suggested that *Homalodotherium* may have had bipedal habits for browsing and defense, but digging behavior was not supported by her results.

Notoungulates: a menagerie all in themselves

6.1. Typotheria Hegetotheriidae *Paedotherium*. One of the few meridiungulates to survive the Pleistocene.

From Elissamburu and Vizcaíno (2005), by kind permission of Asociación Paleontológica Argentina.

Although typotheres are often characterized as rodentlike in overall form, the group encompasses considerable diversity with clades that resemble small deer, wombats or living capybaras (Mesotheriidae), hares (Hegetotheriidae), and hyraxes (Interatheriidae) (Ameghino, 1889; Sinclair, 1909; Bond et al., 1995; Croft, 1999; Reguero et al., 2007). Toxodonts and typotheres are both characterized by a trend to evolve from a generalized masticatory apparatus with complete dentition, lacking a diastema, and brachyodont cheek teeth to forms including, for instance, hypertrophied incisors, simplified crown patterns, and ever-growing (hypselodont) cheek teeth (Ameghino, 1887, 1894; Scott, 1937; Simpson, 1967; Cifelli, 1985; Shockey et al., 2007).

Xenarthrans: the brightest jewel in the South American crown

Xenarthra, a clade traditionally assigned the rank of order (although Magnorder has become increasingly common after McKenna and Bell, 1997), is a disparate assemblage of 31 extant species grouped into 14 genera (Wetzel, 1985; McKenna and Bell, 1997; Anderson and Handley, 2001; Vizcaíno and Loughry, 2008), including the armadillos, the only living armored mammals; the toothless anteaters, an extreme example of the trend toward feeding on ants and/or termites among mammals; and the slow-moving tree sloths, which spend much of their existence suspended upside down beneath tree branches (Grassé, 1955; Nowak, 1999). The group also includes more than 150 fossil genera (McKenna and Bell, 1997; McDonald and De Iuliis, 2008; Fernicola et al., 2008), the majority of which represent two extensive radiations of large-bodied terrestrial herbivores (Fig. 6.3): the ground sloths, the largest of which approach living elephants in body mass (Casinos, 1996; Fariña et al., 1998), and the glyptodonts, massive cousins of the armadillos, with their thick, mostly immobile armor, graviportal limbs, fused backbones, and odd trilobate dentition (Gillette and Ray, 1981; Gaudin, 1999a). This diverse assemblage of living and extinct xenarthrans has traditionally been united on the basis of their restricted biogeographic distribution and by a suite of unusual morphological features. The clade arose in and has been confined to South America for most of its history

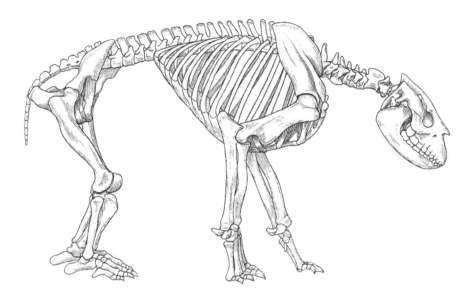

6.2. Reconstruction of the skeleton of *Homalodotherium cunninghami,* based on a specimen housed in the Field Museum of Natural History, Chicago.

Drawing by C. F. Gronemann in Riggs (1937). Reproduced by kind permission of the Field Museum of Natural History.

(Patterson and Pascual, 1972; Simpson, 1980; Pascual et al., 1985), and all but one of the living species is confined to the neotropics (Wetzel, 1985).

As noted by McDonald (2003), the concept of Xenarthra and the features used to diagnose its members has changed considerably from Linnaeus's Bruta. Vizcaíno (2009) summarized the story of Edentata, a name that implies the absence of teeth, although nearly all members that have been grouped under this name do have teeth, though their dentition is typically simplified, reduced, and homodont. Beginning with the French physician and anatomist Félix Vicq-d'Azyr (1792), Edentata included, until a few decades ago, pangolins and aardvarks. In the current usage of Xenarthra for the monophyletic (i.e., natural, meaning it includes a common ancestor and all of its descendents; see Chapter 1) South American clade, there has been a continuing refinement of the selection of morphological characters used to recognize members of the group. Extra articulations between some vertebrae (to which the clade owes its scientific name; *xenarthra* is translated as "strange articulations or joints") and reduction or loss of dentition were among the earliest-used criteria, but now this peculiar group is diagnosed on a suite of apomorphic (or derived, nonancestral) features absent in other mammalian groups (Appendix 2).

Most researchers today support Engelmann's (1985) twofold division of Xenarthra (Fig. 6.3) into Cingulata (armadillos and their extinct kin, pampatheres and glyptodonts) and Pilosa (sloths and anteaters), which is subdivided into Tardigrada (=Phyllophaga=Folivora, the sloths) and Vermilingua (the anteaters). The phylogeny and evolution of the group, both from the point of view of their relationships with other placental taxa and within the group, are discussed below. As well, a brief synopsis of xenarthrans as part of the Lujanian megafauna is provided before the beasts themselves are presented. Appendix 2 presents the more salient skeletal features of Xenarthra, to which the reader may refer during the following discussions.

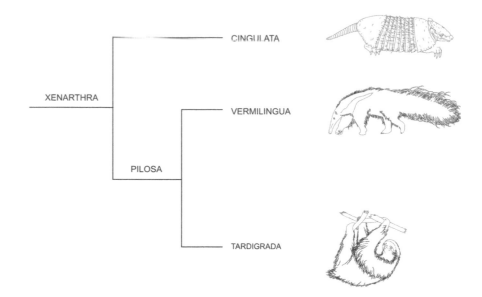

6.3. Relationships among xenarthrans. The main dichotomy is between Cingulata (living and extinct armadillos and their extinct kin the pampatheres and glyptodonts) and Pilosa, including Vermilingua (living and extinct anteaters) and Tardigrada (living and extinct sloths).

Drawing by Néstor Toledo.

Xenarthrans as placentals

The position of Xenarthra among placental mammals was long judged on the apparent archaic flavor of several features of its living members, such as a relatively low body temperature, lack of clear differentiation between the uterus and the vagina, and intra-abdominal position of the testes (Delsuc and Douzery, 2008). For such reasons, xenarthrans were traditionally considered as an early offshoot of Placentalia—that is, that they are basal (though the implication was always that they were "primitive") placentals. Although they have been allied at various times with other groups, both fossil (e.g., palaeanodonts, an extinct group of mammals related to pholidotans, or pangolins, the scaly anteaters of Africa and Asia) and living (e.g., pholidotans and tubulidentates, or aardvarks), most studies do not support such relationships (Delsuc and Douzery, 2008; Gaudin and McDonald, 2008). For a time, the hypothesis of a sister group relationship of xenarthrans to all other placentals (which were placed as a group called Epitheria), reflecting their lowly position among placentals, became fashionable, following McKenna's (1975) analysis, although this view was criticized on morphological grounds and molecular analyses have profoundly altered our understanding of placental systematics.

With regard to relationship to other mammals, molecular-based studies do not support a close relationship between xenarthrans and pangolins. Analyses of mitochondrial DNA sequences have tended to ally xenarthrans with Ferungulata, a clade including carnivorans, artiodactyls, perissodactyls, and cetaceans (Arnason et al., 1997). Studies that used nuclear DNA sequences or a combination of nuclear and mitochondrial DNA sequences position Xenarthra as one of the following: (1) the sister taxon to Afrotheria (which excludes pangolins; Springer et al., 1997; Stanhope et al., 1998; Murphy et al., 2001a,b); (2) the sister group to a clade including all placentals except afrotherians (Delsuc et al., 2001; Eizirik et al., 2001; Madsen et al., 2001); or (3) in a trichotomy with afrotherians and a clade encompassing the remaining placental groups—that is, an arrangement that effectively indicates uncertainty over the relationships among the three groups (Eizirik et al., 2001; Madsen et al., 2001; Murphy et al., 2001a,b). More recently, Prasad et al. (2008) placed Xenarthra with Afrotheria (which is an endemic African clade) in Atlantogenata, a name reflecting its origin on the southern continents that were separated by the spreading Atlantic Ocean. This is the scheme that we follow here. Other studies (e.g., Gaudin and McDonald, 2008) have proposed novel ideas on their relationships, such as a link with carnivorans.

We should note also the possible affinities with the Mesozoic South American fossil mammals known as gondwanatheres and sudamericids. Initial work on these peculiar mammals, represented almost exclusively by dental remains and lower jaws, suggested that they were allied to Xenarthra (Scillato-Yané and Pascual, 1985; Bonaparte, 1986, 1990). The early age of these fossils (Late Cretaceous–middle Paleocene) led to the resurrection of Thomas's (1887) idea that xenarthrans might represent a third radiation of therian mammals, termed Paratheria, independent of Eutheria (placentals) and Metatheria (marsupials). However, this notion was largely abandoned when subsequent work linked gondwanatheres and sudamericids to multituberculates.

Relationships among xenarthrans

Several molecular studies of higher-level relationships among placental mammals have included representatives from all three xenarthran groups (Cao et al., 1998; Van Dijk et al., 1999; Madsen et al., 2001; Murphy et al., 2001a,b), and at least one such study focused largely on the relationships of the groups within Xenarthra (Delsuc et al., 2001; Greenwood et al., 2001). All but one of these studies (Cao et al., 1998) has supported the monophyly of Pilosa. In addition, analyses combining morphological and molecular data support a monophyletic Pilosa (Liu et al., 2001; Delsuc and Douzery, 2008).

VERMILINGUA (THE ANTEATERS)

The comprehensive analysis by Gaudin and Branham (1998) on the phylogenetic (or genealogical) relationships within the highly specialized Vermilingua, based on cranial and postcranial skeletal characters, corroborated Engelmann's (1985) hypothesis that anteaters comprise a monophyletic group. The fossil taxa *Ernanodon* from the Paleocene of China (Ding, 1979) and *Eurotamandua* from the middle Eocene of Messell, Germany (Storch, 2003), have been considered as primitive anteaters related to xenarthrans. Most of the recent interest in vermilinguan systematics has focused on the phylogenetic affinities of *Eurotamandua* (Storch, 1981; Storch and Haubold, 1989). Two subsequent analyses suggested that *Eurotamandua* is not, in fact, a xenarthran. Rose (1999) allied *Eurotamandua* with palaeanodonts on the basis of its forelimb anatomy. Szalay and Schrenk (1998) removed *Eurotamandua* to its own order, the Afredentata, and suggested that it may share a distant relationship with both Xenarthra and Palaeanodonta via an unknown fossil ancestry in the early Cenozoic of Africa. More definitive resolution of this controversy may depend on the discovery of additional *Eurotamandua* specimens, or on a more robust understanding of the relationships among Xenarthra and other groups such as Palaeanodonta and Pholidota. The status of *Ernanodon* has not been satisfactorily resolved. The most pertinent argument against its placement among xenarthrans is its lack of xenarthral joints, which seems a rather telltale absence. It is also unclear whether the skeletal remains of this taxon belonged to a single individual (Horovitz, 2003).

McDonald et al. (2008) remarked on the numerous distinctive modifications of the vermilinguan skeleton and soft anatomy as indicative of adaptive specializations for feeding on colonial insects, including a curved skull base, the presence of a long tongue, and several forelimb and hind limb adaptations for digging. Apart from controversial fossils such as *Eurotamandua*, the fossil record of anteaters is scarce, with the most ancient undisputed remains recovered, surprisingly, from marine sediments of Colhuehuapian age (late Oligocene to early Miocene) of eastern Patagonia (Carlini et al., 1992). Several later species are known from the Santacrucian (McDonald et al., 2008) and later ages. The study of their phylogeny (McDonald et al., 2008) is in preliminary stages, with only a few genera considered, but the three living species are considered to be somewhat distantly related.

TARDIGRADA (THE SLOTHS)

Most people would presume (quite naturally) that the living tree sloths are each other's closest relatives (and would thus comprise a monophyletic group): they share the apparently unique habit of moving upside down, suspended from trees, as well as many other traits compared to all other living (and most extinct) mammals, so there would seem to be no reason to suspect otherwise. Specialists, however, have recognized for decades, following the work of Patterson and Pascual (1972), that this might not be the case, and many researchers consider them diphyletic—that is, derived from relatively distant ancestors. This would mean that the similarities between the two genera of tree sloths were derived independently (i.e., convergently, as was noted in Chapter 4 for many of the sabertooth specializations of sabertooth carnivorans and a marsupial).

Gaudin (1995:674) asserted that the question of tree sloth monophyly versus diphyly is "inextricably linked" to questions of how the various clades of ground sloths are related to one another. Several studies subsequent to Webb (1985b) have provided data relevant to resolving these questions, but all suffer either from limited taxonomic scope (De Iuliis, 1994; Höss et al., 1996; Poinar et al., 1998; Greenwood et al., 2001; White and MacPhee, 2001) or a limited character base (Gaudin, 1995). Gaudin's (2003) comprehensive cladistic analysis, based on craniodental data and including representatives of all the major extinct sloth subgroups and the two tree sloth genera (Fig. 6.4), corroborated the alliance of *Choloepus* with megalonychids (as suggested by Patterson and Pascual, 1972) and placed *Bradypus* as a sister taxon to all other sloths (in contrast to most previous views that allied this genus with megatheriids). If Gaudin is right, it means all sloths, including all fossil sloths as well as *Choloepus*, are more closely related to each other than they are to *Bradypus*. A similarly comprehensive analysis by Pujos et al. (2007) also nested *Choloepus* among megalonychids but was less certain on the sister group position of *Bradypus* to all other sloths. Nonetheless, it too

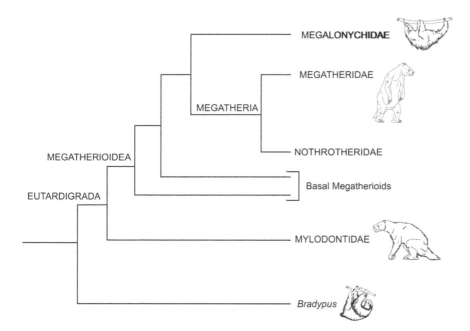

6.4. Relationships among Tardigrada (=Phyllophaga = Folivora). Gaudin's (1995) cladistic analysis based on the morphology of the auditory region incorporates the most comprehensive taxonomic sample, including representatives of all the major extinct sloth subgroups plus the two tree sloth genera. This analysis corroborated the alliance of *Choloepus* with megalonychids but placed *Bradypus* as sister taxon to all other sloths.

Drawing by Néstor Toledo.

corroborated a dipyletic origin of the living tree sloth genera. Molecular-based investigations of sloth relationships used DNA obtained from skin samples of the extinct genus *Mylodon* (Höss et al., 1996; Poinar et al., 1998; Greenwood et al., 2001) and from coprolites attributed to the extinct genus *Nothrotheriops* (Poinar et al., 1998; Greenwood et al., 2001). The conclusions of those studies vary in regard to the position of the modern sloths, but a consensus has emerged on the diphyletic origin of the modern tree sloths. The relationships among the main sloth clades, including the fossil sloths, is presented in Fig. 6.4.

CINGULATA (ARMADILLOS AND KIN)

Gaudin (2003) noted that the cingulates have been less intensely studied, particularly with respect to phylogenetic relationships, than pilosans. Fig. 6.5 presents a phylogeny of the cingulates. Although recent efforts by Gaudin and McDonald (2008) on cingulates generally and Fernicola (2008) on glyptodonts have done much to address the shortcomings of earlier morphology-based studies, much remains to be learned about the evolutionary history and phylogenetic relationships among cingulates. Vizcaíno et al. (2004a) proposed that the dental peculiarities that principally characterize armadillo-like xenarthrans (which are presumed to represent the ancestral xenarthran condition) are related to an ancestral adaptation to insectivory and that they represent a strong phylogenetic constraint that restricted, or at least conditioned, adaptations toward other alimentary habits. However, the great evolutionary diversity reflects a number of adaptive possibilities ranging from myrmecophagy (feeding on social insects) to carrion feeding to predation among the animalivores, and from selective browsing to bulk grazing among herbivores, as well as omnivores. Whereas armadillos developed varied habits, mostly animalivorous but also including omnivores and herbivores, pampatheres (Pampatheriidae) and glyptodonts (Glyptodontidae) were herbivores.

Several morphofunctional and biomechanical studies reevaluated previous hypotheses solely on the basis of comparative morphology. Although in some cases these hypotheses were refuted (carnivory in peltephiline armadillos, Vizcaíno and Fariña, 1997), in others they were corroborated (carnivory in armadillos of the genus *Macroeuphractus*, Vizcaíno and De Iuliis, 2003; herbivory in eutatines, Vizcaíno and Bargo, 1998; pampatheres, Vizcaíno et al., 1998; and glyptodonts, Fariña and Parietti, 1983; Fariña, 1985; Fariña and Vizcaíno, 2001) or refined (different kinds of herbivory in eutatines, pampatheres and glyptodonts; Vizcaíno and Fariña, 1997; Vizcaíno and Bargo, 1998; Vizcaíno et al., 1998; De Iuliis et al., 2000; Fariña and Vizcaíno, 2001; Vizcaíno et al., 2011b). The morphological and adaptive diversity suggests a more extensive early history of the group than that reflected by current systematic schemes. Analyses have also revealed that some cingulates have evolved mechanical solutions that are neither shared by closely related taxa nor have current analogues that can be used as models to investigate and interpret adaptations of lineages without living representatives (Vizcaíno et al., 2004a), as in the cases of peltephilines (Vizcaíno and Fariña, 1997) and glyptodonts (Fariña and Parietti, 1983; Fariña, 1985; Fariña and Vizcaíno, 2001).

Xenarthrans as part of the Lujanian Megafauna

The xenarthrans had enormous presence in the Lujanian. Here we provide a brief synopsis of the main species presented below, as well as note some that we do not discuss. The xenarthrans covered include representatives of Tardigrada and Cingulata, for these clades include large mammals. We do not include the anteaters, which were neither particularly large nor well represented in the fossil record during the Lujanian. In addition, many of the smaller cingulates, mainly armadillos, are not discussed, although in their case, it is due to their size, as they are abundantly represented as fossils.

GROUND SLOTHS

In contrast with only two genera of tree sloths that currently live in the forest canopies of tropical American forests, sloths were common mammals during the Tertiary and Pleistocene of South America. Once the Isthmus of Panama began to emerge several million years ago (see Chapter 5), sloths invaded North America (as well as the Caribbean islands, but via another route; see Chapter 4). Of the Lujanian megafauna, all of the main sloth clades—Mylodontidae, Megalonychidae, Nothrotheriidae, and Megatheriidae—were represented by quite large beasts. We can thus be confident that they did not share the arboreal lifestyle of their modern relatives and so refer to them as ground sloths. Their remains are abundant and indicate that they were common and widespread faunal elements of the South American Plio-Pleistocene. Among Mylodontidae were *Glossotherium*, *Mylodon*, and *Lestodon*, all very large sloths, included in Mylodontinae. The other main mylodontid clade comprises the scelidotheriines (Scelidotheriinae), including *Scelidotherium*, *Scelidodon*, *Catyonyx*, and *Valgipes*. Although not as large as the mylodontines, they were still large mammals, weighing hundreds of kilograms, making them medium-sized sloths. Their lobate

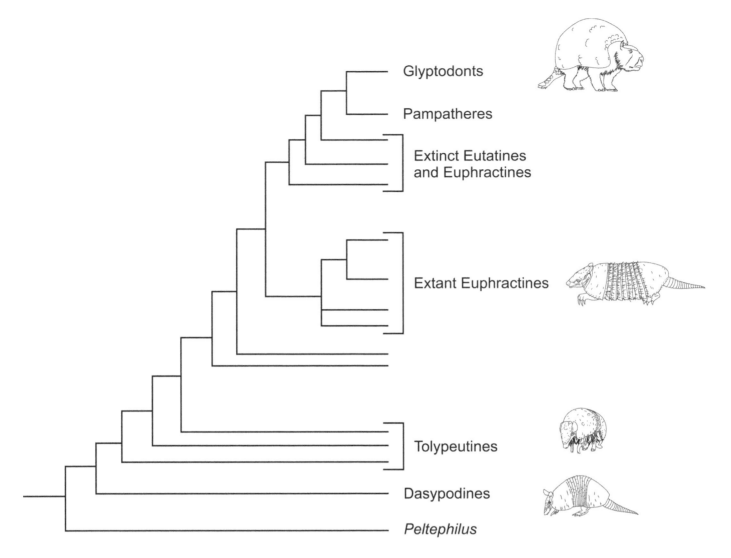

teeth with flattened grinding surfaces is general of mylodontids (McDonald and De Iuliis, 2008). The main differences between Mylodontinae and Scelidotheriinae is that in the latter, the skull is narrow and elongated, and the dentition is transversely compressed (McDonald and Perea, 2002; Dantas and Zucon, 2007). Megatheriidae include the true giants among sloths, with two well-known genera, *Megatherium* and *Eremotherium*. Megalonychidae, represented by *Ahytherium* and *Australonyx*, were less widespread in South America, although they were abundant in North America and the West Indies. They were thought to be quite rare until recently. Nothrotheriidae comprise a relatively less diverse group, known mainly through remains of *Nothrotherium*. Although reasonably large as mammals, nothrotheriids did not reach the gigantic proportions of their close kin such as *Megatherium*, *Eremotherium*, and *Lestodon*.

PAMPATHERES AND GLYPTODONTS

These two groups, now extinct, represent the giants among the cingulates (armadillos, pampatheres, and glyptodonts), a name inspired by the presence of defensive armor encircling much of the body (cingulate is derived from the Latin *cingula*, meaning "belt," "band," or "sash"). Pampatheres

6.5. Phylogeny of Cingulata. Glyptodonts and pampatheres are united in a single group that is allied to extinct forms traditionally classified among the clades Eutatini and Euphrachtini, which are hence paraphyletic groupings. Tolypeutines and the varied dasypodines are placed in a more basal position, but the Miocene specialized *Peltephilus* appears as the sister group of the remaining cingulates.

Modified from Fernicola et al. (2008). Drawing by Néstor Toledo.

6.6. Reconstructions of armored vertebrates. (A) The ankylosaur *Edmontonia* (reconstruction by Mariana Ruiz Villarreal); (B) The turtle *Meiolania* (reconstruction by Heinrich Harder, 1914); and (C) a *Glyptodon* (reconstruction by Mauro Muyano).

superficially resemble armadillos, as indicated by the movable bands that separate the scapular (anterior) and pelvic (posterior) shields and the form of the skull, and for a long time they were generally classified as Pampatheriinae among the armadillos (Daypodidae). However, it appears they are more closely related to glyptodonts (Fig. 6.5; Gaudin and Wible, 2006; Fernicola, 2008; Fernicola et al., 2008). This was suggested by Patterson and Pascual (1968, 1972), and both Gaudin and Wible (2006) and Fernicola (2008) found strong support linking pampatheres and glyptodonts in a clade that Gaudin and Wible (2006) referred to as Glyptodonta. Features of the detailed anatomy of the ear region and masticatory apparatus are among the characteristics that indicate their strong phylogenetic relationship.

Glyptodonts are among the most striking of the already stunning Lujanian mammals. They are easily differentiated from armadillos and pampatheres on the form of the skull and, more obviously, on armor that forms an essentially immobile carapace (that is, it lacks movable bands). In relation to this, their vertebral column attained a degree of fusion far greater than that in any other mammal, and they resemble turtles and ankylosaurs in this regard (Appendix 2; Fig. 6.6). As their highly derived anatomy would suggest, glyptodonts have followed a separate evolutionary history for at least 35 million years.

The presence of such an exoskeleton, as well as the large size of the Pleistocene species, made them excellent candidates to be ever-present as fossils. The first megafaunal fossil specimen collected was a carapace found in the banks of the Carcarañá River, near Santa Fe, Argentina, by the English Jesuit missionary Thomas Falkner (1774:9), who was amazed by its large size (he described the "shell," composed of "hexagonal ossicles," as about 3 yards wide) and referred it to an armadillo, which was not far off the mark; indeed, Pasquali and Tonni (2008) suggested that this was likely the same material on which Owen (1839) erected the genus *Glyptodon*.

Another early glyptodont find, this one by local residents, resulted from an expedition led by Teodoro Vilardebó (Box 6.1). The outcome of the expedition, carried out with Bernardo Berro, Dámaso Larrañaga's (see Chapter 2) nephew and later the president of Uruguay, and the French consul in Montevideo, who serendipitously happened to be the naturalist Arsène Isabelle (Onna, 2000), was indeed remarkable, as a glyptodont carapace and many bones were recovered. These were published in 1838 in the only medium then available to them: the local newspaper, *El Universal*. In this report, the authors demonstrated themselves to be beyond the level of unsophisticated amateurs. As noted by Onna (2000:65), the publication demonstrated Vilardebó's and Berro's "formation, erudition, and updated knowledge" (our translation). Although the name given to the yet undescribed animal, *Dasypus antiquus*, is not considered valid because they were unable to follow the formalities of systematics and nomenclature, it emphasized that the beast was related to armadillos. The remains were sent to the Muséum national d'Histoire naturelle in Paris and was later identified by the Swiss zoologist Pictet (1844–1846) as *Glyptodon clavipes* and by Gervais and Ameghino (1880) as a new species, *Glyptodon principalis*, which is no longer considered valid.

Teodoro Vilardebó (Box Fig. 6.1) was one of the first physicians in the emerging republics of the Río de la Plata and, like his colleague Muñiz (see Chapter 2), was also a fervent patriot and competent paleontologist. He was born in Montevideo in November 1803, then still part of the Río de la Plata viceroyship of the Spanish crown, and during his rather short life (he died while carrying out his duties during the epidemics of yellow fever in March 1857), he managed to obtain his degree from the University of Paris, to become a member of the Brazilian Instituto Histórico e Geográfico when he lived in Rio de Janeiro, and, in the few years he spent in his native country, to preside over the committee that created the Uruguayan Museo de Historia Natural in 1837. In this year, a short time after Darwin visited the region (without, unfortunately, having attempted to meet the local naturalists), Vilardebó organized the first recorded expedition to collect paleontological remains from Pedernal Creek, which today lies a short hour's drive from downtown Montevideo, but was then a much more isolated place.

Box Fig. 6.1. Teodoro Vilardebó (1803–1857), physician and one of the early South American naturalists. This picture was one of the first daguerreotypes ever taken in South America, converted into a photograph after Vilardebo's death.

Whereas the Uruguayan naturalists had to rely on the local newspaper to publicize their finds, Owen (1839), making use of his Old World advantages, soon produced the first formal description of a glyptodont, *Glyptodon*. It was included as a note at the end of the tenth chapter of the comprehensive account of the Río de la Plata region by the diplomat and naturalist Sir Woodbine Parish (1839) on many aspects of the geography of the new republics of the Río de la Plata, based on the information he had gathered as chargé d'affaires of the British crown in Buenos Aires. Owen (1839) named the genus as different from that of modern armadillos, an approach that remains valid still, after "the regularly fluted or sculptured form of the tooth . . . (γλυφω, *sculpo*)." He refrained from providing further detail, because in those times educated people knew the classical languages, but the reason why he preferred *glypto* instead of *glypho* (which is the closer transliteration) is because it is the form of the past participle of the Greek verb γλύφειν (*glyphein*, "to hollow up" or "carve," and, by extension, "sculpture"), therefore meaning "sculptured," as he remarked.

Glyptodontidae was created several decades later by the long-term keeper of zoology of what was then called the British Museum (Natural History), John Edward Gray (1869), to include the following genera: (1) *Glyptodon*, created by Owen (1839); (2) *Panochthus*, named by Burmeister (1866); (3) *Hoploplorus*, created by Lund (1939), and with *Schistopleurum*, proposed by Nodot (1857), the director of the Muséum d'Histoire Naturelle of Dijon, France, as a synonym for the latter. Gray (1869:387) recognized the following main characteristics: "dorsal shield entire, not revolute, immoveably affixed to the upper part of the very large pelvis." This author then proceeded to incorporate in the description Burmeister's (1864) proposal on the existence of a ventral shield that led the German scholar to name the group Biloricata ("double shelled")—and such a shield, weaker and much smaller than the carapace, was described by Burmeister and later

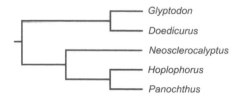

6.7. Phylogeny of the glyptodonts described in this chapter.

Drawing by Sebastián Tambusso.

also by Lydekker (1887). Finally, Gray (1869:387) ventured that the head was probably "contractile under the dorsal shield," much the way many turtles are capable of retracting the head under the shell.

Parish (1839:xviii) himself was amazed by the paleontological finds, noting in the introduction of his work that his "inquiries for fossil remains" led to "the examination of the monstrous bones . . . by learned individuals," which, in turn, proved "indisputably that the vast plains in that part of the world, at some former period, the further history of which has not been revealed to us, were inhabited by herbivorous animals of most extraordinary dimensions, and of forms greatly differing from those of the genera now in existence." Glyptodonts were among the most prominent of those "monsters," and we are fortunate to witness the huge increase in our knowledge and understanding of these amazing beasts. The efforts of such researchers as Ameghino (1884, 1889), Burmeister (1874), Scott (1903–1904), Kraglievich (1932), Castellanos (1931, 1932, 1959), Paula Couto (1947, 1957), and Hoffstetter (1958), as well as the work of several other more recent workers, have documented the extensive radiation that produced a vast array of species throughout much of the Cenozoic. They deserve a book all to themselves, but we will have to be content with presenting a few of the most characteristic Lujanian forms in our Bestiary, although others, such as *Lomaphorus, Plaxhaplous, Neuryurus,* and *Neothoracophorus,* have been named. The sequence in which the glyptodonts are treated in the Bestiary (Fig. 6.7) essentially follows Fernicola (2008) and Porpino et al. (2010).

The beasts

Boreoeutheria
Laurasiatheria
Carnivora
Feliformia
Felidae
Machairodontinae
Smilodon—sabertooth

One of the most recognized extinct predatory mammals, *Smilodon* is related to the modern cats, though it is not, as generally assumed, a tiger. Two late Pleistocene species of *Smilodon* are usually recognized, S. *populator* (Fig. 6.8), restricted to but widely distributed in South America east of the Andes, and S. *fatalis,* widely distributed in North America, but appearing also along the coast of South America west of the Andes. Although an iconic mammal of the North American Pleistocene and linked in the public's mind to the La Brea Tar Pits of California, the first remains of these cats were found in South America and named by Lund (1842), a giant among early paleontologists. A third species, S. *gracilis,* smaller than the later species, is known from Pliocene to middle Pleistocene remains from North America (Kurtén and Werdelin, 1990; Turner and Antón, 1997; Rincón, 2006).

The mass of the somewhat smaller *Smilodon fatalis* was estimated as 350 kg by Anyonge (1993). That of *Smilodon populator* was estimated as 400 kg on the basis of scale models (Fariña et al., 2005), about 350 kg on the basis of allometric equations (Fariña et al., 1998), and 405 kg on the

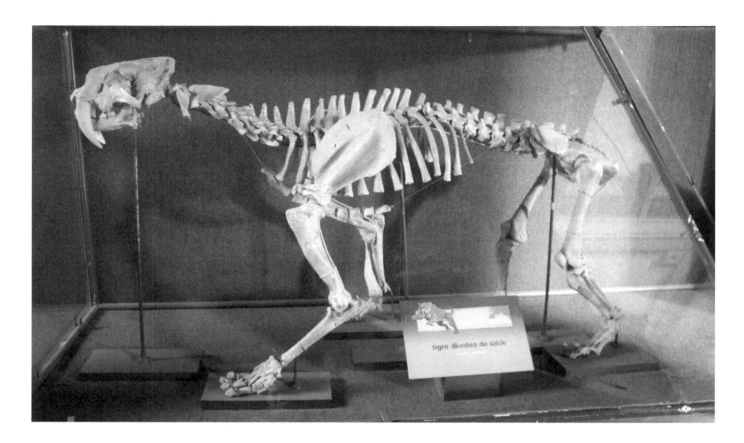

6.8. Skeleton of *Smilodon populator* mounted in the Museo de La Plata, Argentina, by kind permission of that institution. Total length of the specimen: nearly 2 meters.

basis of Anyonge's (1993) equations. Such size rivals that of the great North American lion *Panthera atrox*, making *Smilodon populator* among the largest felids to have ever walked the earth. The North American lion, incidentally, is an interesting taxon as well. There is no doubt that it was a large felid and that it did enter South America, but there has been considerable speculation in the literature over its systematics, with opinions alternating between considering it a species, *Panthera atrox*, distinct from the extant lion, *Panthera leo*, and a subspecies of the latter, with the traditional idea that these species, along with the Eurasian cave lion, *Panthera spelaea*, formed a tightly knit group (for example, Nagel et al., 2003). More recently, however, Christiansen (2008) argued that *P. spelaea* and *P. atrox*—clearly as a separate species—are not even true lions, but fall as successive outgroups to a lion (*P. leo*) and leopard (*P. onca*) clade (see Chapter 4).

With its relatively short legs, *Smilodon* does not appear to have been well adapted for chasing down its prey; instead, it was an ambush hunter, as Muñiz proposed a century and a half ago (see Chapter 4). The results of Anyonge's (1996) analysis relating limb bone dimensions and habits in several carnivores suggested a most likely interpretation of *Smilodon fatalis* (morphologically similar to *S. populator*) as an ambush hunter, with ambulator being the next likely interpretation. Most studies of *S. fatalis* (Palmqvist and Arribas, 2001) concur that it was an ambush predator like most modern felids (with the exception of the cheetah). Also, the forelimbs are extraordinarily robust, with the humerus stronger than the femur, a characteristic probably related to the functions of seizing and killing prey with the forelimbs (Muñiz, 1845; Akersten, 1985; Anyonge, 1996).

In extant felids, the canine, particularly, and incisor teeth are normally used in a lunging killing bite and have circular sections to withstand

6.9. The skull of a young glyptodont found in Arizona is pierced by two oval perforations that were probably inflicted during an attack, likely by one of these felids.

From Gillette and Ray (1981), by kind permission of the Smithsonian Institution Scholarly Press.

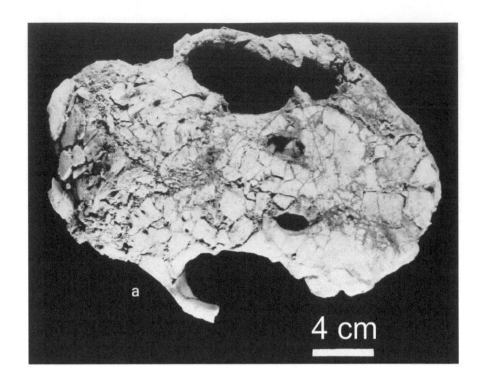

4 cm

unpredictably directed bending stresses produced by the twisting actions of struggling prey (Farlow et al., 1991; Meachen-Samuels and Van Valkenburgh, 2009a, 2009b). This would also be true of other felines with conical teeth. On the other hand, sabertooth mammals, with their highly specialized, elongated canines of elliptical section (Biknevicius and Van Valkenburgh, 1996), probably used a different style: the teeth would have had to penetrate tissues without the prey moving in such a way that might break them. Such habits would have required extremely strong forelimbs to seize and immobilize prey. It has also been suggested that the long canines would not have been suitable to strike in bony regions, such as the skull and cervical vertebrae, at least in the case of adult prey. This would have been less of a problem for juvenile prey (Fig. 6.9), whose skulls would not yet have been fully formed, but even so, the ferocity of attacks may have produced rather uncomfortable accidents, with animals possibly getting locked to death and even perhaps even fossilized as such (but see Box 4.4). Another possibility for delivering critical bites may have been to slash through the neck ventrally, severing the trachea and blood vessels, causing severe loss of blood and inducing shock. Similarly, fatal wounds may have also been applied to the abdominal region (Turner and Antón, 1997).

The dentition in carnivores (that is, any predominantly meat-eating animal, whereas members of the placental mammal clade Carnivora—cats, dogs, bears, and so on—are referred to as carnivorans; see Chapter 4), and cats in particular, is specialized for cutting and slicing flesh and tendons. This involves a reduction in number and narrowing of the cheek teeth, those behind the canines. Some of the cheek teeth, particularly in carnivores such as dogs, remain robust, however, to function in bone crushing. A pair of teeth, the fourth upper premolar and first lower molar, termed carnassials, are particularly specialized for slicing and become elongated, bladelike structures (see Chapter 4; Appendix 1). These specializations, particularly loss of teeth and bladelike form of the carnassials, are taken

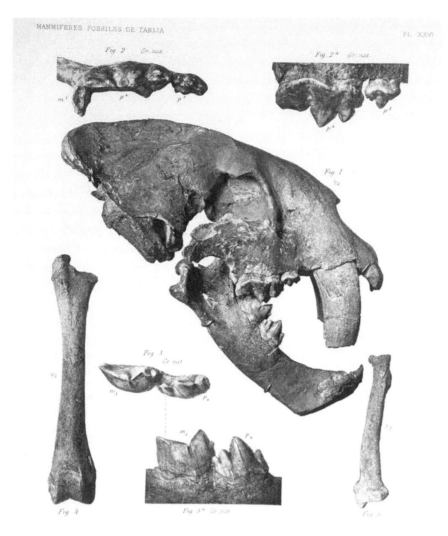

Smilodon

6.10. Remains of *Smilodon* from Tarija, Bolivia. The central figure (labeled Fig. 1) is of the skull and lower jaw. Although their tips are broken, the great size and power of the canine teeth can be easily appreciated (see also Fig. 4.15). The carnassial teeth that form the main slicing and shearing surface of the dentition are shown just as they are engaging, as would have occurred at the beginning of jaw closing. The main carnassial pair is formed by the fourth upper premolar and first lower molar, which are tall, elongated, and bladelike teeth. The upper carnassial, at the top of the figure and labeled p4, is shown in occlusal and side views. The lower carnassial, at bottom central, is labeled m1. The catlike morphology, taken to an extreme in sabertooths, also involved reduction and loss of the remaining cheek teeth—for example, there are no posterior grinding molars (compare with the form in a more doglike form such as the creodont *Sinopa* in Fig. 4.13). Also shown are the femur (lower left) and the radius (lower right).

From Boule and Thévenin (1920).

to extremes in the sabertooth cats (Fig. 6.10), effectively precluding bone crushing.

Akersten (1985) suggested that the North American species *Smilodon fatalis* might have been a social hunter. Akersten based this on the high number of entrapped mammals in the tar pits of Rancho La Brea, which this author interpreted as being the consequence of responses to distress calls that prey animals might have uttered when struck down. Such responses are typical of social hunters but not of solitary hunters (Carbone et al., 2009). Although such evidence is not available for the South American species, its inferred presence in a closely related species is suggestive.

Some aspects of the paleobiology of *Smilodon populator* were considered early on by Muñiz (1916) (Box 6.2). Fariña et al. (2005) found that the individuals of *Smilodon populator* were able to perform strenuous locomotor activities, such as fast running. Very large herbivores escape predation by virtue of their size, at least as adults (Owen-Smith, 1988). In Africa and Asia today, an adult body mass of 1000 kg seems to be the boundary between those mammals that are susceptible to predation and those that are not. Particularly interesting is the case of the giraffe (*Giraffa camelopardalis*). Male giraffes have an adult body mass larger than 1 tonne and do not usually fall victim to predation, whereas females seldom reach 1 tonne and are sometimes successfully attacked by lions (Owen-Smith, 1988). A similar phenomenon—that is, a threshold for escaping predation by size alone—might have been present in the Lujanian, especially with regard to juveniles, although perhaps that threshold may have been somewhat higher. *Smilodon* may have been capable of bringing down larger prey than could a conical-toothed feline (Akersten, 1985), such as lion or tiger.

Box 6.2   **Aspects of the paleobiology of *Smilodon populator* by Muñiz (1916)**

> La admirable macicez y la magnitud de las extremidades, las asperezas de las fuertes inserciones tendinosas que las cubren; las crestas humerales de donde parten las potencias musculares movientes de la mano y de los dedos; las del radio a que adherían las que sirven a la pronación y supinación; la firmeza, en fin, del ensamblaje braquial; el anchuroso pecho; el irresistible poder de los cinco dedos unguiculados, sin igual en su género; el de sus uncinados o cortos incisivos, y el más terrible aparato de colmillos de dos filos de raro y especial tamaño; señalan inequívocamente al lacertoso y terrífico rey de las selvas, el primer monstruo de la tribu felina; el más indómito e infatigable en la caza; el más feroz y audaz en sus ataques bruscos e insidiosos.

> [The admirable stoutness and magnitude of the extremities, the coarseness of the strong tendinous insertions that cover them; the humeral crest from which arise the powerful musculature moving hand and fingers; and of the radius, to which attached those effecting pronation and supination; the stability, all told, of the brachial apparatus; the broad chest; the irresistible power of the five unguiculate fingers, unmatched in kind; and that of its uncinate (hook shaped) or short incisors, and the most terrifying set of double-edged, oversized canines; point unequivocally to the strong and terrifying king of the jungles, the first monster of the feline tribe, the most indomitable and indefatigable in hunt: the boldest and most ferocious in its sudden and deceitful attacks."]

Muñiz advanced, with poetic flair (although our translation may lack some of his language's flavor), two levels of interpretation for *Smilodon*, the first strictly morphological, on the importance of the forelimb for hunting and immobilizing its prey, a hypothesis widely supported by more recent research (Akersten, 1985); and the second, at a higher level of abstraction, on its admirable endurance, which is contradicted by recent views. On the one hand, Muñiz presented *Smilodon* as an "indefatigable" animal, i.e., a predator capable of long chases, that attacked, at the same time, by "sudden and deceitful" methods. Modern interpretations point instead to an ambush hunter that used its forelimbs for subduing prey and thus imparting an accurate, mortal blow with its long and sharp, but perhaps easily breakable, canines (Biknevicius and Van Valkenburgh, 1996).

Boreoeutheria
Laurasiatheria
Carnivora
Caniformia
Ursidae
Arctotherium—South American giant short-faced bear

Today there is only a single ursid species in South America, *Tremarctos ornatus* or the spectacled bear, a relatively small herbivore that lives in the great forests of the Andean slopes. This bear and its kin are the short-faced bears, or Tremarctinae, comprising several genera and species, including the North American giant short-faced bear, *Arctodus simus*, already dealt with in Chapter 4. In South America, the genus *Arctotherium*, which is widely distributed during the Pleistocene but which has not so far been recovered west of the Andes, encompasses the extinct tremarctine bears.

Five species of *Arctotherium* have been recognized (Soibelzon et al., 2008; Soibelzon and Schubert, 2011), although only A. *wingei* (Brazil, Bolivia, and Venezuela, where it survived until the early Holocene), A. *bonariense* (Argentina), and A. *tarijense* (Argentina, Bolivia, Chile, and Uruguay) are known from the Lujanian, whereas A. *angustidens* (early to middle Pleistocene of Argentina and possibly Bolivia) and A. *vetustum* (middle Pleistocene of Argentina and possibly Brazil) are known from earlier deposits.

Like other bears generally, their diets were probably omnivorous, including active hunting, scavenging, and prey theft (Soibelzon and Prevosti, 2007). The North American *Arctodus* has been proposed as having had a rather carnivorous diet. Its colossal mass, skull morphology, and teeth and limbs suggest that it was capable of hunting large prey, perhaps even subadult proboscideans (Christiansen, 1999). However, Figueirido et al. (2010) suggested other alimentary habits, as discussed in Chapter 4. *Arctotherium* may also have been able to hunt large prey, although the South American forms have generally not been regarded as large as *Arctodus simus* (e.g., by Christiansen, 1999), but some individuals of the Ensenadan *Arctotherium angustidens* may have attained the size of *Arctodus simus* and even surpassed it (Figueirido and Soibelzon, 2009; Soibelzon and Schubert, 2011).

The body size of the smaller A. *tarijense* was given as about 400 kg by Soibelzon and Schubert (2011). That of A. *bonariense*, based on a specimen in the Museo Argentino de Ciencias Naturales, is estimated to have been on the order of a few hundred kilograms, similar to that of the large grizzlies of North America (Christiansen, 1999). Fariña et al. (1998) estimated its mass as slightly above 300 kg, although only nine equations could be applied to this specimen, rendering the results less reliable. Soibelzon and Tarantini (2009) estimated an average size of 200 kg for this species. The smaller A. *wingei* probably weighed about 105 kg (Soibelzon and Schubert, 2011). Soibelzon (2004) estimated the body size of A. *angustidens* as between 273 kg and 1124 kg, with an average of 671 kg, by using Anyonge's (1993) equations for limb dimensions. As Soibelzon (2004) pointed out, *Arctotherium* species apparently exhibited remarkable sexual dimorphism, which might affect the estimations, and the use of craniodental features yields underestimates. However, the recent report of an extraordinarily large specimen of this species suggests the likelihood that it could achieve far greater size: Soibelzon and Schubert (2011), on the basis of equations that used the more reliable predictive measurements of the humerus and radius, settled on a likely size of 1700 kg. This would put A. *angustidens* in a league of its own, making it not only the largest known bear, but "probably the most powerful terrestrial carnivoran of the late Cenozoic" (Soibelzon and Schubert, 2011:72).

Pathologies have been reported for the remains of these bears (Soibelzon and Prevosti, 2007), especially in A. *latidens* but also in A. *angustidens*, including tremendous wear of the occlusal surface of the molars that in some cases left the pulp cavity exposed. Moreover, some teeth are found broken as a consequence of having chewed hard material, probably bones, and cavities are also recorded, perhaps due to the consumption of carbohydrate-rich food, such as honey (Ferigolo and Berman, 1993) or fruits. In the case of A. *angustidens*, Figueirido and Soibelzon (2009) noted that the damaged teeth may have resulted from chewing on bones, but they suggested that its mandibular anatomy demonstrated that this species was neither a hypercarnivore nor a specialized scavenger. Given the ecological conditions, with abundant prey and few competitors, these authors considered it reasonable to suppose that this short-faced bear was capable of preying on megamammals and of scavenging carcasses as the opportunity arose, and thus that its diet possibly did include a considerable amount of meat, as suggested by Soibelzon and Prevosti (2007). On the other hand, bears eat a wide variety of hard foods other than bones, such as bark, hard fruits, and roots, that might also have damaged teeth. The overall assessment of this bear by Figueirido and Soibelzon (2009:10) was that it "was mainly omnivorous, and had ecological opportunities and morphological capabilities to feed on carcasses or meat whenever these resources became available."

In one of the Pleistocene burrows treated in Chapter 7, three bear specimens, assigned to *Arctotherium angustidens*, were collected (Soibelzon et al., 2001). This remarkable find, including articulated skeletons of an adult female and two juvenile males for which there is no evidence of postmortem transport, was interpreted as a family that had invaded the cave and died when the cave suddenly collapsed.

Boreoeutheria
Laurasiatheria
Cetartiodactyla
Camelidae
Camelinae
Palaeolama—large llama

In Lujanian times, llamas and guanacos were much more widespread than their restricted modern range in the Andes and Patagonia, and they were common mammals on the plains (Fig. 6.11). The typically recognized genera are *Palaeolama* and *Hemiauchenia*. Their taxonomy, however, has proved problematic (López Aranguren, 1930; Cabrera, 1932, 1935; Hoffstetter, 1952; Cardozo, 1975), and the most recent review (Guérin and Faure, 1999) recognizes the presence of only one genus, *Palaeolama*, formed by two subgenera, *Palaeolama* and *Hemiauchenia*, with the species referred to below and their geographic distribution remaining valid. These authors accepted that *Palaeolama* is found in North America with the Pleistocene species P. (*Palaeolama*) *mirifica*, P. (*Hemiauchenia*) *vera*, P. (*Hemiauchenia*) *blancoensis*, and P. (*Hemiauchenia*) *macrocephala*, and in South America with the species P. (*Hemiauchenia*) *paradoxa*, P. (*Hemiauchenia*) *major*, P. (*Palaeolama*) *weddelli*, and P. (*Palaeolama*) *aequatorialis*. They

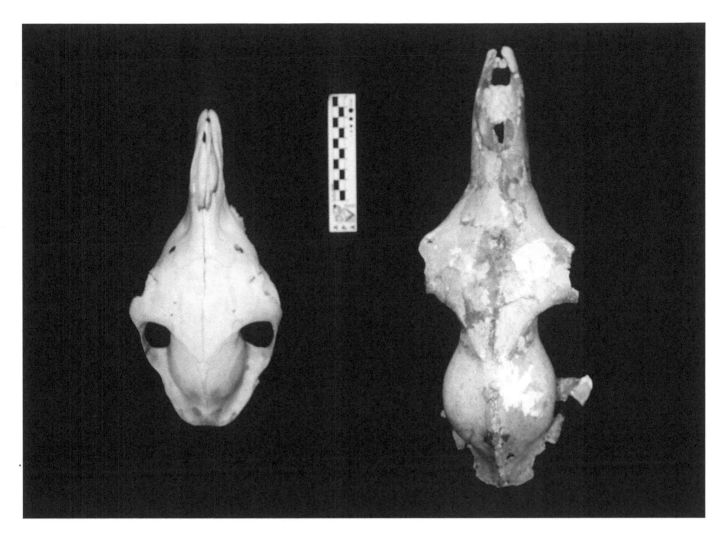

also erected two new species, *Palaeolama (Palaeolama) hoffstetteri* from Bolivia, and *P. (Hemiauchenia) niedae* from Brazil.

The species of *Palaeolama* and the former genus *Hemiauchenia* have been considered browsers (MacFadden et al., 1994), grazers (Webb, 1974; Menegaz and Ortiz Jaureguizar, 1995), and mixed feeders (Guérin and Faure, 1999), although no attempt to establish these habits has been undertaken, apart from MacFadden and Shockey's (1997) study, mainly based on isotopic evidence. *Palaeolama* was the giant of the group, reaching a size not much smaller than that of their living relatives, the camels of the deserts and steppes of Africa and Asia. Its estimated mass is 200 kg (Fariña and Czerwonogora, unpublished data).

Boreoeutheria
Laurasiatheria
Perissodactyla
Equidae
Equinae
Hippidion—South American horse

Given how common horse riding is today on the great plains of the Americas, it is almost inconceivable that horses haven't always been here. Yet we know that Hernán Cortés was lucky in his predilection for horses,

6.11. Skull of *Palaeolama* (right) compared to that of *Lama guanicoe*. Scale bar equals 10 cm.

Photograph by Sergio F. Vizcaíno by kind permission of the Museo de La Plata, Argentina.

for we are told that his native adversaries were so frightened at the sight of the double-headed monster that they gave up on the spot. Although the past five centuries—a mere instant for a paleontologist—have heard the hooves of domesticated *Equus caballus* galloping over American soil, other species of horses were quite common in this continent a few hundred centuries earlier. As an integral part of the fauna that arrived during the great interchange, two genera of extinct horses inhabited South America, including *Equus*, the modern horse but of the subgenus *Amerhippus*, and the endemic Plio-Pleistocene genus *Hippidion* (Alberdi, 1987; Alberdi and Prado, 1993; Prado and Alberdi, 1996). According to these authors, the latter represents a monophyletic (i.e., natural) group that must have diversified somewhere in northern South America or perhaps in Central America.

All South American horses became extinct a few millennia before the European colonizers reintroduced the domestic species, which has adapted to this continent as well as its extinct native relatives had done. *Hippidion* was more robust and large-headed that the modern species (see the reconstruction in Plate 8, a beautiful watercolor by Cabrera). Among the cranial characteristics that differentiate *Hippidion* is that the nasal bones were largely unattached for a great part of their length, as is clear from Fig. 6.12. Its mass was estimated as about 500 kg by Fariña et al. (1998) by using allometric equations. Alberdi et al. (1995 b) estimated the mass of *H. principale* as 460 kg by using a different set of equations. The mass of another Lujanian equid, *Equus* (*Amerhippus*) *neogeus*, was apparently not particularly different from that of the modern horse, *E. caballus* (Prado and Alberdi, 1994).

Fariña et al. (2005) assessed the athletic capabilities of *Hippidion* and compared it with the two living zebras (*Equus zebra* and *E. burchelli*). All of them show high values of the indicator of athletic capacity, which is consistent with their observed or inferred fast galloping gait. Furthermore, the strength of their humerus (upper forelimb bone) is substantially greater in the anteroposterior direction than transversely. On the other hand, the femoral (thigh bone) sections of the three species are fairly circular. Therefore, and different from the case of *Macrauchenia* (see Chapter 7), *Hippidion*, like its modern relatives, seems likely to have primarily relied on fast running without swerving, as its humeral dimensions are much stronger anteroposteriorly than transversely.

As for the habits of South American horses, Alberdi and Prado (2004) stated that even though both genera arrived at different times, they produced ecologically similar types, perhaps as a consequence of a shared phenotypic plasticity. Generally speaking, the species of the genus *Hippidion* are more stoutly built and are less clearly adapted to open environments than those of *Equus* (*Amerhippus*), which seemed to have been more cursorial, based on the length of the leg bones, especially the metapodials, i.e., metacarpals (the element or elements between the wrist and the fingers) and metatarsals (between the ankle and the toes).

As can be seen in Fig. 6.12, *Hippidion* exhibits a narrower frontal region of the skull as well as largely unattached nasals. These have been interpreted as an adaptation to more closed habitats, such as parks or woody steppes. Also, *Hippidion* bears teeth with a relatively lower degree of enamel folding and less hypsodont (ever-growing) than the teeth in *Equus* (*Amerhippus*),

6.12. Skull of *Hippidion*. Note the nasal bones largely unattached for a great part of their length. Scale bar equals 10 cm.

Photograph by Sergio F. Vizcaíno by kind permission of the Museo de La Plata, Argentina.

which suggest a less silicon-rich diet. Congruently, biogeochemical data suggest that the species of *Equus* (*Amerhippus*) had a more grazing diet (MacFadden et al., 1996, 1999; MacFadden and Shockey, 1997).

> Boreoeutheria
> Euarchontoglires
> Rodentia
> Myomorpha
> Hydrochoeridae
> Hydrochoerinae
> *Neochoerus* — giant capybara

Although it is known from remains that are too fragmentary (Fig. 6.13) to permit a reliable reconstruction, we know enough of the Pleistocene giant capybara *Neochoerus* to say that it is similar to its living relative, except that it was at least twice its size. Its estimated mass is between 100 and 150 kg. The habits of this interesting taxon have not been thoroughly studied, but they might have been similar to those of the living capybara, *Hydrochoerus hydrochaeris* (which means, rather erroneously, given our modern systematic understanding, "water hog"). Probably less evident than for other mammals, being a rodent imposes its own phylogenetic constraints. Although diverse and variable in morphology, caviomorph rodents parallel but do not achieve the extreme adaptations of other specialized mammals such as running ungulates, with their long legs, light hooves, and long, elastic tendons (Elissamburu and Vizcaíno, 2004), making very large rodents interesting creatures that managed to achieve large sizes without the specializations observed in other large mammals.

Although the Pleistocene cabybara *Neochoerus* was much larger than its living kin *Hydrochoerus*, there were perhaps even larger, though older,

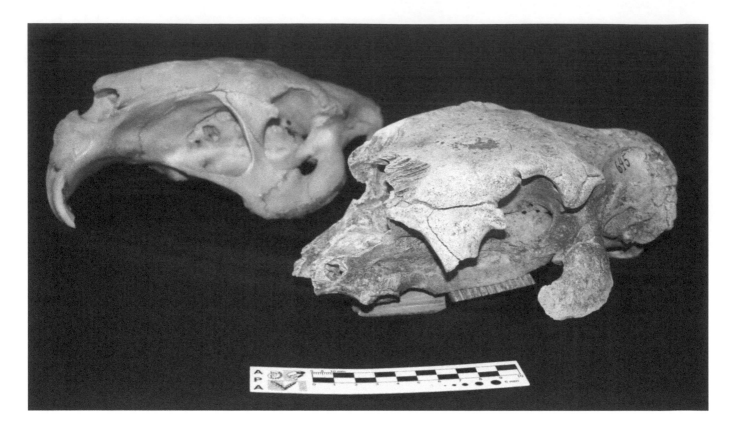

6.13. Fragment of skull and mandible of *Neo-choerus*. Even though this individual was a juvenile, its large size is clearly appreciated compared to the complete skull of modern capybara. Scale bar equals 10 cm.

Photograph by Sergio F. Vizcaíno by kind permission of the Museo de La Plata, Argentina.

extinct rodents, although also known from fragmentary material. Among them was the giant chinchilloid rodent *Phoberomys* from the Miocene and/or early Pliocene of Argentina, Brazil, and Venezuela. Sánchez-Villagra et al. (2003) and Horovitz et al. (2006) obtained values for *Phoberomys* as high as 700 kg, a size comparable to that of buffalo (Alexander, 2003).

Other very large rodents include the Pliocene Chapadmalalan *Chapalmatherium* (about 200 kg) and *Telicomys* (about 600 kg) (Vizcaíno et al. 2012). *Chapalmatherium* is a hydrochoerid, also related to extant capybaras. *Telicomys* is a dinomyid related to the pacaranas, 10–15-kg living animals that are apparently capable of climbing trees. The Pliocene–early Pleistocene dinomyid *Josephoartigasia monesi* was proposed as the largest known rodent, with a body mass of about 1000 kg (Rinderknecht and Blanco, 2008), although there has been some controversy on the statistics and calculations, as well as on the use of only measurements along the main axis of the skull. All these factors might potentially have yielded overestimations (Millen, 2008). Also, the age of this species cannot be precisely established because it is reported to have been found in a fallen boulder, with the additional problem of the material having been lost and refound after it was collected. Finally, the mass estimations are based on rather fragmentary material.

Atlantogenata
Afrotheria
Proboscidea
Gomphotheriidae
Cuvieronius, "*Stegomastodon*," and *Haplomastodon*

As noted in Chapter 4, at least two and possibly three genera of New World proboscideans were present during the South American Pleistocene, in addition to the possible early presence of the Amazonian late Miocene gomphothere *Amahuacatherium peruvium* (see Chapter 5). The recognition of these proboscidean groups in South America was essentially established by Cabrera (1929b), as can be seen in Plate 9. We noted in Chapter 4 that one group of Gomphotheriidae is considered as the New World proboscideans, and that members of this group entered South America. We noted in Chapter 5 that representatives of Elephantidae and Mammutidae did not enter South America. The South American gomphotheres are short jawed and short faced, as are some North American gomphotheres, and in contrast to the long-jawed gomphotheres of North America. Among the features are a shortened mandibular symphysis (chin region) and loss of permanent mandibular incisors (Ferretti, 2008). The South American gomphotheres share several prominent features that suggest that they are closely related and form a distinct clade among New World gomphotheres. Among these features, we note a relatively high cranium and wide forehead (Fig. 6.14), and the absence in adults of lower tusks, as well as more technical anatomical characters (e.g., the premaxillae, the bones bearing the incisors or tusks, are large and poorly pneumatized, meaning that they contain small air sinuses; there is only a single infraorbital foramen, a passageway in the bone of the cheek region; Ferretti, 2008). Being proboscideans, the South American gomphotheres were massive and stoutly built beasts. The estimated mass of "*Stegomastodon*" *platensis* is more than 4 tons, based on allometric equations (Fariña et al., 1998). Mass estimates for the South American species are not available. "*Stegomastodon*" *platensis* appears to be on average somewhat larger than *Haplomastodon*, whereas the remains of *Cuvieronius* from Tarija (Bolivia) suggest a stoutly built animal, though somewhat smaller at shoulder height than "*S.*" *platensis*. These observations, however, are not based on rigorous analysis, and size differences between these proboscideans may eventually be demonstrated as being unimportant, given individual variation and sexual dimorphism (M. P. Ferretti, personal communication).

Alberdi and Prado (1995) agreed with Simpson and Paula Couto (1957) in recognizing two genera and three species: *Cuvieronius hyodon*, *Stegomastodon waringi*, and *S. platensis*. However, some researchers have questioned this scheme (e.g., Ferretti, 2008, 2010; Mothé et al., 2010) and suggested, as noted by Ferretti (2008, 2010), the recognition of a third genus, *Haplomastodon*, for *Stegomastodon waringi*. As well, Ferretti (2010) suggested that, even though *H. waringi* (here the concern is the specific epithet *waringi*) is in widespread use, the proper name for this species is *Haplomastodon chimborazi*. Given that the rules governing the naming of species, as codified by the International Code of Zoological Nomenclature (1999) specify that the oldest available name has priority over subsequent names, it indeed appears that Ferretti is correct. In addition, it is not clear that *Stegomastodon*, also known as a North American genus (as is *Cuvieronius*, incidentally), is truly present in South America: Ferretti's (2010) analysis suggested that the South American form included as *Stegomastodon* is not closely related to the North American *Stegomastodon*, and thus does not

6.14. Skull and mandible (from Tarija, Bolivia) of the South American gomphotheriid *Haplomastodon chimborazi*. (A) The skull, with its massive tusks, is shown in anterior (or front) view in the main image. The central region of the skull is dominated by the large openings into the nasal cavities. In proboscideans, the openings are farther back than usual in mammals, indicating the presence of a trunk. The form of these large skulls, minus the tusks, as is often the case in fossil specimens, may have appeared similar to a giant human skull, possibly contributing to myth of the Cyclops, giant humanlike creatures with a single eye in the middle of their forehead. The images at top left and right show the detailed texture of the tusks; the central lower image is of a section through a tusk. The mandible, or lower jaw, sits below the skull; here it is reproduced at half its scale. For reference, one of the isolated tusks would be about 150 cm in length. (B) The main figure shows the skull and mandible in lateral (side) view. At bottom left the skull and mandible are shown in anterior (front) view but at a different perspective than in A. At top right, the skull and mandible are shown in posterior (back) view. At bottom center and right, the mandible is shown in occlusal (from the top) and lateral (side) views, respectively. Note the presence of only a single tooth in each half of the mandible, which is a typical proboscidean pattern. Top center shows the symphysis (chin region) in ventral (underside) view.

From Boule and Thévenin (1920).

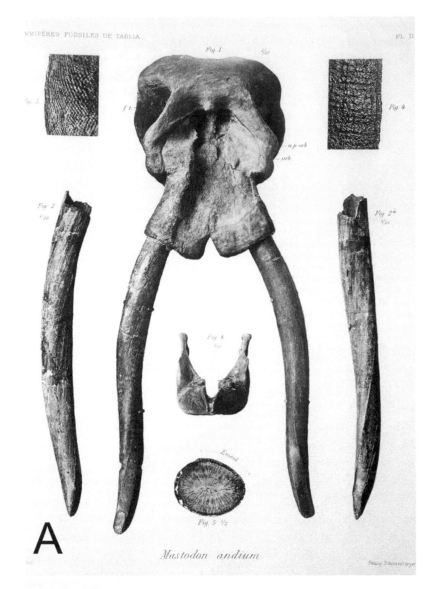

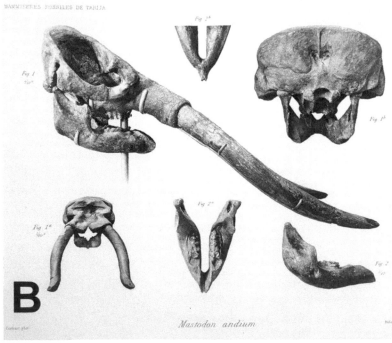

belong in this genus. Although there has been the suggestion that the South American *Stegomastodon* should be placed in the genus *Notiomastodon*, Ferretti (2008, 2010) preferred to retain the species tentatively in *Stegomastodon*, but as "*Stegomastodon*" *platensis* to underscore its taxonomic uncertainty (its proper generic name has not been worked out as of this writing). We will follow, for convenience, the scheme proposed by Ferretti (2008, 2010) in recognizing *Cuvieronius hyodon*, "*Stegomastodon*" *platensis*, and *Haplomastodon chimborazi* as the South American proboscideans.

As usual in paleontology, the taxonomy of these gomphotheres is more complex and convoluted than these taxonomic wrinkles imply. Although we need not delve too deeply into this matter, we may mention that the treatment of these gomphotheres extends nearly as far back as the beginning of the science of paleontology. *Cuvieronius hyodon*, named by Fischer de Waldheim (1814) originally as *Mastotherium hyodon* on the basis of a tooth from northern Ecuador, was the first proboscidean scientifically named from South America, although the same specimen had been treated by Cuvier (1806) as "Mastodonte des Cordilières." Cuvier (1824) named it *Mastodon andium*, but Fisher's name takes priority. Despite this priority, subsequent and ongoing research indicates that the original tooth cannot be distinguished from those that researchers have come to recognize as *Haplomastodon*. Because the abundant remains subsequently recovered from Ecuador may be referred to *Haplomastodon*, but none surely to *Cuvieronius*, it is likely that the tooth described by Fisher and Cuvier in fact belongs to *Haplomastodon* rather than *Cuvieronius*, and thus that the latter was never actually present in Ecuador (Ficcarelli et al., 1995; Ferretti, 2008). This would then mean that the naming of the taxa is indeed doubtful. To maintain stability, there has been a general consensus to retain *Cuvieronius*, though not its original meaning.

As for the relationship of the South American gomphotheres, Alberdi and Prado (1995), following such authors as Cabrera (1929b) and Simpson and Paula Couto (1957), have considered that they share close ancestry with such Old World forms as *Anancus* and *Sinomastodon*. On the other hand, several authors (e.g., Shoshani, 1996; Ferretti, 2008) have found no support for such a relationship, and Ferretti (2008) suggested a more logical (as well as morphologically supported) relationship with the North American long-jawed gomphothere *Rhynchotherium*.

Gomphotheres were fairly widely distributed in South America (Fig. 6.15). *Cuvieronius hyodon* was restricted to the Andes south of Ecuador during the middle and late Pleistocene (although on the basis of the controversial dating of sites in Tarija, Bolivia, there is doubt on the middle Pleistocene presence, and it may be considerably younger; Coltorti et al., 2007; Ferretti, 2008). *Haplomastodon chimborazi* is known from later Pleistocene to early Holocene localities across tropical South America, including the Andean region and possibly also Chile, whereas "*Stegomastodon*" *platensis* is recorded from the pampean region of Argentina and perhaps Uruguay (Ferretti, 2008). As we noted in Chapter 5, the proboscideans have generally been considered as having entered South America from North America at a relatively late date, during the Ensenadan Age (about early to middle Pleistocene). Ferretti (2008) noted that the presence of South American gomphotheres represents a single dispersal event, with the

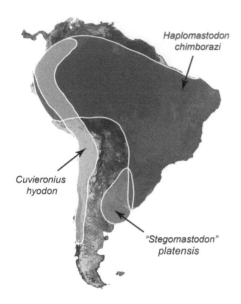

6.15. Distribution of the species of South American gomphotheres.

Based on Alberdi and Prado (1995). Drawing by Sebastián Tambusso.

entrance of *Cuvieronius* into South America followed by a rapid diversification that gave rise to *Haplomastodon* and *"Stegomastodon" platensis* (in this sense, South American *"Stegomastodon"* is distinct from North American *Stegomastodon*). This scenario for the entrance and diversification of the South American gomphotheres is the one generally recognized. However, as explained in Chapter 4 and alluded to above, the earliest presence of gomphotheres in South America may extend back much further than usually supposed. The remains of the Amazonian *Amahuacatherium peruvium* were dated as about 9.5 Mya (late Miocene) by Campbell et al. (2010). This is controversial because the fossils of this beast have also been considered as a variant of *Haplomastodon*, a Pleistocene gomphothere. If Campbell et al. (2010) are correct, then we would have to rework our traditional ideas on the gomphothere radiation in South America. We may, however, not need to radically alter our view on the essential outline of the story—that the *Cuvieronius*, *Haplomastodon*, and *"Stegomastodon"* clade represents a fairly recent (i.e., late Pliocene–Pleistocene) dispersal and radiation, as opposed to accepting *A. peruvium* as the earliest representative of the Pleistocene clade. This latter possibility seems unlikely; it is difficult to believe that such an extended lineage of very large individuals (and hence potentially excellent candidates to be found as fossils) left no remains during the intervening time periods, or that, equally unlikely in our view, we have been unable to recover any fossils of such large beasts. It may be that *A. peruvium* was an early proboscidean South American immigrant but that its lineage became extinct, and that the Pleistocene clade represents a second immigration event. We will need more evidence to tell this tale.

The paleoecology of South American gomphotheriids has been investigated by several authors (e.g., MacFadden and Shockey, 1997; MacFadden et al., 1996; Sánchez et al., 2004). Sánchez et al. (2004) analyzed the carbon and oxygen isotopic composition of enamel, dentine, and bone for all three species from different localities. Among specimens that can be clearly assigned to species (and following the taxonomic scheme outlined above), these authors suggested that *Cuvieronius hyodon* specimens from Chile apparently fed exclusively on C3 plants, whereas those from other localities had mixed C3–C4 dietary preferences. *"Stegomastodon" platensis* specimens from Buenos Aires Province (Argentina) relied mainly on C3 plants, whereas those of *Haplomastodon chimborazi* indicate mixed C3–C4 feeders. The general impression is that these proboscideans were capable of a varied diet, depending on availability. Mothé et al. (2010) studied the population structure of *Haplomastodon chimborazi* from Águas de Araxá, Minas Gerais, Brazil. Their results suggested a stable population that lived in an open environment.

The gomphotheres were the only proboscideans that entered South America, even though Mammutidae and Elephantidae ranged into southern Central America. Sánchez et al. (2004:159) suggested that the "most likely explanation for the absence of *Mammut* and *Mammuthus* in South America is that they were highly specialized feeders with habitat preferences not represented in the Panamanian land bridge." However, there have been reports that Elephantidae also entered South America. Elephantid remains from Guyana were reported during the 1930s, but their identification as elephantid was questioned. The specimen was subsequently

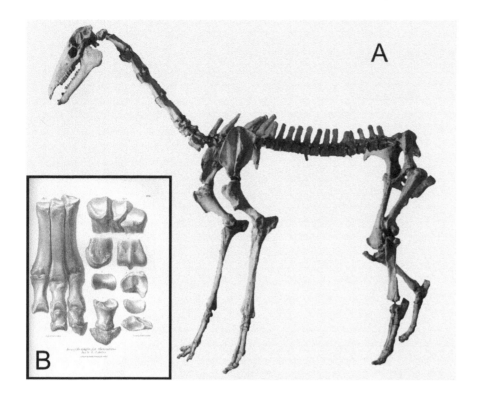

6.16 A) Complete skeleton of *Macrauchenia* in lateral view. B) Detail of the anatomy of the manus after Owen (1839). Skeleton length: 3 meters.

lost, and claims of an elephantid presence in South America were generally forgotten. However, Nascimento et al. (2010) reported a new elephantid specimen, a partial tooth, from late Pleistocene deposits in the state of Rondônia (Brazil). Although this report is an abstract (and thus further analysis and rigor are required), elephantid teeth (see Chapter 4) are difficult to misidentify, and so this report is intriguing. Among the surprises is that Rondônia is far south, by 3000 km, of other localities yielding elephantid remains, suggesting that the elephantid presence cannot easily be attributed to the accidental meanderings of one or a few individuals. Nascimento et al. (2010) suggested that the previous absence of elephantid remains from South America may be due to their having been restricted to the forested Amazonian region, where fossils are not easily formed or recovered.

> Atlantogenata
> Meridiotheria
> Meridiungulata
> Litopterna
> Macraucheniidae
> *Macrauchenia*

Litopterna comprised a now entirely extinct, splendid group of mammals that were among the ancient inhabitants of South America since the early Cenozoic (Pascual et al., 1965). *Macrauchenia* (Fig. 6.16), meaning "big neck," was first described by Owen (1838), on the basis of limb bones and vertebrae collected by Darwin in Puerto San Julián in Santa Cruz Province, Argentina, in 1834. The first-named and best-known species, *M. patachonica*, was the last representative of the entire group, possibly with the tropical form *Xenorhinotherium* (Cartelle and Lessa, 1988),

6.17 Complete skeleton of *Toxodon* as displayed in the exhibition of the Museo de La Plata.

Courtesy Museo de La Plata.

although the validity of this genus was questioned by Guérin and Faure (2004b). They vaguely resembled living camels, but the retracted position of the large, elliptical nostrils of *Macrauchenia* suggests the presence of a trunk. This, or the alternative proposal of a sphincter to close the nostrils, must have been useful for enduring dust and sandstorms in the arid regions that these animals inhabited (Lavocat, 1958). However, Sefve (1924), interpreting the retracted nostrils in an entirely different way, suggested that they were evidence of an aquatic existence. In any case, it seems that this animal must have been adapted to the climates of different latitudes, from southern Chile to northeastern Brazil (a region reconstructed as more humid than today, Guérin and Faure, 2004b) and even the coast of Venezuela (Hoffstetter and Paskoff, 1966; Hoffstetter, 1968; Guérin and Faure, 2004b). Also, they seem to have coped with the environments of the somewhat high altitude (~2000 m) of Tarija (Hoffstetter, 1986) and nearly 4000 m with the species *M. ullomensis* of other regions of Bolivia near La Paz (Ulloma, Charaña, and Ayo Ayo).

Its estimated mass is about 1 tonne (Fariña et al., 1998), while Fariña et al. (2005) obtained an estimate of 1100 kg with a scale model. To avoid predation (especially by the sabertooth *Smilodon*), this large litoptern seems to have been particularly adapted to swerving behavior (Fariña et al., 2005), as discussed in Chapter 7.

Atlantogenata
Meridiotheria
Meridiungulata
Notoungulata
Toxodonta
Toxodontidae
Toxodontinae
Toxodon

Toxodon (Fig. 6.17) and its related tropical form *Mixotoxodon* are further examples of the last representatives of a glorious lineage that is today completely gone. Pleistocene toxodonts are among the most interesting examples of the multitudinous notoungulates because they were about the size of rhinoceroses and hippopotami and probably shared some of the habits of these large living mammals. *Toxodon* has also been blessed by fate as being one of the first, if not the first, major sources of inspiration for Darwin's (1859) theory of evolution, as indicated by his comments in his diary:

> The Toxodon, perhaps one of the strangest animals, ever discovered: in size it equalled an elephant or megatherium, but the structure of its teeth, as Mr. Owen states, proves indisputably that it was intimately related to the Gnawers, the order which, at the present day, includes most of the smallest quadrupeds: in many details it is allied to the Pachydermata: judging from the position of its eyes, ears and nostrils, it was probably aquatic, like the Dugong and Manatee, to which it is also allied. How wonderfully are the different Orders, at the present time so well separated, blended together in different points of the structure of the Toxodon! (Darwin, 1839:59)

The skull remains used by Owen (1840) to erect this genus had been bought by Darwin (1839:112) in western Uruguay for 18 pence (equivalent to £5.60 today, according to the Web page Rate Inflation, or a bit less than US$10), thus saving it from being destroyed by children honing their stone-throwing skills.

Its upper incisors were strongly arched, hence the name *Toxodon* ("arched tooth") bestowed by Owen, whereas its lower incisors were horizontally arranged. Their great lateral expansion gave the lower jaw a giant, spadelike appearance. Owen did not seem impressed by the animal's intelligence: he judged that the direction of the occipital foramen, then used as a measure of how smart a person or animal was, was nearly the opposite of humans. The estimated mass of *Toxodon*, on the basis of allometric equations by Fariña et al. (1998), is over a tonne. A scale model was also used to estimate the mass of *Toxodon platensis* (Fariña and Álvarez, 1994), and this, too, turned out to be 1100 kg. Jerison (1973) reported the same estimate.

Atlantogenata
Xenarthra
Pilosa
Tardigrada

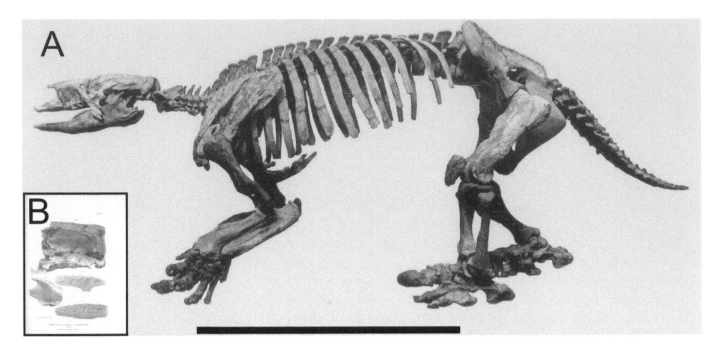

6.18 A) Complete skeleton of *Scelidotherium* after Lydekker (1894). B) Detail of skull and teeth after Owen (1839). Bar: 1 meter.

Courtesy Museo de La Plata.

Mylodontinae
Scelidotheriinae
Scelidotherium

There were several scelidothere genera, some of which are discussed further below. *Scelidotherium leptocephalum* (Fig. 6.18) is a commonly recovered scelidothere, closely related to S. *parodii*. Its estimated body mass ranges between 600 kg and 1 tonne, according to Fariña et al. (1998). Bargo et al. (2000) used a volumetric estimation and obtained a figure of 850 kg from a scale model and 830 kg from a computer model.

Vizcaíno et al. (2001) discussed several Pleistocene burrows in the Mar del Plata area (Buenos Aires Province, Argentina) and interpreted some of them as paleoburrows built by fossil animals on the basis of morphological patterns, transgressive boundaries in relation to the sedimentary units (i.e., they extend through rock units of different age), and the presence of claw marks on the walls and ceilings of the burrows. The paleoburrows are discrete structures of similar size, several meters in length, and with subrounded cross sections. Their diameters range from 0.80 to 1.80 m, and the width generally exceeds the height. Some remain open and only partially filled with sediment. These structures occur in Pleistocene deposits containing fauna referable to the Ensenadan and Lujanian Ages. The evidence identifying *Scelidotherium* and *Glossotherium* as the probable excavators of the paleoburrows is presented in Chapter 7. One of the burrows preserved the skeletons of three individuals of the short-faced bear *Arctotherium*, as noted above.

The remains of late Pleistocene ground sloths, including scelidotheres, have been frequently found preserved in natural caves and rock shelters, which suggests that use of caves, for whatever reasons, was a common behavioral pattern in ground sloths (as explained more fully in Chapter 7). One possible use was to escape large-sized carnivoran predators, such as the sabertooth cat *Smilodon* and the short-faced bear *Arctotherium*.

These mammals were possible predators of these sloths, if not of the adults then at least of the juveniles; and we have already noted that *Arctotherium* occasionally also made use of such burrows. Another possibility involves climatic and physiological factors. Sloths may have used burrows to avoid alternatively excessively cold or warm climatic conditions. During the early Pleistocene, the climate of the pampas was warmer than today, but the last part of the Pleistocene was mainly colder and much more arid, and a dry steppe developed in this area. In such an environment, the sloths may have needed a warmer place to breed, or even to survive, during the colder season.

> Atlantogenata
> Xenarthra
> Pilosa
> Tardigrada
> Mylodontidae
> Scelidotheriinae
> *Catonyx* and *Valgipes*

Two other scelidothere genera are considered here not only because they are part of the Lujanian megafauna, but also because they serve as examples of how taxonomic practice can lead to widespread confusion. The two genera are treated together because their taxonomic histories are so intertwined: to discuss one without numerous references to the other would be impossible. The story of these genera is a classic case of tortuous taxonomic mixups. This is fairly common in paleontology, but the situation with these scelidotheres was so convoluted that it led researchers to believe in the existence of things that weren't there. The story was unraveled by Cartelle et al. (2009), whose conclusions have produced a profound reinterpretation of the mylodontid (and megalonychid) sloth fauna in Brazil. Ironically, the solution to the taxonomic problems was provided by P. W. Lund over 150 years ago and corroborated nearly 100 years ago by H. Winge; we've had to go a long way to get back essentially to the same place.

Lund (1846) concluded that two Pleistocene scelidotheres (which he had initially considered as megalonychid sloths) had existed in Minas Gerais, Brazil. Winge (1915) reviewed Lund's material and came to the same conclusion, though these authors referred to these scelidotheres by different names, none of which turned out to be valid (Cartelle et al., 2009). Lund recognized *Scelidotherium owenii* and *S. bucklandi*, and Winge *S. magnum* and *Catonyx giganteus*. Another element of this story includes Gervais's (1874) erection of *Valgipes deformis* on a partial calcaneum (heel bone), also from Minas Gerais, that Winge (1915) agreed represented a scelidotheriine. So far, apart from some possible confusion over names, so good.

The problems began with Reinhardt (1875), who based *Ocnopus laurillardi* on a tibia–fibula (fused bones of the shin region; Appendix 2) that Lund (1846) had assigned to *S. bucklandi* and a tooth of a juvenile giant ground sloth (Megatheriidae) that Lund (1842) had correctly identified as a megatheriid sloth. Paleontologists have long realized that Reinhardt erred with regard to the tooth. Next came Hoffstetter (1954b), who really caused turmoil. Hoffstetter determined that Winge had erred in recognizing *Scelidotherium magnum* (which had been described from Argentina) for the Brazilian material, which should have been assigned to *Scelidodon cuvieri*. Hoffstetter was correct in this instance (although this species is now placed in the genus *Catonyx*). Hoffstetter did not stop here. He considered Winge's *C. giganteus* invalid because it was based on an erroneous association of remains: the skull, mandible, and teeth belonged to the scelidotheriine *C. cuvieri*, whereas the postcranial skeletal remains belonged to a megalonychid. Among these postcranial remains was the tibia–fibula that Reinhardt (1875) had attributed to *Ocnopus laurillardi*. Hoffstetter (1954b) thus proposed a new combination, *O. gracilis*, known only from the postcranial remains that Winge (1915) had illustrated and two isolated megalonychid-like molariforms (i.e., cheek teeth) that Hoffstetter assigned to the species. The teeth were no doubt meant to reinforce the megalonychid nature of the postcranial remains, but they were not associated with the postcranial remains (that is, there was no reason to suppose they belonged to the same individual). The peculiarities of the alleged megalonychid *O. gracilis* prompted Hoffstetter (1954b) to propose a new subfamily Ocnopodinae.

Moreover, Hoffstetter reflected on the partial calcaneum (heel bone) on which Gervais had erected *Valgipes deformis*. Noting that its overall ax-shaped tuber (the actual end of the heel) resembled that of *Megalonyx*, he also considered it to be an odd megalonychid. Hoffstetter's opinions were unfortunately accepted by subsequent paleontologists, who labored for nearly 60 years under the mistaken impression of the past presence of strange megalonychids in Brazil. This has been frustrating, particularly for workers exploring Brazilian deposits hopeful of finding more remains of these strange megalonychids.

This episode in the history of vertebrate paleontology shows that even great paleontologists can be wrong; it also illustrates that their mistakes, like their good work, may have widespread influence. Another great paleontologist was swayed by Hoffstetter's authority. During the 1970s, C. de Paula Couto wrote a manuscript describing an incomplete (though truly megalonychid) skull from São Paulo (Brazil) and assigned it to *O. gracilis*. Here was the missing skull of this species, confirming Hoffstetter's view of its megalonychid affinities. Fortunately, Paula Couto's manuscript was never published.

One final twist to this story is that recently Guérin and Faure (2004a) erected a new scelidothere species *Scelidodon piauiense* on the basis of a skull, mandible, and humerus from Piauí (Brazil). The assessment of Cartelle et al. (2009) of the scelidothere situation in Brazil was based on numerous recently uncovered remains, and they determined that there were two Brazilian Pleistocene species, as Lund and Winge had initially believed on the basis of considerably less material, and swept away any ideas on odd Brazilian megalonychids (though one further point needs to be related, but more on this below). The valid names for these scelidotheres are *Catonyx cuvieri* and *Valgipes bucklandi* (which includes, of course, the partial calcaneum). *Scelidodon piauiense* turns out to be a synonym of the latter species.

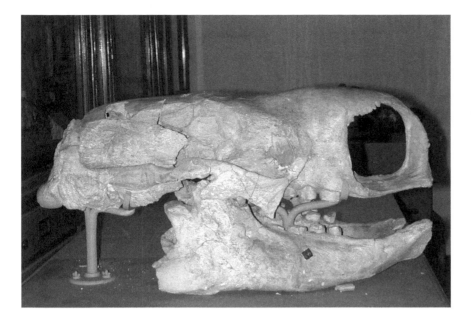

6.19. Lateral view of the skull of *Mylodon darwinii* kept in the Zoologisk Museum in Copenhagen. Total skull length: 61 cm.

Atlantogenata
Xenarthra
Pilosa
Tardigrada
Mylodontidae
Mylodontinae
Mylodon

Another kind of sloth is named *Mylodon darwinii* in honor of the great evolutionist. To *Mylodon* belong the skin and feces found in southern Patagonia. Embedded in the skin were numerous dermal ossicles, which had previously only been found associated with the skeleton, so their position in life had been speculative. A lateral view of the skull housed in the Zoologisk Museum in Copenhagen, collected in Argentina, is included in Fig. 6.19. Christiansen and Fariña (2003) estimated its mass to be from 1 to 2 metric tonnes.

As mentioned above, *Mylodon* is known to have occupied, at least seasonally, colder environments farther southward, near the southern extreme of continental Patagonia. Physiology provides a possible explanation for this behavior. Living xenarthrans have low body temperatures, low basal rates of metabolism, and high thermal conductance, characteristics that influence the geographical limits of their distributions (McNab, 1985). Mylodontid ground sloths probably also had a low metabolism, as suggested by their relatively small tooth occlusal surface area (Vizcaíno et al., 2006; see Chapter 7). Hibernation is not known to occur in living xenarthrans, but the Patagonian armadillo, *Zaedyus pichiy*, enters a state of torpor during winter, when the availability of insects, one of its main food resources, declines dramatically.

The hide of *Mylodon*, found in a cave in southern Patagonia, is one of the most remarkable fossil remains of this already astounding Lujanian fauna (Fig. 1.9). The story of its discovery, narrated by Hauthal et al. (1899), is worth recounting. In 1895, the German landowner Hermann Eberhardt and several of his guests were exploring caves near Última Esperanza Sound when they came across a portion of hide. A year later, the great Norwegian explorer Otto Nordenskjöld visited the cave and collected another piece of hide, as well as claws and fur balls. He took them to the Natural History Museum of Stockholm, where they were studied by Einar Lönnberg (1899). In 1898, part of this material was sent back to South America, to the Museo de La Plata, Argentina. Previously, that institution had shown interest in the remains and had sent an expedition to the area. This party included the prominent Argentinian naturalist Francisco Pascasio Moreno, the zoologist Emile Racowitza, geologist Rodolfo Hauthal, and engineer Luis Álvarez, who had already been in the cave with Eberhardt. Although they apparently never entered the cave, they took the hide collected by Eberhardt to the Museo de La Plata. This specimen was later described and presented by Arthur Smith-Woodward in a meeting of the Zoological Society of London (Moreno and Woodward, 1899). It is possibly the specimen currently housed in the collections of the Natural History Museum in London, as Eberhardt's skin specimen has never made it back to the Museo de La Plata.

Hauthal began the first systematic excavation of the cave in 1899 and produced the collection currently housed in the Museo de La Plata, including a piece of hide preserving evidence of having been artificially cut, as well as bones of other mammals and human-made tools. This assemblage led scientists of the day to consider the possibility of an interaction, distinct from a predator–prey relationship, between humans and the fossil fauna. As implied by the name *Grypotherium domesticum*, erected erroneously by Roth (1899) for the *Mylodon darwinii* remains found in the cave, the distribution of elements encountered at the fossil site was considered evidence that they had been intentionally arranged to produce a primitive (by modern standards, at any rate) enclosure, a pen of sorts, where ground sloths were corralled as though they were giant cattle. They based this novel point of view on the fact that the dung attributed to *Mylodon* was located in an area that could have been easily restricted. Moreover, grass was found arranged in such a way that suggested it had been stored there for feeding the animals. Although the site is no longer interpreted as evidence for domestication, the hypothesis formulated more than a century ago is an interesting example of how our scientific knowledge expands.

Martinic (1996) and Borrero (2001) reviewed the available evidence. The accepted interpretation today is that the rocks forming the enclosure were not intentionally placed in position but fell from the cave ceiling at various times over the past 12,000 years. Beneath them are still intact stratigraphic sequences providing evidence that *Mylodon* and other mammals used this and other caves in the area as natural shelters. Moreover, this might have been true for early South American Paleoindians, as suggested by the archaeologic evidence, which also indicates that in this remote part of the continent, humans must have scavenged on *Mylodon* and other megamammals rather than having developed specialized techniques for hunting them (Borrero, 2001, 2009).

Atlantogenata
Xenarthra

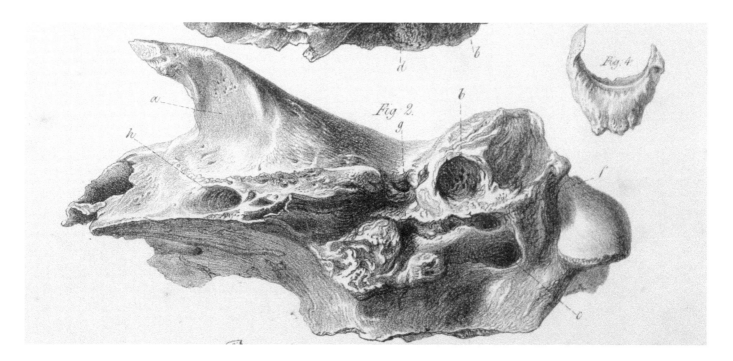

Pilosa
Tardigrada
Mylodontidae
Mylodontinae
Glossotherium

Similar to *Mylodon*, but somewhat larger and with a shorter, wider rostrum, the first remains of this genus were recovered by Darwin from the same Arroyo Sarandí in Río Negro, Uruguay, where the holotype of *Toxodon* had been found. The remains were studied by Owen (1840), who erected *Glossotherium* for them, on the basis of the posterior half of a skull. In his *Beagle* diary, Darwin (25 November 1833; Keynes, 2001:203) noted that "We heard of some giants bones, which as usual turned out to be those of the Megatherium—With much trouble extracted a few broken fragments." Owen (1840) derived the name *Glossotherium* ("tongue beast") from the presence of a large, almost circular depression, about 2.5 cm in diameter, that lies close to the ear region on the underside of the skull (Fig. 6.20). Owen knew that this depression served for the attachment of the bony framework, the hyoid apparatus, supporting the tongue, and he realized that the beast had been endowed with an enormous tongue, a supposition supplemented by the very large size of the cranial passage for the nerve that controls the tongue. In combination with other skull features, Owen made a strikingly accurate prediction, given that he had only a fragment of the skull to work with, on the form of the beast's limbs (though our current thoughts on its dietary habits differ). Here is Owen's (1840:57–58) description:

> The importance of this surface [that for the jaw joint] in the determination of the affinities of a fossil animal has been duly appreciated, since the relations of the motions of the lower jaw to the kind of life of each animal were pointed out by Cuvier; but yet we should be deceived were we to establish, in conformity with the generalization enunciated by Cuvier, our conclusion, from this surface, of

6.20. The name *Glossotherium* ("tongue beast") was derived from the supposed presence of a large tongue, in part based of the large (nearly 2.5 cm in diameter), almost circular depression (labeled b) lying close to the ear region on the underside of the skull. The depression received the proximal end of the hyoid apparatus, the bony structure that supports the tongue.

From Owen (1840: fig. 2,b, pl. XVI).

the nature of the food of the extinct species under consideration; for the glenoid cavity is so shaped as to allow the lower jaw free motion in a horizontal plane, from right to left, and forwards or backwards, like the movements of a mill-stone; and, nevertheless, I venture to affirm it to be most probable, that the food of *Glossotherium* was derived from the animal and not from the vegetable kingdom; and to predict, that when the bones of the extremities shall be discovered, they will prove the Glossothere to be not an ungulate but an unguiculate quadruped, with a fore-foot endowed with the movements of pronation and supination, and armed with claws, adapted to make a breach in the strong walls of the habitations of those insect-societies, upon which there is good evidence in other parts of the present cranial fragment, that the animal, though as large as an ox, was adapted to prey.

Let us explain the significance of this passage and Owen's methods. His prediction on the morphology might seem nothing short of miraculous, but in the final analysis, his prediction was pedestrian indeed. Owen was careful to point out that he was relying on Cuvier's method of the correlation of parts, dealt with in Chapter 2, but also that he did Cuvier one better in concluding that the glossothere had claws rather than hooves. We venture to suggest that Owen was being a bit of a show-off. Owen tipped his hat to Cuvier's influence (i.e., the correlation of parts), but in the same breath, he suggested that although Cuvier might conclude that the form of the jaw joint would predict a cowlike animal, his own deeper understanding was that the glossothere's front limbs would have claws rather than hooves. Owen lets on that he surmised this from the fact that the beast's tongue must have been enormously large, much more so than in any typical herbivorous mammal (e.g., a cow), and that it must therefore have used its tongue the way an anteater would; thus it would have required claws to break into ant and termite nests. He was right about the claws, but Owen might have been being a sneak—he knew full well that the glossothere would have claws, not because of its large tongue, but because he was entirely aware that he was dealing with a xenarthran (he classified it as such, after all, though as Edentata), and he knew that xenarthrans, living and extinct, have claws rather than hooves. Had science not then known about sloths, particularly giant sloths (remember that *Megatherium* was already famous and that Owen himself had thoroughly described the beast), then predicting claws on the basis of the depression for the hyoid apparatus alone would have been a real feat. On the diet, however, Owen was incorrect, according to our modern view, for *Glossotherium* is considered herbivorous rather than myrmecophagous, as pointed out in Chapter 7.

The taxonomic history of *Glossotherium* is complex and intricately tied to that of *Mylodon darwinii* (Fernicola et al., 2009). We recount the story as another example of just how finicky paleontologists have to be in getting their (scientific) names right. Owen (1842) erected the species *Mylodon robustus* on remains including a nearly complete skull; he assigned to *Mylodon darwinii* (which was based on a dentary) the cranial fragment that he (Owen, 1839) had previously described as *Glossotherium*. Reinhardt's (1879) detailed description of a fossil sloth skull and mandible from Pergamino (Buenos Aires Province) recognized that (1) the mandible was similar to that described as *Mylodon darwinii* by Owen (1840), (2) its skull features were sufficiently distinct as to suggest generic separation from *Mylodon robustus*, and (3) the cranial fragment originally assigned by Owen (1839) to *Glossotherium* was closely allied generically to *Mylodon robustus*.

Not recognizing *Mylodon darwinii* as the type of the genus, Reinhardt (1879) proposed the new genus *Grypotherium*, in which he placed the dentary assigned by Owen (1839) to *Mylodon darwinii* and the specimen from Pergamino as the species *Grypotherium darwinii*; and he recognized *Mylodon robustus* as the type species of *Mylodon*. Ameghino (1889) accepted the generic differences noted by Reinhardt (1879), but considered, as Owen (1842) had before, that the cranial fragment of *Glossotherium* and the dentary of *Mylodon darwinii* belonged to the same species, and so included these specimens—as well as the skull and mandible from Pergamino—in *Glossotherium darwinii*, given that in this scenario *Glossotherium* has priority over *Grypotherium*. Smith-Woodward's (1900) revision of Darwin's South American fossil sloth collection concluded that the cranial fragment originally assigned to *Glossotherium* was congeneric with the specimen assigned to *Mylodon robustus* by Owen (1842) and that the dentary of *Mylodon darwinii* and the specimen from Pergamino, described by Reinhardt (1879), were conspecific. However, because Smith-Woodward did not recognize *Mylodon darwinii* as the type species of *Mylodon*, he resurrected Reinhardt's (1879) *Grypotherium*.

Kraglievich (1928) modified the taxonomy of these taxa on the basis of a detailed revision of the group. This author held that the root of the problem was the rejection of *Mylodon darwinii* as the type species of *Mylodon* and the lack of agreement on the assignment of the cranial fragment assigned to *Glossotherium* by Owen (1839). Once Kraglievich (1928) had established *Mylodon darwinii* as type species of *Mylodon*, *Grypotherium* fell as a synonym of *Mylodon*. Kraglievich (1928) agreed with Reinhardt (1879) and Smith-Woodward (1900) that the cranial fragment of *Glossotherium* was congeneric with *Mylodon robustus* but not conspecific with it. Consequently, and with the general understanding that *Mylodon robustus* was generically distinct form *Mylodon darwinii*, Kraglievich (1928) revalidated *Glossotherium*, but with two species, i.e., *Glossotherium robustus* and *Glossotherium uruguayense*, the latter including Owen's (1839) cranial fragment. These nomenclatural conclusions were accepted by Cabrera (1936), although in this author's revision of the species of *Glossotherium*, he recognized its two valid species as *Glossotherium robustum* and *Glossotherium lettsomi* (Owen), with *Glossotherium uruguayense* a synonym of the latter. (*Glossotherium lettsomi* was originally assigned to *Pleurolestodon lettsomi* by Gervais and Ameghino, 1880, based on observation of a skull exhibited in the Natural History Museum, London, that had been labeled by Owen as *Mylodon lettsomi*; see Ameghino, 1889). Esteban's (1996) review of Mylodontinae considered *Glossotherium lettsomi* (sensu Cabrera, 1936) a synonym of *Glossotherium robustum*, so that the cranial fragment collected by Darwin in Uruguay is currently assigned to the latter species.

Christiansen and Fariña (2003) estimated the body mass of a gracile morph of *Glossotherium robustum*, originally described as a species of the genus *Mylodon*, *M. gracilis*. Fariña et al. (1998) estimated the mass of *Glossotherium robustum* as between 1 to 2 metric tonnes by using allometric equations. On the other hand, Bargo et al. (2000) used a graphic model and obtained figures of 1500 kg from a scale model (see Fig. 7.1) and 1200 kg

6.21. Lateral view of a skull of *Lestodon armatus*. Its large caniniform tooth gives it the appearance of a predator, from which its name, meaning "thief tooth," is derived. Scale bar equals 10 cm.

Image courtesy of Susanna Bargo, with permission of the Museo de la Plata.

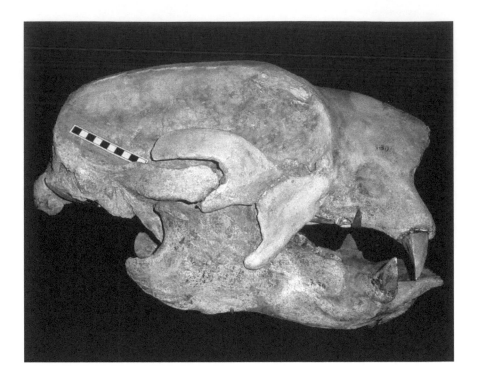

from a computer model. The result of the mass estimation by Christiansen and Fariña (2003) of the gracile morph of *Glossotherium robustum*, by using allometric equations based on craniodental and limb bone measurements from three specimens, was between 500 kg and 1000 kg. McNab (1985) proposed that cold-temperature tolerance of some ground sloths in North America (including *Glossotherium*) probably stemmed from a thick fur coat and continuously available food. However, only *Mylodon* (Moreno and Woodward, 1899) and *Nothrotheriops* (Lull, 1929) are known to have possessed a good thick fur coat.

> Atlantogenata
> Xenarthra
> Pilosa
> Tardigrada
> Mylodontidae
> Mylodontinae
> *Lestodon*

Lestodon was one of the real giants during Lujanian times—among sloths, only the giant megatheres were larger. Even though xenarthrans lack true incisors and canines, the first tooth in *Lestodon*, caniniform in shape, is displaced far forward in the jaw and protruded as a small tusk (Fig. 6.21). Fariña et al. (1998) estimated its mass at more than 3 tonnes. Bargo et al. (2000) obtained masses of 4100 kg with a scale model and 3750 kg with a computer model. Again in this case, some measurements yield exaggerated estimates, as is the case of the almost 38 tonnes obtained based on the transverse diameter of the femur. On the other hand, this rhino- to elephant-sized beast had a tibia (shinbone) as short as that of a 200-kg mammal.

A curious point about *Lestodon armatus*, the only valid Pleistocene species of the genus (Czerwonogora and Fariña, 2011), is its distribution.

Its remains are commonly found in Uruguay, east of the Río de La Plata, where *Megatherium americanum* is much less abundant. The latter, on the other hand, is common in Argentina, west of the estuary; *Lestodon* is less common. Clearly, the distribution of these taxa suggests some degree of competitive exclusion, or perhaps a subtle ecological difference between the pampas on either side of the estuary. These considerations have yet to be addressed in the scientific literature. Another interesting paleobiological feature is the potential social organization that these mammals might have had, according to evidence furnished by their massive ear ossicles (Blanco and Rinderknecht, 2008), which imply the use of low-frequency sounds for long-range communication as in elephants, and the preservation of several skeletons of varied size together (Arribas et al., 2001; Fariña and Castilla, 2007) in the site of Arroyo del Vizcaíno (see Chapters 4, 8, and 9).

The big caniniform tooth present in many specimens, but particularly in the larger ones, led the author of the genus, the French paleontologist and entomologist Paul Gervais, to name it "thief tooth," implying it was, or at least looked like, a predator. Another genus with one species, *Mylodonopsis ibseni*, has been reported for the northeastern Brazilian state of Bahia (Cartelle, 1991). It is quite similar to *Mylodon darwinii*, with a bony bridge linking the nasals and premaxillae, but it retains a small caniniform tooth (absent in *M. darwinii*) set in a wider muzzle (Cartelle, 1999).

Atlantogenata
Xenarthra
Pilosa
Tardigrada
Megatheriidae
Megatheriinae
Megatherium

Among the largest of sloths, and together with gomphotheres and the more tropical giant sloth *Eremotherium*, the species *Megatherium americanum* is one of the largest members of the Lujanian megafauna. It may also have been the largest bipedal mammal (see Chapters 5 and 7). It had the powerful claws characteristic of Lujanian sloths, but in this beast, they were really exaggerated. It has been suggested that the claws were used to strip bark from trees, which it would then eat. Other possible uses are explored in Chapter 7. Dámaso Larrañaga, one of the earliest South American naturalists, worked on this mammal, as did several other important early scientists, such as Cuvier, Owen, and Ameghino—indeed, its importance to the early development of paleontology has been recounted in Chapter 2. Its great size, particularly given its early discovery, provided an obvious name for it, for *Megatherium* means "great beast." Fariña et al. (1998) obtained highly variable mass estimates, between about 3 to 6 tonnes, by using allometric equations. Casinos (1996) obtained a figure of 3800 kg by using scale models.

Reconstructions of *Megatherium americanum* usually portray this giant ground sloth with a thick furry coat (for example, as in Figs. 6.22 and 7.10). This appearance is based largely on the fact that most mammals possess a good amount of hair covering their bodies and the demonstrated

6.22. Reconstructions of *Megatherium america-num* usually portray this giant ground sloth with a thick, furry coat.

Drawing by Smit in Hutchinson (1893).

presence of hair in the sloths *Mylodon* and *Nothrotheriops*. However, the possibility that *Megatherium*, given the implications of its large body size and geographic distribution, may have been largely hairless is considered in Chapter 7.

Megatherium americanum is one of several Pleistocene *Megatherium* species described in the literature. Several other smaller species, such as *M. medinae*, *M. tarijense*, and *M. sundti*, are clearly distinct from *M. americanum* and almost certainly valid species. They are known mainly from the western part of South America such as Chile (De Iuliis, 1996, 2006) and Bolivia (although *M. americanum* is also known from this country), as is the earlier and even smaller *M. altiplanicum* (Saint-André and De Iuliis, 2001). However, many other species have been named for remains that almost certainly belong to *M. americanum*. The plethora of names is a reflection largely of past practices in vertebrate paleontology, where almost any scrap of bone was deemed fit for naming a new species and justified on such dubious grounds as geographic provenance (e.g., from a different country, or even different provinces within a single country) or minor morpho-logical variation—a bump of bone here, a bigger groove in a tooth there. Such typological methodology is no longer viewed as justifiable under our modern biological paradigm, with the understanding it has given us on population biology and the existence of considerable individual variation within species. Unfortunately, there are still elements of typological think-ing among researchers. De Iuliis (1996) used the information on variation gleaned from the large samples of *Eremotherium laurillardi* (Cartelle, 1992; De Iuliis, 1996) to support a similar range of variation in the large-sized

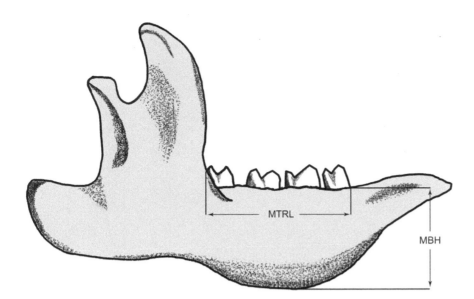

6.23. Figure of the mandible of *Eremotherium* illustrating the variables used for the hypsodonty index, calculated as MBH/MTRL × 100, where MBH is mandibular body height and MTRL is mandibular tooth row length.

Drawing by Sebastián Tambusso.

Megatherium remains, mainly from the Pleistocene of Argentina, essentially on the basis of a series of remains from Luján, the classic as well as type locality for *M. americanum*. Among the synonyms of this species are *M. lundii*, *M. gaudryi*, *M. parodii*, and *M. gallardoi*.

However, Brandoni et al. (2008) argued for the latter's validity, even though all the features purportedly validating this species can be found in other remains that undoubtedly belong to *M. americanum*, except possibly a relatively lower mandibular ramus and unfused premaxillae, all features that are based on only a single skull. These were features that De Iuliis (1996) recognized but suggested were more likely the result of intraspecific variation. Certainly there is some doubt with regard to the depth of the mandible, given that the recognition of this character is based on the hypsodonty index (HI), defined as follows: (maximal mandibular depth/tooth row length) × 100 (Fig. 6.23). The lower HI in the single *M. gallardoi* skull is in part due to the way tooth row length is measured. For all other *M. americanum* skulls, the length was measured from the teeth, but the aggregate alveolar (the alveoli are the sockets in the jaw bones in which the teeth are set) length was used in *M. gallardoi* (because its teeth are not preserved). This overestimates the length of the tooth row and hence underestimates the relative depth of the mandible. The minimum HI in other *M. americanum* specimens is 92. Brandoni et al. (2008) reported a value of 87, and HI as calculated by measurements provided by De Iuliis (1996) is 90. However, if we allow for the fact that these values are based on alveolar tooth row length, then HI approaches a value of 91 for the *M. gallardoi* type (that is, reducing tooth row length by about 4 mm over a total length of about 225 mm). The point is, what are we to make of such small differences, in part resulting from measurement error, when the range of HI based on all other *M. americanum* specimens is between 92 and 112? As for the other purportedly diagnostic feature of the type of *M. gallardoi*, although premaxillae are usually found fused to the skull in *M. americanum*, this is not always the case, and in such instances, it should be assumed that they were unfused and thus lost after death. In the case of *M. gallardoi*, the premaxillae happen to be unfused, but fortunately they were found with the skull. We have not introduced these aspects of *Megatherium* as arguments for resolving its taxonomic status—this is not an appropriate arena—but rather to provide a detailed example of the sorts of systematic issues that paleontologists deal with and the difficulties in trying to resolve them.

Atlantogenata
Xenarthra
Pilosa
Tardigrada
Megatheriidae
Megatheriinae
Eremotherium

As noted above, *Eremotherium laurillardi* rivaled and perhaps surpassed *M. americanum* in mass. Its estimated mass is also about 4 tonnes, but its limb bones are longer than those of *M. americanum* (Fig. 6.24). It has only three digits in its hand, with the first and second strongly reduced, as opposed to four in *Megatherium*, and its skull is less robust (Fig. 6.25). It is found in lower latitudes, so it may be considered a tropical megathere. So far as is known, this genus is never found south of the Brazilian state of Rio Grande do Sul, whereas *Megatherium* occurs in Uruguay but not in Brazil. The range of this species was vast—so much so that *E. laurillardi* has been dubbed the Panamerican giant ground sloth. From its southern limits in Brazil, its range extended well into North America, as far north as the U.S. state of New Jersey.

As with *M. americanum*, numerous species (e.g., *E. mirabile, E. carolinense, E. cucutense*) were erected on remains of *E. laurillardi*, seemingly one for each country in which its remains were encountered. Cartelle (1992), De Iuliis (1996), and Cartelle and De Iuliis (1995, 2006) analyzed the large samples of this giant ground sloth from two principal localities from nearly opposite ends of its range: Toca das Onças (Minas Gerais, Brazil) and Daytona Beach Bonebed (Florida, USA), as well as the sparser remains from localites between and beyond these two sites (such as Venezuela, Ecuador, Mexico, and Panama, as well as other Brazilian and U.S. states). These authors argued for their conclusion that there was a single Panamerican species mainly on the basis of the fact that the two largest collections could not be distinguished on either size or morphology. In fact, the Toca das Onças collection alone subsumes nearly all metric and morphological variation reported in the literature or observed by these authors.

Cartelle and De Iuliis (1995) also corroborated the hypothesis of marked sexual dimorphism in this species (an idea first proposed by Cartelle and Bohórquez, 1982), establishing a variation of nearly 35% in linear measurements among both adult and juvenile members of a single population. This was based on the astragalus (an anklebone), which is among the most frequently recovered skeletal elements in the sample recovered from Toca das Onças. Among adults, linear dimensions vary by nearly 35% between the largest and smallest astragali. Juvenile astragali exist that are about the same size as the largest adult and smallest adult astragali, so that the difference between these juvenile astragali is also nearly 35%. Of course, there are also many juvenile astragali that are considerably smaller that the smallest adult astragalus. Thus, nearly identical proportions exist among the

6.24. The great size of the giant megatheriine ground sloths is easily appreciated in this image of a mounted skeleton of *Eremotherium laurillardi* (from Daytona Beach Bonebed, Florida, USA), with a child sitting beneath it.

Courtesy of Daytona Museum of Arts and Science.

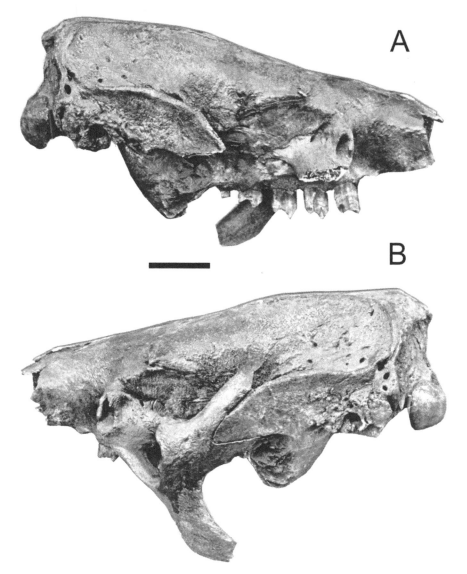

A

B

6.25. A skull of *Eremotherium laurillardi* from Minas Gerais, Brazil. In the top figure, showing the skull in right lateral view, the zygomatic bone is missing, revealing the robust dentition. Each of the first four molariforms has sharp front and back crests separated by a deep V-shaped valley. The bottom figure of the same skull in left lateral view preserves the zygomatic bone with its long descending process. Also missing from the skull are the premaxillae, the bone at the front of the palate. The scale bar represents 10 cm, so the length of this skull as preserved is nearly 70 cm, but it would reach nearly 80 cm with the premaxillae in place.

From De Iuliis (1996). Courtesy of Cástor Cartelle.

two groups of larger—presumably mainly male—and smaller—presumably mainly female—adult-sized individuals, but the groups include both adults and juveniles. Although the elements tend to fall into these groups, there is no distinct separation between them, so that a continuous size range exists. Although there are fewer of the other skeletal elements (though they still comprise reasonably large samples), the same pattern may be identified: they tend to fall into large-sized and small-sized groups, each of which includes adult and juvenile remains (Cartelle and De Iuliis, 2006). Significantly, a nearly identical range of morphological variation exists within the large-sized and small-sized groups, so that there is no morphological justification to support their specific distinction. An almost identical range of size variation was documented by De Iuliis and Cartelle (1999) for an earlier species of the same genus, *E. eomigrans*, known from Florida, so it seems likely that a profound sexual dimorphism may have been typical of giant-sized megatheres.

Despite the data presented by Cartelle and De Iuliis in their various publications, Guérin and Faure (2000) argued for the recognition of a dwarf species of *Eremotherium* based on remains in Brazil, to which, they suggested, the name *E. laurillardi* should properly be applied. However, as noted by Cartelle and De Iuliis (2006), these authors misinterpreted two fundamental points of sloth morphology. The basis for this dwarf species is a maxilla that is not of an adult individual but rather a (fairly young) juvenile. This maxilla fits easily within the ontogenetic trajectory established by Cartelle and De Iuliis (2006) for the Panamerican giant sloth.

The second point is that Guérin and Faure pointed to the existence of small-sized, parallel-sided molariforms as indicative of adults of their concept of a dwarf species. This argument rests on the belief that molariforms of juvenile megatheres are tapered at their tops and those of adults are parallel sided. However, as Cartelle and De Iuliis (2006) pointed out, and as had been reported earlier in the literature, the molariforms are tapered only in young juveniles, but parallel sided in older juveniles; in other words, a parallel-sided molariform is not necessarily an adult condition.

An interesting condition in *Eremotherium laurillardi* is that the skin was probably embedded with numerous dermal ossicles, as is the case in *Mylodon*. Although skin has not been reported for *Eremotherium*, several of its skeletal elements have been found in situ with numerous small, circular, and mound-shaped ossicles resting atop the bones (Cartelle and Bohórquez, 1986).

Atlantogenata
Xenarthra
Pilosa
Tardigrada
Megalonychidae
Ahytherium

The megalonychids are a diverse clade mainly from Central America and the West Indies, where they are abundantly represented and highly diverse (e.g., *Meizonyx*, *Acratocnus*, *Parocnus*, *Megalocnus*, and *Neocnus*), and from which the oldest certain member of the clade has been recovered

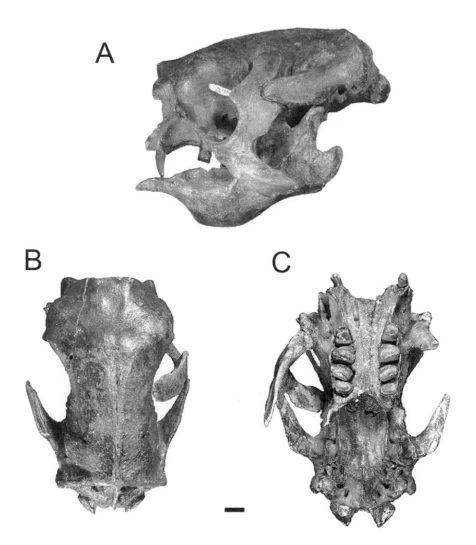

A

B

C

6.26. Skull and mandible of *Ahytherium aureum*. The skull and mandible are shown in left lateral view in the top image, whereas the skull only is shown in dorsal view (bottom left) and ventral view (bottom right). Among the features that characterize this megalonychid sloth from Brazil are its short but wide rostrum (muzzle). Scale bar = 2 cm.

Images courtesy of François Pujos. Fossil in the collection of the Museu de Ciências Naturais da Pontifícia Universidade Católica de Minas Gerais, Belo Horizonte, Brazil.

(Cartelle et al., 2008). They are reasonably abundant, though much less diverse in North America (e.g., *Megalonyx* and *Pliometanastes*). They have a long history in South America, known mainly from Argentina. They were reasonably abundant in the Miocene but later became scarce (following the classification that excludes the nothrotheriids from megalonychids).

Three genera, *Pliomorphus*, *Megalonychops*, and *Diabolotherium* (the latter a climbing sloth from Peru; Pujos et al., 2007), seem reasonably well established, but several other genera have erroneously been considered as megalonychids until recently (Cartelle et al., 2008, 2009). Two of these, *Ocnopus* and *Valgipes*, were noted above as having long been considered odd (and as it turned out, nonexistent) megalonychids. A third genus, *Xenocnus*, falls into the same category. *Xenocnus* was erected by Paula Couto on a partial unciform (one of the wrist bones) of the Panamerican giant ground sloth (*E. laurillardi*) that he mistook for the astragalus (an anklebone) of yet another odd megalonychid (Cartelle et al., 2008). This seems to be a glaring blunder, given that Paula Couto must have been familiar with remains of *E. laurillardi*. Although this is yet another example that shows that even the great paleontologists are not immune to error, we wonder whether in this instance Paula Couto's judgment was clouded, conditioned as he must have been by the (supposed) existence of odd megalonychids in Brazil. Paula Couto thought that *Xenocnus* was such an

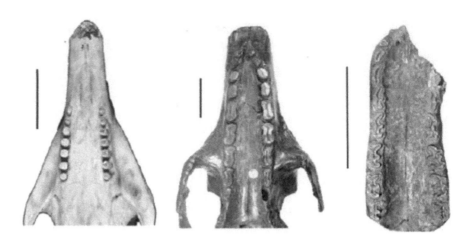

6.27. Diagram illustrating the form of the teeth among cingulates (scale bar = 5 cm). The typical armadillo pattern of rather simple, peglike, and circular to oval teeth is represented by the armadillos *Dasypus*. The teeth of pampatheres and glyptodonts are more complex, with pampatheres having mainly bilobate teeth and glyptodonts having mainly trilobite teeth.

From Vizcaíno et al. (2009) by kind permission of Paleobiology.

odd megalonychid that it too, like *Ocnopus* and *Valgipes*, deserved its own subfamily, the Xenocninae, which he formally erected.

Cartelle et al. (2008, 2009) sorted out these problems on Brazilian scelidotheres and megalonychids. There were no odd megalonychids in Brazil, though it does seem that they remained rather rare in the Plio-Pleistocene of South America. These authors (2008) reported a new megalonychid species and genus, *Ahytherium aureum* (Fig. 6.26), which is a perfectly typical megalonychid. It is smaller and more gracile than the well-known *Megalonyx jeffersoni* from the Pleistocene of North America. Its skull is markedly inflated dorsally, with a notably wide and shortened rostrum. As well, it has a remarkably distinct tail, with vertebrae that widen progressively until approximately midlength, then taper to the tail tip. A second megalonychid, *Australonyx aquae*, was reported from Brazil by De Iuliis et al. (2009), but its remains are less well known that those of *Ahytherium*.

Atlantogenata
Xenarthra
Cingulata
Glyptodontia
Pampatheriidae
Pampatherium and *Holmesina*

Pampatheres (Fig. 6.27) are larger than dasypodid armadillos, with *Pampatherium typum*, the smallest of the Pleistocene representatives, reaching 95 kg (Abrantes et al., 2005) and with an estimated maximum transverse body diameter at about 80 cm (Gervais and Ameghino, 1880). In comparison, the Pleistocene armadillo *Propraopus* was only about 50 kg (Fariña and Vizcaíno, 1996), and the living species *Priodontes maximus* is 45 kg (Fariña and Vizcaíno, 1997). Among the more notable differences between armadillos and pampatheres is the form of their teeth: whereas all the teeth of armadillos tend to be homodont (i.e., all of similar form), simple, peg shaped, and oval in section, the more posterior teeth of pampatheres, central image, are bilobate (Fig. 6.27).

The taxonomy and systematics of pampatheres are unsettled and require revision. One of the main reasons for the confusion lies in the imbalanced reliance on osteoderms. For example, Edmund (1996) attempted to recognize genera and species solely on the patterns of osteoderm size and ornamentation. In one sense, such a practice is inevitable, as osteoderms of pampatheres (and of cingulates generally) are rather common fossils in South American Plio-Pleistocene deposits, whereas their skeletal remains are much less abundant. There has been particular confusion with the Pleistocene pampatheres, which are usually classified either in *Pampatherium* or *Holmesina*. Some authors, such as Cartelle and Bohórquez (1983), proposed that all species should be included in *Pampatherium*, and hence recognized *P. paulacoutoi*. However, Cartelle (1992) transferred this species to *Holmesina*, but later (Cartelle, 1994b) reverted to *P. paulacoutoi*. The problem is compounded by the exceedingly few examples of good skull and mandible remains from any single South American locality that might provide insight into the extent of intraspecific variation. De Iuliis and Edmund (2002) noted that osteoderms, especially from different localities and in the absence of associated skull and mandible material, are unreliable in generic and specific recognition. There is considerable variation in size and ornamentation among osteoderms of a single individual. Thus, association of additional skeletal elements is highly desirable and necessary for confident taxonomic assignment.

To this end, Scillato-Yané et al. (2005) suggested that a combination of osteoderm and craniodental features, as well as size, allows the recognition of the two genera and five species of Pleistocene pampatheres. These authors recognized *Pampatherium* and *Holmesina*, with the following species: *P. humboldti*, *P. typum*, *H. majus*, *H. occidentalis*, and *H. paulacoutoi*. The osteoderms of *Holmesina* bear a conspicuous central figure, delimited by two lateral furrows, which Scillato-Yané (1982) considered a plesiomorphic (or ancestral) condition for all cingulates (Fig. 6.28). The central figure may be a narrow ridgelike protuberance or more rounded. The osteoderms of *Pampatherium* are nearly flattened and without a central figure in *P. typum* and slightly sculpted in *P. humboltdi* (Figs. 6.28 and 6.29). Measurements of the long bones indicate that *H. majus* is the largest of the pampatheriids, whereas *H. paulacoutoi* is the most robust (Scillato-Yané et al., 2005). *Pampatherium typum* is the smallest and least robust of the pampatheriids. In South America, *H. occidentalis*, as its name implies, is known from the northwestern part of the continent, including Colombia, Ecuador, Peru, and Venezuela. *H. majus* is known with certainty only from Brazil, whereas *H. paulacoutoi* is reported from Brazil and Argentina. *P. typum* had a wide distribution that extended through parts of Bolivia, Brazil, Paraguay, and Argentina, and *P. humboldti* is known from Brazil and Uruguay (Scillato-Yané et al., 2005).

The phylogenetic and biogeographical histories of *Pampatherium* and *Holmesina* were discussed by Scillato-Yané et al. (2005). Interestingly, these authors suggested that the *Holmesina* lineage likely differentiated in North America from more generalized pampatheres (ultimately, of course, from South America), either from an earlier representative of *Holmesina* or the older genus *Kraglievichia*. *Holmesina* then reentered South America, likely along a western coastal route, as evidenced by *H. occidentalis*, the other species of the genus differentiating later in South America. In North America, the genus is represented by *H. septentrionalis*. *Pampatherium*, on the other hand, underwent its main differentiation in South America, with the earlier genus *Vassallia* considered the sister group of *Pampatherium*. The latter genus also spread north, represented by *P. mexicanum*.

1 cm

6.28. Osteoderms of *Holmesina paulacoutoi*, bearing a conspicuous central figure, delimited by two lateral furrows. The smaller osteoderm, on the left, is typical of those composing the carapace over the shoulder and hip regions. Often termed "buckler osteoderms," these were the fixed elements of the caparace. In pampatheres, three rows or bands of movable, or imbricating, osteoderms (the number is usually higher and more variable in armadillos) lay between the shoulder and hip regions. The longer osteoderm here is from the movable band region. The front end of these osteoderms (toward the top in the figure) comprise a short, smooth, and nearly squared surface that lay and moved against the underside of the longer rugose region of the osteoderm in front of it. The figure below it is a section through the rugose region.

Based on Scillato-Yané et al. (2005).

6.29. Osteoderms of *Pampatherium*. The surface of the osteoderms are nearly flattened compared to *Holmesina*. The several examples of buckler osteoderms, the smaller elements on the right side, demonstrate the variability in their shape.

Photograph by Sergio F. Vizcaíno by kind permission of the Museo de La Plata, Argentina.

Among the more interesting aspects of pampathere paleobiology is that they may have been avid excavators of burrows. There is little doubt that the armadillos *Eutatus* and *Propraopus* built the larger burrows (about 1 m in diameter) found in several areas of southern South America, but the much larger pampatheres may also have excavated some of them. Edmund (1985) noted that the presence of moveable bands in the carapace gave the skeleton more flexibility than in glyptodonts, which almost certainly did not excavate burrows (see Chapter 7). Vizcaíno et al. (2001) summarized the research postulating the possibility of pampatheres as excavators. For example, whereas Quintana (1992) proposed *Eutatus, Propraopus,* or *Pampatherium typum* as the probable excavators of a paleocave near Mar del Plata (southern Buenos Aires Province, Argentina), Imbellone and Teruggi (1988) and Imbellone et al. (1990) tentatively attributed some burrows near La Plata (northeastern Buenos Aires Province, Argentina) to *Eutatus seguini* or *Pampatherium typum,* and Bergqvist and Maciel (1994) attributed large crotovines (paleoburrows that have been filled in by sediment) in Rio Grande do Sul (southern Brazil) to the pampatheres *Pampatherium* and *Holmesina,* and to *Propraopus.* More recently, Buchmann et al. (2010) attributed several other such structures to *Propraopus* or *Eutatus* and provided a schematic reconstruction of a paleoburrow (Fig. 6.30). Although the possibility of pampatheres as excavators has received support, largely on the form of the burrows, we should note that anatomical details suggest we should be cautious about such claims: the relative length of the olecranon process (the proximal end of the ulna, or funny

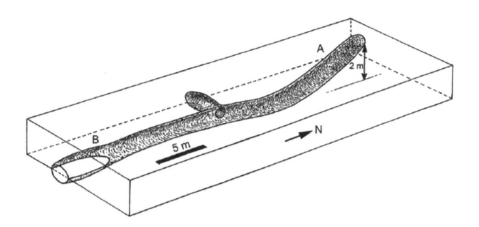

6.30. The partial reconstruction of a paleoburrow system found in Rio Grande do Sul, Brazil, demonstrates how extensive such structures could be. Buchmann et al. (2010) suggested that this paleoburrow was likely constructed by a large armadillos such as *Propraopus* rather than by pampatheres.

Reproduced by kind permission of F. S. Buchmann.

bone) of pampatheres is considerably less than in the fossorial armadillos. A long olecranon process improves the moment arm of the triceps muscle, which acts as an extensor of the forearm, and within mammals, this design is appropriate for diggers. Even among some unspecialized clades of mammals, species that are known to dig more frequently show relatively longer olecranon processes than those that do not dig (Vizcaíno et al., 1999). Abrantes et al. (2005) analyzed limb proportions of *Pampatherium* using several indices developed for armadillos (Vizcaíno and Milne, 2002). The results led them to conclude that *Pampatherium* was not particularly well suited for digging behavior.

The dietary habits of Pleistocene pampatheres were investigated by De Iuliis et al. (2000), who studied the masticatory apparatus of *Pampatherium typum, P. humboldti,* and *Holmesina paulacoutoi.* Comparison of the apparatus with those of other pampatheres suggests a trend toward increasing ability to process resistant vegetation from *Holmesina* through *Pampatherium.* The mechanical design of the apparatus among these pampatheres is nearly identical; the main differences lie in features associated with the musculature, and thus the primary means for differentiation in masticatory function, which is correlated with diet, were changes resulting in differential force input. Although the taxonomic status of some specimens is equivocal, indications are that the Plio-Pleistocene paleobiogeographic distribution of pampatheres is correlated with masticatory function (and hence diet), with *P. typum,* the species best adapted for grinding coarse vegetation, occurring in the more arid Pampean regions of South America. At the other extreme, *H. occidentalis,* known from deposits near the current Peru–Ecuador border, an area of humid lowlands during glacial maxima, was the least suited to coarse vegetation, although it was still a capable grinder. Other species lie between these extremes.

Atlantogenata
Xenarthra
Cingulata
Glyptodontia
Glyptodontidae
Glyptodon

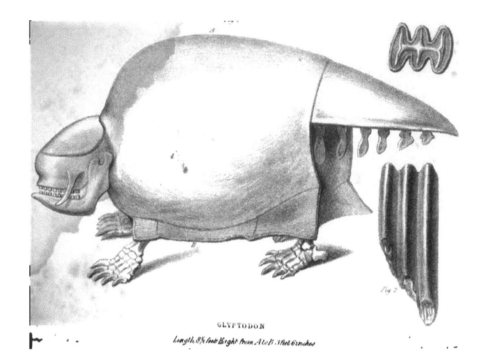

6.31. This early illustration of *Glyptodon,* though crudely executed, shows the main characteristics of this group of xenarthrans: the presence of a cephalic casque (or head shield), the long descending process (a downward bony extension) of the zygoma (the cheek region), the peculiar sculptured trilobate teeth, and the essentially immobile carapace.

From Owen (1839).

It should come as no great surprise to learn that the remains of glyptodonts, particularly of *Glyptodon,* were among the first fossils collected in the Río de la Plata region, and are those most commonly found among Lujanian megamammals. The latter circumstance is largely the result of the paleontologically recent times during which these beasts existed and probably their abundance, but also to the heightened potential for fossilization produced as a consequence of the high number of osteoderms (as many as 1500) that composed the carapace of each individual

In the brief note first describing *Glyptodon* (with the first described species, *G. clavipes*), Owen (1839c:178) was already able to report several other singular features of glyptodonts: that the vertebral column was "from the neck to the sacrum . . . altogether" and the ilia (the main hip bones) united the sacrum (the part of the vertebral column joining the hips) in "one single and immovable piece." As well, the great anatomist noted the presence of a cephalic casque (or head armor), the long descending process (a downward bony extension) of the zygoma (the cheek region), and the peculiar "sculptured" teeth (Fig. 6.31). A few years later, Owen (1845a) identified several other species: *G. reticulatus, G. tuberculatus* (now known as *Panochthus tuberculatus*), and *G. ornatus* (currently *Neosclerocalyptus ornatus*).

As in all glyptodonts, the tail was completely armored (Fig. 6.32), but it lacked a proper caudal tube formed by many ankylosed (or fused) terminal bony rings. Instead, the tail was covered by individual bony rings.

This genus has been found in several South American countries (Argentina, Bolivia, Brazil, Chile, Paraguay, Peru, Uruguay, and Venezuela). The estimated mass of the largest specimen of *Glyptodon clavipes* is about 2 tonnes, as based on a scale model (Fariña, 1995). The smaller *Glyptodon reticulatus* must have had a body mass of nearly 1 tonne, according to the estimates made by Fariña et al. (1998), who used allometric equations on cranial and postcranial measurements.

6.32. Carapace and tail of *Glyptodon clavipes*. A readily identifiable feature of this genus is that the osteoderms of the tail are arranged as individual rings that are movable relative to each other, rather than being ankylosed together to form a solid caudal tube. Total length: approximately 3.5 meters.

From Lydekker (1894).

Atlantogenata
Xenarthra
Cingulata
Glyptodontia
Glyptodontidae
Doedicurus

Owen (1846) described yet another glyptodont specimen. Although he had only the distal portion, or caudal tube, of a very large tail, Owen was sufficiently impressed with its widened extremity, which reminded him of big club, that he erected a new species for it, *Glyptodon clavicaudatus*, the latter derived from the Latin, meaning "club-tailed." Nearly three decades later, Burmeister (1874) reclassified this species, distinguishing it from those of the genus *Glyptodon* by its osteoderms, which lack ornamentation and instead are pierced by two or more rather big holes, and, of course, the monumental caudal tube. Burmeister saw in the latter a general similarity to a pestle, from which its name, *Doedicurus*, is derived: in Greek, δοῖδυξ means "pestle"; and ουρά refers to "tail"). The caudal tube of *Doedicurus* is even more formidable than in other glyptodonts, reaching about a meter in length, but this is just the length of the caudal tube; the entire tail was considerably longer (Fig. 6.33). The surface of the widened end of the caudal tube bears large depressions to secure cornified and perhaps spiky structures. The thick osteoderms of this genus are not as firmly fused to each other as in other glyptodonts.

Doedicurus was allied to *Glyptodon* in a clade termed Glyptodontinae on the basis of several technical details of the skull, such as the relative positions of some of the foramina (or passages through the bones of the skull), a circular orbit (the depression that houses the eyeball), and the shape of the occipital condyles (the bony skull protuberances that form a joint with the first vertebra of the spinal column). Together with *Plohophorus*, *Glyptodon* and *Doedicurus* constitute Glyptodontidae (Fernicola, 2008; Fernicola et al., 2008).

6.33. Carapace and tail of *Doedicurus*. The figure clearly depicts the huge and highly domed carapace and massive caudal tube of *Doedicurus*. Note also the large terminal figures at the end of the tube that probably served for the attachment of scutes. Total length of carapace and tail about 330 cm.

From Dana (1896).

Doedicurus remains have been recovered from Argentina, southern Brazil, and Uruguay. Its mass was estimated by allometric equations as more than 1 tonne (Fariña et al., 1998) and as 1400 kg by scale models.

Atlantogenata
Xenarthra
Cingulata
Glyptodontia
Panochthidae
Panochthus

The name coined by Burmeister (1866–1867) for the genus he considered different from *Glyptodon* is based on the hellenization of the specific epithet that Owen (1845a) erected for the species "*G.*" *tuberculatus*. *Panochthus* comes from the Greek πᾶν, "everything" or "all," and ὄχθος, referring to an eminence like a hill or a riverbank, and also a tubercle in leprosy: each osteoderm of the carapace of *Panochthus* (except for those in marginal regions) is covered by little tubercles, as shown in Fig. 6.34.

As was obvious to Burmeister, who was able to conduct a more thorough analysis of this species because he had more complete remains available, *Panochthus* is distinguished by its highly domed carapace, which has a long pelvic region (Fig. 6.35). The caudal tube of *Panochthus tuberculatus* is almost as large and thick as that of *Doedicurus clavicaudatus*, but its extremity is not expanded. It also bears a series of elliptical rugose regions, termed lateral figures, along its lateral surface that probably mark the attachment of cornified pads or spines, but again, with or without such structures, it must have been a powerful weapon (see Chapter 7). In contrast to the depressed surfaces in *Doedicurus*, the lateral figures of *Panochthus* bear conspicuous, conical tubercles. *Panochthus* is currently known from Argentina, Bolivia, Brazil, Paraguay, and Uruguay. The animal's estimated mass is over 1 tonne (Fariña et al., 1998). According to Fernicola et al. (2008), *Panochthus* and *Hoplophorus* form a sister group relationship.

Atlantogenata
Xenarthra
Cingulata

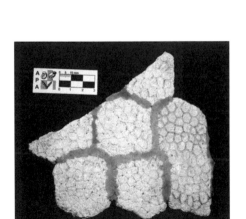

6.34. Osteoderms of *Panochthus*. The osteoderms of this genus each bear several tubercles. The individual osteoderms of this specimen were glued together with mastix, an adhesive that was commonly used in paleontological preparations during the late nineteenth and early twentieth centuries. Mastix was composed of resin, beeswax, paraffin, and plaster. The resin was originally obtained from a shrub native to the Mediterranean region. Scale bar = 3 cm.

Courtesy of the Museo de La Plata, Argentina.

6.35. The massive carapace and caudal tube of *Panochthus.* Note the several lateral and terminal figures on the latter. Total length of carapace and tail about 2.6 meters.

From Lydekker (1894).

Glyptodontia
Panochthidae
Hoplophorus

Along with *Dasypus antiquus* and *Glyptodon* (noted above), *Hoplophorus* (named after the familiar ancient Greek infantry; Box. 6.3), with its species *H. euphractus,* were the first names applied to glyptodont remains. Lund (1839) named it on the basis of material collected in the valley of the Rio das Velhas, Minas Gerais, Brazil. Perhaps because the author lived in the isolation of the nearly uncharted territory of the Lagoa Santa region in the Brazilian countryside, perhaps because he wrote in Danish, or perhaps because of a combination of the two (see Chapter 2), it was *Glyptodon* rather than *Hoplophorus* (which, in our opinion at least, conveys a more vivid image) that became the basis on which the whole group came to be identified (recalling that *Dasypus antiquus* was not considered valid). The first assignment of these genera to a separate clade (as distinct from Dasypodidae) was proposed by none other than Darwin's bulldog, Thomas Henry Huxley (1863, although it was presented before the Royal Society in January 1864), who named it Hoplophoridae. However, it was Glyptodontidae, created by Gray in 1869, that became anointed by researchers who worked on the group (although today we consider the entire group of glyptodonts as Glyptodontia; Fernicola, 2008).

The genus erected by Lund (1839), a large-sized glyptodont known from Brazil and Bolivia (Porpino et al., 2010), is characterized, following Paula Couto (1957), by a rather long and narrow skull compared with other glyptodonts. The long bones tend to be longer than in *Neosclerocalyptus,* and the armor is less thick than in other genera with large individuals. The osteoderms of the carapace tend to resemble those of *Panochthus* but with a nearly flat central figure. In the dorsal part of the carapace, they tend to be rounded rather than pentagonal or hexagonal, as in *Panochthus* and *Neosclerocalyptus.* Along the sides, however, they are pentagonal or hexagonal in all three genera (Porpino et al., 2010). The caudal tube is as long as the rest of the tail, with two large elliptical figures on each side, one of which

6.36. Caudal tube of *Hoplophorus* in dorsal (A), lateral (B) and ventral (C) views. These images clearly show the large elliptical figures of the caudal tube, each bearing a conical tubercle (as also occurs in *Panochthus*). At the terminal end of the tube are two rows of enlarged rounded plates, each with a conspicuous central figure, that lie on the top of the club between the distal figures. Scale bar = 10 cm.

Composed by Sebastián Tambusso from pictures kindly provided by Kléberson Porpino.

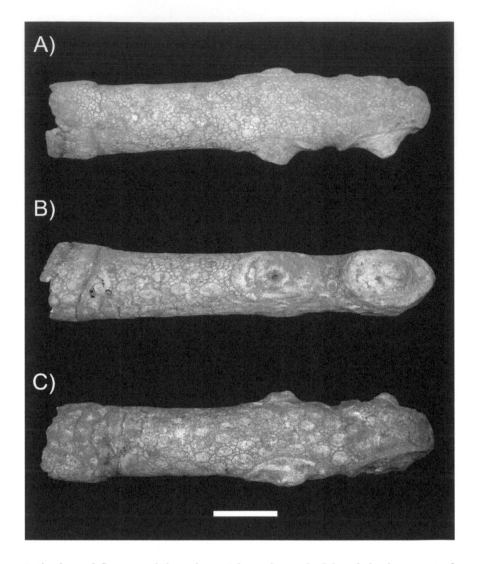

is the lateral figure and the other, right at the end of the club, the terminal figure. As in *Panochthus*, the figures bear conical tubercles (Fig. 6.36). Two rows of enlarged rounded plates, each with a conspicuous central figure, lies on the top of the club between the distal figures.

An interesting specimen of *Hoplophorus euphractus* includes a partial carapace associated with a tail club. The carapace has a hole, with at least two osteoderms lost, and Cartelle (1994b) hypothesized on the reason for their loss as follows. After its death, the remains of the individual were transported by water into a cave and then deposited. When recovered, the remains were found as they had been deposited, with the carapace lying above the tail. The carapace had become filled with heavy sediment with the passage of time and pressed against one of the spiky, cornified spikes, not yet decomposed, that probably attached to one of the figures on the tail club, until the spike pushed through and dislodged the missing osteoderms. It's certainly a plausible story, as well as a good example of deductive reasoning that allows us some insight on the condition of fossil remains.

Porpino et al. (2010) considered the phylogenetic relationships of *Hoplophorus*. These authors indicated a close relationship between *Panochthus* and *Hoplophorus* on the basis of several features, among which we may note that the last seven teeth are trilobate, the terminal figures of the caudal tube are separated dorsally (at the top), and a conical tubercle on the lateral and terminal figures is present.

What's in a name?   Box 6.3

Is there any rhyme or reason for the often strange or obscure names given to biological taxa? Well, yes, there usually are reasonably sound and logical, though not always obvious, reasons for how scientists come up with names to christen new groups of organisms. Naming privileges belong entirely to the scientist (or scientists) first describing a taxon, and the logic behind the name is usually part of the formal description. Paleontologists seem to be especially interested in such singular, some might say trivial, matters. We'd venture to say that our profession is particularly afflicted by an attraction to the arcane.

Let's consider some examples. Names are often based on a feature that strikes the scientist's imagination (e.g., *clavicaudatus*, "club tail," or *Megatherium americanum*, "big beast from America"—again, meaning the entire land mass stretching from Cape Horn to Alaska). Taxa may be named after a person, often a famous scientist (e.g., *Mylodon darwinii*, named after Charles Darwin), a once-famous scientist, the person who found the specimen on which the name is based (e.g., the no longer valid species *Mylodon listai*, named after the traveler Ramón Lista), a friend or family member (notably, the Eocene marsupial genus *Caroloameghinia*, named by Florentino after his brother), or a place (e.g., the astrapothere genus *Uruguaytherium*). One of the more, ahem, intriguing examples we've come across is the name of a trilobite (a group of extinct marine arthropods) species. Ludvigsen (1979a,b) erected *Ceraurus maewestoides* (though now assigned to a different genus) on the basis of its apparently curvaceous resemblance to Mae West, the U.S. actress and sex symbol famous especially during the 1950s and 1960s, thereby contributing to her immortality: centuries from now, when Mae West, the person, will have probably long been forgotten, her name will live on in the annals of science as a moniker for an extinct marine invertebrate. It is a consolation of sorts, we suppose. Ludvigsen's (1979:36) choice of language is revealing: "A *Ceraurus* with a forwardly expanding and smooth glabella [part of the head between the eyes] possessing a pair of large mammilate swellings on the anterior lobe and, behind these, four tubercles arranged in a square." However, Ludvigsen was just one among many in paying homage to Mae West. For example, a popular item of the surrealist movement was Salvador Dalí's Mae West Lips Sofa (1937); World War II Allied aircraft crew dubbed their inflatable, vest-style life preservers "Mae Wests" (presumably on their form and the rhyming of "breast" with "vest"); and "Mae West" is the term for a parachute malfunction, a partial inversion that contorts the canopy into a brassiere shape.

Getting back to xenarthrans, the name *Hoplophorus* is derived from the Greek ὅπλον, which was the shield, and more broadly the armor, of the infantry, better known as hoplites, of ancient Greece, and φέρειν, "to carry or bear," so the name thus means "armor bearer." Other taxon names are based on *Hoplophorus* that make sense only if one attempts to follow the logic of the scientists involved, usually by slogging through piles of old journal articles. As told by Simpson (1980), Ameghino (1887) created another genus, *Palaeohoplophorus,* "old armor bearer," based on an older glyptodont specimen that he considered possibly ancestral to *Hoplophorus,* and that, also in 1887, Ameghino created yet another genus, *Propalaeohoplophorus,* meaning "before the old armor bearer," on even older material that he considered possibly ancestral to *Palaeohoplophorus* (incidentally, a similar set of derivatives also exists for the anagrammatical *Plohophorus,* another glyptodont genus). Certainly, it's a neat story. But in telling it, Simpson, as so often happens with taxonomic matters, made several errors (not to denigrate Simpson—our admiration for him is made abundantly clear in Chapter 2, but even the best of paleontologists make mistakes, and given Simpson's stature, his are more likely than not to be propagated). In the first place, Ameghino named the first genus in 1883, not 1887; next, he actually spelled it *Palaehoplophorus*—there is no "o" after the "ae" diphthong, almost certainly because the "h" is silent in Spanish, Ameghino's native language, and he might have wanted to avoid a duplicated "o" sound that would have resulted with a doubling of this vowel; and lastly, he spelled the second genus *Propalaehoplophorus* (again, without the "o"). The latter was formally emended, illegitimately, by Castellanos (1932) to *Propalaeohoplophorus,* which, as "palaeo," makes more sense grammatically. Several prior authors, presumably not noticing the original spelling (e.g., Scott, 1903–1904), had already spelled it *Propalaeohoplophorus.* As noted, however, the emendation does not count, even though it might be better, more appropriate, or even what the original author might actually have intended (although this does not seem to be the case here: Ameghino consistently spelled both names without an "o" in subsequent publications, for the reason noted above), because the rules of taxonomic nomenclature clearly state that the oldest published name (given that a few other conditions are met) has priority. In addition to all this, Ameghino's names reflect concepts that we no longer consider to be accurate. Modern research does not support the ancestor–descendant relationships that Ameghino used as a basis to coin the names. No matter; we are stuck with them. This is a good thing, though. The rules try to maintain stability, and one of the ways to accomplish this is to adhere as strictly as possible to the priority criterion. Otherwise we'd waste time trying to agree on arbitrary name changes, and as must be clear from the preceding sentences, we already spend enough time on names.

Atlantogenata
Xenarthra
Cingulata
Glyptodontia
Panochthidae
Neosclerocalyptus

6.37. Carapace, tail, and cephalic shield of *Neosclerocalyptus*. The carapace of this genus is lower and flatter than of the other genera considered here. Total length of carapace, tail and cephalic shield about 2 meters.

From Lydekker (1894), who mistakenly assigned it to Lomaphorus.

Neosclerocalyptus (Fig. 6.37) is the smallest of the Lujanian glyptodont genera, with an estimated mass of about 300 kg (Fariña and Vizcaíno, 1995). Owen (1839, 1845a) first described it as *Glyptodon ornatus* on the basis of a few bits of a carapace collected near the Rio Matanza, then about 20 miles south of the city of Buenos Aires. The remains included four or five osteoderms, of smaller size than *Glyptodon clavipes*, the species that Owen had already described. This author also noted that the external surface of the osteoderms was relatively smoother than in *G. clavipes* and the central disc (his words; we'd call it figure today, as we do here) smaller as compared with the peripheral discs, of which there are seven on each osteoderm, forming a rosette pattern (Fig. 6.38). Its tail was protected by a caudal tube, smaller than that of *Hoplophorus*, that likely served in defense (Fig. 6.39). The lateral figures are smaller than those of *Panochthus* and *Hoplophorus*, do not have a markedly raised crestlike edges, and do not bear a conical tubercle.

Burmeister (1871) transferred Owen's *Glyptodon ornatus* to the genus *Hoplophorus*, which had been erected by Lund (1839) on the basis of *H. euphractus*. The latter is thus the original, or in more technical parlance, the type species of *Hoplophorus*. Ameghino (1891b) replaced this generic name with *Sclerocalyptus* (from σκληρός, "hard," and καλυρτὸς, "veil," or more generally, "cover"), a decision that contravened the rules of zoological nomenclature: *Sclerocalyptus* is therefore a synonym of *Hoplophorus*. A century later, however, researchers realized that *H. euphractus* is not sufficiently closely related to "*H.*" *ornatus* to share its generic name (Simpson, 1945; Paula Couto, 1957). Therefore, because *Sclerocalyptus* is not available (because it is a synonym of *Hoplophorus*—recall that *H. euphractus* is the type, and so long as the generic name of the type is valid, its synonym cannot be pressed into service), a new name was required. Paula Couto (1957) therefore proposed *Neosclerocalyptus*, so the species (formerly assigned to *Glyptodon*, then *Hoplophorous*, then *Sclerocalyptus*) became *Neosclerocalyptus ornatus*. Lydekker (1894) added his share to the confusion by describing and figuring the carapace and skull of this species under the genus *Lomaphorus*.

6.38. Osteoderms of *Neosclerocalyptus*. The rosette pattern with relatively large peripheral figures is typical of this genus. Scale bar = cm.

Photograph by S. F. Vizcaíno, by kind permission of the Museo de La Plata, Argentina.

6.39. Caudal tubes of several glyptodonts. Differences in the patterns of figures can be observed.

Composed after Lydekker (1894).

Doedicurus

Panochthus

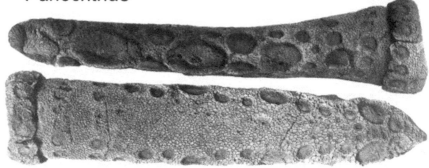

Neosclerocalyptus

Apart from the intricate story of its nomenclature, we may note that additional remains of *Neosclerocalyptus* are now known (Fig. 6.40) from several countries in South America (Argentina, Brazil, Bolivia, Chile, Paraguay, and Uruguay) and provide us with a better understanding of its anatomy, including its characteristically and greatly expanded nasal cavity.

Porpino et al. (2010) included *Neosclerocalyptus* in Panochthidae as the sister group to the clade formed by *Panochthus* and *Hoplophorus*. Among the characters that tie these three glyptodonts together, we may list the following: a relatively short distance between the back of the mandible to the front of the last cheek tooth; the width of the external nasal aperture (opening) is greater than the width across the occipital condyles (the processes at the back of the skull that articulate with the vertebral column); the infraorbital foramen (a passageway in the bony cheek region) is clearly lower than the orbit (the depression of the skull housing the eyeball); the lacrimal foramen (a passageway associated with the apparatus that produces tears) is within the orbit, rather than outside it; and the descending process of the zygomatic (a downward extension of the bony cheek region) is relatively long.

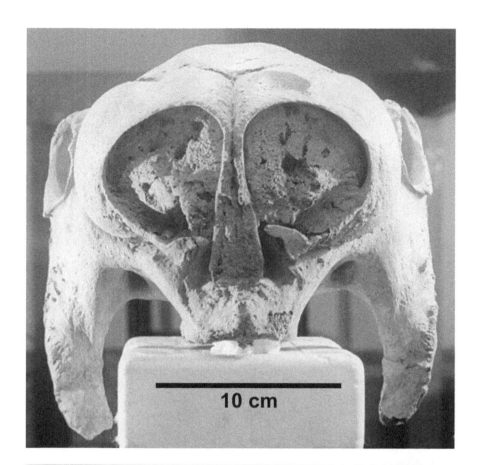

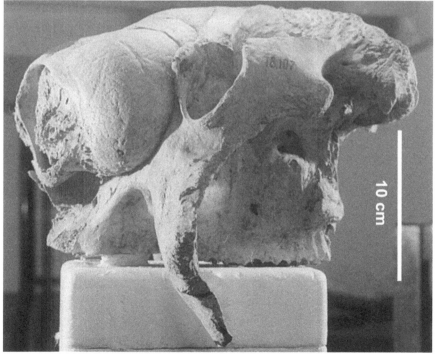

6.40. The skull of *Neosclerocalyptus* in (A) anterior and (B) lateral views, demonstrating the expanded nasal region in this genus.

Image courtesy of S. F. Vizcaíno, by kind permission of the Museo de La Plata, Argentina.

7.1. Scale models of giant sloths used in Bargo et al. (2000). From left to right, *Lestodon armatus, Glossotherium robustum,* and *Scelidotherium leptocephalum.*

Classical mechanics has been instrumental in shaping our views on the habits of extinct vertebrates. That this should be so might at first seem counterintuitive; after all, it is not unreasonable to expect that most animals—in the broad sense, say of fishes or the great cats—have always had the same general ways of life, and thus that understanding those of the present would provide ample insight for making sense of those of the past. We may confidently use the present as a guide to the past, but only when we are reasonably sure that the extinct forms are closely similar morphologically, physiologically, and phylogenetically to the extant forms we use as comparators. However, many of the animals now gone from existence were not, except in a general sort of way, like those still living, either in habits or ecological relationships; in more technical terms, we say that there are no modern analogues to guide us. For such cases, the use of modern forms as comparators often produces unsatisfactory or puzzling reconstructions, and paleontologists must resort to other means.

The most familiar example is that of the large nonavian dinosaurs. Only a few decades ago, these great beasts were generally viewed as being big, slow, dumb, lumbering, and, except for their great size, rather biologically uninteresting creatures. This impression largely came from combining their massive size with the characteristics of their closest modern nonavian relatives, the cold-blooded lizard- or alligator-like reptiles, which, it is now obvious, are not suitable models for dinosaurs. An overreliance on such simplistic transference of characteristics from the living to the long dead not only produces incorrect results, but also makes for uninteresting biology: it condemns the past to be just like the present. Our views on dinosaurs have changed tremendously since the 1960s. These dynamic beasts, now considered to be particularly intriguing examples of biological evolution (although a biologist would probably consider any group just as intriguing), were one of the first groups to be analyzed comprehensively by the tools offered by mechanics, beginning with the classic works of Alexander (e.g., 1983, 1985, 1989).

The use of such an approach involves the application of the principles of physics and engineering to paleobiological subjects in analyzing such aspects as locomotion, mastication, bone strength, and muscle reconstruction in the formulation of explanatory hypotheses on the lives of extinct beasts. Many of the fossil species that were part of the South American megafauna, and particularly the odd extinct xenarthrans, fall into the same category as the dinosaurs: classical mechanics combined with creative paleobiological approaches have done much to elucidate their otherwise puzzling morphologies. In this chapter, examples are provided of how paleontologists put Newton's ideas (especially on mechanics, but also on thermodynamics) to work in helping us, Darwin's followers, infer the habits of this remarkable assemblage of mammals. In this chapter, we examine some of the more resourceful efforts aimed at providing a better understanding of the life and habits of several of the extinct South American megamammals.

How do we tell a giant? A note on body size

Body size, which may be inferred for fossil mammals, is an essential part of the appeal of paleontology and a remarkable influence on an animal's life history, perhaps the single most important variable in terms of impact on all aspects of an organism (Peters, 1983; Schmidt-Nielsen, 1984; Damuth and MacFadden, 1990). Virtually all life traits are decisively influenced by or correlated with body size, including, for example, metabolism (Kleiber, 1932), limb bone dimensions and biomechanics of locomotion (Alexander, 1985, 1989; Alexander et al., 1979a; Fariña et al., 1997), population density and home range (Damuth, 1981a,b, 1987, 1991, 1993; Lindstedt et al., 1986; Nee et al., 1991; Reiss, 1988; Swihart et al., 1988), behavior and social organization (Jarman, 1974), and proneness to extinction (Flessa et al., 1986; Lessa and Fariña, 1996; Lessa et al., 1997). This is especially true in the case of very large mammals, as noted by Owen-Smith (1987, 1988). Ever since the earliest descriptions of the first collected specimens, the members of the Lujanian land mammal fauna, especially its xenarthran representatives, have impressed researchers and laypeople alike with their large size (Cuvier, 1804; Owen, 1838; Darwin, 1839; Burmeister, 1866–1867, 1874, 1879; Ameghino, 1887, 1889; Lydekker, 1894; Kraglievich, 1940; Patterson and Pascual, 1972; Simpson, 1980). Indeed, it prompted Darwin's (1859) correction of Buffon's assertion on the lack of vitality of South American forms.

Estimating body size, therefore, is a critical first step for investigating many of the biological parameters of these strange and spectacular species. Fariña et al. (1998) built up a database of mass estimates for several of these Lujanian species, including seven xenarthrans (three glyptodonts and four ground sloths) and six other mammals (a notoungulate, litoptern, perissodactyl, and proboscidean, and two carnivores); these taxa were selected as subjects of study because of the relative completeness of their skeletal remains. These authors obtained several skeletal measurements for these beasts and plugged them into the appropriate allometric equations for cranial, dental, and limb skeletal elements, as well as total body length data, as worked out for modern mammals by Anderson et al. (1985), Janis (1990), and Scott (1990). More recently, it has been complemented by a multivariate approach (De Esteban-Trivigno et al., 2008), in which the equations generated imply consideration of weighed variables of the limb skeleton. There is also an interesting general outcome from this study, important to the issue of body size and its influence on anatomy; because the database used, which was obtained from ungulates and xenarthrans, yielded five selected equations that were able to predict body mass in species from other, distant groups (rodents, carnivores, hyracoideans, and tubulidentates), the authors suggested the presence of a complex common allometric pattern for all quadrupedal placentals.

Another aspect to be stressed is that approaches that use one single method obviously have specific shortcomings. To avoid such inevitable biases, Fariña et al. (1998) compared these estimates with those obtained by the method used by Alexander (1985, 1989) to generate dinosaur body mass estimates for some of the species that compose the fauna treated in this book. The latter author's method is based on Archimedes' principle (Box 7.1).

Archimedes' principle   Box 7.1

An object immersed in a liquid experiences an upward force that equals the volume of liquid displaced by the object. Consider an object of weight W (mass = W/g, where g is the acceleration of gravity) and density ϱ immersed in a fluid of density ϱ'. The volume of the object is W/rg, so the weight of fluid it displaces is $W\varrho'/\varrho$. The net downward force on it, W', is given by the equation

$$W' = W - (W\,\varrho'/\varrho)$$

The scale models (Box Fig. 7.1) were submerged in water, so ϱ' equals 1000 kg m^{-3}. For *Glyptodon clavipes*, commercial models available through the Natural History Museum of London were used, but for those taxa lacking readily available models, models were specifically made (Box Fig. 7.1 and Fig. 7.1). Then the volume of water displaced by each model was weighed. The mass of a model, obtained from Eq. (1), was multiplied by the cube of the linear proportions (usually about 1/40). These calculations assume that the density of an animal is close to that of water, 1000 kg m^{-3}. This is not at all an unrealistic assumption; animal matter is made up largely of water. A precision balance was used to assess the mass of the model, with potential error being less than 0.5 g. After that, the volumes of the mammals were obtained by multiplying those of the models by the cubes of the linear ratios, as they are, for instance in the case of *Glyptodon*, 40 times longer, 40 times wider, and 40 times higher. Therefore, the final error introduced by multiplying by the scale should be less than 32 kg. Because the animals of this fauna had masses measured in hundreds of kilograms and tonnes, this source of error was not regarded as relevant.

Box Fig. 7.1. Scale model of a glyptodont (*Panochthus*) used for estimating body mass. It is posed on a real carapace, although one of a different genus (*Glyptodon*).

Now that technology is so ubiquitous, this procedure can be complemented by computer-generated models, from actual measurements on mounted skeletons, such as in Henderson (1999) and Bargo et al. (2000). Allometric equations describe how a variable (bone length, for instance) varies with body mass. These equations, which are based on relationships in modern organisms, have also been widely used to predict body masses from bone and tooth measurements (Damuth and MacFadden, 1990), although caution has recently been urged by Packard et al. (2008) against excessive extrapolation. In the cases of the mass estimates of the megafauna reported here and first published by Fariña et al. (1998) and Bargo et al. (2000), we claim they are valid, as a few dozen equations based on measurements of different anatomical regions (teeth, skull, long bones, and total length) were used and averaged, thus minimizing the risk of sampling wrong measurements. In other words, if an animal has larger teeth, longer skull, and wider femora (thighbones) than another, it is likely to be larger as a whole. The equations used and their source are listed in Appendix 3.

In Table 7.1, modified from Fariña et al. (1998), the mass estimations obtained by these equations are listed for several of the members of the megafauna. In many cases, some equations could not be used as a result of the unavailability of the measurement or as a result of the peculiarities of the anatomy of the appropriate species. In the case of the xenarthrans, the results showed a great dispersion, probably because the equations were obtained from stereotypical herbivorous mammals, the modern ungulates (Janis, 1990; Scott, 1990), and then applied to the peculiar xenarthrans. As usually results for xenarthrans, the use of allometric equations generated from a database of modern mammals (mostly perissodactyls and artiodactyls) yields curious predictions when applied to this phylogenetically distant clade. A body mass of only 40 kg is obtained for the very large *Glossotherium robustum* when the length of the tibia is used as the estimator, and more than 20 tons in the case of the transverse diameter of the femur. (Other, more extreme cases are mentioned below.) This suggests that as a result of the morphological peculiarities present in this group, estimates obtained from one or a few measurements must be regarded with caution.

The results for *Smilodon populator* are the basis for the claim made in Chapter 6 that the South American sabertooth, along with *Panthera atrox*, were among the largest felids to have ever walked the earth. This is hardly surprising because the accompanying herbivores, potential prey of these large felids, are also extraordinary in terms of size.

Most, if not all, Lujanian ground sloths were truly gigantic. The exception was *Scelidotherium leptocephalum*, the smallest of them, with an estimated body mass between 600 kg and 1 tonne, according to Fariña et al. (1998). Bargo et al. (2000) used a volumetric estimation and obtained a figure of 850 kg from a scale model (Fig. 7.1) and 830 kg from a computer model. De Esteban-Trivigno et al. (2008), by means of multivariate statistics, found it to have been positively svelte, at 580 kg. *Glossotherium robustum*, in turn, qualified as megafauna. Fariña et al. (1998) estimated its mass as between 1 to 2 metric tonnes by using allometric equations. Bargo et al. (2000) used a graphic model and obtained figures of 1500 kg from a scale model (Fig. 7.1) and 1200 kg from a computer model. Christiansen and Fariña (2003) estimated the body mass of a gracile morph of *Glossotherium robustum*, originally described as a species of the genus *Mylodon*, *M. gracilis*, by means of allometric equations that were based on craniodental and limb bone measurements from three specimens. The result obtained was between 500 kg and 1000 kg, which is close to the value of 750 kg obtained by a computer-generated model (a technique called double graphic integration analyses; Henderson, 1999) on a virtually complete skeleton. These results are congruent with the possibility that *G. robustum* may have displayed sexual dimorphism, with the gracile morph representing the presumably smaller female. Sexual dimorphism was identified for the similar North American mylodont sloth *Paramylodon harlani* by McDonald (2006), but in this species, the differences occur in skull proportions and caniniform wear pattern rather than overall size. Christiansen and Fariña (2003) estimated the mass of *Mylodon darwinii* as from 1 to 2 metric tonnes. The fourth mylodontid, *Lestodon armatus*, is one of the real giants during Lujanian times. Fariña et al. (1998) estimated its mass at more

Table 7.1. Mass estimations of several species of megafauna[a]

Species	No. of Equations	Arithmetic Mean	Geometric Mean	Median	Mode	Maximum Value	Minimum Value
Smilodon populator	27	352	328	347	350	744	127
Scelidotherium leptocephalum	39	1119	617	633	724	4059	21
Glossotherium robustum	38	1713	891	1041	1448	20092	40
Lestodon armatus	40	3397	1784	1918	2896	37706	205
Megatherium americanum	44	6265	2903	2543	2896	97417	524
Glyptodon reticulatus	43	862.3	403	457	362	7005	31
Panochthus tuberculatus	43	1061	528	701	724	9088	22
Doedicurus clavicaudatus	37	1468	613	708	512	10472	3
Stegomastodon superbus	23	7580	4311	2831	4096	56606	1458
Macrauchenia patachonica	66	988.1	830	781	1024	2843	123
Toxodon platensis	58	1642	1187	1191	724/2896	6795	213
Hippidion principale	66	511	476	483	512/500	993	193

[a] From Fariña et al. (1998). Arithmetic mean is the simple average, obtained by summing up the values and dividing by the number of cases. Geometric mean is a slightly more complicated calculation, with more robust results. It is obtained by multiplying the values and taking the *n*th root of the product (with *n* being the number of cases), and its results are less likely to be exaggerated overestimates. The median is described as the value separating the higher half of a sample from the lower half and the mode is the most frequently occurring value.

than 3 tonnes (Table 7.1). However, some measurements yield obviously exaggerated estimates, as is the case of the almost 38 tonnes obtained based on the transverse diameter of the femur. On the other hand, this rhino- to elephant-sized beast had a tibia as short as that of a 200-kg mammal. Still, those estimates were corroborated by using other approaches: Bargo et al. (2000) obtained masses of 4100 kg using a scale model (Fig. 7.1) and 3750 kg using a computer model, while De Esteban-Trivigno et al. (2008) used multivariate statistics to obtain a result of 3600 kg.

Among the largest of sloths, and together with mastodons and the more tropical giant sloth *Eremotherium*, the species *Megatherium americanum* is one of the larger members of the megafauna. It may also have been the largest bipedal (perhaps only facultatively—see Chapter 5) mammal known. Fariña et al. (1998) obtained diverse mass estimate values, between about 3 to 6 tonnes. Casinos (1996) obtained a figure of 3800 kg. The phylogenetic remoteness of this group from those used to construct the database from which the allometric equations are generated occasionally produces ridiculous estimates. For instance, the transverse diameter of the femur of *Megatherium* cannot be used for a realistic estimate of the body mass of this ground sloth because the value obtained is 97 tons—higher than the estimated mass of the largest dinosaur (Mazzetta et al., 2004).

Among cingulates, the body mass of *Pampatherium* was estimated at 200 kg by scaling down glyptodonts and assuming geometric similarity with them. Glyptodonts were larger, with *Glyptodon reticulatus* reaching a body mass of several hundred kilograms, according to the estimates of Fariña et al. (1998), and the largest specimen of *Glyptodon clavipes*, estimated by a scale model as reaching about 2 tonnes (Fariña, 1995). The mass of *Panochthus tuberculatus* was estimated as 1100 kg on the basis of a scale model by Fariña (1995). For *Doedicurus clavicaudatus*, Fariña et al. (1998) estimated mass as more than 1 tonne on the basis of allometric equations and 1400 kg on the basis of scale models. Even the armadillos were giants among their own kind: *Propraopus* and *Eutatus* were estimated as about 50 kg (Fariña and Vizcaíno, 1997; Vizcaíno et al., 2003), similar to the living giant armadillo *Priodontes maximus* (approximately 45 kg), and much beyond the mass of other armadillos.

The proboscidean "*Stegomastodon*" (we have already warned the readers about this generic name; see Chapter 6) was respectful of its heritage: at more than 4 tonnes, it reached the gigantic size common for its group. The strange South American ungulates were also giants, with *Macrauchenia patachonica* at about 1 tonne based on allometric equations (Fariña et al., 1998) and 1100 kg based on a scale model (Fariña et al., 2005), and *Toxodon platensis* slightly exceeding 1 tonne, by a scale model estimate (Fariña and Álvarez, 1994) and 1100 kg by allometric equations (Table 7.1). Jerison (1973) also reported an estimate of 1100 kg.

The South American horse *Hippidion principale* was assessed at about 500 kg by Fariña et al. (1998) by means of allometric equations. Alberdi et al. (1995b) estimated its mass as 460 kg by using a different set of equations. Clearly, the two estimates are in agreement for members of this species. The mass of another Lujanian equid, *Equus (Amerhippus) neogeus*, was similar to the 400 kg of an average modern horse, *E. caballus* (Prado and Alberdi, 1994). The large llama *Palaeolama* exceeded the modern representatives of the group at 300 kg, and the same is true for the extinct South American caviomorphs, which achieved the largest size of all rodents (Sánchez-Villagra et al., 2003; Rinderknecht and Blanco, 2008; Millen, 2008). The Pleistocene capybara *Neochoerus* was similar to its living relative, except that it was at least twice its size, with mass estimated at about 150 kg.

How athletic were glyptodonts?

Heavily armored and as enormous as they often were, the glyptodonts—the tanks of the South American Plio-Pleistocene—would probably be at the bottom of most people's list of athletic animals. To suggest that they were would surely raise many eyebrows; yet various aspects of these apparently ponderous beasts indicate that their paleobiology was much more interesting than traditionally thought. The ability to perform strenuous locomotor activities, such as jumping and galloping, may be assessed through the study of the long bones of the extremities. During strenuous activities, such elements are subjected to enormous stresses, and activities that require frequent bending are the ones most likely to cause fractures through fatigue. The strength of these bones may be studied through the application of beam theory, where a long bone is likened to a beam. A parameter Z, termed the section modulus, is obtained, as shown in Fig. 7.2. This quantity is related to the shape and area of the cross section of the bone, measured at a distance x from its distal extremity (Fig. 7.2). For instance, if the section were circular and with a diameter of 2 cm, its area would be πr^2, or 3.1416 $\times$ 1 cm $\times$ 1 cm = 3.1416 cm^2, and its section modulus Z in any direction, according to the formula given in Fig. 7.2, would be 0.7854. For an elliptical section measuring 4 cm along its major axis and 1 cm along its minor axis, the area would be the same as the previously considered circular section (3.1416 $\times$ 2 cm $\times$ 0.5 cm = 3.1416 cm^2), but its section modulus Z measured along its major axis would be larger, 1.5708 cm^3. On the other hand, the section modulus Z of that elliptical figure measured along its minor axis would be smaller: 0.3927 cm^3. In other words, if you want to break an object

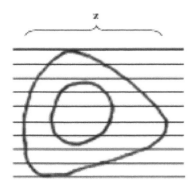

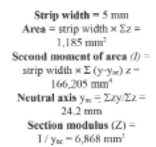

Strip width = 5 mm
Area = strip width × Σz =
1,185 mm²
Second moment of area (I) =
strip width × Σ (y-y_ne) z =
166,205 mm⁴
Neutral axis y_ne = Σzy/Σz =
24.2 mm
Section modulus (Z) =
I / y_ne = 6,868 mm³

7.2. Calculation of the section modulus (Z). A cross section of a long bone, typically taken at midlength, is divided into slices (in this case, of 5 mm) whose distance (y) from an chosen origin and area are determined by multiplying the distance by its measured width (z), excluding the width of the marrow cavity. The second moment of area (I), which is the summation of the cumulative areas multiplied by their squared distance to the neutral axis, is also obtained. The neutral axis (y_ne) is the longitudinal (horizontal and perpendicular to the plane of the drawing) surface that, if the bone is bent, remains of constant length—that is, it is not strained or compressed. It is calculated as the ratio of the summation of the slice cumulative areas by the summation of the slice widths. The ratio of the second moment of area by the neutral axis is the section modulus (Z), which has units of volume. A formula for approximative calculation of the second moment of area of an hollow ellipse is $I = (\pi/64) (D_y^3 \times D_z - d_y^3 \times d_z)$, where D and d are the diameters of the external and internal ellipses, respectively.

Illustration by Sebastián Tambusso, based on material kindly provided by Professor Robert McNeill Alexander.

Strip N°	y mm	z mm	zy mm	y-y_ne	(y-y_ne)²z
1	2.5	13	32.5	-21.7	6122
2	7.5	27	202.5	-16.7	7530
3	12.5	12+12	300.0	-11.7	3285
4	17.5	10+13	402.5	-6.7	1032
5	22.5	10+17	607.5	-1.7	78
6	27.5	10+21	852.5	3.3	338
7	32.5	14+24	1235.0	8.3	2618
8	37.5	37	1387.5	13.3	6545
9	42.5	17	722.5	18.3	5693
Σ		237	5742.5		33241

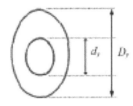

Fast calculation

$$I = (\pi / 64) (D_y^3 \times D_z - d_y^3 \times d_z)$$

with an elliptical section, you may want to try to do so across its thinnest dimension (although this is intuitive enough!).

We then calculate its relationship with the percentage, *a*, of the weight, *mg* (i.e., mass × the acceleration of gravity), that each pair of extremities supports (Fig. 7.3). Thus, a strength indicator, also termed the indicator of athletic capability, Z/amgx, is produced, which gives an idea of the athletic capability of a land animal. Briefly, this concept is defined as the capability of an animal to perform strenuous activities without risk of bone failure. Its units are the inverse of those for pressure, which is intuitively easy to understand if the strength indicator is viewed as the force that the bone is capable of supporting per unit area. Higher values imply a better athletic capability: a land animal with long, thin bones that support a larger weight is more prone to fractures than one with the opposite characteristics. In fossils, this indicator is useful because one can infer the extent to which an extinct organism could perform strenuous activities without excessive risk of fractures.

When this indicator is calculated for living tetrapods, as in the classical study by Alexander (1985), it turns out that those animals with high values are able to perform more athletic activities than those with lower values. As shown in Table 7.2, a living bird (ostrich) and three recent mammals (white rhinoceros, African elephant, and African buffalo) have quite different values for both the humerus and the femur (bones of the upper arm and thigh, respectively). The ostrich shows an indicator of 44 GPa⁻¹ for the

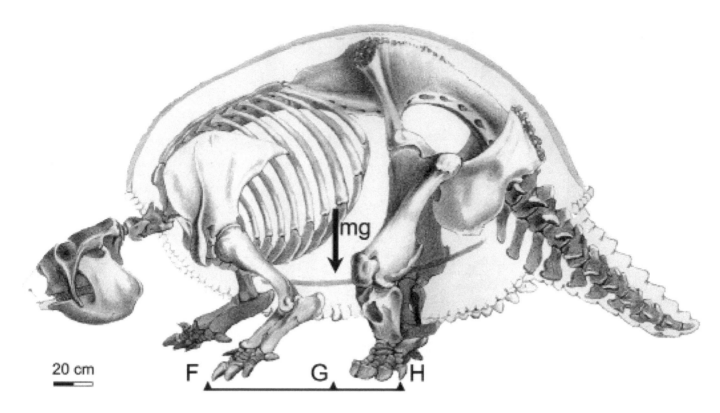

7.3. Reconstruction of a glyptodont showing the position of the center of mass, where the weight was applied. G, projection of the center of mass on the ground; F, forefeet; H, hind feet.

Modified from Fariña (1995), based on a figure of Glyptodon from Burmeister (1874).

femur (the calculation for the humerus is irrelevant because the ostrich is a biped), which reflects its condition of being a fast runner. The white rhinoceros and the buffalo have lower values, but it is obvious that their bones allow them to perform strenuous activities, such as galloping. Finally, the elephant, an animal that cannot gallop, has the lowest values of this set.

Alexander (1985, 1989) applied this approach to assess the athleticism of very large dinosaurs; his results are included in Table 7.2. According to these results, *Apatosaurus* was not capable of galloping, but it could have ambled the way modern elephants do. *Triceratops* must have been as athletic behaviorly as the African buffalo, and hence it may be inferred that its limb dimensions would have allowed it to gallop. The femur of *Tyrannosaurus* has a lower value, comparable to that of the elephant, but Alexander (1989) cautioned against jumping to conclusions because of the anatomical differences between the species considered. Incidentally, the subject of the speed of *Tyrannosaurus*, which is beyond our scope here, has been intensely debated during the last few decades by several researchers, such as Bakker (1986, 1987), Alexander (1996), Christiansen (1998), and Hutchinson and García (2002).

Another interesting case to which such information and methodology may be fruitfully applied is that of the large Pleistocene glyptodonts such as *Glyptodon, Panochthus, Doedicurus,* and *Neosclerocalyptus*, which were analyzed by Fariña (1995). In these taxa, there is a great disparity in the indicator of the femur versus that of the humerus, although this does not occur in the preceding Miocene glyptodont *Propalaehoplophorus* (Table 7.2). This disparity is such that if the entire weight of the body were supported by the back legs, the femur would continue to be stronger than the humerus in a normal quadrupedal posture, as can be seen in the respective values of the indicator of athletic capability.

Table 7.2. Athletic capability in recent and fossil tetrapods[a]

Taxon	Estimated Body Mass (kg)	a_{hind}[b]	*Z/amgx* Femur (GPa^{-1})[c]	*Z/amgx* Humerus (GPa^{-1})[d]
Apatosaurus	33500	0.70	9	14
Tyrannosaurus	7700	1.00	9	—
Triceratops	6400	0.52	19	22
Struthio	40	1.00	44	—
Loxodonta	2500	0.42	7	11
Ceratotherium	750	0.40	26	31
Syncerus	500	0.40	22	21
Propalaehoplophorus australis	50	0.6	64	55
Glyptodon clavipes quadruped	2000	0.60	22	11
Glyptodon clavipes biped	2000	1.00	14	—
Panochthus tuberculatus quadruped	1100	0.56	27	13
Panochthus tuberculatus biped	1100	1.00	15	—
Doedicurus clavicaudatus quadruped	1400	0.57	31	14
Doedicurus clavicaudatus biped	1400	1.00	18	—
Neosclerocalyptus ornatus quadruped	280	0.56	46	22
Neosclerocalyptus ornatus biped	280	1.00	26	—

[a] Data are provided for various dinosaurs (*Apatosaurus, Tyrannosaurus, Triceratops*), the small Miocene glyptodont *Propalaehoplophorus*, and four large Pleistocene glyptodonts (*Panochthus, Doedicurus, Glyptodon*, and *Neosclerocalyptus*). Modified from Fariña (1995).
[b] Fraction of the weight supported by the posterior extremities.
[c] Value of the athletic indicators of the femur.
[d] Value of the athletic indicators of the humerus.

A chain is as strong as its weakest link, and thus an organism is as vulnerable as its weakest extremity. Therefore, it would be useless to increase the strength of one pair while leaving the other prone to accidents. If one breaks an arm, it is not particularly consoling to think that one's legs are safe and can withstand a great deal more. In the same way, a quadruped is much more likely to become a carnivore's dinner irrespective of which bone it breaks. It can be surmised, then, that for the large Pleistocene glyptodonts, it was not only possible but actually convenient to perform strenuous activities of a bipedal nature; otherwise, why would such a discrepancy between the front and back legs exist? The logical thing, after all, would be to distribute weight more equitably on all four limbs, which would result in a much smaller discrepancy, as is observed in *Propalaehoplophorus*. This surprising conclusion requires more evidence before the following counterintuitive image can be appreciated. For example, the muscle structure and the corresponding articulations should be assessed for congruence (Fariña, 1995). On the other hand, a plausible biological explanation should be proposed in trying to understand the reasons why an armored, ton-sized beast must have needed to perform such a difficult activity with its back legs so as to put the bones at risk, and led to such a discrepancy between front and back legs. Further, the question as to why such activity must have been performed in a bipedal mode should be considered.

Let us reflect on the problems that day-to-day reality imposed on these strange creatures. To begin with, it is quite trivial to demonstrate that a glyptodont was capable of standing on two legs. Certainly males had to be able to do it in order to copulate with females. Could reproduction have been this all-important activity? It does not seem to have been. One elementary argument, from an evolutionary point of view, against this idea is that if such robust posterior extremities were needed by males for some sexually related behavior (i.e., they possibly were sexually dimorphic features), then we would not expect to find them in females, in much the same way that female lions lack manes or female moose lack antlers, because such structures are biologically costly to produce and maintain. However, we find that all remains of large Pleistocene glyptodonts have this same femoral characteristic. This is supported by mechanical reasoning as well. *Glyptodon*, for example, which is among the largest glyptodont of all, with a mass of about 2 tons, supported, in a quadrupedal stance, approximately 60% of its weight on its posterior extremities. That this is so may be determined by finding the center of mass and noting at what distance its normal fell relative to the base formed by joining each pair of legs (Fig. 7.3). The closer the center of mass is to any particular pair, the greater the proportion of the weight supported by that pair, as may be confirmed by the limiting case of human beings. The distance in our species from this normal with respect to the posterior pair is nil, and consequently, we humans support all our weight on them. Because the distance of the point of contact of our arms with the ground is generally infinite, the percentage of the weight they support is zero.

When the glyptodont walking on four legs lifted its left rear foot to take a step, all 60% would be resting on the right rear foot. When standing on two legs, each of the two back legs supported only 50% of the total. Therefore, as far as the bones of the legs and arms are concerned, a simple walk on all fours would have proved more of an exertion than reproduction. Having eliminated sex, aerobics would seem to be the answer, but it is difficult to imagine any reason that would have required such very large armored beasts to have had this ability, such as required, for example, in the need to build up speed (yet, even in the case of speed, there is no absolute necessity to accomplish it on two legs).

An aid for an explanation comes from an unexpected source: some carapaces (the armor plating) of glyptodonts preserve fractures with accompanying healing scars, as indicated in Lydekker's (1894) illustration of *Doedicurus* in Fig. 7.4. A reasonable explanation for the cause of these fractures, as considered further below, is intraspecific (i.e., within species) fighting. Given this bit of information, and after considering the movements necessary for delivering a blow with the tail, we can postulate a possible resolution for our puzzlement over the athletic capability of glyptodonts that was initially suggested by the difference in athletic indicators between the front and back limbs.

An animal was confronted by another, the two adversaries face to face. Immediately one stood up and, turning on its right rear leg, contracted the huge muscles on the left side of the tail (or vice versa, of course) in an attempt to strike its opponent with its terrible club. These movements are similar to those executed in karate or other martial arts to deliver a blow using a circular kick.

We would have indisputable evidence of glyptodonts' ability to walk on two legs if we could find fossilized footprints. Unfortunately, no record of such ichnites exists for the larger glyptodonts, and there is only one footprint in the amazing site of Pehuén-Có (see Chapter 5) of a smaller one, perhaps *Neosclerocalyptus* or a young member of the larger species (Aramayo and Manera de Bianco, 1987). Despite the ingenuity for which

7.4. Carapace of *Doedicurus clavicaudatus* (Museo de La Plata, MLP 16-23) with a healed fracture in a laterodorsal position. Unfortunately, this carapace has been broken and repaired several times since the photograph was taken just before the end of the nineteenth century, so we cannot describe the damage from our own observations. The figure shows a roughly circular dent about 35 cm in diameter encircled by a bony callus. Total length of carapace and tail about 3.2 meters.

From Lydekker (1894).

paleontologists are famous, this single print does not tell us much one way or the other.

The animals that walked over the mud at Pehuén-Có, deposited their bones at a nearby location called Playa del Barco. The skeletal remains of glyptodonts are as abundant there as at any other Pleistocene site. How, then, do we account for the fact that they were present in the area but left almost no footprints, in contrast to so many of the other animals?

Applying some elementary principles of physics allows us to approach the problem from a rather singular perspective—and leads to a rather mundane solution. If we calculate the mass of the animal, we can then compare its mass to the total surface area of the four feet, and thus determine the risk of remaining stuck in the underwater substrata so common in the mud of Pehuén-Có. The capybara shows, as would be expected, a good performance in underwater clay substrata (Table 7.3). The pressure its foot would have exerted on the ground is fairly small, suggesting it would have been unlikely to sink far into the mud, as expected for a mammal that walks on such slippery surfaces. Two North American investigators (Gillette and Ray, 1981) proposed a similarity in environmental preferences between semiaquatic capybaras and glyptodonts, but this hypothesis is not supported by the results.

In effect, as Table 7.3 shows, the performance in mud of the large glyptodonts was certainly not much better than that of domestic cattle, and even worse than that of the African elephant and *Megatherium* (with the latter in a bipedal stance), which are animals of much greater size. Additionally, we can easily make an analogy with the performance of cars. The globular carapace of glyptodonts gave them the general appearance of a popular German car. Laterally, the carapace covered over much of the legs, leaving a markedly low clearance. The margins of the carapace could easily make contact with the mud, adding an additional factor that may have contributed to the animal becoming stuck.

It is thus unlikely that the larger glyptodonts had any predilection, in contrast to the hypothesis of Gillette and Ray (1981), for these kinds of substrata because they likely found it difficult to walk on them. Therefore,

Table 7.3. Pressure exerted on the ground of some mammals and dinosaurs[a]

Taxon	Mass (kg)	Foot Area (m²)	Weight/Area (kPa)
Apatosaurus	35000	1.2	290
Tyrannosaurus	7000	0.6	120
Iguanodon	5000	0.4	120
Loxodonta	4500	0.6	70
Domestic cattle	600	0.04	150
Homo	70	0.035	20
Hydrochoerus	70	0.032	22
Giraffa	1100	0.1	110
Megatherium biped	4100	0.7	60
Glyptodon quadruped	2000	0.23	90
Glyptodon biped	2000	0.135	145

[a] From Alexander (1985) and Fariña (1995).

the chances of our finding their tracks there are small—unless, of course, new discoveries turn up that might tell us something different. However, so far, our expectations fit the available evidence.

Fighting glyptodonts

As noted above, glyptodonts might have used their stout tails in an aggressive manner. The tails are usually surrounded by rings of dermal bony scutes, a morphology that suggests considerable flexibility. In many cases, the scutes of the distal end of the tail are fused to form a caudal sheath or tail club, whose function can only be inferred, as it is present in no modern animal. However, this clublike structure is likely to have acted as a weapon, useful either in male–male combat or in defense against predators.

We have mentioned that the glyptodont carapace shows damage attributed to blows from glyptodont tails (Fig. 7.4). If this interpretation is correct, intraspecific fighting might have occurred, as has been proposed for ankylosaurs (Farlow and Brett-Surman, 1997). Lydekker (1894), for example, illustrated a carapace of *Doedicurus clavicaudatus* with a healed fracture in a dorsolateral position. Unfortunately, this carapace has been broken and rebuilt several times since the photograph was taken, so the damage cannot be directly observed. The figure shows a roughly circular dent about 35 cm in diameter encircled by a bony callus, indicating that healing had occurred, so we can be reasonably sure that the condition of this carapace is not the result of postmortem deformation.

The energy of the tail blow may be calculated from muscle action, both of which may be estimated through the principles of physics. As is usual in glyptodonts, the flexible part of the tail in *Panochthus* (Fig. 7.5) is a truncated cone whose volume was measured by Alexander et al. (1999) as 0.105 m³, two-thirds of which (about 0.07 m³) must have been occupied by musculature; most of the rest of the volume is easily attributable to the vertebrae that extended through the caudal tube. The density of muscle in living vertebrates is 1060 kg m⁻³ (Méndez and Keys, 1960), so its mass must have been about 74 kg. To this value, the epaxial trunk muscles (i.e., those above the pelvis) must be added. Alexander et al. (1999) estimated them as about 34 kg. Therefore, the total mass of tail-moving muscle must have been about 108 kg, 54 kg on each side. Muscles yield maximum power between about 20 J kg⁻¹ and 70 J kg⁻¹ (Alexander, 1992), and hence

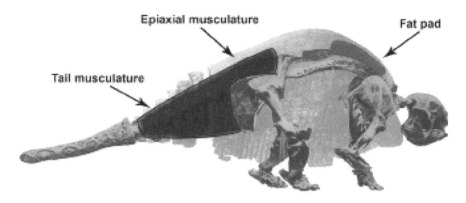

7.5. Diagram of *Panochthus*. The armor has been removed from the near side of the body to show the spaces occupied by the supposed fatty pad, and by the tail and posterior epaxial muscles.

Based on Alexander et al. (1999).

Epiaxial musculature

Fat pad

Tail musculature

Alexander et al. (1999) took an intermediate figure of 50 J kg^{-1} as an estimate of the work likely performed by glyptodont tail muscles, which gives a total amount of energy of $54 \times 50 = 2700$ J. Indeed, this is a doubly conservative estimate, for the animal could have used its legs to move its body as well as its tail in delivering a blow, thus imparting additional momentum (Fariña, 1995), and elastic energy could have been stored before the blow was delivered, as athletes do by activating antagonistic muscles when preparing for a jump (Alexander, 1995). To make a modest allowance for these possibilities, we will increase our estimate of the kinetic energy available for the tail movement from the 2700 J given above to 3000 J. This is a considerable amount of energy—enough to have caused the 40-kg club to move at 12 m s^{-1}, a speed similar to that produced in putting a 7.3-kg shot by a good but not top-level athlete (Ballreich and Kuhlow, 1986).

We may now consider whether this energy was sufficient to have fractured the carapace, especially if the blow was delivered with the center of percussion (or what's known as the sweet spot in a tennis racket or baseball bat), which is the point on an object where a perpendicular impact will produce translational and rotational forces that perfectly cancel each other out, so that the object will transfer all its kinetic energy to the impact (Blanco et al., 2009). Objects are broken by two different phenomena, whose actions are complementary. First, strain energy (i.e., the potential to perform work when the strain is released) must be applied so that it will be stored in the object until some part of it reaches its ultimate stress (Alexander, 1983), and this amount of work is proportional to volume for the same material. It is known as the strain energy capability. Moreover, when an object is fractured, new surfaces are created. The work needed equals the area of these surfaces multiplied by the appropriate work of fracture peculiar to that material. However, the total work required is not necessarily the sum of these components, because strain energy may supply the work needed to create new surfaces. The relevant quantity for specimens that are flexible enough, as were the carapaces under consideration (with many relatively small osteoderms connected by collagen), is their strain energy capability.

Alexander et al. (1999) were faced with the impossibility of testing their results on actual large carapaces. Therefore, they decided to estimate strain energy capability by comparing their result with those obtained from human skulls, which are structures also made of rigid bony pieces connected by collagen. Yoganandan et al. (1995) found that the energy required to fracture human skulls ranged from 14 to 69 J, with an average of 28 J for impact tests. Scaling up this latter figure for the 210-fold larger carapace yields a required energy of about 6000 J. Furthermore, a figure of 1400 J is obtained by comparing the 35-cm-diameter fracture observed in Lydekker (1894) with the 20,000 J m^{-2} obtained by Jaslow's (1990) tests on bars of bone cut from goat skulls. The dent in the *Doedicurus* specimen had an area was about 0.10 m^2, and hence the total length of sutures in that area of carapace would be about 3.5 m. The scutes average 20 mm thick, so the total area of sutures in the dented region was about 0.07 m^2, and the energy required to break them all can be estimated as 20,000 × 0.07 = 1400 J.

As Alexander et al. (1999) noted, their three estimates (the kinetic energy of a blow as 3000 J, and their two estimates of the energy required to break a carapace as 6000 or 1400 J) are crude, but they agree within an order of magnitude. This suggests, as can be expected in an arms race, that not all tail blows may have fractured an opponent's carapace, but not all carapaces withstood tail blows undamaged. Therefore, it seems reasonable to interpret the damage on some carapaces as resulting from fights. Of course, it could be that they were not caused by tail blows. What, then, might have caused the fracture of the carapace? Perhaps it was due to a fall or some heavy object that fell from the sky, but we would be hard-pressed to imagine just what that might have been. It's better, as usual, to have a testable hypothesis; it might be wrong, but at least it gives someone else the chance to show we're wrong.

Carapaces were strongly built at their hind end, as they were firmly sutured both to the ilia and the ischia, bones that make up the hips (Appendices 1 and 2). However, in the thoracic region, the carapaces were not built as solidly; there was no support from the underlying skeleton because the vertebrae and ribs do not contact the carapace. The space in between was improbably filled by musculature, as in glyptodonts the armor was immovable and the dorsal vertebrae are fused (Appendix 2). Thus, Alexander et al. (1999) proposed that this space might have been occupied by a large interscapular (between the shoulders) fat deposit (Fig. 7.5), like that of the camel's hump, which could have had a role as an energy reserve as well as a cushion to partly absorb tail blows. The human heel pad, a structure similar to that proposed for glyptodonts, is compressed at heel stroke to 0.4 of its initial thickness (De Clercq et al., 1994; Aerts et al., 1995).

Digging sloths

We noted in Chapter 6 that pampatheres and several armadillos likely excavated some of the smaller paleoburrows preserved in deposits of several areas of southern South America. The possibility that sloths were responsible for some of the larger paleoburrows, counterintuitive as this might at first seem, is suggested by several aspects of their morphology and of the structures themselves. Among the clues is that digging a tunnel involves a great expenditure of energy, so an animal does not usually bore a burrow much larger than its own body diameter. Thus, although large extinct pampatheres and armadillos such as *Eutatus* and *Propraopus* may have dug many of the paleoburrows, it is considerably less likely that they excavated those of larger diameter. Among other sufficiently large candidates, the glyptodonts may be excluded, because their anatomy is clearly not suited for this activity (Milne et al., 2009). By process of elimination, we may

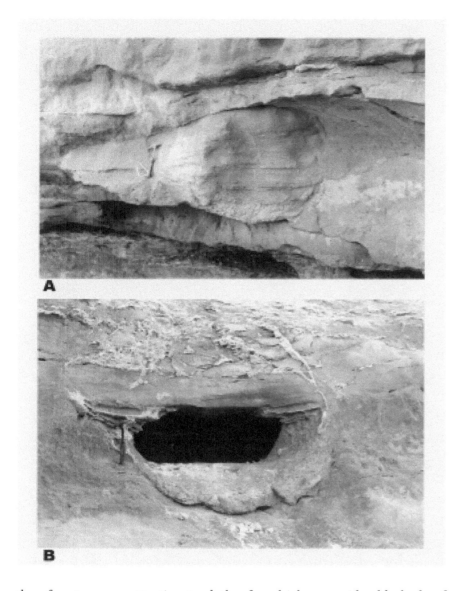

7.6. (A) Sediment-filled paleoburrow in the cliffs near the city of Mar del Plata, Buenos Aires Province. (B) Partially empty paleoburrow located southwest of Miramar, Buenos Aires Province. Both paleoburrows have a transverse diameter of approximately between 1 and 1.10 m.

From Zárate et al. (1998), by kind permission of the Asociación Argentina de Sedimentología.

therefore turn our attention to sloths, for which a considerable body of evidence has been marshaled.

Large paleoburrows and crotovines (see Chapter 6) have recently received renewed interest in the literature. Although in recent years several such structures have been discovered in Pliocene and Pleistocene sediments, especially in cliffs in the area around Mar del Plata, on the Atlantic coast of Buenos Aires Province, the existence of such structures was already known; they were noted in the scientific literature of more than a century ago (by Ameghino, for example). In the 1920s, Joaquín Frenguelli reported the remains of a ground sloth, assigned to *Scelidotherium*, within a cave filled with volcanic ash. This discovery made little if any impact on subsequent researchers, perhaps because it is only just mentioned in an extensive work on the geology of the south coast of Buenos Aires Province (Frenguelli, 1928), or because the absence of modern mammalian diggers of that size prejudiced researchers against such large animals as diggers. The larger paleoburrows have a subcircular section with diameters ranging from 80 cm to over 1.50 m (Fig. 7.6; Zárate et al., 1998). One of these paleoburrows measures at least 40 m (Quintana, 1992) and was probably much longer when it was functional.

7.7. The forelimb of the ground sloth *Glossotherium robustum* as a machine. Input force (F_i), generated by the triceps, and input lever arm (L_i); output force (F_o), produced by the hand on the substrate, through the output lever arm (L_o), i.e., the summed length of the forearm, from the elbow joint, and the hand.

From Vizcaíno et al. (2001), by kind permission of Acta Palaeontologica Polonica.

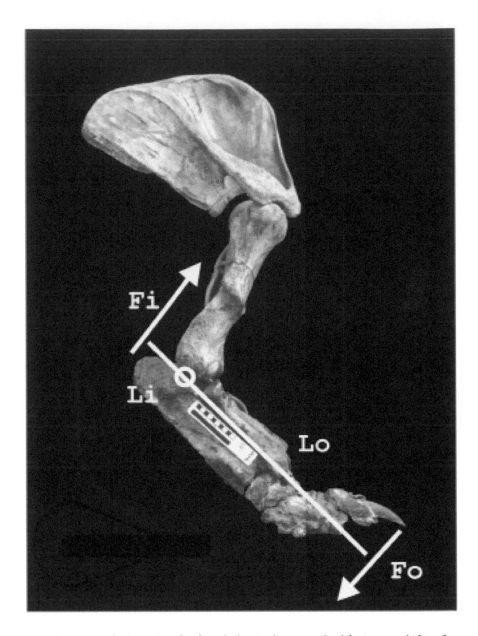

Bargo et al. (2000) calculated the indicator of athletic capability for the bones of the forelegs and hind legs of the mylodontid sloths *Glossotherium* and *Scelidotherium*. Their results indicated that the bones were well adapted to withstand stresses equivalent to, or even greater than, those withstood by large mammals that are able to gallop, such as the African buffalo. However, the proportions and general anatomy of the sloths' limbs, particularly of the hands and feet, make fast locomotion unlikely. The authors, considering the limbs as well as other anatomical aspects and geologic evidence, proposed that digging was the strenuous activity for which the limb bones had been the initial clue. The anatomy of the forearm (Fig. 7.7) clearly supports this hypothesis. The long olecranon process of the ulna (i.e., the part of the ulna behind the elbow joint, commonly referred to as the funny bone; Appendix 1) is three to four times longer than the length required for walking, indicating that the forearm was more appropriately suited for producing great strength rather than speed.

Bargo et al. (2000) and other authors such as Cartelle (1980), Aramayo (1988), and Cuenca Anaya (1995) considered the morphology of the manus

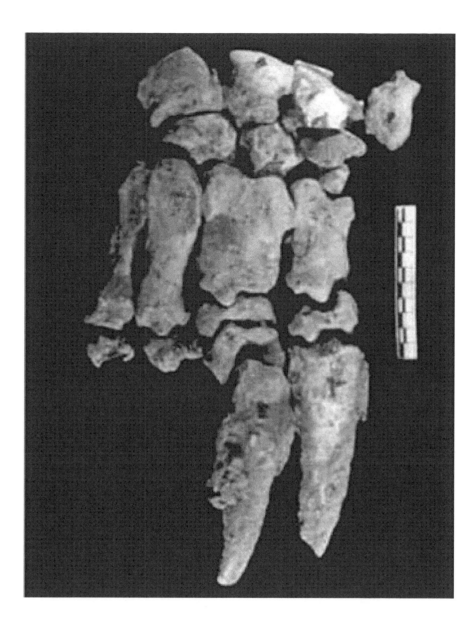

7.8. Right manus of *Scelidotherium leptocephalum,* showing the anatomy of the fingers that are candidates to have made the marks found in the caves. Scale bar equals 10 cm.

From Vizcaíno et al. (2001), by kind permission of Acta Palaeontologica Polonica.

of *Scelidotherium, Glossotherium,* and *Proscelidodon.* The main conclusions reached by these authors are that the carpus (wrist), metacarpus (palmar region), and first phalanx (finger) are closely articulated (Fig. 7.8), forming a shovellike structure. The most remarkable feature is the great development of the second and third digits, especially the ungual (claw-bearing) phalanges, whereas the other fingers are reduced, making the manus ideally suited for digging. As well, Aramayo (1988) indicated that the forearm and manus of scelidotheres were capable of ample flexion, extension, pronation, and supination, and thus must have permitted the movements that are required by digging habits. Finally, Bargo et al. (2000) noted that the animal's center of mass was located mainly over the hindquarters, meaning that when the animal was standing quadrupedally, 60% to 70% of body mass was supported by the hind limbs and 30% to 40% by the forelimbs, the reverse of the situation in modern, nondigging large mammals. This distribution of body mass means that these sloths could easily have assumed a bipedal (though not erect) stance, a posture essential for releasing the forearms to act on the substrate, with weight supported by the hind limbs. Incidentally, this proposed habit for these sloths is not

7.9. The giant living armadillo *Priodontes maximus,* showing how it supports its weight on the hind limbs and on the tail.

Photograph courtesy of Mariella Superina.

entirely conjectural; it is precisely the style used by many mammalian diggers, including their xenarthran relatives, such as in the large living armadillo *Priodontes maximus* (Fig. 7.9), a powerful digger that walks almost permanently on its hind limbs while maintaining the backbone roughly parallel to the ground, the forelimbs flexed, and the hands, turned posteriorly, dragging along the ground (Frechkop, 1949).

In further considering possible excavators, Vizcaíno et al. (2001) noted that the large mylodontid ground sloths *Scelidotherium leptocephalum My-lodon darwinii,* and *Glossotherium robustum* better fit the observed profiles of the larger paleoburrows. For example, *Scelidotherium leptocephalum,* with an estimated body mass of 800 kg, had a maximum transverse diameter of 100 cm and a length of 270 cm, while *Glossotherium robustum* reached 1200 kg in body mass and had a transverse diameter of 120 cm and a length of 325 cm. The diameters of the largest paleoburrows recorded agree much more closely with these dimensions than with those of giant armadillos and pampatheres (Fig. 7.6). As already mentioned, burrowing animals do not build burrows much wider than their own diameter as a way to minimize energy expenditure.

Moreover, the shape of several claw marks preserved on the sides, floor, and ceiling of the paleoburrows fits the form of the hand skeleton of these ground sloths, and this provides perhaps the most convincing evidence supporting these sloths as diggers: the marks are arranged in pairs of parallel grooves about 30 cm long and 3 or 4 cm in width. These coincide with the anatomy and size of the hand of *Glossotherium* and *Scelidotherium.* Both had large claws on digits II and III (corresponding respectively to the index and middle fingers of humans). The marks are too large for the hands of the armadillos *Eutatus* and *Propraopus.* Finally, although *Pampatherium* is larger in body size than the armadillos, it did not possess such large claws, and its three middle fingers were subequal in size (see Chapter 6), so that any marks made by this mammal's hand would have produced three rather than two parallel grooves.

In summary, mylodontid sloths, such as *Scelidotherium* (and also *Glossotherium*) are considered as possible builders for the large late Cenozoic paleoburrows present in the Pampean region. The question that immediately arises, of course, is what were they using the paleoburrows for.

The largest living fossorial mammal is the aardvark (Tubulidentata), with a body mass up to 100 kg, but this value does not necessarily reflect an upper size limit for fossorial mammals. The Australian Pleistocene giant wombat (*Phascolomus gigas*, Marsupialia) has been regarded as the largest fossorial animal that ever lived. It was twice the height and length of any living wombat (Rich, 1990), and if geometric similarity is assumed, it must have had a body mass of about 200 kg to 300 kg, which admittedly is still much less than that of the Pampean mylodonts. Such factors may furnish clues for burrow-digging behavior in some mylodontids.

Vizcaíno et al. (2001) postulated that the upper body size limit for a mammalian digger is determined by environmental and ecological rather than physical factors; it is to these factors that we must look to make sense of digging behavior in these mylodontids, because although we may have evidence that they excavated the paleoburrows, we still have to consider why they would have needed to do so. *Scelidotherium* and *Glossotherium* may, for example, have depended on burrows to escape predation. As noted in Chapter 6, the large-sized carnivorans, such as the sabertooth cat *Smilodon* and short-faced bear *Arctotherium*, were possible predators of these sloths—if not of the adults, then at least of the juveniles. The much larger body size of the contemporaneous mylodontid *Lestodon* probably rendered at least the adult individuals immune to predation, as occurs today among elephants.

Climatic and physiological factors may also have played a role: sloths may have needed caves or burrows to avoid alternating excessively cold or warm climatic conditions. As discussed in Chapter 3, the climate of the Pampas region during the early Pleistocene was warmer than it is today, but toward the end of the Pleistocene, it became cold and dry, and an arid steppe developed throughout the area. In this type of environment, sloths may have benefited from the relative warmth and relatively stable environments of caves, which acted as effective temperature regulators. Perhaps these features were useful in breeding and in rearing their young, and may even have helped them survive the coldest periods of the year. Their living relatives (tree sloths, anteaters, and armadillos) have low body temperatures and low basal rates of metabolism; these influence the geographical limits of their distributions (McNab, 1980). Although we don't know about the external appearance of *Scelidotherium*, it is conceivable that if they did possess a poorer fur coat, unlike that of the high-latitude *Mylodon* (see Fig. 1.9), they might have compensated by using underground shelters, at least during unfavorable seasons.

That sloths made use of caves or burrows is not a particularly radical idea. Fossil remains of ground sloths in natural caves, caverns, and other natural shelters are common elsewhere in the Americas, and are especially abundant in northern Brazil (Lund, 1842; Winge, 1915; Cartelle, 1991). Ground sloths were also recorded in natural caves in cooler regions at higher latitudes and/or altitudes of Patagonia and western North America (Moreno and Woodward, 1899; Scillato-Yané, 1976; Akersten and

McDonald, 1991; McDonald et al., 1996). For example, the remains of another mylodontid, *Mylodon darwinii*, are frequently encountered in Patagonian caves; the hide and fur recovered in the cave of Última Esperanza (Eberhardt or Mylodon Cave), southern Chile, were noted in Chapter 5. There is thus ample evidence suggesting that the use of caves or shelters, for whatever reasons, was a common habit among ground sloths. For North American forms, it was proposed that the relatively stable environments of caves acted as an effective temperature buffer, explaining the frequent occurrence of sloth remains.

But why dig them out, as proposed for *Scelidotherium* and *Glossotherium*? It turns out that on the vast plains of the Pampas region of Argentina, natural caves are restricted to certain sectors of the mountains of Ventania and Tandilia in the south of Buenos Aires Province, and some other mountain ranges west of the pampas, but so far, fossil remains of sloths have not been found in these places. Consequently, if the ground sloths inhabiting this vast plain required closed shelters, they would have had to build them, and these animals were well suited anatomically to have carried out such activities.

Was *Megatherium* a biped?

Many reconstructions of *Megatherium americanum* place the animal in a quadrupedal position, as in, for example, Bru de Ramón's (1784–1786) static pose, and the more realistic image portrayed in the painting in the División de Paleontología de Vertebrados of the Museo de La Plata, both of which are discussed elsewhere in this book. The more active appearance in the painting (shown in detail in Fig. 7.10) suggests that it was comfortable on all four limbs, with its forearms extended. This is in accordance with certain skeletal features of the forelimb. For example, the olecranon process of the elbow, where the forearm-extending triceps muscle inserts, is projected backward. This arrangement is common in terrestrial animals that habitually hold the forearm in full extension, because it maximizes the triceps' leverage. However, analysis of the fossil trackway at Pehuén-Có (near Bahía Blanca, Buenos Aires Province, discussed in Chapter 5) attributed to *Megatherium* (Fig. 7.11), allows consideration of a different posture, one in accordance with another feature of the olecranon process: its short length. A shorter olecranon implies a design for optimizing speed rather than force, including the weight that the extremity has to support. Blanco and Czerwonogora (2003) examined this possibility through consideration of body weight, its distribution, and several geometric variables. They suggested that for the particular instance of making the trackways, *Megatherium* was moving bipedally. Until a comprehensive study of the locomotion of this fascinating animal (which would likely shed further light on its habits) is undertaken, a bipedal locomotory habit for this beast cannot be assumed to be the most efficient. If so, perhaps the tracks hide a trick: if the animal was walking on all fours, the larger steps of the hind limbs might have been superimposed the smaller, previously made tracks of the forelimbs. Indeed, McDonald (2007) advanced just this argument in his analysis of inferred locomotion from the tracks preserved at the Nevada State Prison near Carson City, Nevada. These are the only known tracks of a ground sloth in North America and have been convincingly attributed to

7.10. Detail of the illustration in the División Paleontología Vertebrados, Museo de La Plata, showing *Megatherium* actively walking on all fours.

Image by S. F. Vizcaíno by kind permission of the Museo de La Plata, Argentina.

Paramylodon. This sloth, certainly quadrupedal, left tracks similar to those of *Megatherium* at Pehuén-Có in the sense that there are many more made by the hind than fore feet. McDonald (2007) suggested that when the sloths walked along a more or less straight path, the hind feet covered over the handprints. The few handprints that are preserved apparently coincide with a deviation from a straight path, and this situation seems to occur in both the Nevada State Prison and Pehuén-Có sites.

Some help in understanding the way the animal was walking and how it distributed its weight comes from three hand (or manus) impressions that exist at about the middle of the trackway. These fall 25 cm in front of the anterior border of the pes prints. The maximal depth of the manus prints is 0.14 m, and the maximum depth of the pes prints is 0.26 m. The areas of the manus and pes, according to Aramayo and Manera de Bianco (1987), are 0.0375 m^2 for each manus and 0.398 m^2 for each pes.

The reasoning of Blanco and Czerwonogora (2003) is as follows. If, as derived from the geometry of the posture of the animal on all fours, 70% of the weight was supported by the hind limbs and 30% by the forelimbs and the estimated body mass is about 4000 kg (although higher estimates have been reported), the pressures exerted by each manus and pes in a static quadrupedal position must have been 310 kPa and 68 kPa, respectively. If the depth of the ichnites was directly proportional to the pressure exerted on the substrate, this pressure should have been less for the manus than for the pes, and the ratio between depth for the manus and depth for the pes (0.14 m/0.26 m = 0.54) should have been equal to the ratio between the pressures calculated theoretically, but it is not: P_m/P_p = 310 kPa/68 kPa = 4.56. If the animal was walking bipedally, stopped, and then leaned at the same time on both forelimbs, the manus pressure reduces to 155 kPa (half of the previous value). If it were walking bipedally, the pes pressure would rise to 97 kPa per foot. This latter hypothesis reduces the gap between manus and pes pressure but still maintains a higher manus pressure, in contradiction to the observed depth.

Blanco and Czerwonogora (2003) concluded that *Megatherium* walked in a cumbersome way, exerting a peak force on the substrate greater than

7.11. Our colleague, Susana Bargo (above), steps side by side with *Megatherium*.

three times its body weight, even when it was walking slowly. This might have been a strategy to minimize the cost of transport and to make better use of hind limb bone properties for withstanding the large lateral bending moments that occur in a bipedal walk.

Incidentally, the trackways can also provide another clue to the life habits of *Megatherium*, this being how quickly it might have moved. Blanco and Czerwonogora (2003) calculated the average speed of locomotion along the trackway in Pehuén-Có (Fig. 7.11) as 1.21 ms^{-1} by the method shown in Box 7.2.

Box 7.2   How to estimate the speed of a megathere

The calculation of speed based on a series of fossil footprints may be calculated as based on the method of Alexander (1976), as follows:

$$v = 0.25\, g^{0.5}\, \lambda^{1.67}\, h^{-1.17}$$

where the stride length, λ, is the distance between two successive footprints of the same limb from the same side; h is the height of the hip joint from the ground (h is the sum of the lengths of the femur, or thigh, and tibia, or shin, bones), and g is the acceleration of free fall, 9.8 ms^{-2}. Blanco and Czerwonogora (2003) used the mounted specimen of *Megatherium americanum* (Box Fig. 7.2) in the Museo de La Plata for skeletal measurements and anatomical observations. The pes, or foot, dimensions of MLP 27–VII–1–1 (length 0.85 m and width 0.35 m) were nearly the same as the footprint dimensions measured by Aramayo and Manera de Bianco (1987) at Pehuén-Có (length 0.88 m and maximal foot width 0.48 m). Plugging in the measurements from the specimen into Alexander's formula, the average speed of locomotion along the trackway was found to have been 1.21 ms^{-1} (by summing the speeds between all the available footprints and dividing by the total number of successive footprints), with a speed range between 0.69 ms^{-1} and 1.68 ms^{-1}. In the same trackway, Casinos (1996) obtained a slightly different value of 1.4 ms^{-1}. To put this in perspective, this is about the speed at which humans usually walk.

Aggressive sloth

Fariña and Blanco (1996) tested a particular aspect of the hypothesis that *Megatherium americanum*, one of the two largest Lujanian sloths, had morphological features that are better explained by its forearms reflecting an optimization for speed rather than for strength of extension. If so, such a trait might have been associated with an aggressive use of the animal's large claws. The possibly aggressive use of the claws was assessed through study of the anatomy of the forelimbs and the association of such usage with hunting habits.

Estimating the speed of a part of the body such as the forelimb requires a rather involved series of calculations. Fariña and Blanco's (1996) analysis of the forelimb of *Megatherium* began with a scale model to determine mass and the position of the center of mass of the forearm, which was found to lie 65 cm distal to the humerus–ulna articulation (i.e., elbow joint). The estimated mass of the forearm was calculated as 115 kg. Both these quantities are required because in the study of circular movements, it is necessary to know how much mass is to be moved as well as how it is distributed with respect to the pivot.

Another variable to consider is a quantity called the moment of inertia (I). This is a measure of an object's resistance to changes in its rate of rotation, and for our purposes it requires consideration of both the forearm and hand. The value obtained for I was 49 kg m², a quantity that will be applied later in the series of calculations (Appendix 4).

Having defined the resistance of the forearm of *Megatherium* to rotational movement, we must now consider the force that rotated it. Movements need motors, and muscles are the motors of animals. Muscles move bones by shortening, thus pulling the bone (relative to a joint) to which it is attached. The triceps brachii muscle, which is on the back of the upper arm, is the main muscle involved in rotating the forearm (at the elbow joint) relative to the upper arm, so that the forearm is extended (opening the angle of the elbow joint; the reverse movement flexes or closes the joint, which is accomplished mainly by shortening the biceps brachii and brachialis muscles located on the front of the upper arm). This is the movement, of course, that produces a straight punch, such as a jab, and thus the triceps is the muscle we are interested in.

The triceps inserts on the olecranon process of the ulna (a bone of the forearm; Appendix 1). The olecranon process is that familiar anatomical structure that is soundly cursed when banged hard enough—it is known as the funny bone. (Its colloquial designation is perhaps derived from the mistaken assumption that the aggrieved bone is the humerus, the bone of the upper arm, which in any event is spelled differently from humorous). The triceps inserts onto a wide, rough crest of the olecranon, which is useful in helping reconstruct its volume and hence its possible action. Several decades ago, Hill (1938) described the association of variables involved in muscle shortening. Estimating values for the terms in Hill's equation is the next step in the process, for it allows those values to be plugged into other equations that permit the calculation of maximal and average rotational speed, technically termed angular velocity. From here on, the process of calculating speed of the forearm requires several mathematical juggling acts, where terms from one set of relationships (i.e., equations) are substituted into others. We invite our more mathematically inclined readers to refer to Appendix 4, where the mathematical logic is presented in full detail. Here, we provide a more qualitative account of Fariña and Blanco's (1996) analysis.

The analysis of speed of the forearm of *Megatherium*, once the math has all been done, indicates that the animal could have moved its forearm at something more than 60° per second, which must have delivered a kinetic energy of about 2000 J. This is impressive: it equals that of an object of 20 kg falling from a height of 10 m!

Now that we've demonstrated how powerful a blow could have been delivered, by an analysis of the muscle energy available and how it was transferred to the movement of the forearm, let's consider some anatomical features to see whether that action might have been enhanced. Digits II, III, and IV have large, powerful ungual phalanges (Fig. 7.12), of which the first two are laterally compressed, while that of digit IV is dorsoventrally flattened (De Iuliis and Cartelle, 1993). These would have made formidable

Box Fig. 7.2. Complete skeleton of *Megatherium americanum,* as exhibited in the Museo de La Plata.

Courtesy of the Museo de La Plata, Argentina. Photograph by Pedro Fernández.

7.12. Hand of *Megatherium americanum*. Digits II, III, and IV have large, powerful ungual phalanges. In the two first cases, the phalanges are laterally compressed, while that of digit IV is dorsoventrally compressed.

Photograph by Sergio F. Vizcaíno by kind permission of the Museo de La Plata, Argentina.

weapons to stab prey, daring predators, rival conspecifics, or whatever else the animal might have wanted to destroy.

We might then wonder whether those 2 kJ (2000 J) would have been high enough to penetrate soft tissues. An experiment was performed in which one of the authors (R.A.F.), a then 100-kg, 38-year-old paleontologist with no particular athletic abilities, threw a 1-kg object a height of 2 m into the air by action of his right triceps alone. This means that the kinetic energy developed was 0.02 kJ (20 J). The same subject then stabbed through a 10-cm-thick chunk of domestic cattle muscle (that is, meat) covered with skin and fur with a kitchen knife, again using only the energy produced by his right triceps. It seems reasonable to assume that the 100-fold difference in favor of *Megatherium americanum* more than accounts for the energy consumed in deforming the claws, which are composed of a more flexible material than the steel of the knife. All this demonstrates that *Megatherium americanum* was probably easily capable of stabbing through soft tissues, even if only forearm extension was used. However, we must also take into account that a blow with the claws certainly involved the actions of the muscles of other body segments and thus the momentum imparted by them in delivering a blow, much as in an overarm throw used in several sports (Alexander, 1991), as by a baseball pitcher. Such throws require a coordinated action of the body segments involved with appropriate delays so as to create a whip effect.

Another comparison may be made between the relative length of the olecranon of *Megatherium americanum* and those of other mammals. According to the allometric equations for mammals produced by Alexander et al. (1981), a mammal the size of *Megatherium americanum* should have a triceps moment arm of 26 cm rather than the 12 cm observed. Interestingly, this short olecranon of *Megatherium americanum*, which must have made for a faster claw strike, is in compensation widely flared for the insertion of the triceps. If *Megatherium* were a biped, as suggested by evidence presented in the previous section, then the wide insertion must be interpreted as an evolutionary compromise for not sacrificing too much output force to achieve that higher speed.

The conclusion that *Megatherium americanum* could have been an efficient stabber seems clear, but its implications need to be discussed. Among possible alternative explanations, defensive behavior immediately comes to mind. However, the 10-fold difference in size between *Megatherium americanum* and the largest carnivores, such as the 400-kg sabertooth *Smilodon*, would seem more than sufficient protection to make such a strong weapon redundant. Mammals also display intraspecific aggression, but no evidence of it has been reported for *Megatherium*, despite the recovery of hundreds of its fossil remains; and such behavior, in any event, tends to be less lethal than the massive blow calculated by Fariña and Blanco (1996). We will return to this subject in Chapter 8.

As noted in the Bestiary, reconstructions of *Megatherium americanum* usually portray this giant ground sloth with a thick furry coat. By using the physical principles of energetics, and by comparing this giant ground sloth with a human being, Fariña (2002a) showed that according to its size and the climatic conditions prevalent in at least part of its biochron, a hairless skin may have been a more probable appearance. Indeed, if body shape and metabolism is approximated to that of a person without clothes, this giant sloth would have been about as comfortable at a temperature of a few degrees below freezing as would a person at 27°C. This is because the most important source of heat loss is conduction (Alexander, 1999), and therefore convection and radiation can be considered negligible. The equation that describes the flux of heat loss by conduction is

$$\Phi_{cond} = (k_{skin}/s) \times (T_{inner} - T_{skin}),$$

where Φ_{cond} is the density of heat flow through the skin (i.e., the heat that flows outward per unit area of skin per unit time), k_{skin} is the thermal conductivity of the skin (i.e., the property of a material that indicates its ability to conduct heat), s is the thickness of the skin, and T_{inner} and T_{skin} are the temperatures in the interior of the body and outside the skin, respectively (Fig. 7.13). This equation implies that the thicker the skin and the lower the difference of temperature, the lower the heat flux.

Therefore, if the conduction flux (Φ_{cond} in the equation above) were the same as in a human being, and assuming for both mammals an internal temperature of 37°C (or 310 K, which is the way the equations are actually expressed), the outside temperature (T_{skin} in the equation) for the megathere would have to have been about 54°C less than the internal temperature, or −17° C—well below freezing!

Such a value implies that the environmental temperature could have been several degrees below freezing and the giant sloth would still be in a thermoneutral zone—that is, within the range of temperatures in which an endotherm is comfortable enough not to sweat or shiver. Even during the glacial maximum, the temperature would seldom have approached such low values in the midlatitudes inhabited by *Megatherium*. On the other hand, if the skin were as furry as seen in most reconstructions, the animal would have had to expend vast quantities of water to maintain its temperature, a problematic option for the semiarid environment it must have faced. It should be noted that the calculations provided above are conservative;

Was *Megatherium* furry?

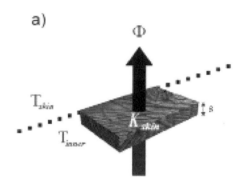

a)

a) In a hairless skin the flow is larger, while b) in a furry skin less heat is lost.

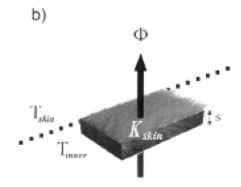

b)

7.13. Flux of heat loss by conduction. Φ_{cond} is the density of heat flow through the skin (i.e., the heat that flows outwards per unit area of skin per unit time), k_{skin} is the thermic conductivity of the skin, s is the thickness of the skin, and T_{inner} and T_{skin} are the temperatures in the interior of the body and outside the skin, respectively.

Illustration by Sebastián Tambusso.

humans have a higher metabolism than that proposed for sloths, which increases our need to protect ourselves from heat loss.

This conclusion may have applied to other giant sloths that inhabited low latitudes (such as *Eremotherium*, as noted by McNab, 1985) or the other midlatitude species mentioned here (e.g., *Lestodon* and *Glossotherium*). However, perhaps the smaller *Scelidotherium leptocephalum* may have needed to shelter in caves. And, of course, this certainly does not apply to the high-latitude species *Mylodon darwinii*, whose furry hide was found mummified in southern Chile.

The strength of a dodging *Macrauchenia*

The litoptern *Macrauchenia patachonica* (whose paleobiology was studied about 90 years ago by Sefve, 1924) had distinctive South American phylogenetic and biogeographical histories. One of its (and probably main) predators, was the sabertooth cat *Smilodon*, whose North American species, *S. fatalis*, was proposed to have been an ambush predator, capable of a short-distance burst of speed (Akersten, 1985). To avoid predation, the large litoptern seems to have been particularly adept at swerving. This is suggested by the fact that its limb bones have indicators of higher transverse than anteroposterior strength (significantly so in the case of the femur), a feature that is also observed in modern swervers, but not so clearly in other fast-running herbivores that do not swerve as much.

When the locomotor behavior of glyptodonts was considered, it was noted that all stresses (including bending components) are particularly important in large animals, as great mass exerts large moments on the bones as a result of inertia alone. Consequently, large animals also use postural adaptations to aid in supporting their body weight (Biewener, 1983, 1989a, 1989b, 1990; Carrano, 1998), maintaining their long bones closer to the vertical plane at larger sizes.

In contrast to other studies on athletic indicators, Fariña et al. (2005) also computed a transverse strength indicator (or transverse indicator of athletic capability, IAC_t). In doing so, the section modulus for bending in a transverse plane (i.e., perpendicular to the fore and aft dimension) was obtained from the appropriate cross section of bone. The rest of the procedure was the same used for calculating the usual, anteroposterior indicator of athletic capability (hereafter referred to as IAC_{ap}). The ratio between IAC_t and IAC_{ap} was then calculated to assess the differential anteroposterior and transverse strengths in many fossil and extant species (Table 7.4).

The departure of the IAC_t / IAC_{ap} ratio from 1 (i.e., a circular cross section) was assessed by subtracting the appropriate value for each bone of each species from 1 and dividing the result so obtained by the standard deviation (a measure of variance about the mean) of this sample. This value was compared with the values given in a *t* table, a statistical model useful when the number of cases is relatively small, to assess the significance of its departure from circularity, in which the ratio equals 1.

According to the results in Fariña et al. (2005), *Macrauchenia* shows a rather high femoral (thigh bone) IAC_{ap} value. Not only is it in the range of galloping mammals, but it's also about 50% higher than expected for its size (Fariña et al., 1997). However, the value of the IAC_t for that bone is even higher, and significantly so (ratio = 1.45). The humerus (upper arm

Table 7.4. Data for mass, distribution, and strength indicators in mammals and bipedal birds[a]

Species	Mass (kg)	a_{hind}	Value of $Z/amgx$ (GPa^{-1})					
			Humerus			Femur		
			Anteroposterior	Transverse	t/ap Ratio	Anteroposterior	Transverse	t/ap Ratio
Smilodon populator	400	0.4	47	31	0.65	19	20	1.17
Puma concolor	47	0.45	30.4	26.2	0.86	23.1	23.3	1.01
Panthera leo	170	0.45	31.0	28.2	0.91	22.4	23.9	1.07
Panthera onca	67	0.45	34.6	32.2	0.93	24.4	30.6	1.25
Panthera pardus	51	0.45	26.6	23.8	0.89	24.1	23.7	0.98
Panthera tigris tigris	145	0.45	26.9	22.0	0.82	18.3	20.5	1.12
Panthera tigris altaica	230	0.45	35.0	25.7	0.73	16.9	17.3	1.02
Uncia uncia	35	0.45	33.2	32.1	0.97	29.6	32.0	1.08
Macrauchenia patachonica	1100	0.35	32	37	1.16	21	30	1.45*
Hippidion principale	400	0.45	55	33	0.60*	53	53	1.00
Equus zebra	150	0.45	44	30	0.69	47	47	1.00
Equus burchelli	136	0.45	46.5	44.2	0.95	53	50	0.94
Gazella thomsoni†	25	0.4	47	47	1.01	39	43	1.13
Connochaetes gnou	150	0.45	36	34	0.94	32	29	0.92
Connochaetes taurinus	150	0.45	44	42	0.96	42	39	0.95
Struthio camelus†	80	1.0	—	—	—	46	55	1.20
Rhea americana†	18	1.0	—	—	—	35	58	1.65*
Pterocnemia pennata†	—	1.0	—	—	—	—	—	1.23

[a] Mass, distribution (a_{hind} is the proportion of the body weight supported by the hind limbs), and anteroposterior (IAC_{ap}) and transverse (IAC_t) strength indicators in several mammals and bipedal birds, both modern (puma, lion, jaguar, leopard, tiger, snow leopard, two zebras, Thomson's gazelle, wildebeest, and hartebeest) and extinct (sabertooth, *Macrauchenia*, South American horse). $Z/amgx$ indicates the strength indicators. The values of the ratios of the transverse and anteroposterior strength indicators that are significantly different from 1 are indicated with an asterisk (*). Modern animals that are observed to swerve are indicated with a dagger (†).

bone), on the other hand, shows larger, though not significantly different (ratio = 1.16), values for the transverse indicators.

In fact, all the values of the indicator of athletic capability obtained for the fossil mammals in the list lie above 20 GPa^{-1}, which in living mammals is associated with the capacity for galloping. Thomson's gazelles and both wildebeests yielded high values for both IACs. Their IAC_t/IAC_{ap} ratios do not differ significantly from 1, although the femora of the gazelle tend to have a larger ratio.

The femora of ratites, bipedal birds that inhabit the southern continents, have unusually high absolute values for both IAC_{ap} and IAC_t, which is a reflection of their capacity for fast running (Alexander et al., 1979b) with femora that are held nearly horizontally, and thus subjected to large amounts of torsional (or bending) stress (Christiansen, 1998; Carrano, 1998). The angle to vertical of the tibiae (shin bones) of ratites is more comparable to the angles at which mammalian long bones are held, and, conversely, the tibia of the ostrich has a strength indicator value similar to the femora and tibiae of fast-moving mammals (Christiansen, 1998). The ratio in the greater rhea was highly significant (ratio = 1.65).

Very large mammals escape predation by virtue of their size, at least as adults (Owen-Smith, 1988). However, the 1000-kg threshold may not have been equally effective, given the very large size of predators such as *Smilodon*. Therefore, it seems likely that *Macrauchenia* must have needed locomotory adaptations to escape predation from them.

As shown by Howland (1974), terrestrial prey can successfully escape from a faster predator if it can swerve on an arc of a circle of smaller radius, and by not dodging too soon to allow the predator to dodge as well and

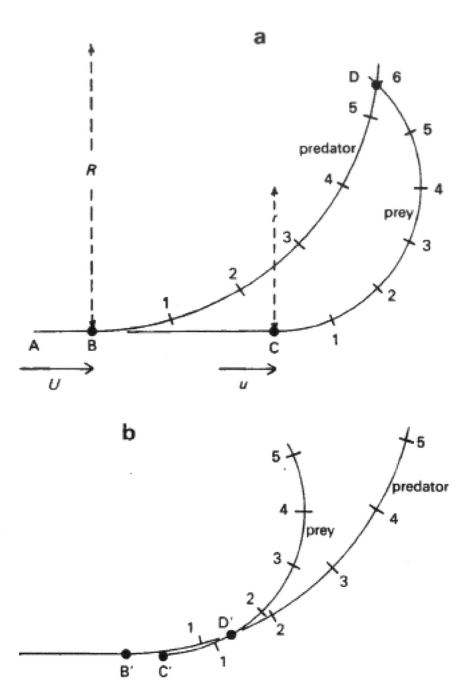

7.14. Diagram depicting the potential paths of a predator chasing slower prey that can turn more tightly. In the top diagram, the prey swerves to the left at position C and continues at speed *u* along a circular arc of radius *r*. The predator responds by swerving at position B at speed *U* along an arc of radius *R*. In a chase where $u = 0.75\,U$ and $r = 0.5R$, the prey arrives at D after 6 units of time and the predator after only 5.4. Therefore, it can intercept and capture the prey at D. In the bottom figure, the prey delays its swerve (at C') until the predator is much closer. The predator now responds by swerving at B'. Given the same quantities for speed and radius as in the top diagram, the arcs now intersect at D', and the prey escapes because it arrives at D' after 1.3 time units and the predator after 1.4 units. So long as the predator and prey continue along a straight path, we may describe how quickly they are moving by referring to their speed (as opposed to velocity, which as a vector quantity includes a speed value as well as direction). When, however, the predator and prey swerve, they change their velocity (because they change direction). A change in velocity (whether a change in speed or direction) is referred to as acceleration. The acceleration of the prey is calculated as v^2/r and that of the predator as V^2/R. The inequality $v^2/r > V^2/R$ (i.e., the acceleration of the prey is greater than that of the predator because, in this case, *r* is much smaller than *R* compared to the speed of the animals) should hold for prey to escape if the coefficient of friction *m* between the feet and the ground is high enough (so that the prey does not slip).

From Fariña et al. (2005), by kind permission of Asociación Paleontológica Argentina.

catch it (Fig. 7.14). An animal running along a circular arc has acceleration (a change in the rate or direction of speed) toward the center of the circle. As noted by Alexander (1982), the acceleration of the prey must be greater than that of the predator for successful escape, provided that the friction between the ground and the feet is high enough and the animal does not slip (thereby making the turn impossible). The theoretical expectation is that during the change of motion from a straight to a circular path, an acceleration results that is related to the rotation of the body in a sagittal plane, and the ground reaction force would produce important transverse bending in the limb long bones.

The humeri of larger mammals should be expected to have higher strength indicator values in the anteroposterior rather than in the mediolateral planes, and to be stronger than the femora, which are held substantially more vertically. This is exactly the observed pattern, although this primarily relates to how the bones are held (their angle to the vertical).

The humerus of *Macrauchenia* is, however, markedly stronger in the mediolateral than anteroposterior plane, an unusual condition in a mammal with elbow flexure and straight limbs. With such a limb posture, bending moments are expected to be stronger in the anteroposterior direction. However, this condition is not reflected in the morphology of the humerus, although the radius (one of the two bones of the forearm) bears an unusual medial flange (Fig. 7.15).

Therefore, Fariña et al. (2005) proposed that swerving and dodging rather than higher speed may have been the strategy that *Macrauchenia* used to flee the attacks of *Smilodon*. Although *Macrauchenia* could have been a rather fast runner, perhaps as fast as or even faster than a rhinoceros, for which a speeds of more than 40 km h^{-1} are cited for short distances (Paul, 1988; Garland and Janis, 1993; Nowak, 1999), *Smilodon* was proposed to have been an ambush predator, whose attack could have been avoided by a sudden acceleration, either forward or lateral, which seems to be a more suitable strategy to escape a short-distance chase. Sudden dodging movement, a behavior found in the ratites (whose femora are congruently stronger in the transverse plane), is corroborated by the maneuverability suggested by the retention of three digits in both pairs of limbs of *Macrauchenia*.

The last feature that needs attention is the unusual medial flange of the radius. Its enlarged size may have been related to the strong pull of musculature toward the sagittal plane of the animal, making dodging more efficient (although this supposition requires further anatomical study).

The posture of *Toxodon*

Springs are structures that are useful for a multitude of purposes: a spring can be fixed to a door to close it; in the suspension system of a car; or in performing neat tricks using a trampoline. Animals long ago discovered the advantages of springs and make use of them in various ways too. The parts of animals that act like springs are ligaments, soft tissue structures that have a variety of properties (depending on the particular functions they are put to) and generally act by linking hard parts, such as bones, together. Mussels use them to keep open the doors of their valves (or shells), which they close with an enormous muscle (a batch of these muscles can be served up as scallops), fleas use them to jump, kangaroos to rebound from the ground and so to travel virtually energy free. The ligament of the nape of the neck, the nuchal ligament, is another example. In this case it is made up of elastin and contributes to keeping the head in position (Dimery et al., 1985).

Scholars have traditionally supposed that *Toxodon* lived in an open environment of arboraceous savanna, in common with the majority of the large Lujanian fauna. Previous reconstructions, derived from those produced at the end of the nineteenth century, portray them with the front limbs bent (Fig. 7.16). As well, they are depicted with their head at a lower level than the first few thoracic vertebrae. With regard to posture, it is difficult to see how this positioning of the legs could be possible, given that a large-sized, ungulate-like animal cannot run or even stand with its

7.15. Right foreleg of *Macrauchenia*, showing the large medial (inward) expansion of the radius. Scale bar equals 10 cm.

Image by S. F. Vizcaíno by kind permission of the Museo de La Plata, Argentina.

legs flexed (such a position, however, may have been possible in very large xenarthrans, given their peculiar anatomy). A telltale sign is the condition of the olecranon, which projects strongly backward, as also occurs in rhinos. This morphology improves the mechanical advantage of the triceps during extension of the elbow joint, suggesting that *Toxodon* held its front limbs extended when standing, as do rhinos.

However, the second characteristic merits greater consideration. At first glance, the head seems to bend too much for the nuchal ligament to support it.

To test this hypothesis, a model was used, similar to the procedure applied in the estimation of the mass of the entire animal (Fariña and Álvarez, 1994). The head and the neck must be cut from the model approximately at the spot corresponding to the articulation between the last cervical (neck) and first thoracic (chest segment of the body) vertebrae. By suspending this portion from two different points, the center of mass is obtained, which must be corrected for by taking into consideration the large contribution of the bone tissue (with a density of approximately 2000 kg m^{-3} and mainly concentrated in the skull and mandible, as opposed to the cervical vertebrae), moving the center to a slightly more anterior position.

The force of the weight acts through this point, tending to make the head and the neck rotate in a clockwise (if the animal is facing to the right, as in Fig. 7.16) direction about the pivot, which is the point at which the last cervical and first thoracic vertebrae articulate. The moment of this force must be compensated by the resistance of the point of insertion of the ligament on the spine at the first thoracic vertebra, which pulls on the neck and head in a counterclockwise direction.

Having made the calculations, F_2 — that is, the weight of the head and the neck together — comes out to about 1200 N; its distance from the pivot is 60 cm. Therefore, the force F_1 — the resistance of the ligament at the level on the spine at the first thoracic vertebra, situated at a distance of 36 cm — must be approximately 2000 N in order for the system to be in equilibrium.

In turn, the force which the nuchal ligament should exert in order to keep the head inclined can be approximated if one knows that the ultimate stress of elastin is 1.8 MPa. In the calculations made above, the force which the ligament must exert at the first thoracic vertebra is about 2000 N. Each square meter of elastin is capable of supporting a force of about 1.8 MN, as already stated in the stress it can withstand.

Before drawing any conclusions, we should remember that all structures are built with a certain margin of safety. If a bridge is built to support a load of 10 tonnes, no engineer in his or her right mind would build it so it would support only exactly that much weight. In general, in view of possible small errors, variations in the load, and nature of the materials used, an engineer will design it to bear much more weight. Usually, it will be built to bear double the maximal load expected. Nature is a marvelous engineer, and the security factor that it builds into its structures is usually on this order of magnitude (Alexander, 1981).

Now we can go back to our calculations and introduce a safety factor of 2. The corresponding diameter of the nuchal ligament, then, should be at least 8 cm. On the other hand, if the head were in a lower position in the diagram, the distance of the center of mass of the head and the neck to

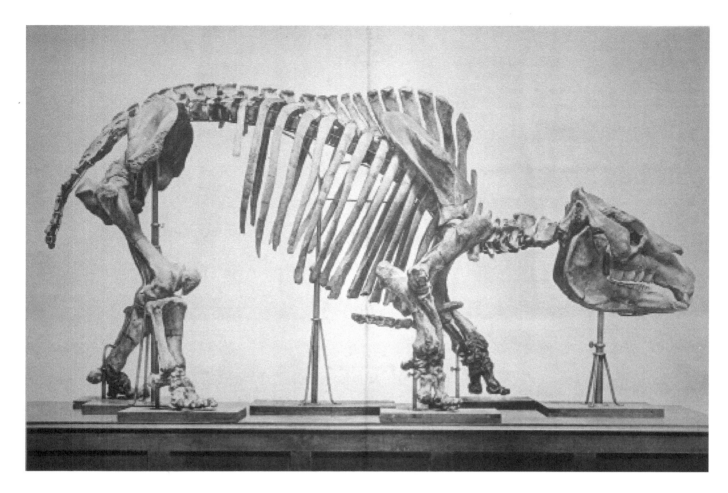

the pivot would be about 35 cm, and from there it would need a ligament of about 5 cm in diameter.

This thickness, although realistic, seems excessive for this ligament. In a deer weighing 20 kg, it is only 1 cm, and it is just a bit larger for a sheep weighing 21 kg. If the animals were geometrically similar, the ligament should be as thick as the cube root of the proportions of the masses—that is, 3.8 cm. Probably, the action of the muscles, which were not taken into account in this model, is sufficiently important to reduce this value even more. All that we can affirm is that it seems to be quite likely that the erect posture of the head was a mechanism for reducing the risk of excessive physical damage in the nuchal ligament.

Here, size matters, because although the weight of a structure increases with the cube of linear measurement, the cross section of a ligament does so only as the square. Thus, it must compensate by growing disproportionately. This growth has a limit, so there comes a point at which the most expedient strategy is to keep the head more erect so as to subject the ligament to less tension. This is what happens in mammals of similar size and shape, such as rhinos and hippos, which do not hold their heads so low.

7.16. Mounted skeleton of *Toxodon platensis*. The high neural spines of the thoracic vertebrae worked as attachments for large muscles that acted as counterpower of possible head-to-head impacts. Total length of the skeleton: 2.7 meters.

From Lydekker (1893).

General speculations on the dietary preferences of ground sloths have been proposed for more than 150 years. Owen (1842, 1860) made remarkable descriptions of the skeletons of the ground sloths *Glossotherium* and *Megatherium*, and presented lengthy explanations on their possible diet

Masticatory function and geometry of the skulls of sloths

and behavior. In the wonderful prose so characteristic of his time, Owen (1842:159–160) wrote:

> The close correspondence between the *Megatherium* and the *Mylodon* [actually *Glossotherium*] in the modifications of the skeleton determining the peculiar forces acting from the hind upon the fore-parts, compels us to infer that they resembled each other in the mode of which they obtained their sustenance; and nevertheless, the difference in the form of the grinding surface of the teeth, as well as in their size and the depth of insertion, obviously indicates some difference in the substances comminuted. . . . On the theory that the Megatherioids subsisted on foliage, it is most natural to suppose that the *Mylodon* and *Megalonyx*, with teeth most closely resembling those of the Sloths, would feed, like them, on the leaves and tender buds; while the *Megatherium*, whose essentially bradypodal teeth were more modified by their arrangement in a closer series, . . . so as concurrently to offer an obvious resemblance to the Elephant's dentition, would be thereby able to bruise the smaller branches, and to masticate these together with the buds and leaves. . . . All the characteristics which coexist in the skeleton of the *Mylodon* and the *Megatherium* conduce and concur to the production of the forces requisite of uprooting and prostrating trees.

Later, Owen (1860:78) stated that "Guided by the general rule that animals having the same kind of dentition have the same kind of food, I conclude that the *Megatherium* must have subsisted, like the Sloths, on the foliage of trees; but that the greater size and strength of the jaws and teeth, and the double-ridged grinding surface of the molars in the *Megatherium*, adapted it to bruise the smaller branches as well as the leaves, and thus to approximate its food to that of the Elephants and Mastodons." Stock (1925) proposed that megatheres, together with megalonychids and nothrotheres, were probably browsers, whereas mylodontids were grazers. Cabrera (1926) discussed the diet of *Megatherium*, rejecting several hypotheses on myrmecophagy or insectivory, and concurred with Owen's statements on a folivorous diet. Winge (1941:364) pointed out that *Megatherium* "has progressed farthest in specialisation as a plant feeder" and that it "must undoubtedly have fed on unusually tough leaves which required much power of mastication."

More recently, McDonald (1987) indicated that the Plio-Pleistocene South American Scelidotheriinae were selective feeders because the long and narrow muzzle was suitable for selecting plant parts. For the North American ground sloths, Naples (1987, 1989) proposed that *Nothrotheriops shastensis* was a selective browser and *Paramylodon harlani* a browser/grazer, instead of a strict grazer; and McDonald (1995) that *Megalonyx* and the Panamerican *Eremotherium* were browsers.

An alternative hypothesis, explored more fully in Chapter 8, to the herbivorous habits of ground sloths was proposed by Fariña (1996), but based on ecological rather than anatomical considerations. Fariña suggested that these beasts may have been opportunistic scavengers, especially *Megatherium americanum*, and Fariña and Blanco (1996) supported this hypothesis, even suggesting that this beast may have been an active predator or a kleptoparasite (i.e., a scavenger that steals its food, forcing predators or smaller scavengers away from carcasses).

The great variation in skull and dental morphology, body size, and proportions among ground sloths suggests that they filled a variety of niches. Their marked differences from other mammalian herbivores in skeletal and

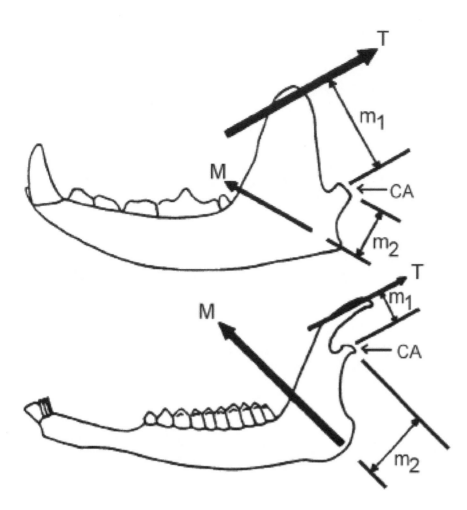

7.17. Mandibles of modern carnivorous (top) and herbivorous (bottom) mammals. CA, cranio-mandibular (or temporo mandibular) articulation; m_1, moment arm of the temporalis muscle (T); m_2 moment arm of the masseter muscle (M). The temporal is more important than the masseter in carnivores, while the opposite is true for herbivores.

Illustration by Sebastián Tambusso based on several sources.

dental anatomy, and the lack of living analogues, make it difficult to interpret the ecology of ground sloths, particularly their dietary habits. Even the extant tree sloths are too specialized to provide good models: with relatively small body masses (less than 10 kg), they are strictly arboreal and almost entirely folivorous, consuming mainly tree and liana leaves.

More rigorous approaches on the way extinct animals masticated and, ultimately, what they ate can be achieved by using biomechanical analyses similar to those applied above to locomotion. Several such studies of the masticatory apparatus have demonstrated correlations among the behavior, diet, and form of the skull, jaws, dentition, and musculature in extant mammals (e.g., Maynard Smith and Savage, 1959; Turnbull, 1970; Moore, 1981; Naples, 1982, 1985; Smith, 1993; Spencer, 1995; Mendoza et al., 2002; Mendoza and Palmqvist, 2008).

In this context, the jaws can be considered as a lever system, with the pivot at the craniomandibular joint, and the masticatory muscles providing the input force, whereas the output force is produced by the teeth on food (Fig. 7.17). The moment arms of the lines of action of the masticatory muscles may be estimated to analyze relationships between bite force and velocity. This procedure was applied to recent mammals (Maynard Smith and Savage, 1959; Turnbull, 1970) and then to fossils with a new geometric model proposed by Vizcaíno et al. (1998) that allows comparisons between fossil and extant mammals.

Bargo (2001b) and Bargo and Vizcaíno (2008) used this methodology with the ground sloths *Megatherium americanum, Glossotherium robustum, Lestodon armatus, Mylodon darwinii,* and *Scelidotherium leptocephalum* (Fig. 7.18). The areas of attachment of the masticatory musculature follow the same pattern, with minor variations in shape or texture. Following these reconstructions and estimation of the moment arm of the masticatory muscles' lines of action, comparison of relative muscle and bite forces, and, more significantly, the analysis of the relationship between bite force and velocity may be analyzed, comparing the proportions of the combined moment arms of the masseter muscle (the jaw-closing muscle on the side of the jaw) and the temporal muscle (the jaw-closing muscle on the side of the skull) with those of the bite. The values of the moment arms (Table 7.5) show little variation among mylodontids, as do the ratios of muscle moment/bite moment, which provides a measure of the relative bite force generated at different points along the tooth row and the bite velocity. *Megatherium americanum* has the highest values all along the tooth row. High ratios indicate strong rather than fast mandibular movements. Hence, the masticatory apparatus of *M. americanum* is designed to generate larger bite forces than those of mylodontids. The means for the posterior teeth and for the whole series provide a measure of the bite force generated at the posterior part of the mandible and its total bite force, respectively. Accordingly, mean values indicate that *M. americanum* has the strongest bite all along the molariform (cheek tooth) series, especially posteriorly, whereas mylodontids have a less powerful bite with little variation of the different points along the cheek teeth.

This analysis was complemented with consideration of functional morphology. In both megatheriids and mylodontids, the shape of the articular surface of the mandible for the skull would have permitted considerable freedom of sideways and front-to-back movements of the mandible. The analysis of tooth wear facets and striations (scratchlike marks made by opposing teeth or coarse food) in mylodontids indicate that the main masticatory movement was in an inward and forward direction. The front and back portions of the zygomatic arch are not fused. This condition might have diminished its ability to withstand the great forces generated by the masticatory musculature during sideways movements, thus suggesting that such movements were not particularly important during mastication. In megatheres, the teeth have two prominent, sharp, and transversely oriented, crestlike ridges (termed lophs) separated by a deep V-shaped valley, which would have been kept sharp by performing essentially vertical movements; this is confirmed by the striations present on the wear facets. Similarly, the zygomatic arch is large and robust, and the anterior and posterior parts may be in contact (or even fused in aged individuals; De Iuliis, 1996), unlike in mylodonts. The stoutly built zygomatic arch, particularly the great development of its descending process, leaves a narrow space between it, the horizontal ramus (or horizontal body of the lower jaw) and, in part, the ascending ramus (or process) of the lower jaw, suggesting a physical restriction to lateral movements.

The mechanical principles of tooth design of mammals in relation to the nature of the food was summarized by Hiiemae and Crompton (1985),

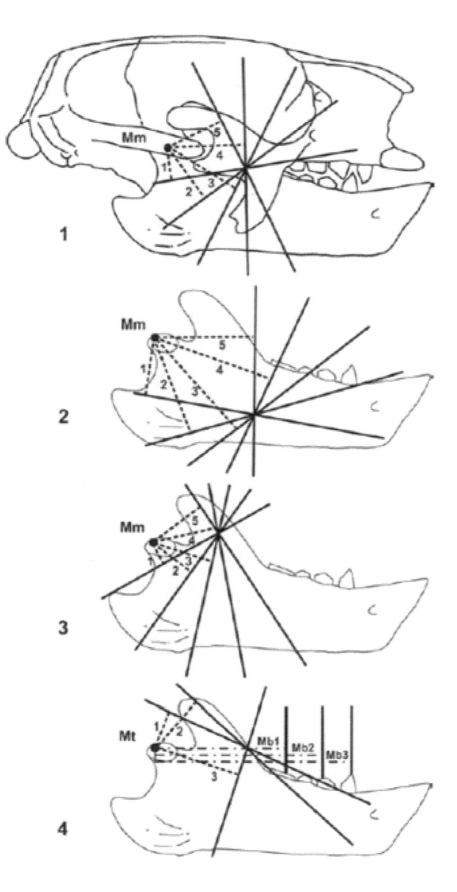

7.18. Geometric model used for the estimation of moment arms of the masticatory muscles in *Glossotherium robustum*. (1) Moment arms of the masseter (Mm 1 to 5) from the middle point of the origin area of the muscle on the zygomatic arch. (2) The same from the anteriormost point of the origin area. (3) The same from the posterior-most point of the origin area. (4) Moment arms of the temporal (Mt 1 to 3), and the bite (Mb 1 to 3, from the posteriormost, middle, and anteriormost tooth, respectively). Once the lines of actions are averaged, a simpler diagram can be constructed, as that of Fig. 7.17.

Illustrated by Sebastián Tambusso based on Bargo and Vizcaíno (2008).

Table 7.5. Summary of mean values of moment arms and muscle/bite ratios[a]

Species	Mm	Mt	Mb$_1$	Mb$_2$	Mb$_3$	r·Mb$_1$	r·Mb$_2$	r·Mb$_3$	X$_1$	X$_2$
G. robustum	28	21	46	62	73	1.07	0.83	0.68	0.95	0.86
L. armatus	29	18	42	61	86	1.13	0.78	0.55	0.95	0.82
M. darwinii	26	15	41	53	65	1	0.77	0.63	0.88	0.8
S. leptocephalum	28	16	41	51	61	1.08	0.87	0.73	0.97	0.89
M. americanum	28	18	28	40	51	1.68	1.18	0.92	1.43	1.26

[a] Calculated by Bargo and Vizcaíno (2008). Mm and Mt, momemt arms of the masseter and temporal muscles, respectively; Mb, moment arms of the bite at the most anterior, middle, and most posterior point of the tooth row; X, mean. Values are in millimeters.

who recognized three basic patterns: (A) a mortar and pestle system suitable for crushing hard and brittle (e.g., nuts) or turgid (e.g., fruit pulp) food; (B) blades to cut soft but tough food (e.g., muscle and skin); and (C) a serial array of low-profile blades acting as a milling machine for tough and fibrous food (e.g., grass). The teeth of ground sloths do not fit easily within any of these patterns, but they can be assigned to an intermediate condition. In mylodontids, the teeth are oval or semioval. They display different degrees of lobation, and the occlusal surface is comparable to Hiiemae and Crompton's pattern A, with some outer hard and sharp cusps that would act as cutting edges during masticatory movements. The combination of this morphology with the evidence that the direction of the mandibular movement is mostly anteromedial indicates that mylodontid teeth would represent an intermediate situation between Hiiemae and Crompton's patterns A and C, i.e., for crushing and grinding. According to Janis and Fortelius (1988), moderately tough and abrasive foods, such as leaves, require reciprocal blades for comminution of food items, a function accomplished by orthal (straight up and down) chewing with bilophodont (i.e., bearing two lophs) teeth. The dentition of *Megatherium americanum* represents a battery of high lophs with sharp, cutting edges, similar to the dentition of tapirs and kangaroos. This morphology would represent an intermediate condition between Hiiemae and Crompton's patterns A and B, i.e., mainly for cutting and crushing. This morphology does not rule out the possibility of processing food with similar physical properties (soft but tough, fleshy), from either plant or animal source, such as muscle or skin (Bargo, 2001a, 2001b).

As is now evident, the two groups analyzed—mylodontids and megatheriids—are morphofunctionally distinct from each other in their masticatory apparatuses.

There is a third approach in searching for evidence on what animals ate. This approach is based on morphology and ecomorphology, which deals with the correlation observed between morphological features and habits in living animals. Recent ecomorphological analyses applied to these ground sloths include three craniodental variables: the dental occlusal surface area (OSA, the sum of the biting area of all the teeth), the relative width and shape of the muzzle, and the height of the tooth crown (hypsodonty, which is a relative measure of the height of the crown). Several authors have produced work along these lines (Vizcaíno et al., 2006; Bargo et al., 2006a,b). The results of these studies offer relevant information that, coupled with biomechanical and functional morphological evidence, allowed Bargo and Vizcaíno (2008) to perform a paleobiological interpretation on the dietary habits of the ground sloths.

Megafauna

The analysis of OSA in xenarthrans by Vizcaíno et al. (2006) supports the proposal that the masticatory apparatus of mylodonts had crushing and, to a lesser extent, grinding functions, and that it was not particularly well adapted for extensive oral food processing. In contrast, the feeding apparatus of *Megatherium americanum* was well designed for generating strong, predominantly orthal movements that were used mainly for cutting rather than crushing and grinding. These authors found that mylodontids have extremely low OSA values in comparison with living herbivorous mammals of equivalent body size, supporting the idea of poor oral food processing. This was probably compensated by high fermentation in the digestive tract, lower metabolic requirements, or both. Surprisingly, OSA of *M. americanum* is the expected (or even higher than the expected) value for a mammal of its size, and much larger than those of mylodontids. It is clear then that *M. americanum* was better suited for oral food processing in the oral cavity, and it most likely had a lower fermentation capability and/ or higher metabolic requirements.

Bargo et al. (2006b) analyzed the relationship between dietary habits and shape and width of the muzzle of the five species of ground sloths considered here, and examined models of food intake by reconstructing musculature and cartilages of the muzzle. According to these authors, ground sloths can be divided in two groups with different feeding behaviors: wide-muzzled sloths (*Glossotherium robustum* and *Lestodon armatus*) that were mostly bulk feeders (i.e., they ingest great amounts of food with each bite; probably roughage and grass eaters), and narrow-muzzled sloths (*Mylodon darwinii*, *Scelidotherium leptocephalum*, and *Megatherium americanum*) that were mixed or selective feeders (i.e., select plants or plant parts; grass and/or tree and shrub foliage). The muscle reconstruction indicates that the upper lip was probably square shaped and not prehensile in wide-muzzled sloths, as in the white rhinoceros, *Ceratotherium simum*. This condition, coupled with the absence of incisors (teeth at the front of the oral cavity), indicates that *G. robustum* and *L. armatus* simply used the upper lip coupled with the tongue to pull out grass and herbaceous plants. Similarly, narrow-muzzled sloths (*M. darwinii*, *S. leptocephalum*, and *M. americanum*) had a thick, cone-shaped, and prehensile upper lip, useful for food intake as in the black rhinoceros, *Diceros bicornis*, to select particular plants or plant parts (e.g., leaves and twigs). It is worth noting that the white rhino is a typical grazer, but its muzzle is relatively narrower than in many mixed feeders. One plausible hypothesis is that in some cases, lip morphology determines functional muzzle width, supporting the need to consider muscular anatomy to complete these discussions when attempting such analyses.

Finally, the comparative study of hypsodonty in Pleistocene ground sloths by Bargo et al. (2006a) suggests that differences in crown height may be explained by a combination of (rather than any single) variables, including dietary preferences (nature of food items), habitat (closed or open, temperate or tropical), and behavior (feeding at ground level or higher, digging). Mendoza and Palmqvist (2007) demonstrated that high-crowned teeth represent an adaptation in ungulates (hoofed mammalian herbivores, such as cows and horses) against tooth wear resulting from the airborne grit and dust accumulated on the herbaceous plants of open environments. However, the absence in xenarthrans of enamel (the hardest outer layer in the teeth of nearly all other mammals), which would make the teeth less durable and subject to faster wear, must be considered as responsible for much of the hypsodonty observed in sloths, as well as in all xenarthrans, obscuring the interpretation of the individual contribution of each these variables. Within mylodontids, *S. leptocephalum* has the highest hypsodonty index (HI, a measure of the relative height of tooth crowns), followed by *M. darwinii*, while *L. armatus* and *G. robustum* have the lowest indices. On the other hand, *M. americanum* has the highest HI, even when compared with other megatheriines (i.e., the Panamerican *Eremotherium* and other *Megatherium* species from north-central and northwestern South America). Determination of the degree to which higher hypsodonty values in megatheriines and mylodontids correspond to feeding on abrasive grasses rather than browsing on foliage (as for living ungulates) is difficult because the proportion of grass in their diet cannot be known. However, one obvious factor in explaining differences in hypsodonty in ground sloths is the increased presence of grit caused by environmental differences resulting from geographic distribution, environmental change over time, or particular habits. For example, the differences in hypsodonty between the megatheriines *E. laurillardi* and *M. americanum* might be explained as adaptations to different environments, as reflected by their geographical distributions. Differences in environment over time, such as from closed to open, were apparently important in North American *Paramylodon* (McDonald, 1995). The presence of digging behavior in *S. leptocephalum* and *G. robustum*, including but not limited to searching for food, may also be relevant. Also, the narrow-muzzled sloths *S. leptocephalum* and *M. darwinii* would have used their muzzles in searching for food. These particular habits must have played a considerable role in shaping the dental characteristics of these sloths, with the important agent being the relative abundance of abrasive soil particles.

Several discoveries of mummified remains and dung of different ground sloth taxa from South and North America have provided additional evidence for the inference of their diets, as summarized in McDonald and De Iuliis (2008). For instance, Moore (1978) states that grasses and sedges are the dominant vegetation identified in the dung of *Mylodon darwinii* found in a cave at Ultima Esperanza. This partially corroborates the morphological information presented above (low OSA values = low metabolic requirements or low-quality food; narrow-muzzled = selective and mixed feeder) for this sloth. The novel application of stable isotope analyses in xenarthrans (by such authors as Coltrain et al., 2004; Kalthoff and Tütken, 2007) and DNA analysis will provide a better understanding of the diets of extinct sloths.

Plate 1. The geologic timescale presents the succession of the divisions of earth time. The basic sequence had essentially been worked out (though there have been refinements) by the first few decades of the twentieth century. The dates (which are also constantly being refined) were superimposed on the existing framework once methods for estimating absolute time were developed, beginning during the 1950s.

Image courtesy of the Geological Society of America.

Plate 2. The Hunterian Museum at the Royal College of Surgeons. Several of the fossils collected by Charles Darwin were housed in this institution, which was overseen by Richard Owen (who may be the person pointing at the skeleton in the foreground). A mount of *Megatherium americanum* is at front left, and a glyptodont carapace is at front right. Watercolor by Thomas Hosmer Sheperd, first published in the *Illustrated London News* in 1844.

Courtesy of the Hunterian Museum, Royal College of Surgeons of England.

Plate 3. An example of the fossil remains found by P. W. Lund. In earlier paleontological publications, such exquisite artwork often accompanied his descriptions (Lund, 1839: pl. XIX).

Fig. 1.

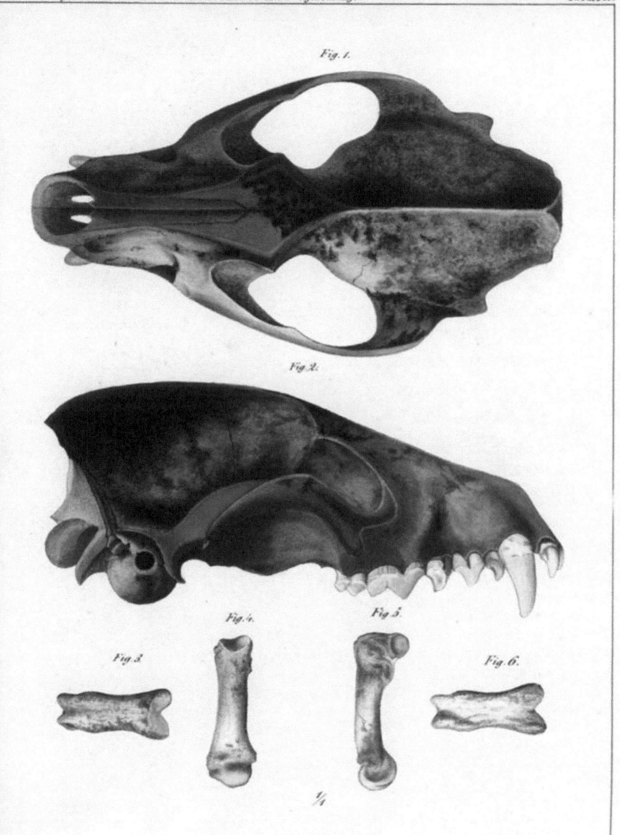

Fig. 2.

Fig. 4.   Fig. 5.

Fig. 3.   Fig. 6.

¼

Plate 4. Map showing biomes of South America.

Modified from Udvardy (1975).

TUNDRA Permafrost
TUNDRA Interfrost
BOREAL Semi-arid
BOREAL Humid
TEMPERATE Semi-arid
TEMPERATE Humid
MEDITERRANEAN Warm
MEDITERRANEAN Cold
DESERT Tropical
DESERT Temperate
DESERT Cold
TROPICAL Semi-arid
TROPICAL Humid
Ice

Plate 5. An image of the Irish elk (*Megaloceros giganteus*) as depicted in Lascaux cave. This representation is interpreted as humans having hunted this mammal in prehistoric times. As noted in the text, this giant deer was not particularly closely related to modern animals referred to as elks, and it was distributed across Eurasia during the late Pleistocene, and thus was not exclusively Irish (although many fine examples of its skeleton have been recovered from Irish bogs). It was among the largest of deer, standing some 2 m at the shoulders, and had the largest antlers, measuring over 3.5 m from tip to tip, of any deer.

Photograph by HTO. Public domain.

Plate 6. (a) Maps of South America during the last glacial maximum (LGM, about 18 kya). At that time, much of the earth's water was locked up in ice, which lowered sea levels in comparison with the present (b), and exposed considerable portions of land.

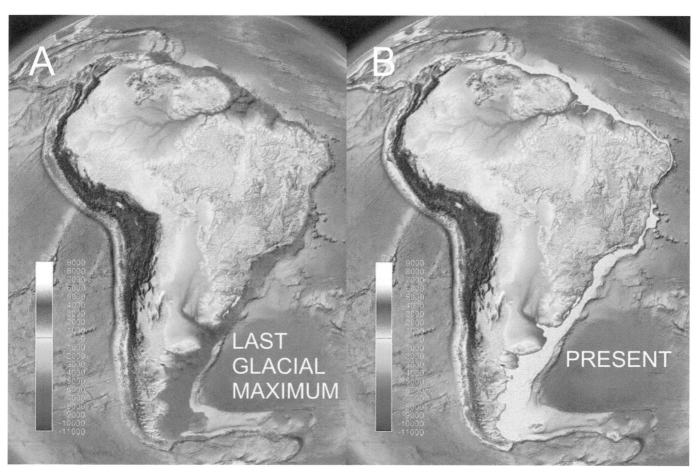

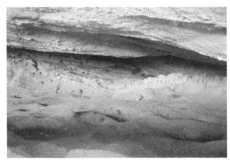

Plate 7. The *Mylodon* cave, in Última Esperanza, Magallanes region of Chile. It is located some 24 km (15 mi) northwest of the city of Puerto Natales. The cave is 150 m above sea level, and it measures 30 m high, 80 m long and 200 m deep. Near the end of the nineteenth century, scientists were fascinated by the discovery in this cave of the skin, bones, and other parts of *Mylodon darwinii* and other extinct animals. Today it is a natural monument and a must-see for tourists staying in Puerto Natales. At the entrance of the monument is a life-sized replica of the *Mylodon* (on the left side of the main figure).

Composed by Sebastián Tambusso after photographs of one of the authors (S.F.V.) and others kindly provided by Luis Borrero and Fabiana Martín.

Plate 8. Reconstruction of *Hippidion*. This water-color by Á. Cabrera was part of an old Museo de La Plata exhibit on the evolution of horses.

Reproduced by kind permission of the Museo de La Plata, Argentina.

Plate 9. Reconstruction of the South American gomphotheriid *"Stegomastodon" platensis*. This watercolor by Cabrera (1929b), who was obviously a competent artist as well as a paleontologist, illustrates two individuals, which he labeled as male (foreground) and female.

Image courtesy of Museo de La Plata, Argentina.

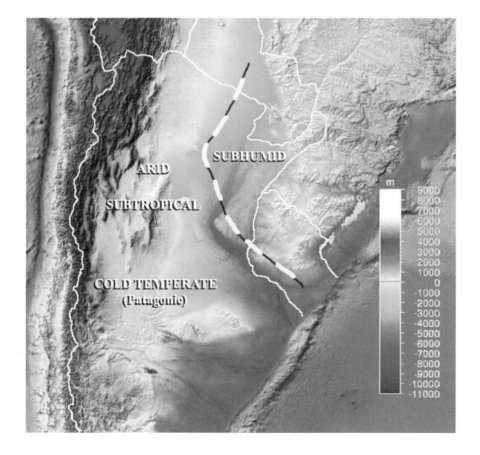

Plate 10. Distribution of climates at the last glacial maximum.

Based on Iriondo and García (1993).

Plate 11. Painting by Luis de Servi (1988) in the main hall of the Museo de La Plata, entitled "Descuartizando un gliptodonte. Escenas de la vida del hombre primitivo" (Quartering a glyptodont. Scenes from the life of primitive man).

Courtesy of Gustavo Politis by kind permission of Museo de La Plata, Argentina.

8.1. Reconstruction of the megafauna. Illustration in the División Paleontología de Vertebrados of the Museo de La Plata of the planned gardens between the Museo de La Plata and zoo.

Image by S. F. Vizcaíno by kind permission of the Museo de La Plata, Argentina.

General Paleoecology

We have already seen how the Pleistocene mammals got to South America, with some having ancestors already established there since the dawn of the Tertiary, notably marsupials, xenarthrans and the native ungulates, others having arrived in Tertiary times by crossing the Atlantic from Africa against overwhelming odds (primates and caviomorph rodents), and still others venturing into southern lands once (or just before) the Panamanian land bridge was formed: sigmodontine cricetids among Rodentia, procyonids, mustelids, canids, ursids, and felids among Carnivora, cervids, tayassuids, and camelids among Cetartiodactyla, equids and tapirs among Perissodactyla, and gomphotheriids among Proboscidea. Many aspects of their proposed habits have been dealt with in previous chapters, so we are now ready to consider how this whole fauna worked together.

Paleoecology deals with reconstructing ecosystems of the past by using data from the fossil record and from its geologic context. It includes the study of the habits of past organisms and their interactions and natural environment. It works on the basis of actualistic, often statistical, models to figure out the habitats of those organisms that we find today as fossils. When dealing with fossils and ancient environments, however, postmortem factors such as temperature and mineral supplies can bias our interpretations, adding another layer of complexity to this discipline when compared with modern ecology. A usual way of dividing the conceptual realm of this subject matter is to consider paleoautecology, which treats the environment from the organism's perspective, and paleosynecology, which studies the large-scale interactions at higher levels, such as communities and ecosystems.

In the following pages, we will summarize the findings and approaches related to the paleoecology of this splendid fauna. Because we have already considered studies on the life habits of particular species, we will focus especially on synecological efforts.

Until recently, the quality of information on this fauna was essentially what could be derived from common knowledge. To get a sense of the perception paleontologists had (and conveyed to other people) of this assemblage, it is easier to consider visual reconstructions of Pleistocene landscapes as opposed to the typical scientific publications, which are cumbersome to track and cite. For instance, the splendid illustration depicted in Fig. 8.1, which adorns one of the entrances of the División Paleontología de Vertebrados of the Museo de La Plata, is enormously significant for several reasons. To begin with, it was conceived at the end of the nineteenth century, just after the city of La Plata had been founded, as an artist's rendition of a plan to embellish the gardens between the museum and the neighboring zoo. This space was envisioned as a place where the

Definition, scope, and background

extinct beasts, whose bones resided in the museum, were brought back to life in the form of life-sized sculptures (and hence why there are pedestrian pathways also depicted). Regrettably, this wondrously grand scheme was never realized, though the reasons why have been lost with time. However, the plan admirably depicts the way those megamammals and their habits were then viewed.

For example, *Megatherium*, in the center background, is represented as furry and walking on all fours along a prairie similar to the pampas that the Ameghinos were familiar with (i.e., before fertilizers). Apart from the general portrayal of a rather luxuriously vegetated plain, it draws attention to the locomotory habits of this giant sloth, considered alternatively as a quadruped or as a biped. It is interesting to note that Cuvier saw it as a quadruped, while Owen (1838) drew it as a biped leaning against a trunk and reaching for leaves, despite Darwin's observations.

In the left foreground, two glyptodonts, likely *Panochthus* and *Doedicurus*, seem cordial in their close proximity; the former seems to be engaged in excavating, an activity based certainly on the habits of its closest living relatives. Although glyptodont anatomy does not, at least in principle, preclude digging behavior, it is unlikely that they practiced such behavior (as discussed further in Chapters 6 and 7).

In the center foreground, a mastodon ambles majestically, and two smaller glyptodonts, *Neosclerocalyptus*, seem attentive to the danger of being trampled by the proboscidean. A similar contemplative attitude is depicted for *Glyptodon*, just to the right of its smaller relatives. Across the pedestrian pathway, one *Macrauchenia* watches warily while another drinks from a shallow pond.

Meanwhile, in the background and from left to right, a ground sloth, probably *Glossotherium*, seems to be leaving the scene, exiting stage left, and a sabertooth is completing a hunt or already consuming its recently hunted prey, a glyptodont turned on its back. In the right background, another sabertooth seems to be stalking a victim just off scene.

Apart from such (quasi) idyllic pastoral images, the views on the way those animals lived were largely unpublished, other than some comments by Darwin (1839) in chapter 5 of his journal of the *Beagle*'s voyage, subtitled "Large Animals Do Not Require Luxuriant Vegetation." Right after having found several specimens of giant sloths, glyptodonts, toxodonts, mastodons, and other large mammals near Bahía Blanca, Argentina, and previously others in Buenos Aires Province and in Uruguay, Darwin made comparisons between the diversity and abundance of Lujanian and African faunas, coming to a conclusion about the relationship between how open the landscape is and the size (or bulk, to use his word) of the inhabiting mammals that differed from the usual view of his contemporaries. Although in India elephants and rhinos dwell in fine forests, Brazilian jungles support mammals on average 25-fold smaller than those in the apparently less fertile African savannas and scrubs, which are populated by an impressive fauna. Likewise, in past South America, Darwin reflected, the large mammals he was finding might not have been related to a richer flora than that of today, for he found "the strongest evidence that there has occurred no great physical change to modify the features of the country," and "yet in former days, numerous large animals were supported on the plains now covered by a thin and scanty vegetation." In the years that followed, knowledge of past climates increased significantly, and we are thus able to judge Darwin's intuitions and observations in the light of current approaches.

Remains of the Pleistocene fauna are common in midlatitude South America, especially in Uruguay and Buenos Aires Province, Argentina; many such fossils had already been collected and housed in museum and personal collections by the end of the nineteenth century. The greatest South American paleontologist, Florentino Ameghino, was the first author to make a comprehensive attempt to systematize the stratigraphy and faunal composition of the Lujanian Age (as in, for instance, Ameghino, 1889). Currently, it is accepted that although older strata also contain large and very large mammals, the splendor of this fauna is revealed especially in the Lujanian. Usually the preservation is good because bones were deposited recently, although taphonomic studies on those giants are only just beginning (Fariña, 2002b). These remains typically occur in loess or reworked loess (deposits due to the action of wind), which is broadly distributed in the region (Panario and Gutiérrez, 1999; Zárate, 2003). Lujanian localities are so ubiquitous that their depiction on a map fitting this page would yield an incomprehensible set of dots.

The fossiliferous Guerrero member of the Luján Formation was deposited during the last glacial maximum (LGM; see Chapter 4). Because of the extensive glaciation in the Andes, the climate was much drier and rather cooler than today in that region (Clapperton, 1983). Different sources of evidence are congruent with this paleoclimatic scenario (Cantú and Becker, 1988; Tonni, 1990). This was precisely Darwin's (1839) intuition about the landscape being less luxuriant than today. However, it seems that it was actually even more arid than surmised by the great naturalist.

Current reconstructions of the period around the LGM show that the Pampean plains underwent intense aeolian (or wind) activity that redeposited large masses of silt and fine sand of periglacial origin, forming a sand sea in the southwestern half of the Pampas, limited by the rivers that trapped those sands. Over the remainder of the area, there is a broad loessic belt (Iriondo, 1990). According to Iriondo and García (1993), climates were shifted about 750 km northeastward relative to present conditions (Plate 10). Hence, Luján would have had climatic conditions similar to those existing today in the north Patagonian locality of Choele-Choel (39°S). Thus, mean annual temperature in the Luján region must have been 2.5–3°C lower and its seasonality more marked, with summers only about 1°C colder but winters up to 4°C colder. More importantly, the aridity must have been higher, with rainfall considerably lower, about 350 mm per year compared with nearly 900 mm currently for the Pampas. The values for the past are approximations, the result of the likely influence of other factors, such as the well-known high quality of the soil of the Pampean region today, and perhaps the effect of the humidity of large local water bodies, such as the Paleoparaná River.

The region between latitudes 33 and 37°S, bounded by the Pleistocene sand fields on the west and the present coast of the Río de la Plata (actually a large estuary) and the Atlantic Ocean on the east, has an extension of

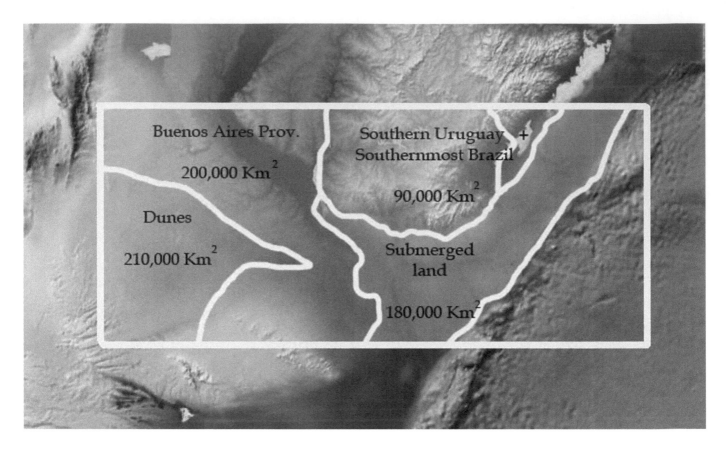

Buenos Aires Prov.

200,000 Km^2

Dunes

210,000 Km^2

Southern Uruguay +
Southernmost Brazil

90,000 Km^2

Submerged
land

180,000 Km^2

8.2. Paleogeographic reconstruction of Uruguay, Buenos Aires Province, and the Río de la Plata and neighboring Atlantic areas during the last glacial maximum (LGM). Available land during the LGM was about 60% greater than that available in modern times.

Illustration by Sebastián Tambusso.

about 300,000 km^2 (Fig. 8.2). Given that sea level was about 120 m (and perhaps more) below the present level during the final phase of the last glacial (Lambeck et al., 2002), the emerged land would have increased the land area to about 480,000 km^2, which represents an increase of about 60%. This has important paleoecological consequences. Current paleogeographic reconstructions (Fig. 8.2) interpret the submerged region as the valley of a large river (the Paleoparaná), today covered by the Río de la Plata, and an immense delta, today submerged by the Atlantic Ocean.

Flora and fauna from the LGM to the Holocene

In this section, we first provide a summary of the studies that address the young Darwin's reflections on the luxuriance of the vegetation in the Lujanian. How are we to begin to answer these reflections, and what do the efforts tell us about the ecosystem that made it possible for such a large quantity of mammalian giants to live together?

As already noted in Chapter 3, information on plants from the LGM is derived from pollen records, sediments recovered from lakes and other bodies of water, and peat and archeological sites (Barreda et al., 2007). The current phytogeographic (i.e., based on the distribution of plants) provinces, the High Andean, Punean, and Patagonic, populated by characteristic assemblages of taxa with defined spatial distributions, are difficult to identify in the Pleistocene. This is the result of the scarcity of pollen records, but the floristic composition does not appear to have changed much in the LGM–Holocene interval, although the evidence available suggests that the vegetation of the central Pampean region composed a psammophytic (i.e., of sandy substrate) steppe and grasslands with few modern analogues.

During the transition between the Pleistocene and Holocene, the eastern Pampas had vegetation that consisted of a bushy steppe, and the Andean region at midlatitude must have had High Andean vegetation (Quattrocchio et al., 1988; Markgraf, 1989; Prieto, 1996; Barreda et al., 2007).

The remains of still-extant mammals (i.e., those for which habitat preference can be safely assigned) belong to species confined to Central and Patagonian faunistic provinces (Tonni, 1985b; Prado et al., 1987; Alberdi et al., 1989). The same can be said about the birds of this age (Tonni and Laza, 1980). According to Cione et al. (2007), and as mentioned in Chapter 4 of this book, the Lujanian started with interglacial conditions 130 kya before the present, which may be inferred from the faunal assemblages (Pardiñas et al., 2004). Its final part coincided with the LGM (18 kya) and the last glacier progression (13 to 11 kya), all of which can be seen in the arid- and cold-adapted fauna (Tonni et al., 1999, 2003). Among their most interesting features, these sediments include evidence of possibly the earliest appearance of *Homo sapiens* in South America and the extinction of the impressive fauna treated in this book, a subject to be dealt with in Chapter 9.

We have previously noted that there are many places where the remains of the Pleistocene megafauna can be found. However, we will focus here on the locality whose name was taken to include the time and the fauna that lived then. The Luján local fauna comprises the fossil remains found in the Guerrero Member of the Luján Formation that were more or less continuously collected since the final years of the eighteenth century. Such collection began even before the independence of the countries in the area of the Río de la Plata, and lasted for 100 years, until the urbanization of the city of Luján rendered it impossible.

Several radiocarbon dates restrict the age of this member to between 10.29 ± 0.13 and 21.04 ± 0.45 kya (Tonni et al., 2003), which includes the LGM (Thompson, 2000). According to dating of the Guerrero Member from another locality (i.e., Arroyo Tapalqué), this fossil assemblage probably represents an average time span of approximately 8000 years (Tonni et al., 2003). One paper on the geology of the Río Luján in the Luján area suggests the possibility of a much longer time span: that the Guerrero Member is composed of two independent stratigraphic beds, with the base of the sequence dated at more than 40 kya (Toledo, 2005), yielding a span averaging more than 20,000 years.

This local fauna has the required grandeur to be the name bearer: although several of its taxa are still in need of revision, as many as 26 herbivorous species have been recorded from the Luján local fauna according to Prevosti and Vizcaíno (2006), although we follow a slightly different classification here. For instance, we have not recognized the diversity of the genus *Toxodon*. We have also retained *Stegomastodon*, although between quotations marks, as explained in Chapter 6. On the other hand, we have replaced *Hemiauchenia*, used by Fariña (1996) and Prevosti and Vizcaíno (2006), with *Palaeolama*, which has no effect on the number of species either.

Composition of the herbivores and carnivores of the paleomammalian fauna

Among these herbivorous species, xenarthrans reach the absolute majority with 14 species: seven glyptodonts (*Neosclerocalyptus migoyanus*, *Neothoracophorus depressus*, *Panochthus tuberculatus*, *Doedicurus clavicaudatus*, *Plaxhaplous canaliculatus*, *Glyptodon clavipes*, *Glyptodon reticulatus*), one pampathere (*Pampatherium typum*), one armadillo (*Eutatus seguini*), and five ground sloths (*Megatherium americanum*, *Glossotherium robustum*, *G. myloides*, *Lestodon trigonidens*, *Scelidotherium leptocephalum*). The list is completed by a rodent (the extinct giant capybara *Neochoerus aesopi*); two notoungulates (*Toxodon platensis*, *T. burmeisteri*); a litoptern (*Macrauchenia patachonica*); a perissodactyl (the horse *Equus* (*Amerhippus*) *neogeus*); six artiodactyls (the tayassuid *Pecari tajacu*, the camelids *Palaeolama paradoxa*, *Eulamaops parallelus*, *Lama guanicoe*, and *Lama gracilis*, and the cervid *Morenelaphus lujanensis*; and a gomphotheriid proboscidean, "*Stegomastodon*" *platensis*). Most xenarthrans qualify as megaherbivores; only a few of them are below the limit of 1 tonne in adult body mass, with *Megatherium* and *Lestodon* reaching several tonnes and *Glossotherium* and most glyptodonts between 1 and 2 tonnes. Competing with the largest, "*Stegomastodon*" honors its proboscidean legacy with an estimated mass of 7500 kg. The species of native ungulates, those belonging to the genera *Toxodon* and *Macrauchenia*, had sizes exceeding 1 tonne. Table 8.1 lists these species and their estimated body mass.

There are five species of Carnivora above the 10-kg size in this local fauna. One was very large: the sabertooth felid *Smilodon populator*, at an impressive 400 kg (see Chapters 6 and 7). Another three were rather large: the extinct ursid *Arctotherium tarijensis* (140 kg, Soibelzon, 2002) and the still living felids *Panthera onca* (jaguar, whose Lujanian individuals attained 120 kg, as discussed in Prevosti and Vizcaíno, 2006) and *Puma concolor* (50 kg, Prevosti and Vizcaíno, 2006). The fifth was the much smaller, extinct canid *Dusicyon avus* (15 kg), perhaps a pack hunter, as are many modern species of canids (Prevosti and Vizcaíno, 2006). The abundance of hypercarnivore canid species in the South American Pleistocene noted by Van Valkenburgh (1991) is not apparent in this local fauna (but see Prevosti and Vizcaíno, 2006), while only a few other carnivorans need be taken into account if the whole late Pleistocene of South America is considered. In accordance with the most recent systematic revisions (e.g., Berman, 1994; Soibelzon, 2002, 2004), Prevosti and Vizcaíno (2006) added other species of carnivorans occurring in Buenos Aires Province to those originally considered for the Luján local fauna: the extinct canid *Canis nehringi* (30 kg, Prevosti and Vizcaíno, 2006) and the ursid *Arctotherium bonariense* (110 kg, Soibelzon, 2002).

The paleoecology of extinct South American large carnivorans is yet to be reasonably well understood, especially in comparison to the studies of North American species. Three of the extinct South American taxa (*Canis nehringi*, *Arctotherium* spp., and *Smilodon populator*) are closely related and similar to North American taxa (i.e., *Canis dirus*, *Arctodus*, and *Smilodon fatalis*, respectively), allowing, in principle, an extrapolation of the paleoecological inferences from the Nearctic forms to the South American ones, although we discussed the habits of *Arctotherium* in Chapters 5 and 6. Fig. 8.3 shows the distribution of the species belonging to Carnivora in a space composed by their estimated size, the possible size of their prey, and the grinding area of their dentition, which is indicative of their

Higher Taxa	Species	Mass (kg)
Artiodactyla, Camelidae	*Palaeolama paradoxa* (extinct large llama)	300
Artiodactyla, Camelidae	*Eulamaops parallelus* (extinct large llama)	150
Artiodactyla, Camelidae	*Lama guanicoe* (guanaco)	90
Artiodactyla, Camelidae	*Lama gracilis* (extinct smallllama)	50
Artiodactyla, Cervidae	*Morenelaphus lujanensis* (extinct large deer)	50
Artiodactyla, Tayassuidae	*Pecari tajacu* (collared peccary)	30
Proboscidea, Gomphotheriidae	*"Stegomastodon" platensis* (mastodon)	7500
Perissodactyla, Equidae	*Equus (Amerhippus) neogeus* (American horse)	300
Notoungulata, Toxodontidae	*Toxodon platensis* (toxodont)	1600
Notoungulata, Toxodontidae	*Toxodon burmeisteri* (toxodont)	1100
Litopterna, Macraucheniidae	*Macrauchenia patachonica* (macrauchenia)	1000
Xenarthra, Dasyposidade	*Eutatus seguini* (large extinct armadillo)	200
Xenarthra, Pampatheriidae	*Pampatherium typum* (pampathere)	200
Xenarthra, Glyptodontidae	*Neosclerocalyptus migoyanus* (small glyptodont)	250
Xenarthra, Glyptodontidae	*Neothoracophorus depressus* (glyptodont)	1100
Xenarthra, Glyptodontidae	*Panochthus tuberculatus* (glyptodont)	1050
Xenarthra, Glyptodontidae	*Doedicurus clavicaudatus* (glyptodont)	1450
Xenarthra, Glyptodontidae	*Plaxhaplous canaliculatus* (glyptodont)	1300
Xenarthra, Glyptodontidae	*Glyptodon clavipes* (glyptodont)	2000
Xenarthra, Glyptodontidae	*Glyptodon reticulatus* (glyptodont)	860
Xenarthra, Tardigrada	*Megatherium americanum* (giant ground sloth—megathere)	6100
Xenarthra, Tardigrada	*Glossotherium myloides* (giant ground sloth)	1200
Xenarthra, Tardigrada	*Glossotherium robustum* (giant ground sloth)	1700
Xenarthra, Tardigrada	*Lestodon trigonidens* (giant ground sloth)	3400
Xenarthra, Tardigrada	*Scelidotherium leptocephalum* (large ground sloth)	1050
Rodentia	*Neochoerus aesopi* (large capybara)	60
Carnivora, Felidae	*Smilodon populator* (sabertooth tiger)	400
Carnivora, Felidae	*Panthera onca* (jaguar)	120
Carnivora, Felidae	*Puma concolor* (puma)	50
Carnivora, Ursidae	*Arctotherium tarijensis* (extinct bear)	140
Carnivora, Canidae	*Dusicyon avus* (extinct wolf)	15

Table 8.1. Body mass in large herbivores and carnivorans in the Luján local fauna[a]

[a] Classification and rounded-up body masses as estimated in Chapter 7.

trophic habits: those with smaller areas are better adapted for consuming proportionately more flesh. The paleoecological information obtained indicates that the Lujanian fauna contained three other carnivorous species, as evidenced by the small grinding area of their dentition: one large hypercarnivorous canid with bone-cracking abilities, a medium-sized canid with moderate carnivorous capabilities (*Dusicyon avus*), and two ursids (*Arctotherium tarijensis* and *A. bonariense*) with rather carnivorous habits.

Taking into account the typical and maximum prey size values obtained from the raw body mass values, as well as those obtained from independent contrasts, and in addition to the estimation of other paleoecological parameters, it becomes possible to establish approximate prey size ranges for the Lujanian carnivorans (Prevosti and Vizcaíno, 2006). Thus, it is possible to identify potential prey for each of these carnivorans within the faunal assemblage of this age (Fig. 8.4). Medium-sized rodents, such as coypo (*Myocastor*) and mara (or Patagonian "hare," *Dolichotis*), and armadillos, such as the hairy armadillo *Chaetophractus villosus*, were probably frequently preyed upon by *Dusicyon avus*, while small rodents and more sporadically larger mammals (e.g., deer, camelids) could have been hunted by this species. The diet of *Canis nehringi* would have comprised mainly

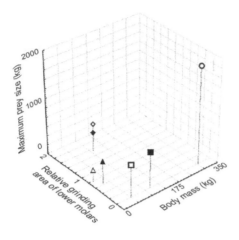

8.3. Three-dimensional distribution of Lujanian large carnivores according to their maximum prey size, relative grinding area of lower molars, and body mass values. White circle: *Smilodon populator;* black square: *Panthera onca;* white square: *Puma concolor;* black triangle: *Canis nehringi;* white triangle: *Dusicyon avus;* black diamond: *Arctotherium tarijense;* white diamond: *Arctotherium bonaeriense.*

From Prevosti and Vizcaíno (2006) by kind permission of Acta Palaeontologica Polonica.

Estimation of population densities for fossil carnivores

middle-sized mammals such as camelids, deer, peccary, armadillos (large and small), and medium- and large-sized rodents. Pack-hunting behavior would have permitted it to prey on large herbivores such as the equid *Hippidion*, the camelids (even the large *Eulamaops parallelus* and *Palaeolama paradoxa*), and small glyptodonts, but it probably preyed only rarely on juvenile megamammals. The diets of *Puma concolor* and *Panthera onca* must have comprised (as they do today) mainly middle-sized mammals, but also including large mammals of up to 600 kg. Given the larger size of the fossil jaguar (Cabrera, 1934; Kurtén, 1973; Seymour, 1993), this species may have fed on somewhat larger prey than those hunted by living individuals. It is also highly probable that it preyed on juvenile megamammals. The size and morphology of *Smilodon populator* indicate that it must have preyed habitually on middle-sized mammals such as large armadillos, equids, small glyptodonts, and large camelids, but it would also have been able to hunt megamammals weighing approximately 1000 to 2000 kg, as well as juveniles of the larger species (e.g., *Megatherium americanum*). According to current paleoecological synthesis, the species of *Arctotherium* were of varied habits (Soibelzon, 2002; Figueirido et al., 2008), and their potential prey would have been in the range of 10–300 kg body mass.

These inferences, along with the distribution of frequencies of herbivore body mass, indicate that all these carnivoran species preyed on species in the 10–300-kg range, but *Canis nehringi, Panthera onca, Puma concolor,* and *Smilodon populator* were able to hunt mammals weighing around 500 kg (e.g., *Equus neogeus*), while only the last species was able to hunt mammals with greater body mass.

The available ecological information on living carnivorans summarized by Prevosti and Vizcaíno (2006) demonstrates the presence of several factors affecting the densities of these mammals in a local scale: climate, prey density and availability, presence of competitors and predators, epidemics, and population genetic diversity. For example, there can be temporal and spatial variations in population densities within a species, and these can be a consequence of the interaction of diverse factors, such as those mentioned by Packer et al. (2005) for the Serengeti lions. These authors found that the density of those top predators exhibit a pattern of variation in an ecologically extended temporal scale (i.e., decades) comprising 10- to 20-year periods of stability punctuated by abrupt increases. This pattern would result from the interaction between prey abundance and vegetation, but would be determined by the population structure. In other cases (e.g., Smallwood, 1997), the variation of intraspecific densities is apparently unrelated to biological causes but may instead be the result of a methodological artifact. Furthermore, the positive relationship between carnivoran density and prey abundance was found for several carnivorans in different habitats (e.g., Bertram, 1979; Handy and Biggot, 1979; Macdonald, 1983; East, 1984; Creel et al., 2001; Fuller and Sievert, 2001; Carbone and Gittleman, 2002; Höner et al., 2005).

Several criticisms and observations on interspecific density–body mass relationships of mammals (including suitability of different regression models, sampling errors and biases, nonlinearity of the relationship,

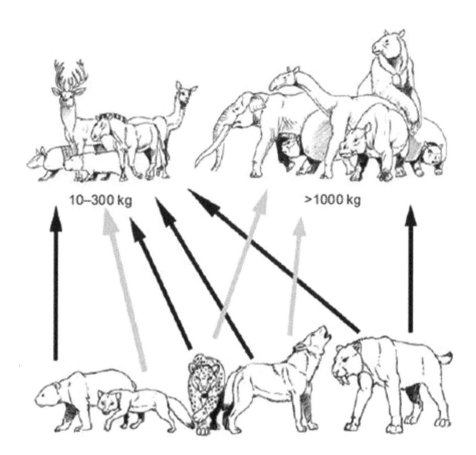

8.4. General trophic relationships between the Lujanian carnivores and their main prey grouped by size classes. Black arrows: frequent prey; gray arrows: occasional prey. The reconstructions are not drawn to scale (see text for body mass estimations). Upper section (from left to right): a pampathere (*Pampatherium typum*), a deer (*Morenelaphus lujanensis*); a capybara (*Neochoerus aesopi*); a horse (*Hippidion principale*); a guanaco (*Lama guanicoe* Müller); a mastodon ("*Stegomastodon*" *platensis*); a glyptodont (*Panochthus tuberculatus*); a litoptern (*Macrauchenia patachonica*); a toxodont (*Toxodon platensis*); and the giant ground sloth (*Megatherium americanum*). Lower section (from left to right): short-faced bears (*Arctotherium bonariense*) and (*Arctotherium tarijense*); a large fox (*Dusicyon avus*); large conical-toothed felid (*Panthera onca*) and (*Puma concolor*); a wolf (*Canis nehringi*); and a sabertooth (*Smilodon populator*).

Drawing by Jorge González in Prevosti and Vizcaíno (2006), reproduced by kind permission of Acta Palaeontologica Polonica.

phylogenetic influence, and use of average values for each species) have been made by authors, suggesting that the observed relationship between the variables is an artifact and as such does not represent a biological pattern (Silva and Downing, 1995; Blackburn and Gaston, 1996; Smallwood and Schonewald, 1998; Griffiths, 1998; Smallwood, 1999, 2001; Silva et al., 2001). These observations argue for caution in the use of allometric equations between living carnivore and herbivore densities and body mass to estimate the densities of fossil communities.

Niche partitioning

The late Pleistocene mammalian fauna is impressive as a result of the quantity of very large herbivores, many of them xenarthrans. Within this group, cingulates are predominant, with several species of glyptodonts, the pampathere, and one very large true armadillo (*Eutatus*). Morphofunctional studies, examples of which were presented in Chapter 7, revealed that the main dietary difference among these cingulates was in the degree of vegetation coarseness that they were capable of processing. The possibility that these differences in capacity may reflect competitive exclusion through niche partitioning among species living sympatrically should be considered—in other words, species living together can avoid competition by adapting to utilize different food resources. This can be achieved by comparing this late Pleistocene assemblage with modern communities of herbivorous mammals including phylogenetically related species. Jarman (1974) clearly established a relationship between body size and feeding style in antelopes (taken as Bovidae excluding Caprinae). Small species are predominantly browsers and tend to be highly selective feeders, relying on specific plants or plant parts. These species utilize diverse diets. On the other hand, the larger species are relatively nonselective grazers and rely on a wide range of grasses, but they may graze and browse. Although feeding strategies among antelopes cannot be applied strictly to cingulate xenarthrans, they provide insight on general dietary style: while the eutatines (up to 50 kg) must have been mainly browsers, the larger pampatheres (up to 200 kg) and glyptodonts (between one and two tonnes) represent increasing degrees of grazing (Vizcaíno, 2000). Such hypotheses are partially borne out by various biomechanical studies (e.g., Vizcaíno et al., 1999; De Iuliis et al., 2000).

Bargo (2001a,b) and Bargo and Vizcaíno (2008), as noted in Chapter 7, studied the masticatory apparatus of the large Pleistocene ground sloths (*Glossotherium robustum*, *Lestodon armatus*, *Mylodon darwinii*, *Scelidotherium leptocephalum*, and *Megatherium americanum*). Extensive morphological analyses, coupled with biomechanical, morphogeometrical, and ecomorphological methods, were used to extract as much information as possible from the specimens. Before proceeding to a consideration of these methods, the use of the terms *browser* and *grazer* deserve attention. Most authors refer to the dietary habits of fossil sloths (and of other mammals as well) as belonging to these categories. In some cases, authors also use the term *mixed feeder*. Conversely, the terms *browsing* and *grazing* are of ambiguous meaning, as noted in studies on living herbivores (Hofmann and Stewart, 1972; Spencer, 1995). These last two terms are used alternatively to describe the mode of food acquisition or the type of food ingested. In other words, *browsing* may refer to selective feeding of any food type, as well as eating dicot (usually arboreous) material; *grazing* denotes grass eating, but is used to mean eating of forbs as well. Ambiguities are possible and do actually happen, as in the case of the mostly grass-eating living cervid *Ozotoceros bezoarticus* (Fig. 8.5), which also takes small portions of nonherbaceous plants. This mammal is a browser from the point of view of food acquisition and a grazer from the point of view of food ingested. Such ambiguity might result in more confusion when the habits of a fossil species are considered.

Hofmann and Stewart (1972) proposed a classification for ruminant ungulates that may help solve such problems. These authors introduced notions such as bulk eaters and roughage eaters (grass eaters, and within them roughage grazers; fresh grass grazers; and dry regions grazers), selectors of concentrate juicy herbage (with tree and shrub foliage eaters, and fruit and dicot foliage selectors), and intermediate feeders (with some preferring grasses, and others preferring forbs, and shrub and tree foliage). This alternative classification was based on the stomach structure and feeding habits of East African ruminants, but it reflects physical properties of the consumed plants: bulk feeders eat low-quality plant material and are wide-muzzled animals that maximize food intake with each bite. On the other hand, selective feeders are narrow-muzzled mammals that select more nutritious small plant or plant parts (Janis and Ehrhardt, 1988; Solounias et al., 1988; Solounias and Moelleken, 1993). From this point of view, the cervid *Ozotoceros bezoarticus* would be a selective feeder, and its long and slender snout clearly correlates with that behavior (Merino et al., 2005).

Solounias and Moelleken (1993) proposed the terms *browser*, *grazer*, and *mixed feeder* to express types of vegetation eaten, not to distinguish

8.5. Pampas deer male (*Ozotoceros bezoarticus*). This species is critically endangered, especially in the southern part of its range. It is one of the main subjects of conservation projects such as Susana González's Pampas deer conservation biology project (Whitley Award winner in 2010).

Photograph courtesy of Susana González.

between selective and nonselective feeders, but this is hardly an option when dealing with fossil species. Attempting to identify the type of vegetation eaten is difficult if behavior is not observable and isotopic analyses are not available. Efforts like the complex multivariate approach by Mendoza et al. (2002) admittedly turned out to be a difficult task, even in the case of those herbivores the model is based on. That work used stepwise discriminant analysis (a statistical approach that allows the consideration of factors one by one, assessing their relative weights in the process as they are being included) of craniodental variables to identify habitat preferences and feeding behavior in artiodactyls and perissodactyls, by first distinguishing between those characteristic of open habitats (subsequently divided into general grazers, fresh grass grazers, and mixed feeders from open habitats) from those characteristic of closed habitats, which were assigned to such feeding categories as browsers (general versus high level), frugivores, and mixed feeders from closed habitats. The results they obtained showed that their algorithms correctly reclassified extant species according to their habitat and feeding preferences, and thus could be of use to infer dietary habits in extinct species. However, the possibility of applying this methodology to phylogenetically distant species (and especially the peculiar xenarthrans) is even more complicated.

On the other hand, the results provided by Bargo and Vizcaíno (2008), coupled with the ecomorphological data, allow the inference of different dietary habits for the most common species of Pleistocene ground sloths. It is also then possible to infer a probable niche differentiation among these species, given that they shared the same habitat. Within mylodontids, *Glossotherium robustum* and *Lestodon armatus*, the wide-muzzled sloths, were mostly bulk feeders, and the lips coupled with the tongue were used to pull out grass and herbaceous plants, which were probably its main dietary items. *Mylodon darwinii* and *Scelidotherium leptocephalum*, the narrow-muzzled sloths, were mixed or selective feeders with a prehensile lip that was used to select particular plants or plant parts. These species could also have used their muzzles to root up food items, such as roots and tubers, much the way hogs do. Mylodontids also have clear forelimb adaptations for digging, using their claws to help search for food. The tooth design of mylodontids, in relation to the nature of food, indicates that teeth were mainly for crushing and grinding turgid and fibrous items, respectively.

According to the muzzle reconstruction of Bargo et al. (2006b), *Megatherium americanum* was probably the most selective feeder, with a thick and strong prehensile lip that was more developed than in the narrow-muzzled mylodontids. This condition probably enabled *M. americanum* to selectively feed on particular plants (shrubs) or plant parts (leaves, twigs, fruits). The dentition was designed mostly for cutting soft but tough items (see Chapter 7). The use of alternative methods (biomechanics, morphogeometry, and ecomorphology) to complement the basic morphological analysis of the masticatory apparatus of forms that have no clear analogues can help advance sound hypotheses on dietary habits. However, it is clear that more evidence (e.g., coprological, isotopic, biochemical, palynological) is required to reconstruct a more accurate understanding of the feeding behavior of this giant ground sloth.

Vizcaíno (2000) concluded that the coexistence of many herbivorous cingulates could have been sustained through a differential exploitation of the resources: eutatine armadillos were mainly browsers, and the larger pampatheres and glyptodonts represented increasing degrees of grazing habits, as mentioned above (Vizcaíno, 2000). This author also suggested that the same scenario might be applicable to other herbivores. Niche differentiation also occurred among Lujanian ground sloths (Bargo and Vizcaíno, 2008), with *Glossotherium robustum* and *Lestodon armatus* as bulk feeders, whereas *Mylodon darwinii* was a mixed feeder. *Scelidotherium leptocephalum* was a selective browser that dug using its muzzle and claws to search for food (perhaps roots and tubers) or browsed on shrubs and grasses at ground level. *Megatherium americanum* was probably a generalized browser—that is, it fed on a mixture of leaves, shrubs, and fruits, although it may have been more selective and capable of a more varied diet.

Stable isotope geochemistry

Elements usually occur in nature with different nuclear compositions, termed isotopes (see Chapter 5). Stable (i.e., nonradioactive) isotope analysis is an approach based on the principles of the fast-growing discipline of biogeochemistry. Along with the analysis of trace elements, it has become over the last few years an important source of information on, among other subjects, the diets of extinct mammals (Palmqvist et al., 2003). Such analyses are useful for paleobiological and paleoclimatological inferences as a result of the particular properties that different isotopes of the same element have.

One of the more relevant uses of this approach is its application to terrestrial plants, which may be categorized into three main groups on the basis of their photosynthetic metabolism: C_3, C_4, and CAM plants, each of them carrying its own proportion (called signature) of ^{13}C.

The main ideas involved are as follows. Photosynthesis is the set of reactions whereby plants use gaseous CO_2 and convert it to sugars, a process known as carbon fixation, or the conversion of gaseous CO_2 into a solid compound. Plants can use different pathways (or reactions) to accomplish carbon fixation. C_3 plants use the Calvin-Benson cycle pathway to fix CO_2 directly. C_4 plants have evolved a second pathway, the Hatch-Slack cycle, which precedes the Calvin-Benson cycle. This pathway is more efficient in its use of CO_2. The designation of these plants is derived from the structure of the first product of carbon fixation: in C_3 plants, this product has a three-carbon atom structure, whereas in C_4 plants, it has a four-carbon atom structure. CAM plants are somewhat similar to C_4 plants in fixing CO_2 as a four-carbon atom structure before it is passed to the Calvin-Benson cycle. The difference is that the preliminary step (which is distinct from that of C_4 plants) occurs at night, thereby allowing CAM plants to keep most of their stomata ("pores" of the leaves) closed during the day so as to avoid water loss by transpiration. CAM stands for crassulacean acid metabolism, which is derived from Crassulaceae, a group of succulent plants in which CAM was first discovered. Many CAM plants can undergo both the C_3 type and CAM pathways, switching to the latter in highly arid conditions. Most plants are C_3 plants, including trees, temperate shrubs, and cool/moist climate and high-altitude grasses, whereas C_4 plants include basically herbaceous, tropical, arid-adapted grasses and warm/dry herbs. CAM plants are generally those that live in arid environments such as deserts, which makes sense, given the need to conserve water in such environments.

All plants, when taking up carbon dioxide from the air to build up organic compounds, sort (or preferentially use) the molecules with the lighter isotope ($^{12}CO_2$) rather than the heavier isotope ($^{13}CO_2$), but C_3 plants take an even lower proportion of $^{13}CO_2$. A measure of the extent to which a particular plant type uses the lighter isotope is expressed as a deviation, designated as δ in parts per thousand (‰) from the normal ratio of the two isotopes in the environment. This is expressed in terms of the heavier isotope, or $^{13}CO_2$, and the deviation is written as $\delta^{13}C_{plant}$. Thus, C_3 plants, which take up less $^{13}CO_2$, exhibit a range of $\delta^{13}C_{plant}$ values between −35‰ in dense forests to −22‰ in open areas exposed to water stress, with an average of −27‰. C_4 plants, on the other hand, which discriminate less effectively against $^{13}CO_2$, exhibit a lower and more restricted isotopic range ($\delta^{13}C_{plant}$ between −16‰ to −8‰), with an average of about −13‰. CAM plants, which are capable of fixing carbon with either photosynthetic pathway, show the whole range of values.

Plants, of course, are eaten by herbivores, which then incorporate carbon (among other substances) into their own tissues. Thus, isotope analysis of animal tissues can help us infer which kinds of plants the animal tends to eat, with the main difference in the deviations discussed above is that an enrichment (or a decrease in deviation) of about +3‰ to +5‰ is observed as the trophic level increases because the consumer differentially eliminates the lighter isotopes. When C_3 plants are consumed by herbivores, thus moving up a trophic level, the collagen, which is used because of its richness in carbon, will exhibit an average deviation value of about of $-23‰$ (with a range of $-31‰$ to $-18‰$), while values for a herbivore that consumes C_4 plants are less negative: around $-9‰$ (from $-14‰$ to $-4‰$). Also, a similar additional fractioning, implying a further decrease in the deviation, is observed in carnivores because we have moved up yet another trophic level: the collagen of a predator that preys upon a C_3-consuming herbivore will exhibit an average $\delta^{13}C_{collagen}$ of $-17‰$, while a carnivore which consumes C_4-grazing herbivores will exhibit a value close to $-5‰$.

In addition to isotopes of carbon, similar analyses that use isotopes of nitrogen are useful indicators of trophic habits. That element is also fixed by plants and passed on to higher trophic levels, with the same process of fractioning that leads to a decrease in the negativity of the values and reaching even positive deviations.

The increase in $\delta^{15}N_{collagen}$ in the collagen of mammals associated with their trophic level is between $+1‰$ and $+6‰$ (average $3.4‰$), although differences are observed, depending on such variables as whether the herbivore consumes N_2-fixing or non-N_2-fixing plants, or whether there exist concentrations of urine as an adaptation to arid environments. Moreover, milk-drinking juveniles of herbivore species may show a high value as a result of the enriched contents of maternal milk.

In the case of extinct xenarthrans, as well as other Lujanian taxa, bone collagen is a good candidate for isotopic analyses (Palmqvist et al., 2003), given the recent age and exceptional state of preservation of this fauna, which minimize the risk of decisive degradation of the bones and its components. In addition, the analysis of bone collagen would allow the possibility of measuring both carbon and nitrogen isotope ratios, which provide valuable indications of the animal's overall diet for the last few years of its life. (The information obtained from tooth enamel carbon isotopes is more limited because it only records the diet during the time the teeth were formed.) The $\delta^{15}N$ composition of collagen has proven useful for estimating the feeding and habitat preferences of extinct mammals, given that each trophic level above herbivore is indicated by an increase in $\delta^{15}N$ of approximately $3.4‰$ (Robinson, 2001). Furthermore, the fact that xenarthrans lack enamel makes it difficult to compare isotope analysis results obtained from teeth with those of other South American indigenous megamammals. For example, for *Toxodon* and *Macrauchenia*, a pure C_4 grazing niche and a mixed feeding diet on both C_3 leaves and C_4 grasses have been inferred, respectively, using the carbon-isotope ratio ($\delta^{13}C$) of tooth enamel (MacFadden, 2000; MacFadden and Shockey, 1997). Given the striking craniodental design of these herbivorous taxa, such biogeochemical inferences will be also testable also by using ecomorphological variables (e.g., hypsodonty index; see Chapter 6) and following a morphological and morphometric approach that studies all the dimensions of the skull and jaws independently of the group to which the animals belong, as proposed by Mendoza et al. (2002) for ungulates.

Some examples are known on the stable isotope contents of late Pleistocene mammals. A North American species of ground sloth has been assessed through it (France et al., 2007), the Rancholabrean megalonychid *Megalonyx jeffersonii*. This animal shows the highest value of $\delta^{15}N$ among the herbivores these authors studied in Saltville, a late Pleistocene site from Virginia, USA, with the exception of juveniles, whose concentration of this heavy isotope was probably enriched by milk consumption. The authors did not rule out the possibility of a small meat component in the *Megalonyx* diet, but they considered it highly unlikely that this animal was a strict meat eater.

Fariña and Castilla (2007) reported surprisingly high values, yet to be thoroughly interpreted, for two 29-kya-old samples of *Lestodon* from Arroyo del Vizcaíno (see Chapters 4 and 9), of $\delta^{15}N$ +10.3‰ and +10.7‰, which is even above the usual range of the carnivores (Palmqvist et al., 2003). This high value has been suggested to be related to a nonruminant herbivorous physiology in a colder and drier environment than the current one (Czerwonogora et al., 2011).

Body size distributions of both modern and ancient mammalian faunas have been extensively studied by paleontologists and macroecologists alike, often through the use of allometric equations (see Chapter 7 for a discussion of this methodology). One of the founders of macroecology is James Brown (1995), whose pioneering research was initially viewed as radical but is now in widespread use. Macroecology is the study of relationships between organisms and their environment at large spatial scales, looking for patterns of abundance, distribution, and diversity (Brown, 1995). Examples relevant to the subject at hand of a macroecological approach are the works of Marquet and Cofré (1999), who studied the size distribution of South American mammals, and Burness et al. (2001), who explored the ecological consequences and the evolution of maximal body size by studying the largest land animals from small islands to continents.

Among those works on fossil vertebrates, Farlow (1993) and Farlow et al. (1995) compared mammalian paleofaunas with dinosaur faunas using cenograms, a methodology that explores the relevance of size distribution. A cenogram is a graphic analysis of the distribution of body sizes of mammals within terrestrial communities, in which the natural logarithm of body mass (a neat mathematical trick that converts values to produce a straight-line graph) is plotted against the rank (from largest to smallest) of each body mass in the fauna. Such methodology was first used by Valverde (1964) to study the pattern of distribution of small-, medium-, and large-sized mammals in living communities, and was later applied to extinct communities (e.g., Legendre, 1986, 1987). After comparison with modern analogues, as MacFadden (1992) explained, cenograms can provide insights for determining the biome that extinct mammalian communities lived in. One important feature of a cenogram is its slope: a species-rich community will tend to have a shallower slope than one where fewer species are present (simply because the species-poor community will have fewer species ranked); and a species-rich community is usually characteristic of humid, forested regions, in contrast to the more arid, open environments sustaining species-poor communities. Another feature of a cenogram is the break point, usually a gap, or limit between

Ecology and size

size categories. The break points tend to be greater in communities living in drier, more open environments.

One conclusion of the earlier cenogram-based studies is that for a wide variety of habitats, each category of terrestrial vertebrates constitutes a separate subcommunity that operates nearly independently of the others, and that they could be compared to equivalent subcommunities from other habitats. This approach was first applied to the study of fossil communities by Legendre (1986), who demonstrated that cenogram shape tends to vary in a predictable way according to paleoenvironmental conditions, as happened in the major faunal turnover event at the Eocene–Oligocene boundary termed La Grande Coupure ("the great break," reflecting the disruption of continuity resulting from widespread extinctions at this boundary), suggesting that the changing body size structure of the European mammalian paleocommunities reflected the global transition to cooler, drier conditions. Cenograms that have a steep slope, basically because of the lack or scarcity of medium-sized (500 to 8000 g) species, indicate an open vegetation structure, while cenograms with a shallow slope suggest a more closed habitat. A shallow decline in species of more than 8 kg, which reflects species richness of large animals, indicates humid conditions, and a greater negative slope in the trend of large herbivores is thought to indicate arid conditions.

Although cenograms have been used by several authors, others, such as Rodríguez (1999, 2000), have been critical of this method, given that it takes into consideration only body size distributions, and assigns communities to some environment on the basis of its ecological structure, as reflected by its body size distribution. Therefore, Rodríguez argues, it is tautological to state that some communities have different ecological structures, because the environment to which any community is proposed to belong is inferred from the ecological structure itself. Another drawback of this approach, and one related to specific problems in applying this and other actualistic approaches to fossil faunas, is that studies using cenogram-based analysis usually include sites with different taphonomic histories without considering the possible effects of biases on the results.

Despite such drawbacks, the results obtained by Croft (2001), relevant to our subject here, are worth considering. This author produced cenograms for eight South American mammalian paleocommunities from the Eocene Barrancan to the Pleistocene Lujanian to compare their inferred paleoenvironmental parameters with the case of modern faunas using the slope for the medium-sized mammals, which is correlated with rainfall, and the gap at 500-g mass, which is correlated with vegetation structure (Fig. 8.6).

Croft's (2001) interpretations agreed in some cases with traditional points of view on the environment those faunas inhabited and shed new light on others. In the case of the Lujanian, the cenogram indicates a dry and open habitat, which is congruent with the presence of a large number of very large animals. According to this author, the relationship between predators and prey is not the result of predator diversity and prey diversity being similarly influenced by one of the habitat variables. Actually, the causal nature of this apparent relationship between predators and prey

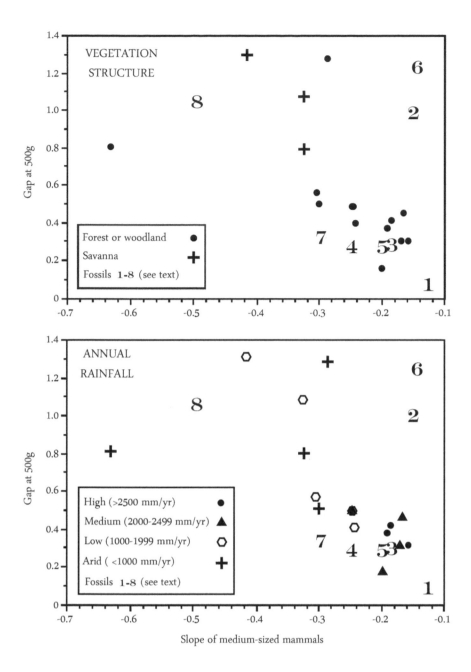

8.6. Bivariate plots of modern and fossil faunas using slope for the medium-sized mammals and size of gap in body size distribution at 500 g. Faunas are grouped by vegetation structure (top) and mean annual rainfall (bottom). Numbers refer to fossil faunas: 1 = Gran Barranca (Casamayoran age); 2 = Tinguiririca (Tinguiririan age); 3 = Salla (Deseadan age); 4 = *Protypotherium attenuatum* Zone (Santacrucian age); 5 = *Protypotherium australe* Zone (Santacrucian age); 6 = Monkey Beds (Laventan age); 7 = Tarija (Ensenadan age); 8 = Luján (Lujanian age).

Drawn by Sebastián Tambusso based on data kindly provided by Darin Croft.

is unknown—that is, whether predator diversity directly depends on the diversity of prey or vice versa.

The mammalian carnivore paleoguild seems to be depauperate, a hallmark peculiarity of Cenozoic South American mammalian faunas. This is a characteristic that may at least in part be explained by the role of large terrestrial predatory birds, the phorusrhacids, in the late Paleogene of South America. However, Croft (2001) claimed that the post–Great American Biotic Interchange (GABI) faunas have predator numbers more typical of modern faunas, in contrast (and perhaps related to the problems discussed by Rodríguez, 1999, 2000) to the conclusions of Fariña (1996) and, to a certain extent, Prevosti and Vizcaíno (2006). On the other hand, the time before the GABI could have had medium-sized mammals in greater numbers than expected because of the low diversity (and, perhaps, abundance) of mammalian predators, a topic that requires further consideration. The greater diversity of predators in the post-GABI fossil faunas is proposed to have potentially resulted from the presence of large placental carnivorans, such as *Smilodon*, preying on large mammals, which would account for the increased predator carrying capacity of faunas at that time.

The plots in the cenogram presented in Fig. 8.7 show the structure of large extinct herbivore Cenozoic faunas and two dinosaur faunas. The Lujanian fauna is the one dealt with in this book. The other large-mammal faunas are the two Australian faunas, including dromornithid birds (called thunder birds or Mihirungs; Murray and Vickers-Rich, 2004) and marsupials and several Old World Plio-Pleistocene assemblages, characterized by proboscideans, rhinos, horses, hippos, cervids, and bovids. The two dinosaur faunas are from particular stratigraphic levels in the Late Cretaceous Dinosaur Park Formation and the Late Jurassic Morrison Formation. It is clear from Fig. 8.7 that the Lujanian fauna is rich in herbivore species and, besides those with dinosaurs, the fauna with the most large-sized species.

Abundance of giants

More than a century and a half had to pass after Darwin for the first sequel: a study of the paleosynecology of Lujanian large mammals undertaken with the modern tools of ecology (Fariña, 1996), which had many derivations referred to in this book, such as the inferences on metabolism through occlusal area and the appearances of ground sloths. The main starting point of that study was the notion that, in the light of modern ecology, such an unusual abundance of large-sized species must have had important consequences for the ecological functioning of the communities involved. As noted in Chapter 7, body size is the most influential single variable of an animal's life history, and this is especially true in the case of the giants. Many ecological traits of mammals that reach a mass of 1000 kg or more have been well documented and discussed by Owen-Smith (1987, 1988).

In the case of the local fauna from Luján, a general ecological relationship between population density and body size was applied (Fariña, 1996), derived from Damuth's (1981a, 1987, 1991) work and based on data from diverse modern ecosystems, as shown in Fig. 8.8. Although there is some variation, a clear pattern between body size and density emerges: there are many more small mammals than big ones. The relationship between size and body mass is easily grasped intuitively; it does not require much specialized biological training to perceive that there will be many more rodents, for example, than elephants in a habitat. The statistical relationship formally describing the relationship is given by Damuth's (1981a) general, empirical equation, which takes the form

$$\log D = -0.75 \log m + 4.23.$$

This equation relates D, the population density in number of individuals per square kilometer, and m, the body mass in grams. Although specific equations valid for a variety of ecosystems were presented by Damuth (1981a, 1987, 1991), the general equation was used for application to the Lujanian fauna as a result of uncertainties resulting from the lack of more precise knowledge of the environment in that region and age. It should be noted that this equation is valid only when a rather high number of species with a wide range of body sizes is considered.

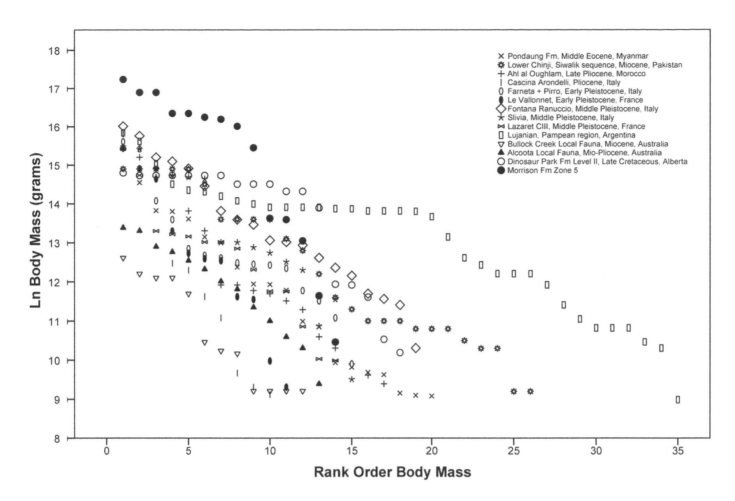

8.7. Cenogram comparing fossil mammalian and dinosaur faunas. Modified from Farlow et al. (MS). Data from Croft (2001), based on Geraads (2006) and personal communication, Montuire and Desclaux (1997), Montuire and Marcolini (2002), S. Montuire personal communication, Morgan et al. (1995), P. Murray, personal communication, Palombo et al. (2005), Palombo and Giovinazzo (2006), Tsubamoto et al. (2005), and M. R. Palombo, personal communication.

Body size has long been known to be related to basal metabolic rate (BMR; Kleiber, 1932; McNab, 1980), which is the amount of energy expended per unit time at rest in a neutrally temperate environment (i.e., when the animal does not expend energy in heating up its body through shivering or in cooling it down by sweating) and during the postabsorptive phase (i.e., when the digestive system is inactive). In other words, it is measured when the energy is used only for the functioning of the vital organs.

To gain insight on the energetic functioning of the Luján community, the basal mass-specific metabolic rate (i.e., the rate of energy expenditure per unit mass; R, in J kg^{-1} s^{-1}) was considered. This quantity differs from the total BMR and varies with body size. Fariña (1996) thus explored how much energy was expended by each herbivore species at rest (for which body mass had been estimated) per unit of mass for every unit of time, using Peters's (1983) equation:

$$\log R = -0.25 \log m + 0.6128.$$

Packard and Birchard (2008) questioned the validity of this equation, derived from that of Kleiber (1932), as a result of the existence of a mathematical artifact that overestimated the metabolic rate of large mammals. A correction was published by McNab (2008) in a noteworthy attempt to provide a more precise description of this relationship in living mammals. McNab gathered data from 639 species of mammals, including their body mass, food habits, climate, habitat, substrate, restriction to islands or highlands, use of torpor, and type of reproduction, which allowed a great refinement

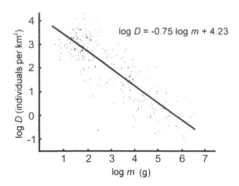

8.8. General relationship between population density and body size in 306 species of mammalian primary producers.

Based on Damuth (1981a).

in the importance of those factors, as well as that of their phylogenetic inheritance. The equation so generated includes several factors for correcting the general relationship and has the mischievous (our apologies, we just couldn't resist) form:

BMR (kJ h^{-1}) = $M I S T C H E F$ 0.062 $m^{0.694 \pm 0.005}$,

where m is the body mass in grams, M for mountains or lowlands, I for islands or continents, S for substrate, T for torpor, C for climate, H for habitat, E for infraclasses, and F for food habits. Note that in many cases the values of these factors clearly depart from 1, so the correction due to habits, habitat preferences, and phylogenetic heritage can be important. Useful as this approach may be (the values of McNab's factors are presented in Table 8.2 for the interested reader), for the Lujanian fauna, it turns out that there is little difference between results obtained using the classic equation of Peters (1983) and those obtained using McNab's (2008) corrections. In addition, as a result of sound thermodynamical reasons related to the loss of free energy as the trophic level increases and to biomechanical reasons related to limb bone strength allometry (Sorkin, 2008), the modern Carnivora are known to be less abundant than their potential prey. Thus, a different equation must be used to estimate their population density, that obtained by Damuth (1993) for African flesh eaters (the symbols are the same as above):

$$\log D = -0.64 \log m + 2.23.$$

Discussion on the use of this formula is given below. The BMR of species belonging to Carnivora is also described by a specific formula, because predators tend to consume more energy than herbivores, even at rest, and that expenditure tends to be even relatively higher as body size increases. Thus, an appropriate equation, also from Peters (1983), was used:

$$\log R = -0.27 \log m + 0.6551.$$

McNab (2008) also published a refinement for this formula, but, as in the case of herbivores discussed above, it yields little difference in the overall results presented here.

The on-crop, or standing, biomass for each species was obtained by multiplying the population density, calculated from Damuth's (1981a) equation, by its body mass. Again, this procedure, which involves the use of allometric equations, must be considered with caution, but in this case, its validity seems justified, given that a good number (26) of species, ranging through three orders of size magnitude, is considered. According to this model, the total on-crop biomass for these species was about 13 tonnes per square kilometer, as originally calculated by Fariña (1996), and about two-thirds of it corresponds to megaherbivores. Using McNab's (2008) corrections, the total on-crop biomass rises to nearly 16 tonnes per square kilometer, of which 11 tonnes (still about two-thirds, thus in agreement with Fariña's original proportional estimate) may be attributed to megaherbivores.

The energy requirements for each species presented in Table 8.3 were obtained by multiplying each species' on-crop biomass by its BMR. However, further factors should be taken into account to have an idea of the energetic requirements of the species under consideration. For instance, the proportion of all vegetable fodder consumed that is assimilated does not exceed 50% of the edible material. On the other hand, the BMR is not

Table 8.2. Coefficients for ecological factors[a]

Factor	Category	Value
M (mountains, lowlands)	Lowlands	0.88
	Mountains	1.00
I (islands, continents)	Islands	1.00
	Continents	1.21
T (torpor and hibernation)	Torpor, hibernation	1.00
	No torpor, no hibernation	1.24
S (substrate)	Burrow, fossorial	0.70
	Tree, caves	0.81
	Terrestrial, aquatic	1.00
H (habitat)	Deserts	0.84
	Xeric, grasslands, savannah	1.00
	Mesic, freshwater	1.19
	Marine	2.37
C (climate)	Tropical, temperate–tropical	1.00
	Temperate, polar, widespread	1.21
F (food habits)	Blood, ants, termites	0.71
	Seeds, leaves, insects, omnivory, worms	0.90
	Vertebrates, fruit, nectar, bulbs, grass	1.00
E (infraclass)	Monotremes	0.58
	Marsupials	0.84
	Eutherians	1.00

[a] Modified from McNab (2008).

a measure of the total energy expenditure of a species, which in modern mammals is typically two to three times higher during the day (Peters, 1983; McNab, 2008). The average actual maintenance metabolism, which accounts for the field energy expenditure (FEE, the average energy an animal expends for its usual functions), was consequently taken to be 2.5 times the basal rate. Adding up the requirements of all the species considered, and converting the units, it follows that they must have needed to consume 1.6 MJ m^{-2} y^{-1}. The megaherbivores alone account for the consumption of more than 900 kJ m^{-2} y^{-1}, or almost 60%. This decrease in the proportion from the two-thirds given for on-crop biomass is because the latter is an instantaneous measurement, whereas the area requirements are consumptions over time (say, per year). Because megaherbivores, as a result of their larger size, have a slower metabolic rate, their share of the energy cycle is smaller when time is involved.

Table 8.4 lists the five large Carnivora species in the Luján local fauna: the extinct canid *Dusicyon avus*, two living felids (jaguar, *Panthera onca*, and puma, *Puma concolor*), the extinct sabertooth felid *Smilodon populator*, and the extinct ursid *Arctotherium*, as in Fariña (1996; but contrast Prevosti and Vizcaíno, 2006). They range in mass from 15 kg (*Dusicyon avus*) to 400 kg in the case of *Smilodon*. As in the case of the herbivores, the on-crop biomass for each species was obtained by multiplying the calculated population density by its body mass. The total on-crop biomass for these carnivores was about 55 kg km^{-2}. Once again, the requirements for each species were obtained by multiplying its on-crop biomass by its BMR (Table 8.4). Summing up the requirements of all the large carnivore species, and converting the units, it follows that they must have needed about 4.2 kJ m^{-2} y^{-1} (that is, the energy needed per square meter over 1 year) as habitat secondary productivity, i.e., the meat available for the predators, to support

Table 8.3. Data for genera of herbivorous mammals larger than 10 kg found in the Luján local fauna[a]

Species	Mass (kg)	Density (individuals km^{-2})	On-Crop Biomass (kg km^{-2})	Mass-Specific Basal Metabolism (J kg^{-1} s^{-1})
Palaeolama paradoxa	300	1.32	397	0.99
Eulamaops parallelus	150	2.23	334	1.17
Lama guanicoe	90	3.27	294	1.33
Lama gracilis	50	5.08	254	1.54
Morenelaphus lujanensis	50	5.08	254	1.54
Pecari tajacu	30	7.45	224	1.75
"Stegomastodon" platensis	7580	0.12	891	0.44
Equus (Amerhippus) neogeus	300	1.32	397	0.99
Toxodon platensis	1642	0.37	608	0.64
Toxodon burmeisteri	1100	0.50	550	0.71
Macrauchenia patachonica	988	0.54	535	0.73
Eutatus seguini	200	1.80	359	1.09
Pampatherium typum	200	1.80	359	1.09
Neosclerocalyptus migoyanus	250	1.52	380	1.03
Neothoracophorus depressus	1100	0.50	550	0.71
Panochthus tuberculatus	1061	0.51	545	0.72
Doedicurus clavicaudatus	1468	0.40	591	0.66
Plaxhaplous canaliculatus	1300	0.44	573	0.68
Glyptodon clavipes	2000	0.32	639	0.61
Glyptodon reticulatus	862	0.60	517	0.76
Megatherium americanum	6073	0.14	843	0.46
Glossotherium myloides	1200	0.47	562	0.70
Glossotherium robustum	1713	0.36	614	0.64
Lestodon trigonidens	3397	0.21	729	0.54
Scelidotherium leptocephalum	1057	0.52	545	0.72
Neochoerus aesopi	63	4.27	269	1.46

[a] Modified from Fariña (1996) and Prevosti and Vizcaíno (2006).

their basal metabolism, if, as noted earlier, an assimilation efficiency of the edible material of 50% is assumed. As in the case of the herbivores, the actual maintenance metabolism was taken to be 2.5 times higher than the basal rate. Therefore, the energy required to sustain their maintenance metabolism was almost 11 kJ m^{-2} y^{-1}.

Productivity and climate

A primary productivity (i.e., the production of organic compounds by plants) of 7300 kJ m^{-2} y^{-1} (Margalef, 1980) is considered high for modern, open-country habitats. Indeed, Uruguay's best present-day natural field, a rich pasture where livestock is raised, reaches a figure for primary productivity of 6600 kJ m^{-2} y^{-1} (Cayssials, 1979; Panario and May, 1994).

If the population density of the Lujanian mammals scaled as predicted by Damuth's (1981a) average equation, about 14% of that primary productivity would have been needed to support the megaherbivores alone, and that figure must have increased to 25% to support all the species larger than 10 kg. In other words, Lujanian megaherbivores, if they had been allowed to roam free in one of the best natural cattle fields of the present-day world, would have consumed an unusually high proportion of the primary productivity, leaving the rest of the species of smaller mammals and other vertebrate and invertebrate consumers in trouble. To put this in perspective, the highest average rate of consumption by large herbivores

Species	Mass (kg)	On-Crop Biomass (kg km⁻²)	Basal Metabolic Rate (J kg⁻¹ s⁻¹)
Smilodon populator	400	0.04	17.6
Panthera onca	120	0.10	11.4
Puma concolor	50	0.17	8.3
Dusicyon avus	15	0.36	5.4
Arctotherium tarijensis	140	0.09	12.1

Table 8.4. Data for species of Carnivora larger than 10 kg[a]

[a] Modified from Fariña (1996).

in modern ecosystems with large mammals is 7.5%, reported for Rwenzori National Park, Uganda, in 1973. This unusually high figure was likely caused by several years of a hunting moratorium, with one result being that it left visitors with the impression that the park was overcrowded by large game. Other African parks show values for this variable between 3% and 6% (Owen-Smith, 1988).

Thus, to assess the situation in the late Pleistocene, at least for the Luján local fauna, but presumably also for others in South America during that age, evidence should be gathered about the primary productivity of the Lujanian habitat. Unfortunately, data on vegetation were scarce when Fariña's (1996) study was undertaken (and, to a great extent, still are; see Chapter 4), and so reconstructions must rely on indirect evidence. Is the Uruguayan cattle field mentioned above a good analogue? Or was the Lujanian habitat much richer, rendering the proportion consumed by the large mammals more balanced? For instance, if the Lujanian plains had been as productive as today's African savanna—about 38 MJ m⁻² y⁻¹ (Margalef, 1980)—the estimated consumption percentage would fall to about 3%, which is an expected, current value.

However, available evidence suggests otherwise, because the climate in the pampas was much drier and rather cooler than today (Clapperton, 1983), according to various lines of evidence (Cantú and Becker, 1988; Tonni, 1990), in agreement with Darwin's intuitions. Thus, primary productivity in the midlatitude Lujanian might have been higher on average than that in today's arid Choele-Choel area, but it does not seem likely that it could have been higher than the best present-day cattle field of Uruguay, and hence it must have been dramatically lower than that of the African savanna.

An interesting possibility that might help explain the apparent overabundance of Lujanian megafauna is that metabolic rates may have been overestimated; we have stressed that this fauna was composed in large part by the strange xenarthrans, whose modern representatives tend to have a low metabolic rate (McNab, 1980). This topic will be carefully developed below, but it may suffice here to state that the members of the Lujanian fauna would need to have been substantially different than assumed in the model to change significantly the results obtained and so far discussed, especially if it is taken into account that there were many species ranging widely in size.

The distinctiveness of the Lujanian mammals is particularly noticeable in the large-sized forms, because most of the smaller mammals found there are members of the modern South American fauna. As already described,

Predators: a second imbalance?

nearly all large species from the Luján local fauna itself are extinct, and a similar proportion of extant to extinct large mammals applies to South America as a whole (Lessa and Fariña, 1996). In current communities, it is usual, despite some variation, that a fauna containing a great diversity of large living herbivores also contains a diversity of large living carnivores (Owen-Smith and Mills, 2008a,b). In the Serengeti, to cite a prominent example of a well-studied modern case, there is a relatively much less diverse fauna of large mammals (four megamammals: elephant, rhinoceros, hippopotamus, and masai giraffe; and six or seven species above the 100-kg limit: buffalo, eland, defassa waterbuck, wildebeest, Coke's hartebeest, Burchell's zebra, and topi). Feeding on them are four predators with masses greater than 50 kg: lion, leopard, spotted hyena, and cheetah. A fifth species, the African wild dog, is a bit smaller than that limit, but it may have compensated with pack-hunting behavior. Admittedly, active carnivores may not be so important in a consideration of predation on the megaherbivores in Luján, at least with respect to adult individuals, because very large animals escape predation thanks to their size alone. Even so, glyptodonts must have experienced some predation, despite their size, complete armor, and defensive behavior (see Chapter 7; Fariña, 1995), as can be surmised by the presence of more than one species in this local fauna, as well as in many others, as predicted by the cropping theory (Paine, 1966). This theory states that in the absence of predation, the most competitively efficient species of a guild would eventually become the only one. On the contrary, predation, when it occurs, is concentrated on this dominant species, given its higher probability of being found. Therefore, predation allows the other species to develop and coexist with the dominant competitor. This also applies in general to megaherbivore genera but is probably restricted to juveniles.

On the other hand, on the basis of available ecological information (e.g., Bertram, 1979; Handy and Biggot, 1979; Fuller and Sievert, 2001; Höner et al., 2005), Prevosti and Vizcaíno (2006) discussed an alternative hypothesis to the imbalance proposed by Fariña (1996), namely that the density of carnivores depends on the density of herbivores (Fig. 8.9). In this view, the possibility exists that a high density of individuals of fewer carnivore species could have been supported (even with changes over an ecologically long time), as is the case in some recent ecosystems, even the Serengeti (e.g., Bertram, 1979; Handy and Biggot, 1979; Macdonald, 1983; East, 1984; Creel et al., 2001; Carbone and Gittleman, 2002; Fuller and Sievert, 2001; Höner et al., 2005). Other elements, including climate, competitors, and predators, might also affect the density of carnivores. For example, Packer et al. (2005) found that the density of Serengeti lions exhibits a pattern of variation in a large temporal scale (decades) comprising 10- to 20-year periods of stability punctuated by abrupt increases. This pattern would result from the interaction between prey abundance and vegetation, but would be determined by the population structure. Even so, the diversity of megaherbivores is so high in the case of the Lujanian that it could possibly have generated a higher diversity of carnivores than expected, which perhaps could have at least filled the role of scavenging carrion feeders or that of a kleptoparasite.

Then again, is it possible that only a few relatively small scavengers, mammals and birds, could have taken advantage of the resource provided

8.9. Schematic representation of the imbalance hypothesis (A) and an alternative hypothesis (B).

Drawn by Jorge González in Prevosti and Vizcaíno (2006). Reproduced by kind permission of Acta Palaeontologica Polonica. In A there are relatively few carnivores compared to herbivores. In B the number of carnivores is relatively more abundant, as discussed in the text.

by dead megaherbivores? If so, they must have been less efficient than a larger carnivore. Large size is more efficient for opening carcasses and cutting tendons, and a large number of individuals is no compensation. Second, smaller scavengers would have had to wait for their food if a larger predator were feeding first, much in the way hyenas have to do until lions are satisfied, unless their numbers are large enough.

As stated above, the metabolic requirements of the five species of carnivorans would have been fulfilled with 11 kJ m^{-2} y^{-1}. The secondary productivity predicted by the appropriate equations in McNaughton et al. (1989) for an ecosystem able to support the herbivore biomass inferred above is at least two and a half times as high. If a growth efficiency, i.e., the proportion of material that passes from one trophic level to the next, of 0.025 is assumed (Peters, 1983), this figure soars nearly fourfold.

The niche of a big carnivore (perhaps a scavenger or a kleptoparasite) was occupied only partially by the sabertooth, jaguar, puma, bears, and canids. This niche might also have been filled by birds, such as vultures, but the available evidence does not support this view. Although the adequacy of paleontological records of birds is usually doubtful, Tambussi et al. (1993) claimed that there are enough remains from the South American late Cenozoic to draw some conclusions. In that study, these authors demonstrated that South American vultures (Cathartidae) are not an exception to the general Cenozoic trend of diversity of birds. Indeed, vultures are less diverse in the Pleistocene than in the late Pliocene, following the same pattern as the rest of Aves.

BMR and phylogenetic heritage

To have a better understanding of how the legacy of a certain fossil species affect traits such as metabolic rate, independent evidence must be obtained. The way bone or teeth grow have been studied to assess this kind of character in theropod dinosaurs (Horner and Padian, 2004), and the size of the brain has also been used as a proxy of metabolic rate in vertebrates (Jerison, 1973). Extinct xenarthrans have been analyzed from the point of view of their occlusal surface area (OSA, Vizcaíno et al., 2006), on the assumption that the relationships among cheek-tooth (i.e., molariforms) OSA relative to body mass are useful in illuminating the possible diet and other biological factors in fossil xenarthrans. Several authors (Janis, 1990, 1995; Janis and Constable, 1993; Mendoza et al., 2002) have noted that OSA is larger in grazers than in browsers among ungulates and kangaroos. For instance, Janis (1988, 1995) stated that monogastric ungulates, such as perissodactyls, have longer molarized premolar rows than ruminant artiodactyls. Janis and Constable (1993) and Janis (1995) proposed that these differences are due to different alimentary strategies in relation with differences in the physiology of digestion. Horses, in comparison with livestock and other ruminants, spend more time chewing, and they also chew more when they eat food with high fiber content.

For their study, Vizcaíno et al. (2006) measured more than 150 specimens from collections housed in different museums, including 47 species of living, mostly herbivorous, mammals of very different sizes of Rodentia, Hyracoidea, Tubulidentata, Proboscidea, Cetartiodactyla, and Perissodactyla, and 24 species of Xenarthra. Among the latter, besides the four living Cingulata, are eight fossil cingulate species, the glyptodonts *Propalaehoplophorus australis*, *P. incisivus*, *Asterostemma depressa*, *Plohophorus* sp., *Panochthus tuberculatus*, *Glyptodon* sp., the pampathere *Holmesina occidentalis*, and the dasypodid *Eutatus seguini*. Three living sloths were included and seven fossil species: the mylodontids *Glossotherium robustum*, *Lestodon armatus*, *Mylodon darwinii*, *Scelidotherium leptocephalum*, the megatherioids *Hapalops* sp. and *Megatherium americanum*, and the megalonychid *Eucholoeops*.

The outlines of the cheek teeth of each individual were digitized in palatal view, and the enclosed area was calculated by appropriate software.

Once obtained, the areas of the individual teeth were summed, and the result was plotted against body mass, then both variables log-transformed to convert the relationship to a linear function, whose regression was then calculated (Fig. 8.10a).

It is evident that the different groups of taxa tend to cluster above or below the line, which shows a slope comparable to that predicted by geometric similarity: most rodents, hyracoids, proboscideans, and perissodactyls (especially horses) fall above the regression line, i.e., those species have larger occlusal areas than expected for their body size, while most artiodactyls (especially ruminants) fall below the regression line, as do living xenarthrans. This is in accordance with Janis's (1988, 1995) conclusion that the monogastric ungulates have longer and more molarized premolar rows than ruminants, reflecting differences in the alimentary strategies and in the physiology of digestion (Janis and Constable, 1993).

Fig. 8.10b compares independent regressions of xenarthrans, both fossil and living, versus the rest of the placental mammals. Both regression lines seem to be parallel and the slopes do not differ significantly, but the regression line of the remaining mammals lies above that of xenarthrans. However, two taxa fall well above the xenarthran regression line; these outliers are the pampatheriid cingulate *Holmesina occidentalis* and the giant ground sloth *Megatherium americanum*. It is also noteworthy that the mylodontid ground sloths fall well below the regression line for xenarthrans.

Fig. 8.10c is devoted to xenarthrans alone and includes independent regressions for cingulates and tardigrades. The regression of cingulates lies above that of tardigrades, and with a marginally higher slope. The tardigrade regression clearly shows a differential distribution of some groups in the same way as in the previous figure, with mylodontids below the line, and *M. americanum* far above it because of its high OSA value (rather than because of a measurement error). The living sloths overlap both regression lines and do not show any particular definitive distribution. In contrast, the cingulates are distributed tightly around their specific regression line, with the exception of *H. occidentalis*, which lies well above it.

These examples indicate that OSA may thus be correlated with the capacity to process food in the oral cavity and suggests some aspects of feeding physiology and metabolism. Xenarthrans have less OSA available for triturating food than other mammals of similar sizes. This may be related to the low BMRs characteristic of living xenarthrans, which fall between 40% and 60% of the rates expected from mass in Kleiber's (1932) relation for placental mammals (McNab, 1985, 2008). This implies that xenarthrans have less energetic requirements than other placental mammals and therefore for a specific type of food require lower intakes than other placental mammals of similar body masses.

Although there is no direct evidence of the rate of metabolism of pampatheres and glyptodonts, indirect evidence, such as the pattern of geographic distribution in North America, indicates that their northern limits were restricted by the combination of low metabolism and high thermal conductance (McNab, 1985). Also, there may be a relationship between OSA and stomach morphology that may help to explain why cingulates fall above the regression line compared to tardigrades. Modern armadillos

8.10. (A) Regression of occlusal surface area (OSA) against body mass for living mammals (n = 125). Symbols: open triangles = Dasypodidae; black triangles = Bradypodidae; gray triangle = Tubulidentata; black circles = Caviomorpha; open circles = Hyracoidea; open squares = Tapiridae; gray squares = Equidae; black squares = Rhinocerotidae; gray diamonds = Cervidae; open diamonds = Bovidae; black diamonds = Giraffidae; gray circle = Hippopotamidae; crosses = Camelidae; gray cross within square = Elephantidae. Dashed lines above and below the regression line = 95% confidence interval. (B) Regression of OSA against body mass of Xenarthra (n = 53) compared with other placental mammals (n = 99). Symbols: circles = other placental mammals; black triangles = Xenarthra; E = other placentals regression line; X = Xenarthra regression line. Dashed lines above and below the regression line = 95% confidence interval. The black triangles contained by the ellipse correspond to the mylodontid ground sloths. (a) Case of the Pampatheriidae *Holmesina*. (b) Two cases of *Megatherium*. (C) Regression of OSA against body mass of xenarthrans (n = 53). Symbols: C = Cingulata; T = Tardigrada; black triangles = dasypodid cingulates; gray triangle = pampatheriid cingulate; open triangles = glyptodontid cingulates; open circles = living tardigrades; gray circle = *Eucholoeops;* open squares = *Hapalops;* black squares = *Megatherium;* black circles = mylodontids. Dashed lines above and below the regression line = 95% confidence interval.

From Vizcaíno et al. (2006) by kind permission of Asociación Paleontológica Argentina.

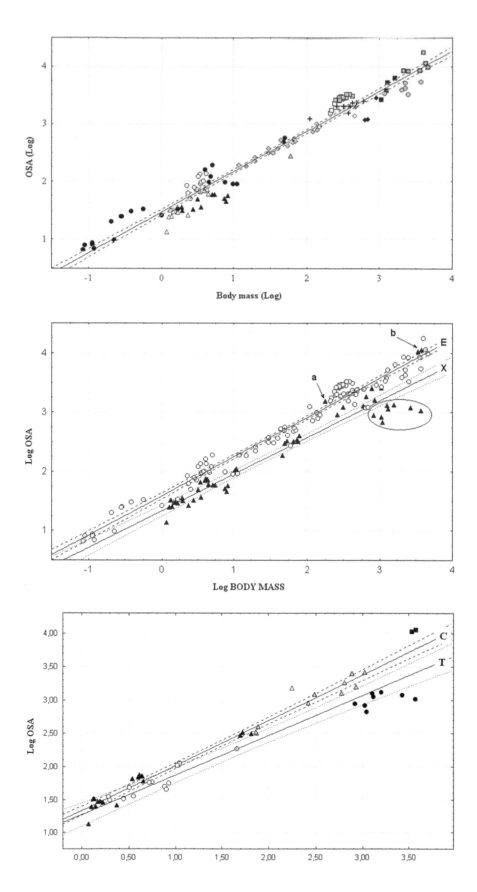

have a simple, saclike stomach (Grassé, 1955) compared to living sloths, which have a chambered stomach (Britton, 1941a, 1941b), and thus presumably do not have the same level of digestive efficiency. Cingulates such as glyptodonts and pampatheres probably also had a simple saclike stomach that lacked chambers and therefore would need a more complex dental apparatus, i.e., lobate teeth and greater OSA, to better process vegetation than sloths.

Among sloths, the extremely low OSA values for mylodontids might reflect poor food oral processing abilities. If this were the case, these ground sloths, in order to maintain diets similar to those of the ungulates of equal body masses, would have been expected to compensate for the low efficiency in food processing by high fermentation capability in the digestive tract, lower metabolic requirements, or both. Interestingly, a similar comparison of herbivorous dinosaur and proboscidean teeth was published several years ago by Coe et al. (1987) to help understand dinosaur–plant ecological interactions and coevolution.

The living tree sloths, *Bradypus* and *Choloepus*, have an extremely large four-chambered stomach, which can be considered equivalent to those of ruminants. Presumably, this might also be true for fossil sloths. Food passage through the digestive tract may take up to 1 week, and hence the digestion and absorption rhythm is also extremely slow (Britton, 1941; Gilmore et al., 2008). Moreover, the generally slow nature of tree sloths, the result of their poor skeletal musculature, also implies lower metabolic requirements than in other mammals (Scholander, 1955; McNab, 1985). Naples (1989) assumed that *Glossotherium* also had a slow rate of food passage through the gut, which might have enabled this sloth to obtain sufficient nutrition from high-fiber foods. Consequently, and also because of the large size in ground sloths, there must have been a long transit time, and hence there may not be as strong selection for the oral mechanical processing of vegetation. Surprisingly, *Megatherium americanum* seems to have followed a different route of specialization with an expected, or even higher, OSA value for a mammal of its size, and much larger if compared with mylodontids.

If a parallel with ungulates (based on Janis, 1995) is assumed, mylodontids might be inferred as foregut fermenters, while *M. americanum* would have been a hindgut fermenter, which in turn reflects lower- and higher-quality nutritional diets, respectively (Alexander, 1996). However, this is not supported by comparison with modern, related forms, as living xenarthrans lack a cecum (an outpocketing of the intestine that functions as the fermentation chamber), and this is not congruent with hindgut fermentation, for a cecum is present in mammals with hindgut fermentation. The case in *M. americanum* must have been one of more intense food processing in the oral cavity than in mylodontids, in accordance with other morphological and biomechanical features; for example, the teeth of *M. americanum* are extremely hypselodont and bilophodont, with the sagittal section of each loph being triangular and sharp edged. This occlusal morphology indicates that the way OSA has been calculated in this study does not reflect the sum of the total area present on the anterior and posterior surfaces of the two lophs (Bargo, 2001a).

The evidence provided by Bargo (2001a) indicates that *M. americanum* could have fed on moderate to soft tough food, probably browsing (including fruits) in open habitats; the possibility of a more varied diet cannot be excluded (Fariña, 1996; Fariña and Blanco, 1996). Fruits and flesh are food items nutritionally richer than most grasses or leaves; richer diets require smaller fermentation chambers (Alexander, 1996), and strict carnivores do not have chambers at all. Frugivory and carnivory imply that flesh and fruit eaters presumably do not require the same degree of mastication, and then the larger OSA in *Megatherium* may have been an indirect way of enlarging the cutting area because cutting does not result in the breakage of cell walls and the release of nutrients as efficiently as crushing and grinding does. Additionally, OSA is assumed to be directly correlated with the quantity of food trapped between the upper and lower tooth rows. Hence, in species that have strong jaw muscles and crush their food between sets of interlocking cheek teeth, as is the case in *Megatherium*, a large OSA is functionally important, increasing the amount of food fragments cut or ground per chewing cycle (Pérez-Barbería and Gordon, 1998).

The great OSA of *M. americanum* would also suggest a BMR similar to that of an extant herbivorous mammal of equal body size. If *M. americanum* had a digestive tract as efficient for processing cellulose as mylodonts, it would have been capable of maintaining a higher basal metabolism compared to mylodonts. The ground sloths, in general, would have had low metabolic rates because of their large body masses (McNab, 1985; Naples, 1989), probably lower than those of ungulates of equivalent size. The skeletal musculature, in contrast to living sloths, was well developed, given their totally terrestrial locomotory habits, and some mylodontids (i.e., *Glossotherium robustum* and *Scelidotherium leptocephalum*) were well adapted for digging burrows (Bargo et al., 2000; Vizcaíno et al., 2001). This adaptation would probably indicate an even lower basal metabolism for these sloths compared with that for *Megatherium americanum*. McNab (1979, 1985) found that in armadillos, low metabolic rates correlate with burrowing habits, probably as a means to reduce heat storage during this activity, which would be even more important for the large burrowing ground sloths.

Cryptic flesh eaters?

It is apparent that, as suggested by Fariña (1996), the Lujanian fauna cannot be properly understood by using modern ecosystems as a paradigm without condemning it to be just like those of the present. However, this is not so much a problem with the paradigm, for when this approach has been applied to other fossil mammalian faunas, they seem to be properly explained. It seems, rather, that the assemblages of the South American late Pleistocene are peculiar beyond our current abilities to cope with it.

For instance, the Rancho La Brea fauna in North America is approximately from the same age as the Lujanian, and it is from a moderately high-latitude habitat as well. Applying the equations given above to the list of species in Stock (1956) and making the same assumptions, this fauna turned out to be easier to sustain than the Lujanian. The metabolic requirements of the megaherbivores would have been fulfilled with a primary productivity of 0.6 MJ $m^{-2} y^{-1}$, which is congruent with the inferred habitat, a conifer parkland (Marcus and Berger, 1984). The consumption efficiency of the large mammals, i.e., the proportion of the primary productivity needed by them, would have been 5.5% of the 11 MJ m^{-2} y^{-1} (Margalef, 1980), which falls within the 3% to 6% range observed in modern examples. Furthermore, given a production efficiency of 2.5%, the secondary productivity must have been 15 kJ m^{-2} y^{-1}, exactly the predicted requirements of the carnivores. Of course this fauna is biased as a carnivore trap, but the bias has to do with the abundance of specimens rather than with the actual species richness. Other studied fossil faunas, such as the Pleistocene Venta Micena in southern Spain (Palmqvist et al., 2003) and the Pliocene Chapadmalalan (Vizcaíno et al., 2004b), are also well balanced.

Even though alternatives were discussed by Prevosti and Vizcaíno (2006), and even though there are examples in modern ecosystems of non-linear ratios between predator and prey species richness, the guild structure of the Lujanian fauna is unusual: estimated excess of herbivores for the primary productivity and scarcity of carnivores, given the high number of large herbivores. There could have been one or more large-bodied species that ate flesh opportunistically. Among these, glyptodonts, capybaras, toxodonts, *Macrauchenia*, large artiodactyl ungulates, and the proboscidean are unlikely candidates because of their inferred paleoautecology (Ferigolo, 1985; Fariña, 1985, 1988; Fariña and Álvarez, 1994; Fariña and Vizcaíno, 2001; see also Chapter 6), or because their living relatives show no preference for animal food. The armadillo-like cingulate *Pampatherium*, if this way of inferring paleoecology is undertaken, might be a potential candidate, because some living armadillos usually eat carrion (Redford, 1985) and even juveniles of other mammals (Newman and Baker, 1942; McBee and Baker, 1982). However, their lobated teeth and structure of their masticatory apparatus strongly suggest that they were browsers and grazers (Vizcaíno et al., 1998; De Iuliis et al., 2000). By a process of elimination, this leaves the ground sloths as the remaining candidates, among which *Megatherium* appears to have been the most likely to have incorporated flesh into its diet.

The strange dentition of this giant sloth, though not typical of a carnivore, is even more atypical for a hard grass herbivore (Bargo, 2001a). It must have been poorly adapted for carnivory if compared with large felids, but they might not have needed to be so efficient. There is no a priori reason to assume that these teeth could not have been used to consume meat, especially if meat was taken in an opportunistic fashion. By way of example, euphrachtine dasypodids manage to use their cone-shaped teeth with a fair degree of efficiency to consume flesh (Vizcaíno and De Iuliis, 2003). Moreover, the strong claws of *Megatherium*, powered by fast-extending forearms (see Chapter 6; Fariña and Blanco, 1996), could have been used to rip carcasses (and perhaps even living animals) and to cut pieces of flesh small enough to be swallowed without much oral processing. Finally, *Megatherium americanum* might have been an opportunistic scavenger that could have driven away a *Smilodon*, which would have had no desire to tangle with a bullying giant, that had captured the prey; this behavior is termed kleptoparasitism.

This proposal offers some relief to an otherwise stressed habitat. A hypothetical flesh eater of 4 tonnes would need to make a kill (or to steal a carcass from a weaker hunter) only every 16 days, if the killing-rate equation given by Peters (1983) can be extrapolated. It would be unwise to make too literal a projection from these equations, but their application suggests that the need for animal food for the largest ground sloths, *Megatherium*, would nonetheless be low.

The diet of the ground sloths has been stated for a long time to be herbivorous (Burmeister, 1879); proposals extolling possible insectivorous or snail-based diets were debunked by Cabrera (1926, 1929c). Other ground sloths are known to have had a herbivorous diet based on their feces, such as the grass-eating *Mylodon* found in Última Esperanza, southern Chile, or the xerophytic-browsing *Nothrotheriops* in North America. However, these are examples of either a marginal environment (southern Patagonia during the ice ages) or a less stressed habitat (southern North America).

In addition, there are many specializations of the digestive tract in herbivorous mammals, including living tree sloths. Species feeding on low-quality, high-fiber forage must develop a fermentation chamber within the digestive system for a symbiotic association with cellulase-producing microorganisms. Large ground sloths would have been able to tolerate low-quality forage because of their large sizes, but in order to be truly effective at feeding on browse at their size, they would have had to evolve modifications to their digestive tract, as all other large herbivores have done. A stomach and intestine partly full of carrion and partly full of poor-quality plant material would not be able to digest either optimally, for they are such different kinds of food (although cecal fermentation should not be ruled out, as explained by Janis, 1976). Thus, if *Megatherium* was eating appreciable quantities of flesh more or less simultaneously with plant material, they must also have been relatively poor herbivores. Bears are a possible analogue, but they never attained the size of the largest ground sloths. Still, the Lujanian fauna's trophic composition was certainly peculiar, and it might have been the appropriate circumstance for a very large herbivore to adopt this kind of diet.

Generally speaking, the choice of diet is related to the potentiality of the morphology, and also to the opportunity of selecting food items. As stated by Peters (1983:106, italics ours): "It should be noted that the distinctions between herbivores and carnivores ought not to be drawn too rigidly. Most, if not all, herbivores will eat meat when available and *often seem to prefer it.*" If the analysis above is correct, availability of meat does not appear to have been a problem in Lujanian times, and after all, "animals eat what they should not," to cite the title of Bozinovic and Martínez del Río's (1996) article.

A fragmentary rib (Fig. 8.11), 26.5 cm long and about 7 cm wide, kept in the collection at the Departamento de Paleontología of the Facultad de Ciencias, Universidad de la República (Montevideo, Uruguay), shows marks that can be interpreted as hard evidence in favor of this hypothesis (Fariña, 2002b). This material drew the attention of our Spanish colleague Alfonso Arribas, from the Museo Geominero in Madrid, who first wondered about the origin of the marks. The description that follows is derived from his personal observations, which he kindly provided. The bone is a middle portion of a (probably fourth) left rib of a very large mammal, likely a giant sloth or a mastodon. Its inner part is composed only of cancellous bone (there is no marrow cavity). Unfortunately, this material lacks data

about its origin, but the sediment still attached to the bone suggests a late Pleistocene origin, while its contents of siliceous plant structures (phytoliths) indicates that the environment in which it was deposited was dry and cool. There are efforts currently under way to identify the chemical signature of the element so its original locality can be recognized.

On its inner face, there are seven shallow marks oriented obliquely to its main axis. Among the possible causes of these marks, inorganic agents can easily been ruled out. Indeed, weathering marks do exist on (and only on) the inner face of the rib, and traverse the other, larger marks, which indicates delayed exposure on the surface before burial, but after the marks were produced, when the bone was left on the substrate in its mechanically most stable position. Moreover, they are small, shallow, split-line cracks probably due to insolation. Furthermore, trampling marks are not observed, which may be explained by the fact that the sediments had a low content of sandy material. A human-made tool, another possible agent, may be ruled out, because marks left by such an object are deeper and sharper (see Chapter 9).

Given their characteristics, the only possible agents likely to have left those marks are mammalian teeth. Indeed, they were not produced by a single instance of contact, but rather show evidence of a repetitive action of a hard material, as well as a general moderate abrasion that could have been the result of the chemical action of the saliva. Among mammals, rodents and carnivorans can be ruled out, especially because of the large dimensions, which exceed those that could have been produced by *Smilodon*, an unlikely agent anyway as a result of its peculiar oral anatomy, which must have made it difficult for the animal to have dealt with such a large bone with its mouth.

The most likely candidate is therefore *Megatherium*, whose transversely bilophodont teeth (Fig. 8.12) have the appropriate size and shape to leave

8.12. Upper image, skull and mandible of *Mega-therium americanum* in lateral view. Scale bar equals 10 cm. Lower image, teeth of the mandible of *Megatherium americanum* in occlusal view. Scale bar equals 10 cm.

From Bargo (2001a) by kind permission of Acta Palaeontologica Polonica.

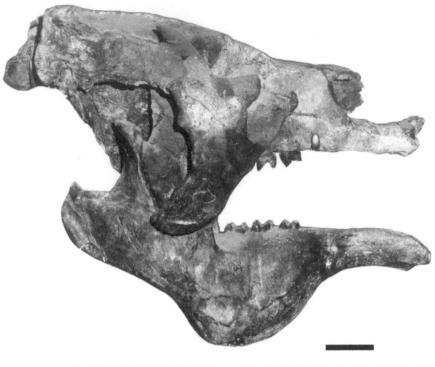

such marks. Indeed, the typical distance between lophs in the teeth of *Megatherium* is about 3 cm, which corresponds exactly to the distance between several pairs of marks (range 2–4 cm). Further, the section of the marks have the proper form and depth, i.e., it is U shaped, wide, and shallow, and the mouth was large enough to house this whole fragment (Fig. 8.12). The most likely explanation for those marks, then, is that the individual of *Megatherium* held the rib fragment with its hands (a normal use of the forelimbs for a biped, even if facultative), with the inner face pointing downward, and gnawed it with the lower teeth. This inferred behavior does not necessarily imply that *Megatherium* was an active carnivore, as the bone could have been scavenged in an opportunistic way to obtain fat or even minerals, as deer do (Sutcliffe, 1977).

This possibility can be further assessed by using McNab's approach on the total daily energy an animal can spend. According to this model (McNab, 2009), the maximal size of vertebrates is limited mainly by the abundance and quality of the resources used to sustain their activities. The maximal daily field energy expenditure (FEE, the amount of energy an animal spends in 1 day while carrying out its usual activities) of an individual varies with a variety of factors, including its mass, mobility, and foods consumed, which render maximal expenditures and body masses greater in herbivores than in carnivores because the resource base for herbivores is greater. Also, FEEs (Box 8.1) of marine mammals are approximately twice those of terrestrial species because the abundance of resources in marine environments permits large masses and high growth rates.

Calculation of FEE   Box 8.1

The FEE of an individual is given by the equation

$$am^b,$$

where a is a coefficient that determines the level at which energy is expended, m is the individual's mass, and b is the power exponent of mass, the value of which is between 0.67 and 0.75 (Nagy, 2005). Therefore, if the maximal individual FEE equals K, the carrying capacity of a certain environment (which is the number of animals the habitat can support throughout the year without damage to the animals or to the habitat), a trade-off occurs between a and m; if a increases, m must decrease, but as a decreases, m may or may not increase, depending on the circumstances in the environment and the characteristics of the species, including its food habits. According to experimental studies (Nagy et al., 1999),

$$FEE = 4.82 \, m^{0.734},$$

where FEE is in kJ day^{-1} and m is in grams. Maximal daily expenditure for terrestrial carnivores is ~1.25×10^5 kJ day^{-1} (McNab, 2009), as this must be the FEE of an 8-ton theropod dinosaur (maximum size as estimated by Christiansen and Fariña, 2004) with a metabolism like the varanids, actively hunting giant lizards whose better-known example is the Komodo dragon. Because mammalian carnivores have faster metabolism—about 50% greater than that of similar-sized herbivores—the appropriate calculation has an intercept of 7.23, and hence the maximum size will be about one ton, which is the size of the largest terrestrial mammalian carnivores known, i.e., the African Miocene creodont *Megistotherium osteothlastes* (880 kg; Savage, 1973).

Xenarthrans of all diets have slow metabolisms—about 76% those of other mammals of the same size and habits (McNab, 2008). Assuming that a potential carnivorous xenarthran had a metabolism that is similarly slower, its FEE should be corrected by the appropriate factor of 0.76. The animal could then attain a larger mass, provided it was not so sluggish as to fail to get its nourishment. Thus, the appropriate intercept would be about 5.5, and hence the maximal body size for a xenarthran with carnivorous habits would be up to 1.7 tons. However, in aquatic environments, there are larger carnivorous mammals, with the maximum size attained by the killer whale (*Orcinus orca*) at 9 tons, a result of the easier locomotion in water and the abundance of marine resources. In the same way, a larger size could be attained by a kleptoparasitic flesh eater in such a meat-rich environment as the South American Lujanian. Such an animal must have had no need to spend precious energy in fast locomotion to catch prey but could have remained much slower in its behavior. A large size can be convenient for such a trophic habit, which depends on bulk to drive original hunters away.

Moreover, it is conceivable that sabertooth cats killed more prey than they consumed, opening the ecospace for carrion eaters.

According to the evidence discussed here, the strongly counterintuitive proposal that ground sloths were at least opportunistically meat eaters was advanced by Fariña (1996) and Fariña and Blanco (1996). In this proposal, *Megatherium* would become the largest known flesh-eating land mammal.

The notion that the review of hypotheses in the light of new evidence is a healthy practice in science applies to these lines of study. For instance, a recent revision of the ursids from South America proposed that during the Late Pleistocene, there were three species of bears (Soibelzon, 2004), instead of one as proposed by Fariña (1996), although only one of them is found in the Luján local fauna. If true, this would force reexamination of the estimates derived from this model of trophic diversity, at least to a certain extent. Similarly, the identification of sexual dimorphism (Christiansen and Fariña, 2003) might diminish the number of herbivore species, although analyses assuming only one species per genus also yielded imbalanced results.

Although much of this chapter is devoted to discussion and explanation of a set of apparently odd paleobiological and paleoecological circumstances (and it is quite clear that there is more than one possible explanation), the whole debate also serves as an illustration of how scientific knowledge is advanced: not necessarily through an orderly, stepwise progression along some previously set line of thought, but with different competing hypotheses all marshaling evidence (sometimes the very same evidence) for their own cause and using some of that evidence in trying to dispute the interpretations presented by other hypotheses. On the other hand, it also shows that scientists do not necessarily have to be, as so often seems to happen, at each other's throats (see Chapter 2 for examples): two of the competing hypotheses in explaining the Lujanian fauna have been proposed by two of the authors of this book.

In any case, there is an indisputable conclusion: this spectacular fauna is weird. Is the cause of its oddness associated with the importance of the presence of the phylogenetically distant, metabolically slow xenarthrans and ancient-styled native ungulates? Or does it have to do with the proposed unexpected habits of *Megatherium*? Whatever the reasons, this amazing fauna excels at being out of the ordinary and still keeps interesting secrets yet to be unveiled.

Among the speculations to be considered as food for thought and future work, it should be mentioned that the ever-growing teeth, a feature present in the large xenarthrans and in toxodonts, have a vital effect on longevity and the prolongation of reproductive activity. A possible scenario, valid as a source of hypotheses for future endeavors, is that the members of this fauna were K strategists, with a relatively low number of individuals per species, some of them with burrowing as a strategy for protection or to conserve water, and with their slow metabolisms being revealed by low OSAs and also by their brain size (see Chapter 9).

The possible energetic imbalance mentioned above can only partly be resolved by the reassessment of the dietary habits of *Megatherium*, and additional hypotheses should be considered, in addition to the already mentioned proposals of low metabolism and differential population density of carnivores. The region between latitudes 33 and 37°S, bounded by the Pleistocene sand fields on the west and the present coast of the Río de la Plata (Fig. 8.2) and the Atlantic Ocean on the east, must have had an extension of about 480,000 km², or 60% more compared to present conditions. This must have had diverse consequences.

First, it makes clear that many of the places now available for pollen or other proxy data were influenced by a higher continentality, as they were farther from the coast than they are today. The samples from Buenos Aires Province that were analyzed for pollen (and on which a psammophytic steppe was inferred; see Chapter 4) would have been, during the LGM and the successive millennia (the so-called tardiglacial), several hundred kilometers from the coastline. They would thus reflect the increased continentality that would have made the climate much more arid, as occurs in modern localities similarly removed from the coast. Today the city of Santa Rosa, the capital of La Pampa Province, Argentina, is about 300 km from Bahía Blanca, the closest point on the coast, and has a rainfall of only 400 mm per year; at the same latitude, but benefiting from the Atlantic's influence, the region that lies along the shore has more than 800 mm of rainfall per year (Hoffmann, 1975). Apart from the already strong regional signal of higher aridity, those samples would thus reflect the increased continentality that would have made the local climate even more arid, as occurs in modern localities similarly removed from the coast. For the Luján local fauna, as defined by Tonni et al. (1985), the same line of reasoning is valid: its primary productivity could have been higher than the average for the region, as it was close to the large Paleoparaná River, which may have increased the available humidity.

Further, current paleogeographic reconstructions interpret the submerged region as the floodplain of that large river (today covered by the Río de la Plata) and an immense delta (now covered by the Atlantic Ocean). These submerged regions would have provided a larger and much more productive territory for the megamammals, based on the presumably dramatic increase in primary productivity due to inferred seasonal flooding and presumably higher humidity, because it was closer to the coast. Even though the general climate, and particularly that of the present temperate region, was much drier than that of the present (Clapperton, 1983; Tonni et al., 1999), current models of rainfall for the tropical part of South America during the pleni- and tardiglacial phases suggest that rainfall was actually higher than today in those regions (Baker et al., 2001), and that large lakes formed then where now only an arid environment can be seen (Chapter 5; Fig. 5.10). The modern Paraná River (and presumably also the Paleoparaná during Pleistocene times) has a substantial part of its basin in tropical latitudes (Fig. 8.13); it therefore must have collected more water than nowadays, making the contrast more intense.

Frequently, bottom-fishing ships collect fossils (Fig. 8.14) from underwater sites in the Río de la Plata and the Atlantic Ocean (thus opening exciting possibilities for submarine paleontology in the near future, as noted

A large river ruling the life and death of large South American Pleistocene mammals

8.13. Modern Paraná River basin, showing the area in the tropical region and in the temperate zone, as divided by the Tropic of Capricorn.

Drawing by Sebastián Tambusso.

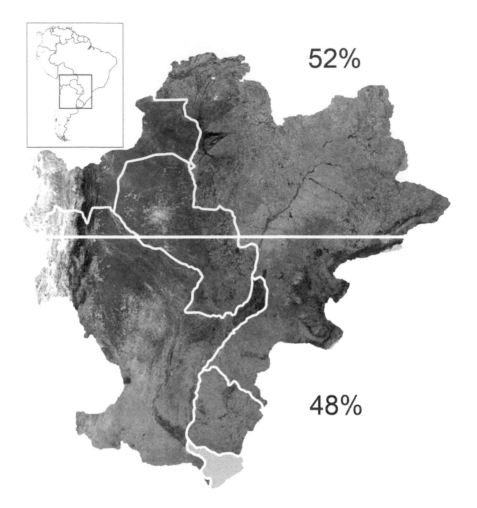

52%

48%

by Fariña, 2002b). All those remains, as well as others found washed up on the beaches, show evidence of not having been reworked and are characteristically permineralized. They all belong to members of the Lujanian fauna, and therefore a late Pleistocene age can be safely assigned, although absolute dating has yet to be attempted.

We may speculate on the possibility that mammals might have experienced seasonal migrations, to and from the occasionally flooded areas, which would have resulted in a more efficient usage of vegetation by the herbivores, such as the impressive seasonal migrations that take place currently between the Serengeti and the Ngorongoro. It is congruent with this notion that a number of different species are usually found in sites of this fauna and that those sites are normally found in fluvial sediments downstream from and near to present-day fords, which presumably coincide with those of late Pleistocene times.

What we want to know next

The complexity of this subject is almost certainly greater than currently conceived. Research on these issues has only begun, and many aspects of this complexity may therefore have been overlooked. The ongoing approaches of the analysis of the mathematical structure of the alimentary nets and counting of specimens will shed new light on these matters.

Other hypotheses will be tested with a number of approaches currently under way or planned for the near future. As usual, teeth will be the

8.14. Permineralized proximal end of the ulna of a small mastodon, collected by fishing boats off La Paloma, southeast Uruguay.

preferred subject for further study of the life traits of this fauna. Preliminary observations of growth lines indicate the potential for reconstruction of many aspects of the life history of these mammals, including the search for seasonal periodicities and growth rhythms, as well as the incipient study of stable isotopes of carbon, oxygen, and nitrogen. This approach, once more results are obtained, will help to determine the dietary niche of *Megatherium* from isotopic comparisons with other members of the South American megafauna. $\delta^{15}N$ values will be also useful for estimating the paleoenvironment, because herbivores from closed habitats exhibit lower $\delta^{15}N$ values with respect to those from open grasslands as a consequence of soil acidity in dense forest and physiological adaptations for concentrating urea in animals inhabiting arid regions (Gröcke, 1997). Finally, the $\delta^{15}N$ values from bone collagen of herbivorous mammals allow estimation of the values of $\delta^{15}N$ in the plants consumed, which is related to annual rainfall and thus can be used for estimating paleoprecipitation (Robinson, 2001).

Another interesting line of research will have to do with further exploring the possibilities of generating diversity. The carnivores may or may not have been able to cope with the impressive diversity of herbivores; this actually could have been related to carnivore diversity having been compensated by higher densities of the species found in the community. If so, this lack of diversity might be explicable in terms of phylogenetic restrictions and contingency, both related to the peculiar history of the Gondwanian heritage of this fauna, evolved in an isolation suddenly interrupted by the mingling of species coming from North America.

Finally, the possibility of a more complete comparison of this peculiar assemblage with dinosaur faunas should be taken into account. Analogies have been advanced as hypotheses between xenarthrans and dinosaurs and other archosaurs. For instance, glyptodonts obviously exploited the same part of the ecomorphological space as ankylosaurs did in the Cretaceous in having been armored and possessed of heavy tails potentially used as clubs. *Armadillosuchus* among crocodiles (Marinho and Carvalho, 2009) also bore heavy armor. Sloths have been compared to the advanced theropod therizinosaurs, such as *Nothronychus graffami* (Zanno et al., 2009). To complete this view, anteaters with their adaptations for digging and tubular snouts have been proposed to parallel the theropod alvarezsaurids (Senter, 2005).

All these subjects open new and exciting possibilities for future research on the unique and bizarre mammals from South America, whose design and evolutionary history captivated such scientific notables as Charles Darwin, Florentino Ameghino, and George Gaylord Simpson, and continues to hold the attention of many of their intellectual descendants, us included.

9.1. Size trends of large-sized South American mammals during the Cenozoic.

Based on Vizcaíno et al. (2012)

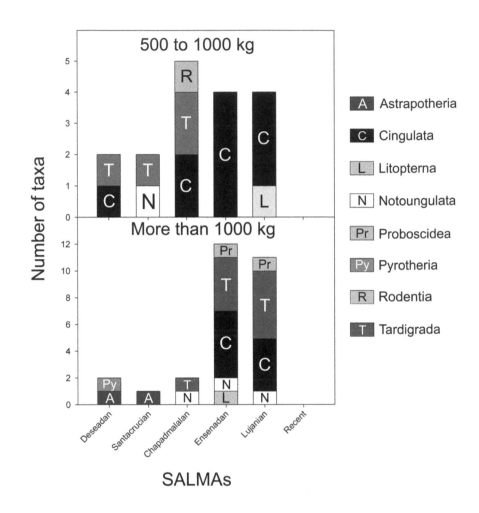

Extinction

It might be said that humans should be as concerned with the extinction of species as much as with death. They are expressions of the same phenomenon at different scales of the hierarchy of life. To add appeal to the subject, the suggestions of extraterrestrial causes, such as asteroid or comet impacts, first to explain the demise of nonavian dinosaurs and then other groups of organisms, are in marked contrast to the more traditional models, such as gradual climate change, advanced to make sense of extinction (Alvarez et al., 1980; Pope et al., 1996). Although science strives for consensus (most scientists today accept that a single extraterrestrial impact scenario was the main trigger leading to the end of nonavian dinosaurs), it seldom achieves unanimity: Keller et al. (2009) found evidence in the Brazos River, Texas, USA, that the extinction occurred over several hundred thousand years at the end of the Cretaceous, a hypothesis recently strengthened by the identification of a second crater in Ukraine (Jolley et al., 2010). In any case, such events serve as strong reminders that a catastrophe from the far reaches of outer space and far beyond our collective human experience has occurred in the geologic past and that, given time, will occur again. The recent controversies and debates over which models best explain the nearly wholesale disappearance of organisms is a main reason why the study of mass extinctions has since experienced a surge of interest, both among scientists and the general public, and new ideas are being developed to understand causes, processes, and patterns.

However, extraterrestrial impacts are not considered by many scientists as the main reason for the devastation most continents suffered during the last several dozen thousands of years, during which large (>44 kg, Martin and Klein, 1984) and very large mammals (>1000 kg, i.e., megamammals) were largely wiped from the face of earth except in Africa and, to a lesser extent, southern Asia (Cione et al., 2009). South and North America were particularly affected, and the splendid fauna this book deals with completely disappeared 10,000 years ago. Although this extinction episode saw the fall of many mammals, it does not fit the typical pattern of mass extinctions because it was only the large mammals that were affected; with the exception of a few smaller mammals, no other animal or plant became extinct (Cione et al., 2009). The various proposals put forth to explain the extinction events that occurred at the end of the late Pleistocene of South America fall into several categories: exclusively or primarily climate driven, exclusively or primarily human driven, some combination of climate and human activity, and extraterrestrial (a relative newcomer to the game, and generally discounted). These will be critically discussed later in this chapter, but we will first deal with one of the suggested biotic causes: the commonly held view of superiority of the North American fauna. This topic specifically addresses the reasons why mammals of North American origin seem to have done so much better than their South American counterparts, and although it also involves extinction, it does not provide explanations for the wholesale disappearance of megafauna from the Americas at the end of the Pleistocene.

Northern superiority?

We noted in Chapter 5 several of the variables that have been used to assess the relative success of the North American immigrants (NAI) versus the South American immigrants (SAI). In that discussion, it was made clear that the manner in which we tabulate the number of clades (ranked traditionally as families) that headed either north or south has a strong influence on our interpretations of determining success. We also made the point that the discrepancies between NAI and SAI essentially disappear if we choose (as we do) to include primates and tree sloths in our evaluation of success in migration. However, migration success is not the sole basis on which earlier interpretations of the inferiority of southern mammals were set forth. The other main factor has to do with what eventually happened to the SAI versus the NAI and the contributions they made to the composition of their adopted homeland at and subsequent to the end of the Pleistocene.

We have noted in other chapters of this book that there was a major extinction episode near the end of the Pleistocene. It has long been recognized that the extinction episode was unequal and that the South American fauna itself (as opposed, in this case, to the SAI) was hit harder and replaced by taxa that had arrived or were descended from the NAI during the Great American Biotic Interchange (GABI). The reason given for this is generally linked, whether or not explicitly stated, to some sort of superiority intrinsic in North American forms. That is, when the two faunas intermingled and had to compete head to head, those from North America won out in the long run because they were inherently better able to adapt and make use of the resources available as a result of their long history of competition and relatively more frequent exchanges with the faunas of Eurasia (in particular) and Africa (Webb, 1976, 1985a). However, discussions on this subject have been based primarily on rough estimates or analyses of pairs of groups that were supposedly in competition with each other, but whose assumptions of ecological equivalence have never been firmly established. The way pairs of groups have been selected — probably unconsciously by most of our otherwise fair-minded colleagues from earlier generations — almost seems in retrospect to have ensured support for a preconceived notion of northern superiority. We are not intending to be overly harsh in assigning such prejudice, but it is no secret, for example, that for much of the past several centuries (and we are not restricting ourselves here to the comments of Buffon and his immediate intellectual descendants), such feelings have not been limited to the realm of the animal kingdom. Many have had (and continue to harbor) misconceptions on the superiority of northern human groups over southern ones — that is, northern societies have, in the last centuries, enjoyed better economic and social successes in part because they have been shaped and conditioned by their more demanding, colder climates: you had better be prepared, or watch out. The idea of better survivability of the northern forms (and here we are back to discussing animals) is based on selective, and not particularly consistent and rigorous, analysis.

Given the magnitude of this phenomenon, the paucity of truly global and quantitative approaches, especially those considering several extremely important factors and evaluation of their relative importance in the study of this episode, is surprising.

The usual evidence for the northern superiority hypothesis is that (1) many more taxa of southern origin became extinct, so that (2) in terms of the overall compositions of modern faunas, South America is heavily represented by NAI (or their descendants), whereas the presence of SAI (or their descendants) in North America is negligible. This is perhaps true if viewed in a global perspective. For example, of the 16 families that were originally North American, three are extinct, whereas four or five (depending on taxonomic vagaries) of 10 South American families (if we exclude the living tree sloths and primates, and recognize nothrotheres as part of another sloth group, as was done traditionally) became extinct. This seems to be rather conclusive factual data in favor of the North American mammals. It might be, except for a few small details. Ah, details, details!

It turns out that we can count three of these North American families as still with us only because they survived on continents other than North America (i.e., Elephantidae, Equidae, and Camelidae). If the problem is considered in this light, then the numbers seem to say something else: eight of 16 North American families became extinct in their continent of origin. Counting those that survived in the Americas as a whole, half of the 16 families of North American mammals became extinct. On the other hand, 10 of the 16 families of South American origin have still-living representatives, including Didelphidae (19 genera with over 90 species), Dasypodidae (eight genera with 21 species), Myrmecophagidae (three genera with four species), Bradypodidae (one genus with four species), Megalonychidae (one genus with two species), Atelidae (five genera with 29 species), Aotidae (one genus with 11 species), Cebidae (two genera with 14 species), Callitrichidae (seven genera with 42 species), and Erethizontidae (five genera with 18 species).

Box Fig. 9.1. Chacoan peccary (*Catagonus wagneri*) in the Saint Louis zoo, Saint Louis, Missouri, USA.

Photograph by Tim Vickers.

Almost extinct   Box 9.1

A case worth mentioning is that of *Catagonus wagneri* (Box Fig. 9.1), of the family Tayassuidae (the peccaries, a family with other living species in both North and South America), which managed to hang on only by the skin of their teeth, as an isolated population in a remote area of Paraguay (Wetzel et al., 1975), found long after the species had been described as a fossil (Rusconi, 1930).

Usually the best such arguments in favor of North American superiority can be made for ungulates and carnivores. The ungulates and carnivores of South American origin, litopterns, notoungulates and other groups, as well as large carnivorous marsupials, are completely gone, their place taken by forms of North American background (Marshall et al., 1982). As we will see, even though these arguments seem convincing, they too are on less than solid ground. In order to understand fully the ungulate part of the story, we need to know that the South American ungulates were already in decline well before North American ungulates arrived. Indeed, by that time, they had been reduced to only five families. Of these, one died out before the Marplatan (that is, before most of the GABI got under way), one during the Marplatan, and a third during the Ensenadan. As Simpson (1980) noted, it appears as though this was a gradual decline in the face of continuously increasing competition, except that the decline had begun before the arrival of the North American imports. Two of the families, the toxodonts and macrauchenids, battled on for two million more years before finally giving up the ghost—about the same time, we should add, that their supposed northern (and "superior") competitors, the mastodons and horses, bit the dust too.

Table 9.1 tabulates the data by genera and groups them in higher categories to facilitate comparisons (cf. Fig. 5.5 and Table 9.1). An advantage of doing so is that we gain resolution and can reassess the interpretations based on tabulations of families. We can gain even greater resolution by listing the species of each genus, which would give us an indication of the numbers of basic biological units, but there are inherent problems with doing this. One of these is that we run into the same taxonomic problems as with higher-level groups: the number of fossil species can vary greatly depending on the author, particularly for those taxa that have not been reviewed by more modern analyses. This can also be true of living species, but given that we have access to much more information (including molecular and behavioral data, as opposed to the almost strictly morphological information for fossil species), we are much less likely to make erroneous taxonomic decisions for living species. Dealing with fossil species is much trickier. Most are known only by their skeletal remains—and partial remains at that. As an example, the giant ground sloth *Eremotherium laurillardi* was until about a decade ago routinely regarded as representing three species, and a decade before that by at least twice this number (see Chapter 6; Cartelle and De Iuliis, 1995; Cartelle and De Iuliis, 2006). Thus, for our purposes, we will mainly concentrate on genera, although we will make occasional references to living species.

The taxa are distributed among similar higher-level taxonomic groupings as in Fig. 5.1 but includes genera. This provides considerably better resolution with regard to which mammals passed from South America to North America (i.e., NAI) and those that passed from North America to South America (i.e., SAI). As with Fig. 5.1, we have included primates and tree sloths as NAI. Before continuing, we should note that a similar effect is in play with regard to taxonomic vagaries: we have and will continue to change our minds about which genera are valid. For example, the carnivoran genus *Felis* includes *Puma* and *Herpailurus*, which are considered distinct genera by some authorities; similarly, *Leopardus* includes *Oncifelis* and *Oreailurus*.

What sort of information may we gather from tabulating the taxa in this format? For the purposes of the following discussion, we will omit Rodentia for the time being. Leaving rodents aside, a simple tally of genera listed in terms of total number that went one way or the other is: total NAI = 27; total SAI = 42. The SAI thus hold an edge here. How might we explain this difference? We could postulate that they were better at migrating, but what would that mean—that they were able to cover greater distances, that

Table 9.1. Taxa involved in the GABI[a]

North American Immigrants

Marsupialia

Didelphidae: *Didelphis*

Primates

Atelidae:

Ateles

Alouatta

Cebidae

Cebus

Saimiri

Aotidae

Aotus

Callitrichidae

Saguinus

Rodentia

Caviidae

Neochoerus†

Hydrochoerus

Erethizontidae

Erethizon

Notoungulata

Toxodontidae

Mixotoxodon†

Xenarthra

Megalonychidae

Meizonyx†

Pliometanastes†

Megalonyx†

Choloeopus

Megatheriidae

Eremotherium†

Mylodontidae

Thinobadistes†

Glossotherium†

Paramylodon†

Nothrotheriidae

Nothrotheriops†

Bradypodidae

Bradypus

Myrmecophagidae

Myrmecophaga

Pampatheriidae

Holmesina†

Pampatherium†

Plaina†

Glyptodontidae

Glyptotherium†

Dasypodidae

Dasypus

Pachyarmatherium†

South American Immigrants

Carnivora

Canidae

Canis

Cerdocyon

Chrysocyon

Dusicyon (in historical times)†

Protocyon†

Pseudalopex

Theriodictis†

Ursidae

Arctotherium†

Tremarctus

Procyonidae

Chapalmalania†

Cyonasua†

Nasua

Mustelidae

Galera (or *Eira*?)

Galicits

Table 9.1., continued.

South American Immigrants

		Lutra
		Lyncodon
		Mustela
		Stipaniciciat
	Felidae	
		Felis (Puma, Herpailurus)
		Leopardus (Oncifelis, Oreailurus)
		Smilodont
Lagomorpha		
	Leporidae	
		Sylvilagus
Rodentia		
	Muroidea	
		Sigmodontinae[b]
		Telicomyst
		Auliscomys
		Bolomys
		Ctenomys
		Cholomyst
		Eumysops?
		Euneomys
		Nectomys
		Scapteromys
		Calomys
		Eligmodontia
		Oryzomys
		Cricetinae[c]
		Reithrodont
		Dankomyst
		Graomys
		Akodon
	Hydrochoeridae	
		Neochoerus (reentered South America over land bridge)
Artiodactyla		
	Tayassuidae	
		Platygonust
		Catagonus
		Pecari
		Tayassu (probably including *Dicotyles*)
	Camelidae	
		Lama
		Palaeolamat
		Hemiaucheniat
	Cervidae	
		Morenelaphust
		Blastocerus (Paraceros)t
		Ozotoceros
		Mazama
		Pudu
Perissodactyla		
	Tapiridae	
		Tapirus
	Equidae	
		Hippidiont
		Onohippidiumt
		Equust
Proboscidea		
	Gomphotheriidae	
		Amahuacatheriumt
		Haplomastodont
		Steogmastodont (probably including *Notiomastodont*)
		Cuvieroniust

[a] Taxa include those that crossed the Isthmus of Panama and those that evolved endemically from ancestors that crossed. Extinct taxa are indicated with a dagger. Although not exhaustive (especially for the South American immigrant rodents), this presents an approximation of genera that are commonly recognized as having participated in the GABI. Data were compiled mainly from Nowak (1999), Vizcaíno et al. (2004a), and Woodburne et al. (2010).
[b] This group is too diverse to provide an exhaustive list of genera, of which there are nearly 90; only a small sampling is given here, including some extinct genera.
[c] Cricetinae are much less diverse than Sigmodontinae; a sample of genera is listed.

they were better at handling the hardships of migration, or that they were more efficient colonizers? These notions might hold water if there was an overwhelming difference in numbers, but the fact that (at least) 27 genera managed the trip north, and that many of them spread widely and underwent significant phylogenetic change does not suggest an inferior group of mammals. As an alternative, perhaps we might consider that there were fewer types in South America to begin with, and most of those present that had a reasonable opportunity to cross the Isthmus of Panama were largely adapted to tropical conditions.

Consider next how many became extinct. Here the numbers are somewhat less straightforward: extinct NAI = 15; extinct SAI = 20. How are we to interpret such numbers? Slightly more than half of the NAI became extinct, slightly less than half of the SAI became extinct—but the point to focus on is that nearly half of either group did go extinct. What sort of biological superiority does this speak to? (A moment's reflection will reveal how silly a question it is in the first place.) Further, more SAI became extinct than NAI. Again, how can we frame an explanation for this result in terms of superiority of one group over the other?

Let's look at things in another way. More than half of the NAI are xenarthrans. This comes as no surprise—we noted in Chapters 5 and 6 that xenarthrans were among the three main South American mammal clades, and that for much of the period over which the GABI occurred, they were probably the most diverse of the three groups. The other two groups, the native South American ungulates and marsupials, had been in decline before the main pulses of the GABI were underway. The primates (one of the ancient immigrants) are the second most represented NAI. This is not surprising because they had diversified by the time of the GABI. It fell to these two groups to represent South America in the North. North America, as described in Chapter 4, contained a much more diverse fauna than South America, and the number of groups invading South America reflects this abundance. Consideration of which of the North American groups fared particularly well in South American shows that it was the carnivorans (both extinct and extant) that are well represented. Again, this is not surprising. As explained more fully in Chapter 5, at the time of the GABI, there were few real mammalian carnivores to speak of. Those that had filled this role, the marsupial borhyaenids, had already suffered a major crash in diversity. They were no match for the incoming carnivorans, but that's because they weren't there.

In terms of extinction among the NAI, the xenathrans suffered most heavily: all but the smaller sloths and cingulates became extinct, as did the only South American native ungulate, *Mixotoxodon*, which was also a large mammal (so nine genera in total). Do we finally have evidence of the inferiority of South American mammals? Hardly. Nearly all of the very large SAI also became extinct; the exception is the tapir, which isn't all that big by megafaunal standards. This includes proboscideans and equids, seven in total. Not only did these become extinct in South America, but they also died out in North America—the land of origin in the case of the equids.

To summarize some of these patterns, once the extinctions have been considered, the apparent success of the SAI may be attributed largely to carnivorans, which are not particularly large, and a few deer and peccary, as well as the single hare. On the other hand, the NAI are largely represented by primates. It is true that they are not nearly as widespread, but the carnivorans and deer are largely temperately adapted, whereas the primates are mainly tropically adapted; we can hardly blame them if the habitat to support them does not extend much farther north than Central America.

We have left out the rodents, and in particular the sigmodontine rodents. Sigmodontines were among the earlier SAIs. What are we to make of this group, poised so strikingly above all other groups in terms of diversity and thus proportional representation in South American, both in the past and today? Is there something about their spectacular performance in South America that is particularly revealing? Certainly, but not with regard to resolving questions of inferiority versus superiority. What sigmodontines have done in South America is what muroids, the group to which they belong, have done just about everywhere else in the world. What it tells us is that they have been as good at speciating in South America as we would have expected them to be. The reasons for their biological proclivity are partly due to the biology and natural history traits of the group: they are small, reproduce quickly and invest in numerous offspring, and exploit varied niches.

A glance at Table 9.1 seems to corroborate the concept of northern superiority, especially if we approach the data with this idea already somewhere in the back of our mind. It's easy to convince ourselves: look at all those genera of SAI compared to the NAI. All those felids, canids, mustelids, ursids, cervids, and gomphotheriids (among others) are much more plentiful than the ground sloths, glyptodonts, myrmecophagids, dasypodids, and didelphids. And the sigmodontid rodents greatly outnumber the caviid rodents.

In the sense of a straight comparison of numbers of genera, our impressions are indeed correct: there are more genera of SAI. But is postulating a northern superiority the only explanation that makes sense of the numbers? In Table 9.1 we consider the significance of these numbers and how they might be interpreted. Here we consider a few other aspects that support these other explanations—for example, leaving out the sigmodontid rodents for the time being (as explained in Table 9.1) facilitates out task. If we want to consider numbers and then deploy them as evidence of biological superiority, it is not equivalent to compare the abundance of carnivoran genera with those of, say, sloths and glyptodonts. If we are going to make comparisons, then we have to compare like with like. Most of the sloths and glyptodonts were megamammals. That many of these mammals tended toward gigantic size was a function of some combination of their phylogenetic heritage and the particular environmental and ecological relationships they found themselves in during much of the latter half of the South American Cenozoic. As Vizcaíno et al. (2012) noted, the trend in size increase culminated in the Ensenadan and Lujanian Ages (Fig. 9.1). If we are going to compare the generic number of sloths and glyptodonts, then we should properly compare them with reasonably similar sorts of mammals—in this case, other megamammals—during similar time periods. How do the sloths and glyptodonts fare against the likes of gomphotheriids and equids (the latter also large and similar in size to some of the smaller giant sloths)? As SAIs, they were not all that spectacularly more successful than the large to very large NAIs. A reasonable tally of large to very large mammalian genera (including one carnivoran, because it really was quite large, although we are trying to avoid carnivorans), arranged by family, would include the following:

South American immigrants (nine genera in all):

Gomphotheriidae: *Haplomastodon, Cuvieronius, Stegomastodon*
Equidae: *Equus, Hippidion, Onohippidion*
Tapirdae: *Tapirus*
Ursidae: *Arctotherium* (at least one of its species)
Camelidae: *Palaeolama*

North American immigrants (eight genera in all):

Megatheriidae: *Eremotherium*
Mylodontidae: *Paramylodon*
Megalonychidae: *Megalonyx*
Notherotheriidae: *Nothrotheriops*
Glyptodontidae: *Glyptotherium*
Pampatheriidae, *Pampatherium, Holmesina*
Toxodontidae: *Mixotoxodon*

This is not much of an advantage. Further, of these, only one, *Tapirus*, survived the end-Pleistocene extinction episode. All the proboscideans and most of the perissodactyls perished. So did the large xenarthrans. On the other hand, and unlike the proboscideans and perissodactyls, xenarthrans are represented in North (as well as South) America by still living representatives: two genera of tree sloths, and one each of anteaters and armadillos. Add to these the five genera of primates, and the numbers no longer seem skewed in favor of the SAI. Granted, there are still more of them, but much of this abundance is attributable to the carnivorans and cervids, which can readily be explained by factors other than superiority (Table 9.1). This ignores the sigmodontid rodents, but their success is explained in Table 9.1 as well.

The carnivorans are among the more notably successful SAIs. For much of the Cenozoic, the South American carnivore niche was occupied by marsupials belonging to a group known as borhyaenids, which disappeared just when placental carnivores entered the scene. This appears conclusive, but appearances do not always tell the whole story. Borhyaenids did coexist with a placental carnivore in the Huayquerian—a carnivoran procyonid, which arrived as an island hopper, but its lineage became extinct. (So much for northern competitive superiority; see Chapter 5 and Fig. 5.5.) However, the canids and felids, clearly the true borhyaenid competitors, appear in the Marplatan, by which time the borhyaenids had already become extinct. It is worth noting that Australia also had a much more diverse marsupial fauna, including very large beasts such as *Diprotodon*, about as large as a rhino, and the giant kangaroo *Procoptodon*, which stood about 3 m high and is honored by Australia with its own stamp (Fig. 9.2). The late Pleistocene extinctions also had a dramatic effect on the Australian fauna, which today

9.2. Australian postage stamp commemorating the giant kangaroo *Procoptodon*.

Drawing by Peter Trusler (2008). Reproduced by kind permission of the Australian Post Corporation.

is severely depleted compared to that of the late Pleistocene. However, the blame can hardly be laid on placentals (except perhaps of the human kind).

To assess quantitatively the proneness to extinction of the Pleistocene South American mammals, Lessa and Fariña (1996) focused on the 122 genera described for the Lujanian land mammal age and classified them according to the following characters, which seemed to have been useful predictors of the probability of extinction:

1. Origin—whether North or South American, and noting where a genus or its ancestor was found before the faunal interchange of the Plio-Pleistocene.
2. Basic trophic niche, or diet—as either herbivore or nonherbivore.
3. Mass, estimated according to various methods (as in Fariña et al., 1998; Vizcaíno et al., 2001, and Chapter 7)—as falling into three categories: from 0 to 1 kg, 1 to 100 kg, and larger than 100 kg.

Subjecting these data to a statistical analysis known as logistic regression allowed investigation of which factors were useful in predicting the probability of extinction and the relative importance of each of these factors. Thus, making use of multivariate statistics, the effects of these factors were considered in one go, and the results showed which of them were valid predictors and which were by-products of the important ones. The results of the analysis indicated that only one of these factors was highly significant and of any predictive value: body mass. Trophic niche—in contrast to the predictions of the theory that claims that predators die out more than herbivores—and geographic origin were not useful.

The successive inclusion of these factors into the analysis, in order of relative importance, permitted an estimation of their effects above and beyond those of any previous variables. Inclusion of the factor of body size explained nearly 55% of the probability of extinction. The effects of the others were not statistically significant, and so the program did not consider them. What is clear from the analysis is that body size is easily the most important factor in explaining extinction in this case, as in many others: the bigger they were, so to speak, the harder they fell. It is so significant, in fact, that it does not make much sense to even think about the possible effects of other factors until the first had already had its way (Fig. 9.3). In effect, this analysis statistically formalizes (and at a much broader scope) the less sophisticated (though intuitive) tallying of genera that we detailed above.

Later analyses (Johnson, 2002) refined this concept, identifying physiological and ecological traits as the main factors influencing proneness to extinction in the late Pleistocene, notably the presence of low metabolic rate. In that study, two general features of the selectivity of Late Quaternary mammal extinctions in Australia, Eurasia, the Americas, and Madagascar were revealed. First, large size on its own was not directly related to risk of extinction; rather, species with slow reproductive rates were at high risk, regardless of their body size and trophic niche. Second, species that survived despite having low reproductive rates typically occurred in closed habitats, and many were arboreal or nocturnal. These features certainly apply to many of the megamammals that became extinct at the end of the Pleistocene, and they also explain the survival of representatives of at least one of the groups: the tree sloths.

In summary, such analyses demonstrate that the classical theory did not have it quite right—at least not in the sense that geographical origin is a valid predictor of the probability of extinction and that the North American forms were superior when the extinction happened. What immigrants were precisely good at was, well, immigrating. Some three million years ago, there were only a handful of immigrants in South America, but now they represent just over half of what is considered, in the geologically short run of time, the autochthonous fauna.

There are various possible explanations for the extinction of the megafauna, and there is no need to postulate that the South American mammals were in one way or another inferior. One lesson we might take away from all this is that there seems to be little to be gained, in this case, by tallying up families, genera, or species to use as evidence of success. Perhaps we should consider refraining from such practice when we set out to formulate hypotheses on the mammalian biogeography of North and South America, or, at the very least, be aware of the limited value of such techniques.

It happens to be that many of the mammals that did become extinct were South American, but more rigorous observation shows that the pattern of extinctions is that the giants (and hence of slow metabolism) were more likely to die out, and most giants were of South American origin. This is really just another way of looking at the discussions on size and slow reproductive biology, but it may make it easier to visualize. That is, the place of origin does not matter. The mammals that were most affected, whether from South or North America, were for the main part the large ones, and the effect on such mammals from either continent was nearly the same. What we are not sure of is precisely why they became extinct, although one, some, or possibly all of the explanations we present here may have had something to do with it. One thing is certain, and is always true of any scientific field of study: further investigation, more research and data, and the generation of new hypotheses are required. There is still

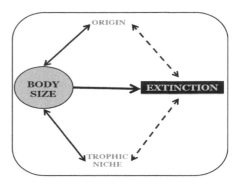

9.3. Diagram representing the importance of some the factors possibly explaining the extinction of South American late Pleistocene mammals. According to Lessa and Fariña (1996), only body size is implicated as a factor of the extinction, with both niche and origin having an indirect influence through their respective correlations with body size. When the factors are considered simultaneously, such as in the logistic regression used, those correlations disappear. As noted in the text, body size correlates with reproductive rate, which Johnson (2002) identified as the main factor of extinction.

Drawn by Sebastián Tambusso based on Lessa and Fariña (1996).

one other thing to clarify (though it is mentioned in Table 9.1), and it may already have occurred to some readers. If the North American mammals were not superior, then how do we explain the fact (and in this case it is undoubted) that the South American fauna has such a pronounced North American flavor, whereas the influence of the southerners in the north is much less important? One of the reasons is likely related to the structure of the original South American fauna, which was of three basic stocks: xenarthrans, marsupials, and South American ungulates.

Let us consider the xenarthrans first. Most of the xenarthrans that disappeared were large forms—the giant sloths, the pampatheres, and the glyptodonts. Nothing mysterious here. There were, however, various smaller sloths, particularly in the Caribbean islands, that did go extinct at the rather late date of 4.4 kya (Steadman et al., 2005). We will keep this in mind for later, but for now, we will note that sloths are still with us, but their range is restricted to tropical rain forests. Other smaller xenarthrans are still with us as well: armadillos and anteaters are both rather widespread.

The marsupials have never been as variable as the placentals and have always seemed to do best when isolated from them. However, some authors propose that marsupials originated in the northern continents, particularly after the finding of *Sinodelphys szalayi* in the early Cretaceous, aged at 125 Mya, of Liaoning Province, China (Luo et al., 2003), then possibly going to North America (but had become extinct there by Miocene times until reinvasion by didelphids, the successful opossums) and South America, and from there to Australia via Antarctica. Certainly they were never able to extensively exploit the large hoofed ungulate niche, although there is at least one exception, the Australian giant herbivore *Diprotodon*.

Vizcaíno et al. (2012) suggested a similar circumstance in South America with regard to the large hoofed ungulate niche. In this continent, sloths and glyptodonts occupied a wide array of the available large herbivore niches. When the artiodactyls and perissodactyls arrived here, their success was limited in the presence of the wide diversity of sloths and glyptodonts, with the result that they achieved relatively little diversification. This was particularly true for the megaherbivore niches, from which the xenarthrans apparently excluded them. Indeed, the relative absence of large hoofed ungulates remains true in the modern South American fauna, with *Tapirus* the only surviving representative. The gomphotheriids were exceptions, but even in this instance, the group is marked by relatively low diversity and taxonomic richness.

The native South American ungulates, a spectacular presence through most of the Tertiary, had already declined by the late Neogene and Quaternary, and only a tiny percentage of the decline occurred in the presence of North American ungulates; some held on to the very end with their North American counterparts. To what, then, should the decline and eventual fall of the native ungulates be attributed? The answer is unclear. Obviously they were unable to adapt to changing environmental conditions, both biotic and abiotic, but this is such a general explanation that it is not particularly satisfying. The point to be emphasized is that their general decline cannot be readily attributed to the superiority of North American competitors, because the decline was already well underway before the North American ungulates arrived. Perhaps the rise and great success of xenarthrans in the megaherbivore niche had something to do with it.

In summary, the South American native fauna at about the time of the GABI basically consisted of a few ungulates and moderately sized carnivorous marsupials, neither particularly diverse and both already under decline, large and diverse groups of xenarthrans, and the caviomorph rodents. This is the stage onto which the North American forms entered. Let us consider how they have fared.

On the whole, they have done quite well, but not all have been equally successful. For example, the insectivores (or, to express it in a better agreement with modern taxonomy, Eulipotyphla; Roca et al., 2004) have barely got a foot in the door, as only five species of shrews inhabit South America, and there they are limited to the northwestern corner's tropical forests and the neighboring Andes (Gardner, 2007). The lagomorphs are widespread, but no more and probably even less so than are, for example, the porcupines and armadillos in North America; and they are not particularly diverse, with only one species present (Robinson and Matthee, 2005).

North American mammals have done best in those niches that the native South American forms had not, for whatever reasons, radiated into by the later Neogene and Quaternary. Thus, the carnivorans, sigmodontid rodents (field mice), and smaller ungulates (mainly cervids) are those that have been most successful. Indeed, some carnivorans achieved their highest diversity in South America, along with the explosive and truly staggering radiation of the field mice (Table 9.1; D'Elía et al., 2005), especially in the cases of two clades that basically evolved in this continent (although they had representatives in North and Central America), namely the South American canids and that of the small South American felids save the yaguaroundi *Puma yaguaroundi* (Prevosti, 2006, 2010; Soibelzon and Prevosti, 2007; Prevosti and Soibelzon, 2012).

The North American forms that have become dominant components of the South American fauna are thus generally those that occupy niches that the South American forms had not already moved into. That is, they faced little, if any, competition—quite the reverse of the idea normally espoused in the superiority hypothesis. Croft (2001) summarized the data on why South American mammals had not filled these niches, although it is admitted there that some inconsistencies in the interpretations (i.e., whether those mammals evolved in habitats similar to those we know today) may result from nonanalogous attributes of South American faunas during much of the Cenozoic. In any case, this circumstance in no way indicates that they were inferior, only that they lacked the appropriate opportunity. Contingency is to be considered the main cause, as in many other cases. To put it in colloquial terms, if you are already committed to being an anteater or sloth, it becomes difficult to enter and exploit the pronghorn or bison niche. However, now that we've raised the point of facing competition, we might consider the other side of the coin: the lack of SAIs' success at exploiting the large hoofed ungulate niche. We noted that this may have been the result of the competition that large North American ungulates faced from the sloths and glyptodonts, which effectively excluded them from this niche (Vizcaíno, 2009; Vizcaíno et al., 2012). Have we here an example where North American taxa were outmatched by well-ensconced native faunal elements?

There is another factor that we should consider as a possible cause of the modern South American fauna and the relative dearth of South American mammals in North America. We have viewed the landmasses in question as two discrete entities: South and North America. From the geologic and geographic perspectives, as discussed in Chapter 3, this is not far from the truth. The two landmasses have largely followed their own geologic histories, and when viewed on a map, they seem to be two distinct regions serendipitously connected by a thin thread of land (and remember that this strip of land rose up between them, rather than having been initially part of one or the other; see Chapter 5).

However, to the animals and plants that must live, survive, and adapt to environments, our convenient subdivisions fail miserably, as ecological regions, discussed in Chapters 3 and 4, must be taken into account if we are to understand the distributions of the flora and fauna. There are various such regions in the Americas, as well as in other continents, but these need not be examined in great detail. For our purposes, it is sufficient to understand that the two continents differ greatly in the distribution of the subtropical to tropical regions as opposed to temperate regions, and that geographical position (i.e., latitude) is one of the main factors in determining climate. Much more of the North American landmass lies in a temperate belt as compared to the South American landmass, which offers a greater tropical region. This is especially true near the very regions where the two continents became connected by the emergent land bridge (Fig. 9.4).

The scenario described below, outlined more thoroughly by S. D. Webb, among others (for instance, in Stehli and Webb, 1985), begins with South America offering greater tropical and smaller temperate regions than North America for much of the Tertiary. Then, during the late Tertiary, uplift of the western Cordillera produced rain shadows and subhumid savanna corridors extending along a north–south axis. Once the two landmasses were linked, these corridors provided a path along which temperate flora and fauna could pass back and forth between them. Most of the mammals that participated in the exchange belonged to groups representing mainly subhumid savanna ecosystems (Webb, 1976).

The formation of the Andes also led to numerous new areas of temperate habitats, particularly in the western part of South America, due mainly to altitude rather than latitude. The unequal temperate areas of the continents thus provide another possible reason for the unequal success of the mingled American faunas. If most of the mammals from South America that had a chance to participate in the exchange were adapted mainly to tropical and subtropical habitats, then we would not expect to find—and we do not find—many of them in most of North America, because the latter provides much more opportunity for temperate-adapted mammals. Similarly, the temperate mammals of North America, an assemblage composed mainly by the carnivorans and cervids, have thrived reasonably well in South America. In the temperate regions beyond the Isthmus of Panama, they found environments that were favorable to their evolution, with many opportunities to fill the empty niches to which they were already adapted.

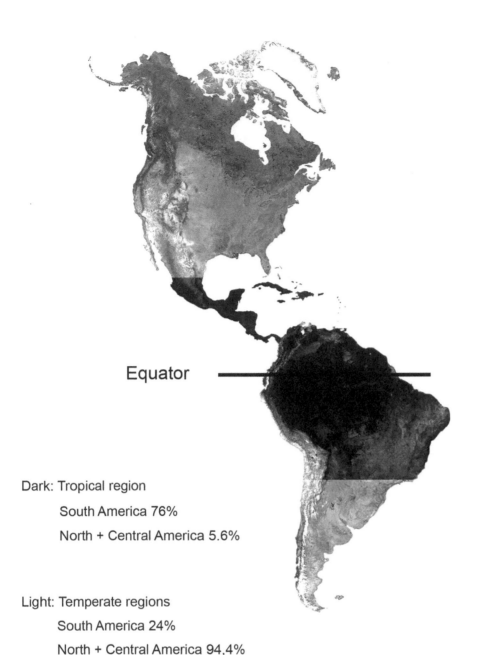

9.4. Tropical and temperate regions in South and North America.

Drawing by Sebastián Tambusso.

Equator

Dark: Tropical region

    South America 76%

    North + Central America 5.6%

Light: Temperate regions

    South America 24%

    North + Central America 94,4%

Thus, with the longtime residents maintaining their own ways (keeping in mind that the larger ungulate niches were largely unavailable to any new-comers in the face of xenarthran domination of these niches) and with the newcomers occupying their parts of the ecospace, the paleoecology of the South American mammalian communities grew in complexity, rather than the native species having been driven to extinction by the immigrants—until extinction affected many of them rather indistinctly, as we shall see, rather than according to whether they were smart or not.

Intelligent primates like ourselves place too much importance to this characteristic in the evolutionary game, and hence we tend to hold other creatures to this standard, to the point of absurdity. Other animals, including other mammals, have eschewed completely the advantages of a brain capable of distinguishing between Monet and Picasso, between a

The role of intelligence

toccata in F minor and "Yesterday," between a play by Pelé and a formula by Einstein.

Rodents, for example, rely on their size and fecundity, producing many offspring very quickly and at various times throughout the year, thus doing their utmost to ensure that enough of them will not be eaten and will produce descendants that will start the process all over again. Elephants, at the other extreme of body size among living terrestrial mammals, put all their chips on growing in size, taking good care of their young while they are defenseless, raising their trunks, and increasing their tusks.

Naturally, however, all creatures need intelligence to a greater or lesser extent, even though it may not be central to the role they have been allotted in the drama of evolution. Thus, viewed as one of many characteristics within the context of the ecological strategies at the disposal of an organism, it begins to make sense (Griffin, 1992). In living animals, intelligence may be judged according to complexity of behavior and ability to learn, with amazing examples of self-recognition assessed in the magpie (Prior et al., 2008). Through this versatility, ethologists appraise the value of this characteristic (Brown and Cook, 2006).

In fossils, as always, matters are more difficult. Yet perhaps because of this, they can also be more interesting. Paleoneurology, mainly through the study of molds both natural and artificial that form inside the cranium, infers the morphology of the brain in vertebrate fossils (Jerison, 1973). For mammals in particular, the relationship of the brain with the cranium that contains it is so close that the region responsible for smell can be clearly identified, and from its relative size, how well developed this sense was can be deduced (Dechaseaux, 1958).

Debate on the value of a quantitative approach to brain size has continued for more than a century. The polemics among French anthropologists over the size of the brain are renowned and led to wagers—to be resolved posthumously, of course, because at the time computed tomographic scans were not available—over the brain size between colleagues in the profession. It is enough to make one pause and reflect on the real evolutionary importance of our inherent intellectual superiority. By the way, the brain of Paul Pierre Broca (Fig. 9.5), one of the bettors, is exhibited today at the Musée de l'Homme in Paris.

Today, although such enthusiasm has not died down completely, the door has been left open for quantitative considerations, but based on relative rather than absolute size. The encephalization quotient gives a measure of the cerebral capacity of a species in relation to its body size, and from this, one can estimate the intelligence of that species. The equation that describes it is as follows:

$$E = k \, P^{\mu}$$

where E is the size of the brain in grams and P is the size of the body in kilograms (Jerison, 1973). Further, k is the constant of proportionality and μ, the allometric coefficient, is the slope of the line in the log–log graph, a neat mathematical trick whose shortcomings have been repeatedly mentioned in this book (Packard and Birchard, 2008), used to turn an exponential curve into a straight line, which is easier to work with.

Both the constants k and μ are almost always determined experimentally in such cases. The results in the case of the encephalization quotient produce a number without any units. It relates the observed value to the expected value (considering the universe of all mammals—or at least those mammals used for creating the allometric equation) for a species of a given body size. A value of 1 means that the species is average; that is, it has a brain size expected for its body size. A value greater than 1 indicates that it is doing pretty well in this regard; and less than 1 that it has to manage by other means, and that you'd never be able to successfully convey to the individuals of that species just how good this book is.

The South American vertebrates offer ample opportunities for this approach. The xenarthrans, along with some marsupials, have the great advantage of belonging to South American lineages while at the same time containing many contemporary examples, as expressed in the pioneering work by Dechaseaux (1958). Apart from the vast field of study present in descriptive and comparative paleoneurology of this odd group of mammals (see Fig. 9.6, showing a digital endocast of *Glyptodon*), some measurements have been carried out in glyptodonts, and the value for the species *Glyptodon clavipes* falls between 0.1 and 0.2—that is, about one-tenth to one-fifth the brain size which a species of its body size should have. If one keeps in mind that armadillos, a group related to the glyptodonts, have a value of 0.4 or higher, one can conclude that the evolutionary strategy of glyptodonts was not based primarily on being particularly perspicacious. Considering the idea of intelligence as one aspect of an entire living system, one realizes that this character was not very necessary for animals that, given their large size and armor, quite efficiently warded off the dangers they faced, and that they did not need much of a brain to choose the patch of vegetation they were going to eat.

In general, South American fossil herbivores fall below the expected range for their brain size compared to the values obtained in a universe of more modern mammals (Jerison, 1973). There is a tendency in the fauna of the globe to increase the variation in relative brain size as we get closer and closer to the present. In simple terms, the regions of the ecospace that remain available as an evolutionary novelty are those that can be occupied by more intelligent creatures.

This process seems to have been much slower among South America mammals, perhaps as a result of their lengthy relative isolation. It would be easy to attribute the extinction of the megafauna to lesser brain development, but this is a topic that requires a detailed discussion. Authors such as Russell (2009) argue that evolution of life is progressive and less contingent than supported by others, notably Stephen Jay Gould. However, the encephalization has to do with low metabolism, an issue already taken into account as the cause of extinction (Johnson, 2002) and other factors, to be discussed, might have had a greater importance.

9.5. The great French anatomist Paul Pierre Broca. French physician, anatomist, and anthropologist. Among his many contributions to the knowledge of the brain, he described a region of the frontal lobe known as Broca's area.

Wellcome Library.

The roles of humans and climate: a mixed bag?

Important to us as biological entities and as cultural beings, the great debate on the human influence on the end-Pleistocene (and present day) extinctions is a decisive crossroads of science and ethics. The idea that humans were responsible for the extinction of the Pleistocene megafauna is an idea extending back more than a century, as recounted by Grayson (1984). The main sides in the debate have defined themselves as supporters

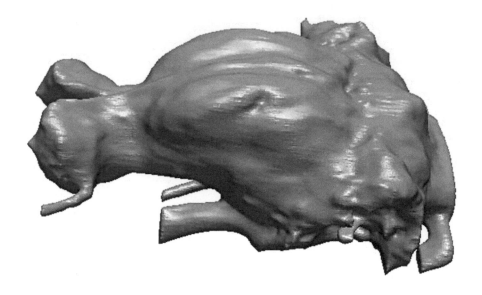

9.6. Digital endocast of *Glyptodon*. Scale bar = 3 cm.

Courtesy of Sebastián Tambusso.

of human-only views (with overhunting as the most radical example) and those who embrace nonanthropogenic climatic change as the cause of the demise of the Pleistocene megafauna. Major contributions of both parties were summarized in a seminal book aptly titled *Quaternary Extinctions: A Prehistoric Revolution* (Martin and Klein, 1984). Since then, a considerable amount of evidence from diverse scientific realms (paleontology, climatology, archaeology, genetics, and ecology, as well as linguistics) has been gathered, and it seems that most authors now support the idea that human contributions to extinctions were important in some continents, but perhaps not solely on the basis of hunting, and that climate also had a role.

In the New World, a high number of mammalian species disappeared, especially those of large and very large body size (Lessa and Fariña, 1996; Lessa et al., 1997). This extinction seemed to have taken place in a short period of time, almost synchronously over only a couple of millennia in North America (Faith and Surovell, 2009) and possibly in South America, although dates are hard to pinpoint (Barnosky et al., 2004). The climatic model implies that swift environmental changes, directly related to the sudden end of the ice age (Lambeck et al., 2002), created ecological conditions that the megamammals could not cope with (Guilday, 1984). Yet another hypothesis postulates that diseases, some possibly carried by humans (or even other immigrant mammals), may have been a main cause of the extinctions (MacPhee and Marx, 1997; Ferigolo, 1999). Indeed, Native American human populations suffered horribly through contact with Europeans because they lacked antibodies and other adaptations against diseases carried to the Americas. It is possible that epidemics were as rampant among mammals during the Pleistocene as for people during the sixteenth and seventeenth centuries. However, it is difficult to explain which sorts of diseases may have affected so many animals belonging to such different groups.

Naturally, the end of the Pleistocene was indeed a time of swift climatic change, with global warming at the end of the ice age at least permitting human expansion and stressing large mammals, especially in the Northern Hemisphere. On the other hand, climate change was a constant throughout the Pleistocene, whereas the arrival of humans in the New World between the extinction episode and the rest of the Pleistocene was a strikingly novel difference. Certainly the climate varied, but it does not appear to have differed much from what happened with the climate and its oscillations between glacials and interglacials in the rest of the Pleistocene (Lambeck et al., 2002; see also Chapter 3). How can we explain the circumstances that came together in the Americas at the end of the Pleistocene in terms of the disappearance of its wonderful fauna? The question all the hypotheses try to answer is: What caused the nearly simultaneous extinction of about 35 megamammalian genera of considerably varied physiology, morphology, and phylogenetic background that lived in distinct habitats over such a wide geographic area? Although this is the general picture we have of what happened to these mammals, it is not clear that the extinction of all 35 genera was a near instantaneous event—which is important in this case (granted that elsewhere in this book we stress that even a few thousand years is an instant, and thus negligible, in the sense of geologic time) if we are to attribute the extinctions to humans. The general reason for favoring humans over environmental factors is the assumption that climatic changes were too varied in intensity, timing, and location to have been the cause of so many extinctions all at once.

Let's begin by examining the hypotheses in further detail, starting with one of the more popular (among both scientists and the general public) recent versions of the human-driven view: the blitzkrieg (or overkill) hypothesis, which has been developed primarily by Martin (e.g., Martin, 1984). The term *blitzkrieg* (German, literally meaning "lightning war," in the vernacular of World War II, meaning a powerful, sudden strike) has been used to describe an overkill so rapid that it resembled the way in which many European countries were overrun by German forces during the war. This overkill hypothesis is driven by observations of several more or less coincident and apparently related events beginning about 12.5 kya (Plate 11).

One of the main events to which the overkill hypothesis is tied is the timing of the colonization of the New World. For much of the middle third of the last century, the idea that a group of humans, referred to as Clovis people (because their artifacts and culture were first recognized near Clovis, New Mexico, USA; Fig. 9.7), settled and spread quickly throughout North America came close to being considered an obvious fact. This idea was based on the then only scientifically imaginable land route into the New World open to the colonizers. During the late Pleistocene last glacial maximum (LGM, as noted in Chapter 3), when the ice sheets that covered much of the Northern Hemisphere (but not Alaska) reached their maximal development, tracts of land that are today submerged became exposed (Fig. 9.8), including the portion between modern northeastern Asia and Alaska, a region known as Beringia (see Chapter 4). Once in Alaska, however, these humans had to wait until sufficient thawing produced an ice-free corridor between the Laurentide and Cordilleran ice sheets (roughly along the boundary between the Canadian provinces of British Columbia and Alberta) that allowed the colonizers to pass south to the interior of North

9.7. Clovis point. Example of a Clovis fluted blade that is 11,000 years old.

Image courtesy of the Virginia Department of Historic Resources.

America. This must have been a harrowing period for these colonizers — waiting in Alaska until the ice melted, then rushing down through a narrow passage 2000 km in length and walled by cliffs of ice to reach south into the heart of North America. Many researchers have pointed out this difficulty, but let's allow the possibility for the time being and continue with our story.

Geologic evidence indicates that the coalesced ice sheets of the LGM began to recede so that by about 12 kya, a narrow but impassable corridor was present, and by about 11.5 kya, the corridor was wide enough to allow passage (Fig. 9.9). The oldest Clovis sites, by comparison, date to about 11.5 kya. The timing couldn't be any better. Many widespread and nearly contemporaneous Clovis sites support a rapid spread across the United States. By current estimates, these early colonizers raced south to reach Tierra del Fuego, the southern tip of South America, just 1200 years later — a remarkably rapid radiation apparently unmatched by any other human group before or since (Meltzer, 2008). Within this framework, the overkill hypothesis has these early people emerging from the ice-free corridor to find themselves in a people-empty continent (though of course they had no knowledge of this) full of wild game. These skilled big-game hunters made quick work of the megafauna, dispatching them with ease in an ever-expanding, lightning-fast wave until they reached the far end of South America. The presence of such readily available and naive wild prey was meat enough and more to spur these hunters ever onward and support the population explosion that peopled the New World. As Martin (1984:1375) put it, "it is difficult to imagine an assemblage of large mammals . . . easier to track, hunt, and exterminate than the numerous kinds of ground sloths and glyptodonts that disappeared down to the last species." Indeed, an elegant model by Alroy (2001) that simulated the timing of events and selectivity against the megafauna outlines how overkill occurred.

The overkill hypothesis apparently fits all the bits of evidence quite nicely, but several other lines of evidence suggest that the hypothesis has serious flaws. One of the main pillars on which the hypothesis rests is that the migration and settling of the New World occurred by way of an overland route, first to Alaska and then through the narrow passage between the ice sheets. For many years, this view, the Clovis first hypothesis, was the most widely accepted view of New World colonization, and for good reason: no other archaeological sites could be shown to be securely dated as older than Clovis. Though there had been several older-claimed sites, none seemed able to withstand the strict standards for secure dating combined with archaeological setting demanded by archaeologists for such an upheaval in thinking. It took many years of excruciatingly patient and thorough efforts before archaeologists generally recognized that at least one older site, Monte Verde (Chile; Fig. 9.8), was really as old as its lead excavator, Tom Dillehay, had claimed it to be (there are other claims of similarly ancient sites, such as Meadowcroft Rock Shelter near Pittsburgh, Pennsylvania, USA). Most estimates securely place Monte Verde at an average of 12.5 kya, and likely as old as 14.5 kya (this is the upper level at Monte Verde; a lower and thus even older level remains controversial). If the glacial geology is correct — and there's no reason to doubt it — then the colonizers obviously couldn't have found themselves in nearly the opposite end of the New World before they had supposedly emerged south of the

9.8. Last glacial maximum (LGM) in South and North America, showing extent of ice sheets and increase in land area.

Drawing by Sebastián Tambusso.

Present day coastline
Coastline at LGM
LGM ice sheet

Sites with human-megafauna interaction:

1 - Fell Cave
2 - Última Esperanza
3 - Monte Verde
4 - Pehuén Có
5 - Campo Laborde
6 - Estancia La Moderna
7 - Luján
8 - Arroyo del Vizcaíno
9 - Tibito
10 - Clovis
11 - Rancho La Brea

ice-free corridor into the interior of North America. As Waters et al. (2011) have noted, there is an emerging view that the Americas were occupied by humans before Clovis, and it is getting more and more difficult to reject this notion. These authors reported that the Debra L. Friedkin site in Texas dates from between approximately 13.2 to 15.5 thousand years ago.

The acceptance of Monte Verde occurred in the final few years of the last century, but several other sites have been proposed as being even older (see Chapter 5). They have not been generally accepted, but if they do turn out to be really ancient, it would require a complete rethinking of the human presence in the New World—perhaps, for example, there was a much earlier and ultimately unsuccessful colonization (but here we are on a much less firmer footing). Even so (and on other grounds), some researchers had begun to question, over the last few decades of the twentieth

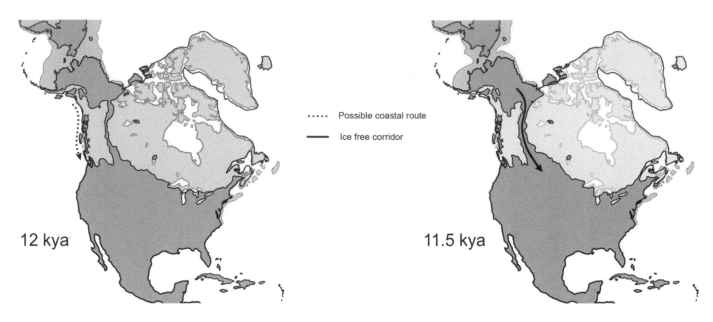

9.9. Possible migration routes for colonizing humans at the end of the LGM. By 11,500 years ago, the ice sheets had retreated sufficiently to open a passable corridor.

Drawing by Sebastián Tambusso based on Meltzer (2008).

century, the route proposed by the Clovis first hypothesis. How, it has been argued, could any group of early people have passed through such a narrow, inhospitable 2000-km corridor between melting glaciers? And if they did, shouldn't there at least be some archaeological evidence of their presence in the area of the corridor? However, there isn't any—at least, none from the critical period, before about 11 kya, when the colonizers were supposed to have been making the trip south. There is archaeological evidence of humans having occupied this region, but it is from considerably later than the critical period, suggesting that the people who left behind the artifacts actually migrated north into the region. Such concerns led researchers to propose other possible migration routes, one being a mainly coastal water passage along southern Alaska and western Canada (Fig. 9.9). Long stretches of this route were free of ice between about 16–14.5 kya, and the coast was probably entirely free of ice beginning about 13.5 kya (Meltzer, 2008). Evidence indicates that humans were present in several of the islands off the coast of British Columbia as early as 10.3 kya and probably even earlier. However, evidence is hard to find: much of the former coast is now submerged, for example. The first colonizers may have actually first entered North America along the western continental United States. How does the evidence from South America fit into this? We have already considered the ancient age of Monte Verde. As well, recent analyses of several sites in Chile and Argentina by Steele and Politis (2009:419) indicated that we can be confident about a human presence in North America and South America close to (but not with certainty much earlier than) 11 kya, implying "the contemporaneous emergence of a consistent and archaeologically-robust human occupation signal at widely-separated locations across the Western hemisphere," although Cione et al. (2009) suggested that this might extend to 13 kya in South America. How can we explain this pattern of occupation? Perhaps it speaks to the possibility of two migratory events, rather than the progressive spread from a single overland route through Beringia (Meltzer, 2008; Steele and Politis, 2009) or the western Canadian or U.S. coast.

Incidentally, there are other fields of study that have had input into this question, among them linguistics and genetics. However, the contributions of these disciplines, although useful and enlightening, have not provided decisive evidence for any particular hypothesis. In any event, the earlier-than-expected presence of first colonizers into the New World weakens the overkill hypothesis.

However, the hypothesis has yet other problems to deal with. One is that there is precious little evidence of kill sites. Grayson and Meltzer (2003) were able to list only 14 well-documented Clovis kill sites, and these included remains of only mastodon and mammoth. If the Clovis did overkill the megafauna, then there should be plenty of kill-site evidence, and not just of mammoth and mastodon. In a strange twist of scientific reasoning, Meltzer (2008) noted that Martin actually considered this lack of evidence as proof of overkill: it happened so fast that that it would be unlikely that the evidence would be preserved in the geologic record—a hypothesis that predicts the lack of evidence! On the other hand, there is ample evidence, beginning just after 11 kya and thus including Clovis times, of extensive bison hunting (and overhunting), yet such intensive and focused activities were not enough to exterminate these beasts, which were waiting in huge numbers until the late nineteenth century for their real troubles to begin (see Chapter 4). What about the model showing how overkill happened? Well, a model is a model. No matter how sophisticated, it is not evidence that an event *did* happen. Much of what a model simulates depends on the variables and values on which it is based—are they relevant and realistic? Given the problems outlined above, we must consider whether whatever is input into the model is matched by evidence (if not, the model can always be adjusted). In this regard, one of Alroy's (2001) inputs was kill sites—but we have pointed out that there are precious few of them, and those in North America preserve only mammoth and mastodons (Meltzer, 2008). What about the kill sites for the several genera of ground sloth, horses, glypto-donts, camelids, musk oxen, and the remaining 35 genera? There's little or no evidence for them. And what if humans did not emerge into the heart of North America from the ice-free corridor 11.5 kya? We have noted above why this scenario may not be reliable.

Despite its problems, the overkill hypothesis continues to have ad-herents, and perhaps with good reason: it is clear that humans hunted mammals, so human predation may indeed have played a part in the extinctions. The question that everyone would like to see answered is, how much? Ironically, overhunting is supported by archaeological evidence, but for guanaco "and deer, which, paradoxically, are those mammals that survived the extinction" (Cione et al., 2009:138). A less brutal view of hu-man predation suggests that although overkill may have been overstated, intense hunting was still a main factor and weakened populations to the point where environmental changes could then act to exacerbate the down-fall of megamammals. The opposite view, that climatic changes weakened populations to the point where human activity was able to deliver the coup de grâce, has also been proposed.

One of the main assumptions of intensive hunting scenarios is that the mammals, never having been exposed to humans, were naive, and so had not evolved antipredator behaviors. Some researchers, in contrast, have considered the many climatic changes that occurred near the end of the Pleistocene to be chiefly responsible for the demise of the mammals.

Variations of this theme contend that climatic changes were paramount, and that human activity, including hunting and habitat alteration and disruption, compounded their effects.

The appeal of humans as the main or sole agents of destruction is understandable based on the overall pattern of human presence in many parts of the world. Other than for Africa and Eurasia, where the fauna and humans and their ancestors coevolved for millions of years (and thus, the reasoning goes, the fauna was able to adapt gradually to the presence of humans), the pattern of human colonization of other continents (the Americas and Australia) and numerous large and small islands suggests that human arrival had quick and devastating effects on the flora and fauna. The pattern generally outlines an apparently stepwise series of extinctions, ranging from slow in the Old World, to quick in Australia and, later in time, in the Americas and on islands (whether prehistoric or historic). Supporters of human-based factors contend that the overall pattern suggests that climate change could not have been particularly important because climate change was occurring, in the background so to speak, throughout much of the Pleistocene. If climate was responsible, then why do we have a staggered pattern of extinction on different landmasses that coincides with human arrival? This thinking has merit, but it depends on being able to determine precisely the timing of the extinctions.

These ideas deserve further elaboration. We will consider first the situations that occurred on islands, small and large, rather than on continental-sized landmasses. It is generally agreed that the first arrival on islands, particularly remote ones, by humans has been followed by swift, devastating, and usually catastrophic consequences for the endemic life of the island. Commonly cited examples are the plights of the moa of New Zealand and the dodos of Mauritius. The idea, then, is that if it happens on islands, then it could have happened on continents and thus can be used as support for extinction on a continental scale. However, extrapolation from island to continent is a real leap, and one that is unsupported. It entails viewing continents as gigantic islands, as suggested by Fiedel and Haynes (2004)—a surprising pronouncement were it to come from professional biologists, because it is true only in the trivial sense that all landmasses are surrounded by water, as Meltzer (2008) has pointed out.

Clearly, as Wroe et al. (2006:62) noted, islands are not just "continents writ small and processes operating at island levels cannot be characterized as scaled-down versions of continental phenomena." There are many examples of this. Grayson (2001), for instance, noted that a combination of several factors was responsible for extinctions on islands. In the case of New Zealand, humans hunted, to be sure, but they also set forest fires that destroyed the natural habitat and imported rats, dogs (which acted as competitors and predators of the native fauna), and parasites. Whereas such features are characteristic of the Holocene island extinctions, those of the Americas and Australia were not accompanied by large-scale "severe anthropogenic landscape modifications, increased human population densities and the introduction of many non-human predators and competitors" (Wroe et al., 2006:62). On continent-sized landmasses, forests may be cut or burned, but entire ecosystems are not destroyed as easily. The early colonizers did not fell all of the trees of all of the forests inhabited by all of the forest-dwelling megafauna. Even modern humans have had a hard time accomplishing this.

One of the main underpinnings of human-driven models of continental extinction is the presence of naive prey that lacked antipredator defensive behaviors. This may characterize island faunas, which had never faced large terrestrial predators and would thus have been vulnerable to any invading carnivore (Wroe et al., 2006). However, the continental megafauna had coevolved with efficient predators, and it has been demonstrated that in the presence of novel or reintroduced predators (following several generations of predator-free existence), potential prey quickly learn and transfer antipredator behaviors to the new or reintroduced predators. Wroe et al. (2006) noted just this pattern for moose faced with a recolonization of wolves, and Meltzer (2008) for the case of African elephants faced with humans. It is thus unlikely that simple extrapolation of patterns of island extinctions to those on continents is a viable strategy. In some cases, there may be factors in common, but the dynamics of island ecosystems, population sizes of native organisms, and severely limited resources suggest extreme caution.

Besides the reliance on island extinction patterns, proponents of mainly human-driven extinction scenarios point to the extinction of mammals in Australia as evidence. Humans arrived on this continent 50 to 43 kya, much earlier than currently accepted arrival times for the Americas (and this early date for Australia is not disputed). The megafauna of Australia suffered, according to Burney and Flannery (2005), an extinction episode coincident with human arrival, and Australia is not a small, remote island (though it is often referred to as an island continent; but then again, all landmasses are ultimately surrounded by water). Thus, as these authors claimed, in 50 millennia of catastrophic extinctions after human contact, the pattern is quite clear: humans arrived in Australia, the megafauna became extinct; humans arrived in the Americas, the megafauna became extinct; humans arrived on islands, the fauna became extinct (though not necessarily megafauna in the latter case). This staggered pattern of extinction episodes does appear to have one thing in common: the arrival of humans. To be fair, these authors did not necessarily view overkill as the only possible mechanism, nor did they discount climatic factors, but they were quite clear that the extinctions were mainly driven by humans through hunting, as well as activities such as rapid overharvesting, biological invasions, habitat transformation, and disease. Climatic factors seemed to these authors to be less important; the Pleistocene was a time of cyclical climatic change, but extinction episodes did not punctuate earlier fluctuations in climate—or so it has usually been claimed.

Presented in this light, humans do indeed seem to have been the main culprits. However, the situation is not as simple as made out by Burney and Flannery (2005). One important difference between Australia and the Americas is that humans arrived in Australia before the LGM. One of the suppositions made by researchers such as Burney and Flannery is that climatic change leading up to the LGM was insignificant compared to previous glacial maxima, such as the penultimate glacial maximum (PGM, about 130 kya, the one that occurred before the LGM), and as previous glacial maxima did not cause major extinctions, then only full glacial maxima, if climate change is to be invoked at all, could be considered as potential causes of extinctions. That is, the climatic changes before the LGM could not possibly be held accountable for the extinction of the Australian megafauna, because they were too small to make a difference (Wroe and Field, 2007). Thus, because the extinctions occurred after humans arrived but before significant climatic change (as would have happened at the LGM) occurred, then the case is fairly clear: they came, they saw, and they slaughtered (and made a mess of other things too). Importantly, this view assumes that all or most of the Pleistocene megafauna was present in Australia when humans arrived. However, neither of these suppositions (that climate was insignificant and all the animals were waiting for humans) is demonstrable; indeed, the reverse seems to be true, according to Wroe and Field (2007:5), who stated that the pattern of extinctions in Australia is more easily explained by a staggered pattern of extinction in which most extinctions predate human arrival, and that the influence of humans was "a minor superposition of broader trends in train since middle Pleistocene times." These authors noted that only 35% of the megafaunal species whose extinction has been attributed to human activity can be demonstrated to have persisted past the PGM, meaning that 65% of these species cannot be securely placed within 80,000 years of firm evidence of human arrival. Further, only 13% (eight species) were demonstrably present at human arrival, and about half of these persisted until the onset of the LGM. The implications are clear: climatic changes over the PGM do seem to have been sufficiently intense to have caused extinctions, so the assumption that previous glacial maxima were unimportant in extinctions during the Pleistocene cannot be maintained; and so, perhaps, we should not minimize the possible effects of the LGM on the American faunas.

That we should not neglect climatic effects is emphasized by the recent efforts of Cione et al. (2009). These authors noted that nearly all the mammals that became extinct were adapted to open areas produced by the dry and cold climate that dominated South America during most of the middle and late Pleistocene. Recall that the Pleistocene was characterized by oscillations between glacial and (shorter) interglacial conditions, resulting in corresponding changes in habitats. For example, modern South America comprises approximately 18% open areas, 15% medium vegetated areas, and 67% closed areas. In contrast, Cione et al. (2003) calculated that during the LGM, these same regions, respectively, comprised 31%, 54%, and 15% (see Fig. 5.9). The changes in climate thus produced corresponding changes in habitats, which in turn led to concomitantly dramatic alterations in the distribution of biomass in South America, as well as other continents (Cione et al., 2009). The periodic interglacial increases in temperature and humidity resulted in drastic contraction of open areas and extreme reduction of the biomass (but not of the biodiversity) adapted to these open areas. Cione et al. (2003) referred to the alterations in biotic trends (i.e., between low and high biomass from open and closed areas) as a zigzag. This alteration was a consistent pattern during most of the Pleistocene, but the end of the Pleistocene (marked by the warming trend following the LGM and a temperature spike near the Pleistocene–Holocene boundary, about 10 kya), coincided with the presence of humans, whose activities disrupted the cycle that had persisted for so much of the Pleistocene. Cione et al. (2003) described this disruption as the broken zigzag hypothesis. Ultimately, claimed Cione et al. (2009), the extinction of large and very large mammals was caused by human foragers, but this event was favored by the climatic changes that led to the contraction of open areas, which led to fragmentation of megamammal populations and obstruction of gene flow. Cione et al. (2009) considered that aspects of life-history traits, such as reproductive biology, of large mammals probably contributed to their demise in the face of the new predator with specialized tools. The age of sexual maturity is an especially important factor, because it indicates how long a female must survive before it can begin to reproduce, and hence is a factor in the rate of offspring production. This threshold for almost all living megamammals is over 10 years of age (Cione et al., 2009). Combined with the long gestation period and low fecundity of most large mammals, it is evident that females would be unlikely to produce more than one offspring per year. Johnson (2002) suggested that an average of less than one offspring per year resulted in a more than 50% chance of extinction for any particular taxon. Thus, one of the critical factors is a low reproductive rate rather than large body size per se; but large mammals generally have low reproductive rates. Cione et al. (2009) proposed that perhaps only moderate and occasional hunting of females (solitary or with offspring) or juveniles would be suffient to cause extinction within a few thousand years.

Critical reviews of at least some of the evidence marshaled to support primarily human-caused extinction scenarios would suggest that we are almost back to square one: it is not clear whether climate or humans were primarily responsible for the extinctions, or whether some combination of the two (and, in such a case, *which* combination) would make for a more reasonable explanation. Let's return to the situation in the Americas. As we have learned from the events in Australia, our confidence in a human-caused explanation would be bolstered if we could be sure that the extinctions did indeed occur (nearly) simultaneously. Meltzer (2008) noted that all were gone (in North America, at any rate) by 10.8 kya, but research indicates that only 16 of the 35 genera can be demonstrated as being alive at about 12 kya. This doesn't necessarily mean that they had gone extinct by then, but we have no evidence that they were alive. We do know that Alaskan bison populations underwent severe contractions about 37 kya, about the same time that evidence points to a warm period leading up to the LGM, which reduced grazing land and increased tree cover, which in turn caused a barrier to movement. As well in Alaska, one species of horse was shown to have gone extinct about 31 kya, another about 12.5 kya, and mammoth about 11.5 kya, with the key event for the last two examples identified as the eventual decrease of grassland following the LGM (Guthrie, 2003, 2006).

This sort of information suggests, at the very least, that the case for synchronous extinctions may be a bit wobbly. However, there is additional evidence, and much of it involves South America, the West Indies, and sloths (recall that sloths were present in the West Indies; see Chapter 5). Steadman et al. (2005) favored the overkill hypothesis to explain the extinction of the ground sloth in the Americas. In terms of timing, these authors claimed that the last appearance dates of these mammals in the Americas and the West Indies coincide with the appearance of human arrival. On the other hand, they argued that climatic changes would be expected (if viewed as widespread and uniform) to have caused near synchronous extinctions throughout the Americas and West Indies, but the last appearance dates were roughly between 12.8 kya for North America, 12.5 kya for South America, and 5 kya for the West Indies, according to Steadman et al. (2005).

Hubbe et al. (2007), however, countered that the data presented by the previous authors were incomplete, particularly for South America. Whereas Steadman et al. (2005) contended that there are no acceptable Holocene last appearance dates for ground sloths, Hubbe et al. (2007) pointed out that secure records are indeed available in the literature, many of which are based on the reliable dating of bone material of ground sloths. For example, some *Megatherium americanum* remains are as young as approximately 7700 years old, and remains of the glyptodont *Doedicurus clavicaudatus* date from just 7000 kya. Several of the South American sites, particularly from Argentina, are archaeological and preserve evidence of use of megamammals by humans, but others, such as at Lagoa Santa, Brazil, show no signs of use either as sources of food or raw materials. In any event, the long lag between the earliest evidence of a human presence in South America, at least 12.5 kya (in Monte Verde, Chile) to the 7.7 kya cited immediately above, speaks against a swift killing spree as envisioned by overkill proponents, but it does fit the hypothesis proposed by Cione et al. (2009). We should note that Steele and Politis (2009) reported that some of the early Holocene dates indicated by Hubbe et al. (2007), specifically those from the Arroyo Seco 2 site in Argentina, are incorrect. However, Steele and Politis (2009) did not make any mention regarding the dates for the Argentine La Moderna (Politis and Gutiérrez, 1998) and Campo Laborde sites (Messineo et al., 2004; Politis et al., 2004), and Borrero (2008) stated that an early Holocene age for the last two sites is still accepted by their excavators. Presumably those dates, which are among the younger dates reported for *M. americanum* and *D. clavicaudatus*, may thus still be considered reliable.

As for the delayed last appearance dates from islands, Hubbe et al. (2007) cited several examples where similar delays have been reported from other islands independent of a human presence, and in those cases, robust environmental factors have been recognized as causative agents of extinction. Again, the evidence of staggered American continental versus West Indies extinctions does not necessarily seem to support an overkill scenario.

We noted above that suggesting a primary role for intensive hunting (though perhaps not as brutal as overkill) is up against the rather sizable obstacle that there is not enough clear-cut evidence of consumption by these ancient settlers on members of the megafauna (Borrero, 2009), but this is not to say that there is no evidence. The presence of several kill sites in North America has been mentioned, but we should also note that several sites in South America suggest a few millennia of coexistence (Plate 11) between the megafauna and early American hunters—time enough for more than one sumptuous feast. Such sites stretch from Fell Cave in the extreme south of the continent, near Beagle Channel, to Tibito (Fig. 9.8) in Colombia. In Chile and northern South America, there are well-documented cases of mastodon hunts and the subsequent consumption of these beasts by a group of humans (Silverman and Isbell, 2008). In the Río de la Plata region, the dramatic reduction of the available area subsequent to the end of the glacial period (an environmental change; see Chapters 5 and 8, and Figs. 5.9 and 8.2), together with the effects of the end-Pleistocene human arrival in the Americas, must have been crucial factors in the extinction of this fauna, at least locally (Fariña, 2002b). The sudden absence of the giants must in turn have induced changes in the environment, so much so that many other herbivores faced depletion of their own food sources. The outcomes of such circumstances may be investigated through research on the impact that elephants and rhinoceroses have on the African vegetation. Johnson (2009) summarized the paleoecological implications of such studies in suggesting that large changes in plant communities may have been triggered by the extinction of the large herbivores. Among the effects of megafauna were the maintenance of vegetation openness and, in wooded environments, the development of mosaics of different vegetation types. Such habitats reverted to denser and more uniform formations after the extinction of the megafauna, and fire frequency increased as a result of accumulation of uncropped plant material. Indeed, understanding "the past role of giant herbivores provides fundamental insight into the history, dynamics and conservation of contemporary plant communities" (Johnson, 2009:2509).

In effect, these large mammals consume such large quantities (never mind the destruction they cause simply by moving about) that their absence leads to profound changes: savannas or prairies are transformed to forests. For example, in the 100 years since elephants have been absent in the National Park of Hluhluwe in Natal, South Africa, three antelope species have disappeared, and the populations of other open-field grazers have been greatly reduced; Owen-Smith (1987) termed this the "pivotal role" of the megaherbivores. In similar fashion, such an occurrence might have taken place in South America, where the extinction of the large mammals may have hastened that of the smaller herbivores, which in turn sealed the fate of the carnivores that preyed on them.

And so it goes, back and forth, in the quest to uncover the main causative factors in the extinction of the megafauna. One side marshals evidence to support one hypothesis, the other side counters with, "Hey, wait a minute, what about this and this and this?" The first comes back with refined arguments, and in the meantime, the other side has found yet other bones to pick. In some ways, the sides are further apart than they were (Grayson, 2007), and the debate has not died down, largely because the available empirical data do not overwhelmingly and convincingly support one or the other side (Kennett et al., 2009). This is the ongoing activity of science, but sooner or later, evidence for or against one of the main hypotheses will be weighty enough for most scientists to either accept or reject it. However, this doesn't necessarily mean that once this happens, its main alternative will be embraced or rejected accordingly.

We noted above the proposal that an extraterrestrial event may have had a hand in the American Pleistocene extinctions. This hypothesis was initially floated in a non-peer-reviewed publication as a supernova, later in a book as multiple comets that impacted North America about 12.9 kya, and finally as multiple comet airbursts/impacts in a scientific journal article by Firestone et al. (2007). This hypothesis attributed several episodes to the proposed extraterrestrial event, such as the onset of the Younger Dryas

9.10. Alfonso Arribas, from the Museo Geo Minero de Madrid, Spain. His observational skills allowed the discovery of the marks on a clavicle of the giant sloth *Lestodon* from the Arroyo del Vizcaíno site.

Courtesy of Alfonso Arribas.

(the cold period during the current interglacial, that took place between 12,800 and 11,500 years ago), ecosystem disruptions, and the extinction of the Pleistocene fauna. Researchers bluntly and quickly pointed out several flaws in this hypothesis. The more serious problems have to do with recognition of the telltale signs of extraterrestrial impacts. For example, an impact crater has not been identified, and the several substances typically recovered from sediments recording impact fallout (i.e., impact markers) do not seem to be those typically observed in such events (or at least not as typically understood by other leading experts). More recently, Kennett et al. (2009) reported the co-occurrence of nanodiamonds (lonsdaleite), soot (in the 12,900-year-old layer), and extinction, which (these authors claimed) is the only known such association other than in the Cretaceous–Tertiary impact layer (i.e., the one widely accepted as responsible for the extinction of nonavian dinosaurs). On earth, lonsdaleite is known only from meteorites and impact craters. As well, the authors countered previous criticisms of missing impact markers by drawing a parallel to the absence of such markers in the commonly accepted impact event called the Tungusta event or airburst. This latter event is generally accepted as a having been caused by the airburst (or explosion, which is still considered an impact) of a comet or asteroid-like meteorite 6–14 km in altitude over the Podkamennaya Tunguska River (Central Siberia) in 1908. The energy emitted by the blast has been calculated at 10–15 MT, which is about a thousand times as powerful as the atomic bomb dropped on Hiroshima. The explosion devastated approximately more than 2000 km² of Siberian taiga, flattening 80 millions trees and setting fire to a great number of trees and bushes in the area (Longo, 2007). Clearly, a similar event over North America about 12.9 kya would have caused a considerable degree of damage to the environment and fauna, and may indeed have been responsible for triggering some of the events postulated by its proponents. But did it happen? As Meltzer (2008) noted, it is still too early for geologists and impact specialists to assess the evidence. If there is one thing we can be sure of, it's that this is already

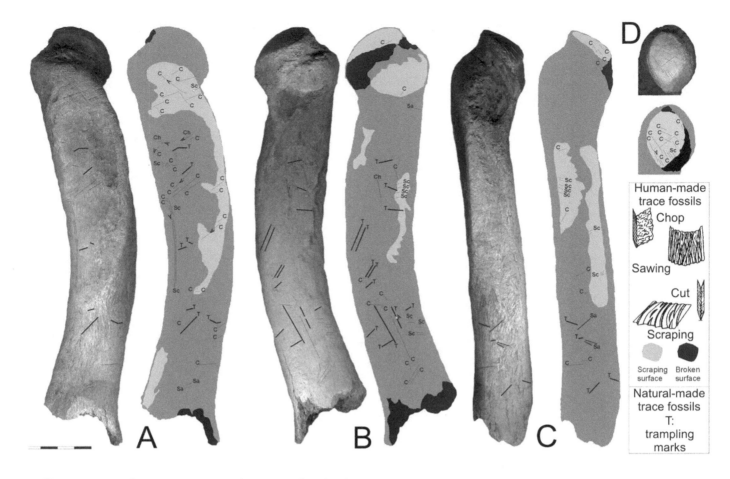

Human-made trace fossils
Chop
Sawing
Cut
Scraping
Scraping surface Broken surface
Natural-made trace fossils
T: trampling marks

well on its way to becoming yet another example of a "here it is/no it isn't" tug-of-war between scientists from opposite sides of the fence.

We mentioned earlier that direct evidence of interaction between humans and megafauna is not particularly abundant in South America, but a recent find may help us view the subject in a new light. Indeed, the presence of humans in this continent and their possible consumption of megamammal meat might have occurred much earlier than previously thought.

The Arroyo del Vizcaíno site was first described in Chapter 4 (and mention of it was made in Chapter 8 for the isotopic contents of the bones) in connection with Dr. Alfonso Arribas's (Museo Geominero, Madrid, Spain; Fig. 9.10) observations of cut marks on remains from this site. The site at Arroyo del Vizcaíno, located in southern Uruguay, preserves fossils of several individuals of the giant sloth *Lestodon*. One of the bones, a clavicle (or collarbone), was observed to preserve 87 marks (apart from the naturally made trampling marks) that had the features of all four types of human-made marks (Fig. 9.11): chopping, sawing, incisions or true cuts, and scraping marks (Arribas et al., 2001).

What a banquet it must have been! A forelimb of *Lestodon* must have represented perhaps more than 100 kg of meat, enough for a big group of hunters to feast for days or even weeks. A small piece of the clavicle (Fig. 9.12) and a sample of an accompanying rib were radiocarbon dated, and they turned out to be much older than expected, between 28 and 29 kya (Fariña and Castilla, 2007). These results imply a much earlier human

Direct human–megafauna interaction

9.11. The four types of human-made marks on a clavicle of the giant sloth *Lestodon* from the Arroyo del Vizcaíno site: chop marks (Ch), sawing marks (Sa), incisions or true cut marks (C), and scraping marks (Sc).

From Arribas et al. (2001), by kind permission of Publicaciones del Seminario de Paleontología de Zaragoza.

9.12. A small piece of the *Lestodon* clavicle, sent for radiocarbon dating.

presence than is usually accepted for both South America and North America, doubling the prehistory in both continents; most dates of the earliest evidence of peopling cluster at about 12 kya. The rather surprising results have generated the expected skepticism among other researchers, and the age of the skeletal remains requires future corroboration, the evidence for which is being gathered as of this writing. Among the implications of the possible early age of these remains is that humans and the New World megafauna would have coexisted for a much longer period of time than previously believed, suggesting that the megafauna may not have been as naive as has been claimed.

The results discussed here suggest a first attempt at colonization before the climate became too harsh for humans to thrive, as it approached the LGM, which took place 18 kya. What happened, then, to this group of humans that lived in such a southern location and at such an ancient time? The answer (pending, of course, corroboration of the results by new finds) is that we have no idea. They may have been scarce, and become extinct or moved to tropical latitudes when the climate deteriorated. All the other, and much younger, evidence of ancient human people may represent a second wave of immigrants that claimed the land only when the climate started to ameliorate in the last few millennia of the Pleistocene, once the LGM was over.

The marks show that an early population interacted with species of the megafauna, but we cannot determine whether they actually hunted large prey or simply took advantage of a lucky find of big, recently dead animals. In any case, they obviously did not drive them quickly to extinction. The human impact could have been a decisive force in the extinction, but perhaps through the action of that possible later wave of colonizers, who might have taken advantage of improved climatic conditions that must have favored the growth of their populations. The recent resumption of excavation of the Arroyo del Vizcaíno site (Fig. 9.13) will, we hope, provide evidence to help answer such questions.

As Barnosky et al. (2004) noted, the late Pleistocene extinction event was spread over more than 50,000 years throughout the entire world. These authors viewed it as the accumulation of diachronous, shorter-term pulses that took place on a regional basis. Those pulses seem to have been amplified by the interaction of both biotic (especially humans, but also other invasive species) and abiotic (climatic) drivers.

We claimed above that this matter is of special interest because it deals with subjects relevant for modern conservation biology. The present-day biosphere shows the consequence of the extinctions of the past, a phenomenon whose long-lasting causes have not ceased but only increased their effects: one species (*Homo sapiens*) is now so successful that it may turn out to be a victim of its own triumph, having created simplified, more vulnerable food webs in most, if not all, ecosystems. On the other hand, the megafauna, once splendid all over the world, in diverse climates and environments, is today but a scattered array of a few species in Africa and tropical Asia.

Barnosky et al. (2004) suggested what they termed "productive future directions," with their views about the need for further contributions, both moving on from the old architecture of the debate (which should not be

seen as the clashing of humans-only versus climate-only supporters) and improving of the quality of the data available from all the disciplines involved. We have endeavored here to follow their suggestions and make our contribution to the subject.

9.13. Wednesday, 9 March 2011, the first day of renewed excavation of the Arroyo del Vizcaíno site. One of us (R.A.F.) is in the foreground waiting for the next sediment-laden bucket to arrive.

Epilogue: Lessons from the Deep Past

E

In this book, we have undertaken a journey through the wonders of the South American megafauna, not only because the fauna is intrinsically interesting in itself, but also to provide examples of how paleontology manages to overcome the paucity of remains—meager scraps, really—that have been left to us to interpret the history of past life. Through our adventurous and often tortuous path, we discussed the never-ending game of science, the ephemeral truths that shed transitory light on our vast ignorance; we paid homage to those giants on whose shoulders we, as a community rather than individually, stand (to paraphrase Newton and several other less famous scientists) by giving a short account of their lives and accomplishments; we set the geoecological stage on which our drama unfolded by summarizing the history of South America as a land mass and the life it had supported, giving special attention to one of the main biogeographic phenomena in vertebrate history, the interchange of mammalian faunas between North and South America that occurred with the emergence during the Pliocene of a land bridge that connected these continents after their long isolation.

Within this framework, we presented the several and varied beasts that comprised this unique fauna, the most impressive assemblage of mammals ever gathered in terms of the number of giants living together, their morphological and ecological diversity, and the peculiar signature of their ancient heritage. But we would have fallen short of our goal had we simply presented what we have learned about them (as is typically the case in strictly nontechnical narratives) without having also provided an account of how we have arrived at our knowledge; and so we have endeavored to let non-paleontologists in on our secrets. We have thus also explored the methods, traditional and new, sophisticated and elementary, that have allowed us and many of our colleagues to share the intellectual joys of venturing into the growing realm of knowledge on the habits of those superb animals and how the whole group of them worked together as an ecological unit. In doing so, we fare no better than the mythological ancient Greek heroes who tried to slay the Hydra, guardian of the underworld, by severing its many heads, one by one. But whereas the ancient heroes were dismayed at the results of their labors, with the mythical creature immediately replacing one severed head with two, we scientists are fortunate, rewarded as we are with pleasure and increased wonder for our efforts in slaying our ignorance and doubts: resolution (or not) of one problem enlightens us, but it also raises new questions, often revealing a depth of ignorance that we had not yet perceived. Such multiplication in the subjects and course of study brings renewed interest and motivation. This is our lucky fate, and one we hope (actually, know) will continue for us, our colleagues, and our intellectual descendants.

However, the megafauna is no longer with us—and hence paleontology is the science that studies it and the reasons for its demise. Was it the rapid climatic and ecological changes wrought by the melting of the vast Pleistocene glaciers that caused their extinction? Were they swept away by as yet unknown diseases? Perhaps it was us, or at least our ancient kin, that hunted them to extinction. Were they cut down by an impact caused by an extraterrestrial body? Or was their fate sealed by an unknown or unthought-of agent to be unveiled by future scientists? Whatever the reasons, the magnificent beasts are now all gone, and we invite our readers to think about the ethical issues inherent in the possibility that they might have been driven to extinction by us. This very much remains an entirely plausible reason, for the early New World people should not be idealized as some sort of *bon sauvage*. They must have had all the miseries and glories inherent to the human condition, nothing more and nothing less. They, however, had the excuse of ignorance. No emails, no cell phone calls arrived from other parts of that ancient world to warn the hunters that the fauna was locally extinct in many other places and that they would be wise to take good care of the remaining beasts and exploit them rationally. No ecological models were advanced to promote sustained development in the Pleistocene. Perhaps there is, after all, a marvelous world in which humans can coexist with wildlife, but we suggest that such a world resides in the future and certainly not in the past, in knowledge and not in ignorance.

Thus, if this unique, this wonderful, this splendid megafauna present in South America just before and during the great ice ages—if that assemblage of amazing Pleistocene beasts fosters interest, stimulates curiosity, and instructs us, then perhaps, even though they ceased roaming the South American plains one hundred centuries ago, they may, in a certain and deep sense, not be completely gone.

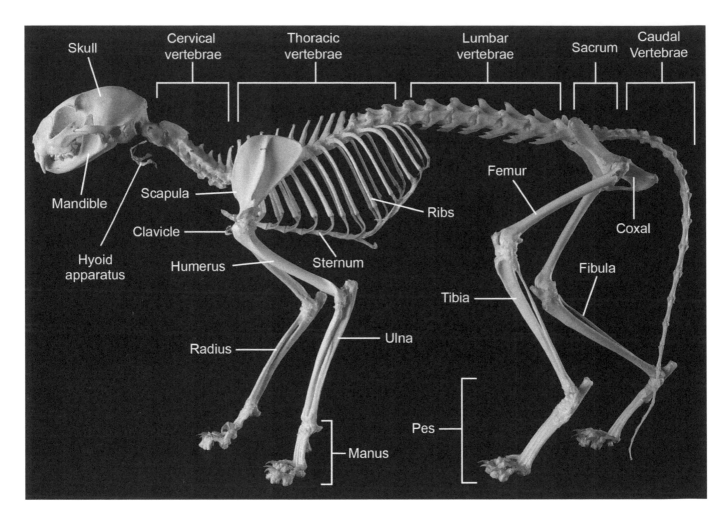

Appendix 1.1. Skeleton of the cat, lateral view.

*Image © Dino Pulerà reproduced by kind
permission of Elsevier Inc.*

The skeleton of vertebrates forms a bony internal framework that serves several functions, among which are that it supports the body, offers protection for vital organs, participates in ventilation of the lungs, provides attachment sites for musculature that together with the bones produce movement, and serves as a store for several minerals. In some vertebrates, such as turtles and some xenarthrans (Appendix 2) and dinosaurs (mainly ankylosaurs), bony structures, such as osteoderms, may be widely distributed and associated with the integumentary system (the skin) to provide external defensive protection. The bony osteoderms are covered by keratinized (i.e., made from essential the same material as nails and hooves) scutes, which often closely mimic the form of the bony element.

The skeleton includes numerous bones (Appendix Fig. 1.1), and while the number and types of bones remain reasonably constant among mammals, their shapes and proportions change, largely in response to functional differences. This primer (adapted mainly from De Iuliis and Pulerà, 2010, with contributions from Getty, 1975) provides a basic overview of the mammalian skeleton, allowing the reader a familiarity with it parts, their main functions, and the functional implications of some of the differences among mammals. It is based primarily on the skeleton of the cat, a mammal with which we are all reasonably familiar, but we include some skeletal elements of other mammals, especially the sheep, beaver, horse, and cow, also all reasonably familiar, to provide a wider scope. This allows a means for assessing and appreciating both the similarities and differences, as well as for evaluation of the functional adaptations reflected by their form (that is, the relationship between form and function).

The skeleton is subdivided primarily into the axial skeleton and appendicular skeleton. The axial is the central part and includes the cranial skeleton (that of the head), the vertebral column (or spine), the ribs, and the sternum (or breastbone). The appendicular skeleton includes the bones of extremities (forelimbs and hind limbs or, more simply, the arms and legs), as well as the elements that connect them to the axial skeleton: the pectoral girdle (shoulder blade) in the case of the arm and the pelvic girdle (hips) in the case of the legs. Often, as for example in paleontology, the skeleton is subdivided for descriptive purposes into the cranial (that of the head) and postcranial (the rest of the body) skeleton, as we do here.

The cranial skeleton includes the skull and the lower jaw, or mandible, as well as the hyoid apparatus, the bony structure that mainly supports the tongue. The skull of mammals, as with other vertebrates, protects and supports the brain and sense organs and is used in food gathering and processing. It may be conveniently divided into a facial or rostral region,

Cranial skeleton

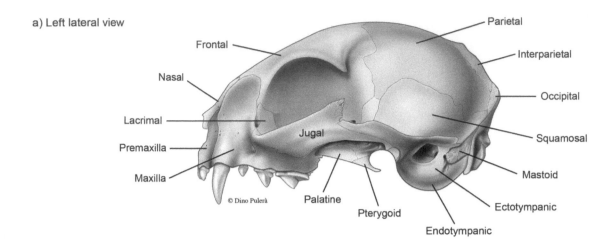

a) Left lateral view

Frontal

Parietal

Interparietal

Nasal

Occipital

Lacrimal

Jugal

Squamosal

Premaxilla

Maxilla

Mastoid

© Dino Pulerà

Palatine

Ectotympanic

Pterygoid

Endotympanic

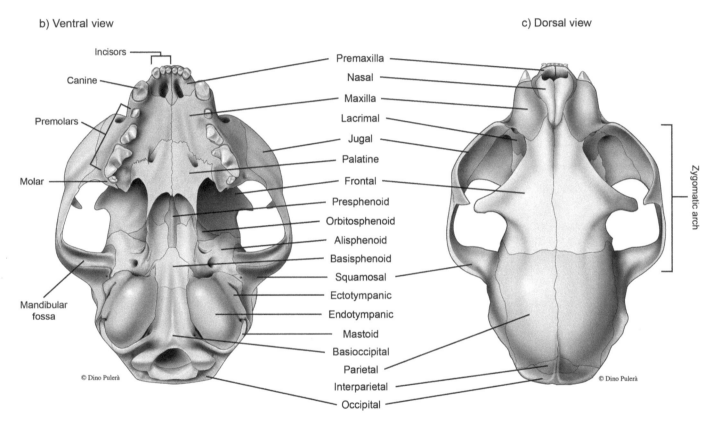

b) Ventral view

c) Dorsal view

Incisors

Premaxilla

Canine

Nasal

Premolars

Maxilla

Lacrimal

Jugal

Palatine

Molar

Frontal

Presphenoid

Orbitosphenoid

Alisphenoid

Basisphenoid

Squamosal

Ectotympanic

Endotympanic

Mandibular
fossa

Mastoid

Basioccipital

Parietal

© Dino Pulerà

Interparietal

Zygomatic arch

© Dino Pulerà

Occipital

Appendix 1.2. Skull of the cat in (a) left lateral view, (b) ventral view, (c) dorsal view.

Images © Dino Pulerà reproduced by kind permission of Elsevier Inc.

including the nose, orbits, and upper jaws, and a cranial region, including the braincase and ear. In adults the skull is a single structure formed from various centers of ossification (regions of bone development). These centers expand and their edges generally fused together so as to obscure their boundaries. However, it is still possible to note the position and form of the bones.

The great diversity of mammals is reflected in the form of their skulls. Among the more extreme modifications are the elongated, tubular, and edentulous skulls of anteaters, the enormous and highly pneumaticized skulls of elephants, and the massive skulls of baleen whales, in which the skull seems little more than gigantic and strutlike upper and lower jaws linked to the back of the skull to allow support of and feeding with baleen. Such differences in form mainly reflect evolutionary modifications for different feeding habits, but other factors (such as digging behavior) may also

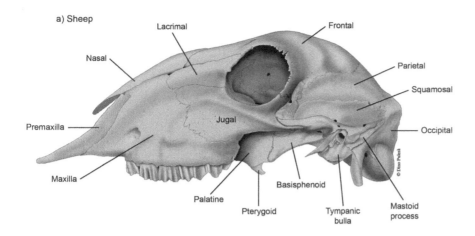

a) Sheep

Lacrimal

Frontal

Nasal

Parietal

Squamosal

Jugal

Premaxilla

Occipital

Maxilla

Palatine

Basisphenoid

Pterygoid

Tympanic bulla

Mastoid process

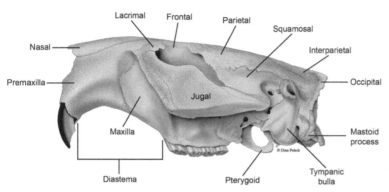

b) Beaver

Lacrimal   Frontal   Parietal

Squamosal

Nasal

Interparietal

Premaxilla

Occipital

Jugal

Maxilla

Mastoid process

Diastema

Pterygoid

Tympanic bulla

Appendix 1.3. Skull of (a) the sheep and (b) the beaver in left lateral views.

Images © Dino Pulerà reproduced by kind permission of Elsevier Inc.

play a role in shaping the form of the skull. Indeed, the jaws, particularly the mandible, and the rest of the skull may need to respond to different pressures. The mandible is clearly affected by feeding behavior, but the upper jaw and the skull may face conflicting demands, including the size of the brain, the support of antlers or horns, and defensive behavior.

The skull of the cat is illustrated in Appendix Fig. 1.2, and of the sheep and beaver in Appendix Figs. 1.3 and 1.4. At the anterior (or front) end of the skull is the naris, a large opening leading into the nasal cavity. In life the opening is separated by a cartilaginous septum, so that two external nares or nostrils are present. The orbits are the large cavities that house the eyeballs. Posterior to each orbit is an expansive cranial wall. This surface is the temporal fossa, which covers the brain and serves as the origin for the temporal musculature, which helps to close the jaw. The ventral margins of the orbit and fossa are marked by the laterally bowed zygomatic arch, a bridgelike extension of bone that serves mainly for the origin of the masseteric musculature, another group of jaw-closing muscles. The zygomatic bone forms most of the zygomatic arch. Its front end is referred to colloquially as the cheekbone. The presence of another bridge of bone should be noted in the sheep (and occurs as well in cows and horses), this being the postorbital bar that encloses the orbit from behind. It is formed by processes of the frontal and zygomatic bones.

Just posterior to (or behind) the zygomatic arch is the external auditory meatus, the opening and passageway leading to the middle and inner ears. The posterior surface of the skull, the occiput, contains the large foramen

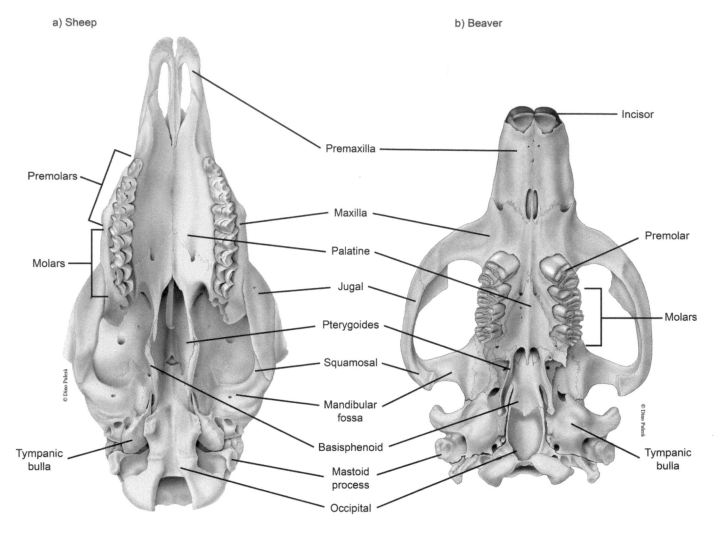

a) Sheep

- Premaxilla
- Premolars
- Molars
- Maxilla
- Palatine
- Jugal
- Pterygoides
- Squamosal
- Mandibular fossa
- Basisphenoid
- Tympanic bulla
- Mastoid process
- Occipital

b) Beaver

- Incisor
- Premaxilla
- Maxilla
- Premolar
- Molars
- Tympanic bulla

Appendix 1.4. Skull of (a) the sheep and (b) the beaver in ventral views.

Images © Dino Pulerà reproduced by kind permission of Elsevier Inc.

magnum through which the spinal cord passes to the brain. On the ventral (underside) of the skull, the flat region between the teeth is the hard palate. At its posterior end is the median internal opening of the nasal passage, termed the choana. Farther back in the cat are the paired, oval, and strongly convex tympanic bullae. The bulla forms the floor of the middle ear. It is especially well developed in carnivorans, but not all mammals (the sheep, for example) have such a large osseous bulla; and some do not have a complete bony covering for the ear.

Nearly all the bones of the skull are paired. Those forming most of the roof are, from front to back, the nasals, frontals, and parietals. A small interparietal is wedged between the posterior ends of the parietals. Usually in older individuals the interparietal fuses to the surrounding bones. The sagittal crest is usually a sharp, raised structure on the skull roof, but may be blunted and represented only by a rugose ridge. The occipital bone forms the occiput, the skull's posterior surface, and contributes to the basicranium (or base of the cranium). Its most notable features are the foramen magnum and the occipital condyles, which articulate with the first vertebra and support the head on the neck. On either side is a blunt jugular process, which serves as a site for muscular attachment.

It is clear from comparing the figures of the skull that all of the bones (and their parts) so far described, though differing in shape, form, and proportions, are similar in position and relationship to one another. We may, however, begin to note more significant differences at the front of the skull: the premaxillae are fairly small bones that help form the hard palate and into which, in the cat and beaver, are implanted the front teeth or the upper incisors. In the cat, there are three incisor teeth in each premaxilla, and they are small and rather conical. In the beaver, there is only a single incisor, and it is a large, curved tooth. The sheep lacks upper incisors altogether. The functional implications of these differences will be discussed below. The maxillae are larger bones that hold the rest of the teeth and form most of the hard palate. In the cat, the large, curved canine is the first tooth in the maxilla, and the following three teeth are premolars. Phylogenetically, they represent premolars 2–4 (P2–P4; Appendix Box 1 provides an explanation of tooth designations). Cats lost P1 during their evolution and a gap, the diastema, is present between the canine and the small, peglike P2. P3 is a larger, triangular tooth that is followed by the long, bladelike P4. Note the small, final tooth, molar 1 (or M1), set transversely and covered in lateral view by the back end of P4. The bladelike P4 is the main meat-shearing tooth, and referred to as the carnassial. It meets the bladelike m1, the carnassial of the lower jaw. The teeth of the other mammals are considerably different. The beaver lacks the canine and the diastema is much longer. The sheep also lacks a canine so that all its teeth are cheek teeth. In both the beaver and sheep the cheek teeth are broad rather than sharp-edged as in the cat. The cheek teeth of the sheep include three premolars and three molars, and those of the beaver one premolar and three molars. In these mammals, the cheek teeth are all fairly similar to each other: the premolars have become molarized (i.e., have evolved a molarlike appearance). In the beaver and sheep, the occlusal surfaces of the cheek teeth have multiple crestlike ridges. Such morphology is also present in horses and cows, and molarization is especially prominent in horses. One might wonder, given that they all look similar, how we can tell which is a premolar and which is a molar (Appendix Box 1). In addition to the molarized cheek teeth, horses may have a tiny P1 (referred to as the wolf tooth) in front of the cheek teeth. As well, teeth may vary by sex. For example, the canines are present in male horses, but in females, they are absent or rudimentary.

The condition of the dentition in the vast majority of mammals, as might be surmised from the preceding descriptions, is termed heterodonty; i.e., the teeth are specialized according to their position in the oral cavity, with the ones farther back tending to be more complex. Homodonty describes the condition where the teeth tend to be generally similar morphologically and functionally, such as the typically simple, peglike teeth of armadillos (and other vertebrates including most reptiles). The teeth of nearly all mammals are composed of three substances: dentine forms most of the tooth, with a layer of cementum covering the roots and enamel covering the crown (in some cement is also present on the crown). Enamel is not only the hardest part of the tooth, but also of the skeleton. As well, most of the teeth (Appendix Box 1) are usually replaced once, a condition termed diphyodonty, in contrast to monophyodonty, the condition in a few mammals (including nearly all xenarthrans; see Appendix 2), in which there is only one set of (usually ever-growing) teeth (many nonmammals, such as reptiles, are characterized by polyphyodonty, in which teeth are replaced

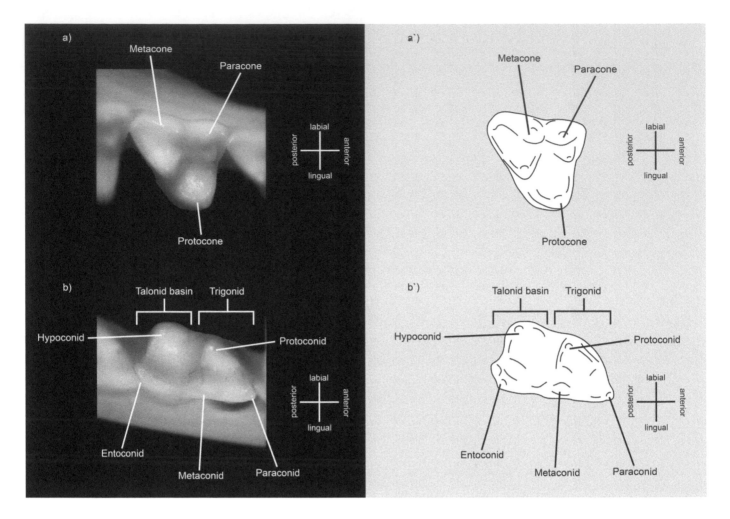

a)
Metacone
Paracone
labial
posterior   anterior
lingual
Protocone

a')
Metacone
Paracone
labial
posterior   anterior
lingual
Protocone

b)
Talonid basin   Trigonid
Hypoconid
Protoconid
labial
posterior   anterior
lingual
Entoconid
Metaconid   Paraconid

b')
Talonid basin   Trigonid
Hypoconid
Protoconid
labial
posterior   anterior
lingual
Entoconid
Metaconid   Paraconid

more or less continuously as needed). In terms of morphology, the ancestral pattern typical of mammalian molars is referred to as tribosphenic (Appendix Fig. 1.5), in which three main cusps (bumpy or pointed projections) are borne by the occlusal surface of upper and lower teeth. This arrangement is termed a trigon in the upper teeth, and a trigonid in the lower teeth. The latter also have a lower flatter surface behind the trigonid termed the talonid (or heel). Numerous specializations have evolved from this basic tribosphenic pattern, from simplification, such as the laterally compressed, bladelike teeth of felids, to increased complexity, with an increase in the number of cusps that may have crests joining them together to form more or less continuous structures termed lophs (or lophids in the lower teeth) as in the teeth of horses and elephants.

Appendix Box 1

How to differentiate teeth

We begin by considering a common nomenclature, followed by designations for the position of teeth (there is more than one system in use). The teeth are referred to as incisors, canines, premolar, and molars. The upper teeth are abbreviated with uppercase letters I, C, P, and M, and the lower teeth with lowercase letters i, c, p, and m. These letters are followed by a number indicating the position of each type of tooth. The ancestral pattern for adult placental mammals, in each half of both the upper and lower jaws, is three incisors, one canine, four premolars, and three molars (44 total teeth). Thus, the third upper incisor is written I3; we begin counting at the midline of the jaw and work our way toward the back teeth. The complete dental formula is given as follows: I3/i3, C1/c1, P4/p4, M3/m3. In marsupials, however, the ancestral pattern is I5/i4, C1/c1, P3/p3, M4/m4. It is important to remember that these represent ancestral patterns; the number of each type of tooth varies, and each species has its particular dental formula, although many mammals may share a particular formula. For example, in the evolution of humans, our dentition has been reduced (i.e., we have lost several of the teeth present ancestrally) to I2/i2, C1/c1, P2/p2, M3/m3. This gives an indication of the number of teeth but not which of the ancestral teeth have been lost or retained. Figuring this out requires knowledge of the phylogenetic history of the species under discussion, but specialists can usually decipher this fairly readily, at least with respect to dentition. In humans, the two incisors correspond to the first two, so we have lost ancestral I3/i3, and the premolars correspond to P3–4/p3–4, so we have lost the first two ancestral premolars.

The names of the different types of teeth require further explanation. The incisors are those teeth in the upper jaw implanted in the premaxilla bone; incisive bone is an alternative name. The canine is the first tooth in the maxilla (or on the suture between the premaxilla and maxilla). The lower incisors are those that occlude against the upper incisors; the lower canine occludes against the upper canine. Ancestrally, the teeth are arranged so that a lower tooth occludes slightly in front of the corresponding upper tooth, except in the case of the incisor teeth. Thus, the lower canine bites in front of the upper canine (the back surface of the lower occludes against the front surface of the upper—this is easily observable in a pet like a cat or dog).

The canines are followed by the cheek teeth, either premolars or molars. Ancestrally, and in many mammals still, the premolars tend to be less complex than the molars—that is, they are morphologically simpler, being smaller and having fewer bumps and ridges. However, the morphological pattern is not the distinguishing criterion. Rather, whether a tooth is a premolar or molar depends on whether the tooth has a juvenile precursor tooth; and here we must discuss the dentition of juveniles. In placental mammals, a juvenile erupts a set of teeth that lasts until a particular age. The teeth erupt sequentially as a wave, starting as a rule at the front and continuing toward the back. Each of these teeth, when its time comes, is shed, essentially in sequential order, and replaced by an adult tooth. The juvenile teeth are termed deciduous (because they fall out, as do the leaves of a deciduous tree) or milk teeth, and the adult teeth are termed permanent or adult teeth. To designate juvenile teeth, paleontologists usually use an abbreviation such as dI1 or dp4 (i.e., deciduous upper incisor three or deciduous lower premolar four). All of the juvenile teeth up to and including the last premolars are replaced: juvenile incisor one is replaced by permanent incisor one, and so on, until the final juvenile premolar is replaced by the final permanent premolar. Molars, however, are not replaced. By the age when the last few premolars are beginning to be replaced, the first molars are starting to erupt, and the latter end their eruption sequence near the age that a mammal reaches its adult size, when the oral cavity has grown

Appendix 1.5. The teeth of vertebrates, particularly of mammals, play an important role in phylogenetic and functional studies. The usual condition in most mammals is heterodonty, i.e., the specialization of teeth according to their function and position in the oral cavity. Thus, with variations between placentals and marsupials, there are usually are two to five incisors at the front in each half jaw, followed by a canine, and then several cheek teeth. The occlusal surfaces of teeth tend to become increasingly more complex in posterior teeth, particularly of the molars, the more distal cheek teeth. The ancestral arrangement of cusps and other features in marsupials and placentals is traditionally referred to as the tribosphenic molar, a term derived from the Greek *tribos* ("friction, rubbing, wearing") and *sphenos* ("wedge"), in reference to the essentially triangular (wedge-shaped) main elements of the tooth that are arranged to produce wearing and grinding between their occluding surfaces. The main cusps, conelike protuberances of the occlusal surface, form, as noted, a triangular configuration. In the upper tooth, the cusps are termed the protocone, paracone, and metacone, outlining the triangular trigon (additional intermediate cusps may be present but are not labeled). Comparable features of the lower teeth are denoted by the suffix -id, and so the main cusps are the protoconid, paraconid, and metaconid, outlining the trigonid. The surface of the tooth outlined by the main cups is basined. Lower teeth also have a prominent talonid or heel (though the term is derived from the Latin *talus*, "ankle"), a lower portion extending posteriorly from the trigonid. The talonid is ringed by several cusps (such as the hypoconid and enotconid) that enclose another basin (the upper teeth have a talon, but it is a much smaller extension posterior to the protocone). The orientation of the trigon and trigonid are reversed, so that the protocone is lingual and the protoconid is labial. In occlusion the triangles formed by the three principle cusps are wedged between each other, with the protocone occluding against the talonid basin.

Illustration by Sebastián Tambusso.

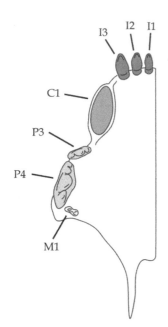

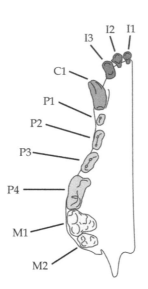

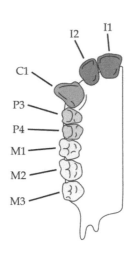

to become sufficiently large to accommodate them. In humans, the last of the molars erupt near the time when we reach physical maturity, so they are called wisdom teeth. In a sense, the molars may be viewed as part of the first (or juvenile) eruption wave; they follow in the sequence established for the deciduous teeth but are not later shed. This is what distinguishes premolars from molars; the latter do not have juvenile precursors.

Continuing on with the remaining bones, which are also reasonably similar among mammals, we may note that, in side view, the maxilla contributes to the rostrum and the edge of the orbit, where it contacts the delicate lacrimal bone, which contains the lacrimal foramen. The infraorbital canal is the large passage at the front end of the zygomatic arch. Just in front of the orbit on the wall of the rostrum, the beaver has a concave surface or fossa that serves for the attachment of part of the masseter musculature, as described below. The palatine bones complete the hard palate. Posteriorly each palatine forms, together with the pterygoid bone, a thin, vertical plate of bone that helps form the pterygoid blades or flanges.

In ventral view, the posterior end of the palatine contacts a complex, composite element, the sphenoid, consisting of various ossifications. The main part, or body, is the basisphenoid. The pterygoid processes of the basisphenoid are the portions that contact, on each side, the palatine and form the posterior end of the pterygoid blades. The narrow, hooklike posterior projection at the end of each pterygoid blade is the hamulus, to which a palatal muscle is attached. In the beaver this region is modified and two processes, the medial being particularly large, are present. The main part of the basisphenoid helps form the basicranium, which underlies the brain. The lateral, winglike portion of the basisphenoid is the alisphenoid. The presphenoid is the elongated and narrow bone lying between the pterygoid blades. The presphenoid also has a lateral extension, the orbitosphenoid, exposed in the ventral orbital wall. The connection between the presphenoid and orbitosphenoid is concealed by the palatine. Several large openings or foramina (sing., foramen) pierce the skull just above the pterygoid blades. These conduct nerves and vessels between the cranial cavity and structures, such as muscles, external to the skull. In the cat there are four foramina, termed, from front to back, the optic canal (mainly for the optic nerve passing from the eyeball to the brain), orbital fissure, foramen rotundum, and foramen ovale. The pattern of these openings varies; for example, in the sheep there are three, as the orbital fissure and foramen rotundum merge to form the foramen orbito-rotundum.

The temporal bone consists of three components that are fused together: the squamous, petrous, and tympanic portions. The squamosals are the large, flat bones that contribute to the lateral wall of the braincase. Ventrally the squamosal sends out a projection that forms the posterior part of the zygomatic arch. The petrous portion of the temporal includes the petrosal and mastoid bones. The petrosal contains the inner ear and is best observed in a sectioned skull. The tiny bones or ossicles of the middle ear are the malleus, incus, and stapes. The mastoid bone is the only part of the petrous portion that is exposed externally and its ventral portion forms the mastoid process, which is small in the cat but can be large (and you can feel it protruding just behind your ear.) The hyoid apparatus attaches to this region of the skull.

The tympanic portion of the temporal includes the rounded, oval tympanic bulla, which is formed from two ossifications. The ectotympanic mainly forms the ring of bone (to which the tympanum or ear drum attaches) surrounding the external auditory meatus, but contributes to the bulla ventrolaterally. The rest of the bulla is formed by the endotympanic. As already noted, not all mammals have an ossified floor to the middle ear.

The mandible or lower jaw is formed on each side by a single bone, the dentary (Appendix Fig. 1.6). Left and right dentaries articulate anteriorly at the mandibular symphysis, and in some mammals are solidly fused together. As with the skull, although there is a strong similarity among the several mandibles, there are also easily noticeable differences on the form and proportions of its various parts. These differences will be noted below as well, once the mandibles are described. The horizontal part of the dentary, in which the teeth are implanted, is the body. The cat has, as in the upper jaws, three incisors on each side, which are followed by a canine and then the cheek teeth, including p3, p4, and the elongated, bladelike m1. The m1 is the carnassial of the lower jaw, so the carnassial pair is P4/m1. The lower dentition of the sheep and beaver closely resemble the upper dentition in number and form (the beaver's lower incisor is smaller), with the major difference that there are lower incisors and a canine in the sheep. These teeth are all similar, so that the canine has become incisiform. Although not illustrated, the dentition of the cow is similar to that of the sheep, whereas the horse has incisors in both the upper and lower jaw. As noted above, a rudimentary p1 is also generally present in horses. In addition, male horses have lower canines.

The posterior part of the dentary is the ascending ramus, which has three processes. The coronoid process is the largest, extending dorsally. The

Appendix Box 1.1. The upper dentition of (a) a sabertooth, (b) a dog, and (c) a chimpanzee, showing variation in tooth number, type, and form. Images not to scale. Abbreviations: C, canine; I, incisor; M, molar; P, premolar. In the sabertooth, a highly carnivorous mammal, the dentition behind the canine is strongly reduced, with P4 very elongated and bladelike. In the dog, a less strictly carnivorous mammal but one that can chew bones, the cheek teeth are largely retained. The P4 is less markedly elongated, and the molars behind it are large teeth suitable for crushing bones. In the chimpanzee, some reduction of the dentition has occurred, but the cheek teeth form a battery of flattened teeth that are suitable for a more generalized, omnivorous diet. Chimpanzees primarily eat vegetation but can (and do) also eat meat.

Illustration by Sebastián Tambusso.

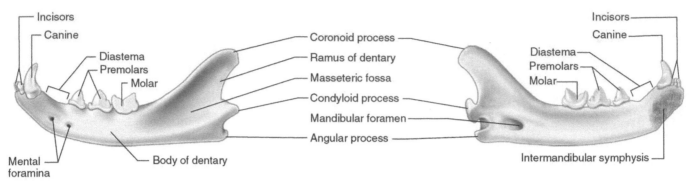

Incisors
Canine
Diastema
Premolars
Molar

Coronoid process
Ramus of dentary
Masseteric fossa
Condyloid process
Mandibular foramen
Angular process

Incisors
Canine
Diastema
Premolars
Molar

Mental
foramina

Body of dentary

Intermandibular symphysis

(a) Left dentary, lateral view

(b) Left dentary, medial view

Appendix 1.6. Lower jaw of the cat in (a) lateral and (b) medial views.

Images © Dino Pulerà reproduced by kind permission of Elsevier Inc.

temporal muscle inserts mainly on its dorsolateral and medial surfaces. The masseteric fossa, the large, triangular depression on the lateral surface of the coronoid process, serves as part of the insertion area for the masseteric musculature. The transversely expanded condyloid process, forms the lower half of the temporomandibular (or jaw) joint. The angular process is the projection of the posteroventral part of the dentary. Its medial surface serves as the insertion site for the lateral pterygoid muscle and its lateral surface for part of the masseteric musculature.

Form versus function

We may readily appreciate the differences in the form of the several skulls described in terms of their function; indeed, most of them are due to adaptations reflecting differences in dietary habits. This is not surprising, given the skull's importance in procuring and processing food. The cat represents a carnivore (and a strict one at that), and many of its features reflect its carnivorous habits. Of the main jaw-closing muscles, for example, the larger size of the temporalis muscle over the masseter is mirrored by the relatively large temporal fossa, as does the relatively large coronoid process (on which the temporalis mainly inserts) compared to the smaller angular process (on which part of the masseter muscle inserts) in the mandible.

The sheep and beaver represent different herbivore types, but several of their features reflect the processing of vegetation (more precisely, of food with certain physical properties as opposed to all kinds of vegetation, as other plant eaters — primates, for instance — have other mechanical requirements and the feeding apparatus is different from that typical of the rodent and the ruminant artiodactyl described here). In herbivores the temporalis muscle is generally smaller than the masseter, and the temporal fossa is likewise relatively smaller; conversely, the angular process is larger than the coronoid processes. Also, the condyloid process of the mandible is higher in the herbivores, a position reflecting a more dorsal position of the jaw joint. The effect of this difference is that in the cat the joint is approximately level with the tooth row, so the teeth can be engaged sequentially, in a scissor-like manner, which is ideal for cutting and slicing. In the herbivores, the position of the joint allows the cheek teeth to be engaged almost simultaneously, thus deploying the long grinding and crushing batteries suited for processing vegetation.

The form of the cheek teeth is also important. The cat's teeth are mainly narrow with a single sharp edge, a suitable mechanism for cutting and slicing flesh and tendons as the teeth pass adjacent to each other in a scissorlike manner. Not all carnivorous mammals have such strictly blade-like teeth. A less restricted diet is reflected by the dentition of canids (dogs and their relatives), which retain a sharp-edged carnassial pair but also have crushing molars; and the dentition of bears reflects an omnivorous diet.

The cheek teeth in herbivores tend to be morphologically similar to each other and molarized, with relatively wide and complex occlusal surfaces bearing several sharp-edged ridges or lophs (again, not all herbivores, such as primates, have this pattern). The lophs are composed of enamel, which is the hardest substance forming a tooth, whereas dentine lies between the lophs. The dentine, being softer, wears more rapidly, thus leaving depressions between the lophs. As the lophs move past each other, they act like rasps to grate food between them. The main difference between the beaver and sheep is that the ridges are oriented transversely in the former and anteroposteriorly in the latter, reflecting differences in jaw movement, but the overall rasping effect is similar (see ventral view in Appendix Fig. 1.2 and Appendix Fig. 1.4).

The need to process relatively coarse vegetation results in considerable wear and, to cope with this, herbivores have evolved tall (or hypsodont) teeth, whose sockets in the jaw bones (technically, alveoli) are deep, and thus the maxilla and mandible have become deeper to accommodate the taller teeth. This effect is especially evident in grazing mammals, such as sheep, cows, and horses; indeed, the orbits have been displaced posteriorly to accommodate the roots of the upper teeth. In some mammals (e.g., some rodents and rabbits) the roots of the cheek teeth remain open and the teeth are ever-growing (or hypselodont).

Also characteristic of herbivorous mammals is that the lower jaw has a considerable range of movement, either from side to side or backwards and forwards, enabling the lophs of the cheek teeth to move past each other. In contrast, carnivores have a mandible that tends to be restricted mainly to orthal (up and down) movements. The form of the jaw joint reflects these movements. The condyloid process of the cat is rounded and transversely widened, and it articulates with a concave and transversely wide mandibular fossa, a joint that restricts lateral movement. In the sheep, the condyle and fossa are nearly flat, allowing considerable side-to-side movement. The beaver, and rodents in general, have specialized mandibular movements: the lower jaw functions in two distinct positions. With the mandible in posterior position, the cheek teeth occlude to crush and grind, but the tips of the incisors do not meet. The mandible, however, may be pulled forward so the incisors can meet to gnaw and clip vegetation (and even to dig, in some cases), in which case the cheek teeth are not fully engaged. These possibilities are reflected in the form of the jaw joint: the condyloid process is rounded and the mandibular fossa is elongated, allowing for both transverse and propalinal (anteroposterior) jaw movements. The rodent dentition increases masticatory efficiency by having upper and lower tooth rows nearly equidistant from each other, whereas in the cat and sheep (and most other mammals) the distance between the lower tooth rows is narrower than in the upper, so that the upper and lower teeth of only one side can occlude at any one time. In the beaver, however, both side upper and lower teeth meet at the same time, so mastication occurs on both sides through anteroposterior movements of the jaw, which also explains its transversely oriented lophs.

At the front of the oral cavity, the incisors are used to obtain food, whether nipping prey or grasping vegetation. The cat possesses the full number of incisors for placental mammals and the canine is a prominent, pointed tooth. This makes good functional sense. Incisors are typically used to capture and manipulate prey (but may serve other functions, such as grooming), and the canines in particular to stab and pierce in subduing and killing prey. These actions require a quick biting movement from a relatively wide gape. The temporalis muscle, given its position, is most mechanically efficient for initiating such movements in producing a strong bite at the front of the jaws, thus explaining the tendency of a large temporalis in carnivores.

Herbivores have less need for a powerful temporalis, as they do not need to subdue and kill prey. Rather, the emphasis, particularly in grazing forms, is on constant mastication, mainly a function of the cheek teeth, for which the masseter muscle is ideally positioned. The postorbital bar noted in the sheep (present also in horses and cows) is likely an adaption for absorbing stresses during mastication. The upper incisors and canines are absent in the sheep, and this is generally true of ruminant artiodactyls. Functionally they are replaced by a hard pad of fibrous tissue with a thick, horny covering (the dental pad or plate), against which the lower teeth act. The males of some antlerless ruminants, such as the mouse and musk deer, retain a large canine for fighting and display.

The beaver (and rodents generally) has only a single pair of enlarged, curved, gnawing, and ever-growing incisors in the upper and lower jaws, the other incisors and canines having been lost during evolution. Enamel is present only on their anterior surfaces. As it is harder than the dentine posterior to it, the enamel wears more slowly, resulting in the self-sharpening, chiseled edge characteristic of rodents. These teeth, as mentioned earlier, can be engaged by pulling the lower jaw forward. To this end, rodents (except for the mountain beaver, *Aplodontia rufa*, which is not a true beaver) tend to modify the masseter musculature so that at least one of its subdivisions arises from the rostrum, in contrast to the typical mammalian condition in which the masseter arises mainly from the zygomatic arch.

An anterior attachment of part of the masseter produces a more nearly anteroposterior line of action, increasing the muscle's mechanical advantage in drawing the mandible forward. There are several patterns for such reorganization of the masseter. The beaver represents the sciuromorphous condition, in which the anterior part of the superficial masseter arises from the anterior surface of the zygomatic arch and the lateral surface of the rostrum—the depression that serves this origin was noted above. The other patterns are termed hystricomorphous and myomorphous, which were also noted in Chapter 5. In the hystricomorphous pattern (e.g., porcupines and guinea pig) the anterior part of the medial masseter arises from the anterior surface of the rostrum and extends through a usually greatly enlarged infraorbital foramen. In the myomorphous pattern (e.g., rats and

a) Skull of Bos taurus

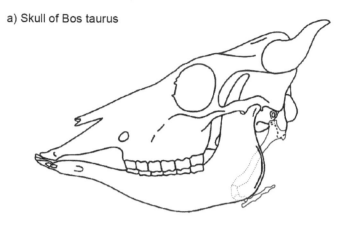

Appendix 1.7. (a) Skull of an artiodactyl showing the position of the hyoid skeleton. (b) Hyoid elements.

Illustration by Sebastián Tambusso, modified from Pérez et al. (2010).

b) Hyoid apparatus

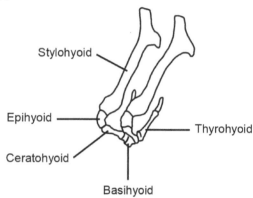

Stylohyoid

Epihyoid

Ceratohyoid

Thyrohyoid

Basihyoid

mice), the anterior part of the superficial masseter arises from a markedly enlarged anterior portion of the zygomatic arch, and the anterior part of the medial masseter arises from the rostrum and extends through a moderately enlarged infraorbital foramen.

Lastly, we may consider the size of the diastema, which tends to be more extensive in many (though not all) herbivorous mammals. A longer diastema may be a function of the longer snout and jaws, but the reason for this is not clear. Perhaps the diastema is functionally important in mastication by providing additional space for the tongue to manipulate food, although the longer snout may also reflect a need for a herbivore to select food with the front of the oral cavity without injuring its eyes. As well, separation of the incisors and the cheek teeth improves the specific function of both kinds of teeth: the sharper, cutting front teeth are relatively far from the pivot (the jaw joint), where speed is enhanced over force; and the crushing check teeth are closer to the pivot, where force in enhanced over speed. In rodents, the long snout also helps accommodate the extremely large incisors. These teeth extend far back into the skull and mandible, curving posteriorly through the rostrum. In the mandible, the incisor curves within the ventral part of the dentary and extends back beyond the cheek teeth.

The hyoid apparatus is composed of several small bones (Appendix Fig. 1.7). It sits in the throat at the base of the tongue and supports the tongue and muscles of the larynx (the voice box). It is composed of a median bar, the basihyoid. There are two pairs of horns or cornua that extend from the basihyal. The lesser cornua (sing., cornu) are the longer (in the cat and most mammals; the terminology, however, is based on the condition in humans), anterior pair, whereas the greater cornua are the smaller, posterior pair. Each lesser cornu consists of a chain of four ossicles that curves anteriorly and dorsally to attach to the skull. These are the ceratohyoid, epihyoid, stylohyoid, and tympanohyoid (the latter is typically a soft tissue structure not usually preserved in skeletal preparations). The greater cornu is formed on each side by the thyrohyoid, which articulates with the thyroid cartilage of the larynx. These elements are similarly arranged in most mammals. In large herbivores such as cows and horses that use the tongue to gather food, the basihyoid sends out a stout lingual process forward into the tongue for further support.

Postcranial skeleton

Vertebral column

The vertebral column is composed of a series of movable bones, or vertebrae (Appendix Fig. 1.1). The column is an important structure in support and locomotion and has evolved in mammals into five distinct regions. In the neck are the cervical vertebrae, which are followed by the rib-bearing thoracic vertebrae in the trunk. Next are the lumbar vertebrae, and then the sacral vertebrae, which are fused into a solid structure that articulates with the pelvis (or hip bones). Lastly, the caudal vertebrae support the tail. An opening, the intervertebral foramen, is present on each side between adjacent vertebrae for the passage of a spinal nerve.

Each vertebra generally possesses several parts (Appendix Fig. 1.8a,b). The centrum (plural, centra) or vertebral body, forms the main support of the vertebral column. The neural canal, through which the spinal cord passes, lies dorsal to the centrum. The neural canal is enclosed laterally and dorsally by the neural arch. Extending from the arch are several processes that serve largely for the attachment of musculature that control and stabilize the spine. A neural process projects dorsally from the neural arch and a transverse process projects laterally on either side. A vertebra makes contact with each preceding or succeeding vertebra at three points. One is at the centrum. Two other points of contact occur at the front and at the back of each vertebra via a pair of anterior and a pair of posterior projections. The former are termed prezygapophyses, while those at the back are the postzygapophyses. The prezygopophyses of vertebra articulate with the postzygapohyses of the vertebra in front of it.

The cat has seven cervical, or neck, vertebrae, as do most mammals. The first two, the atlas and axis, are specialized and markedly different from the remaining five as a result of their function in support and movement of the head. The atlas (named after the mythological giant Atlas, who was charged with supporting the sky on his shoulders) articulates in front with the occipital condyles of the skull and permit mainly up and down or nodding movements of the head on the neck. Posteriorly, the atlas articulates with the axis, and the joint between them allows mainly rotational movements of the head. The axis has an expanded neural process. The remaining cervical vertebrae each possess a relatively slender neural process that increases in height from the third cervical, in which the process is strongly reduced, to the seventh cervical.

a) Thoracic vertebra,
left lateral view

b) Thoracic vertebra,
anterior view

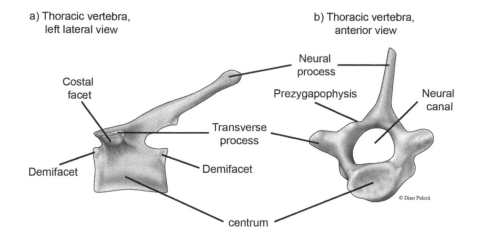

Costal facet

Neural process

Prezygapophysis

Neural canal

Transverse process

Demifacet

Demifacet

centrum

© Dino Pulerà

c) Sacrum, dorsal view

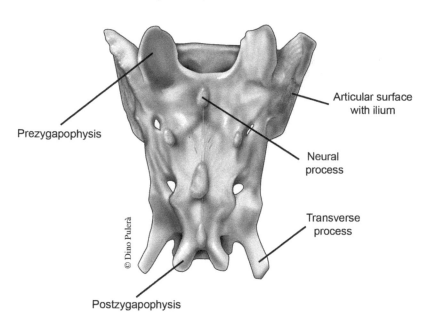

Prezygapophysis

Articular surface with ilium

Neural process

Transverse process

Postzygapophysis

© Dino Pulerà

Appendix 1.8. Vertebrae of the cat: thoracic vertebra in (a) left lateral (anterior or front toward the left) and (b) anterior views; (c) sacrum in dorsal (from the top) view. The sacrum, which articulates with the hips, is a fusion of three (in the cat) vertebrae. The individual vertebrae may be recognized by counting the neural processes.

Images © Dino Pulerà reproduced by kind permission of Elsevier Inc.

The cat usually has 13 thoracic vertebrae, those articulating with ribs, for which there are facets not present on other vertebrae. The capitulum (or head) of a rib typically sits between and articulates with two successive centra, for which most thoracic vertebrae possess demifacets. These are anteriorly and posteriorly paired facets, so that a vertebra typically has four demifacets. The demifacets of successive vertebrae form (on each side) a complete or "full" articular surface for the capitulum of a rib. Thoracic vertebrae also have a costal facet on the ventral surface of each transverse process for articulation with the tuberculum of the corresponding rib.

Cats usually have seven lumbar vertebrae, which are characterized by their large size and pleurapophyses, the elongated, bladelike processes that sweep anteroventrally. A pleurapophysis represents a fusion of a transverse process and an embryonic rib. The sacral vertebrae, of which the cat normally has three, are fused together to form the sacrum, which links the spine to the hips. Although they are fused together, several of its parts can be recognized, such as the neural processes and the pleurapophyses, which

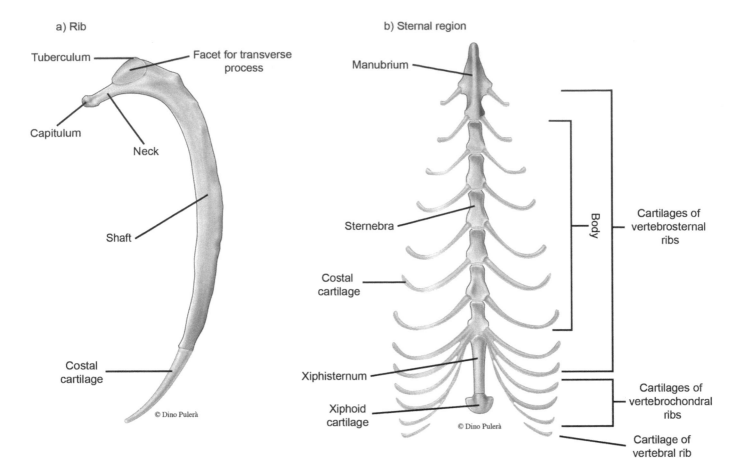

a) Rib

Tuberculum

Facet for transverse process

Capitulum

Neck

Shaft

Costal cartilage

© Dino Pulerà

b) Sternal region

Manubrium

Sternebra

Costal cartilage

Xiphisternum

Xiphoid cartilage

© Dino Pulerà

Body

Cartilages of vertebrosternal ribs

Cartilages of vertebrochondral ribs

Cartilage of vertebral rib

Appendix 1.9. Components of the rib cage of the cat: (a) a rib, with its costal cartilage; and (b) sternum and costal cartilages. The curved ribs and sternum contribute to formation of the rib cage (the thoracic vertebrae also participate; see Appendix 1.1), a bony structure that surrounds and protects organs and functions as well in ventilation of the lungs.

Images © Dino Pulerà reproduced by kind permission of Elsevier Inc.

are expanded anteriorly and posteriorly and fuse with each other (Appendix Fig. 1.8c). The caudal vertebrae, the smallest of the series, usually number between 21 and 23 in cats. The caudals tend to become progressively smaller and less complex posteriorly. The anterior caudals have zygapophyses, and neural and transverse processes, but the more posterior ones are elongated, cylindrical structures consisting almost entirely of the centrum. Many of the caudals bear hemal arches, small V-shaped bones on the underside, that enclose the hemal canal, through which caudal blood vessels pass. Among mammals the number of vertebrae is variable (but few depart from seven cervicals). For example, the horse has 18 thoracics, six lumbar, five sacral, and generally 18 caudals, the pig 14 or 15 thoracics, six or seven lumbar, four sacrals, and 20–23 caudals.

Ribs

The ribs (Appendix Figs. 1.1 and 1.9a) help form a strong but partly flexible cage that protects vital organs (e.g., heart and lungs) and participates in moving air into and out of the lungs. Each rib articulates dorsally with the vertebral column and ventrally with a costal cartilage. The ribs, of which there are 13 in the cat but usually 18 in the horse (the number matches the number of thoracic vertebrae) are subdivided into vertebrosternal, vertebrochondral, and vertebral, based on the attachment of their costal cartilages onto the sternum or breast bone. In the cat, the first nine ribs are vertebrosternal (or "true" ribs), meaning that the costal cartilages attach directly to the sternum. The next three are vertebrochondral (or "false" ribs) because their costal cartilages attach to the costal cartilage of another rib; that is, they do not attach directly to the sternum. The last rib is a vertebral rib because its short cartilage does not gain access to the sternum. It is thus also known as a floating rib. A floating rib (typically the last) is usually but variably present in other mammals.

Ribs differ in length but are generally similar in being curved, slender, rodlike bones. The capitulum or head bears surfaces for articulation with the demifacets of the thoracic vertebrae. This is followed by a short, constricted neck and then by the tuberculum, which has a facet for articulation with the costal facet on the transverse process of a thoracic rib. A tuberculum is absent in the final two or three ribs.

Sternum

The sternum consists of a series of individual elements (Appendix Fig. 1.9b), the sternebrae, of which there are usually eight in cats, as well as horses and dogs, but 7 in cows and 6 in pigs. The front-most sternebra is the manubrium. In the cat, the next six sternebrae form the body of the sternum. The last is the elongated and tapering xiphisternum, and the xiphoid cartilage attaches to its free end. The costal cartilages of the vertebrosternal ribs attach directly to the sternum.

Forelimb

The forelimb is attached to the rest of the body by the scapula or shoulder blade (Appendix Fig. 1.10). The forelimb consists of proximal, middle, and distal segments: the brachium (arm), antebrachium (forearm), and manus (hand or front paw). The scapula is a flattened, triangular bone, with its lateral (or external) surface bearing a prominent scapular spine. Ventrally the spine ends in the acromion process. The latter articulates with the clavicle or collar bone in mammals that possess a well-developed clavicle, such as humans but not cats or horses (in which it is entirely absent). The other end of the clavicle, when well developed (hence not in the cat or horse), articulates with the sternum. Just dorsal to the acromion process is the posteriorly projecting metacromion process. The spine rises prominently and separates the supraspinous fossa from the infraspinous fossa, which are both fairly smooth surfaces and mainly contain muscles that help move the brachium. The glenoid fossa is the smooth, concave surface at the apex for articulation with the humerus, the bone of the brachium. The delicate coracoid process projects medially from the anterior margin of the glenoid fossa. The medial surface of the scapula bears the subscapular fossa, a relatively flat surface onto which other arm-moving musculature mainly attaches.

The humerus is the bone of the brachium (Appendix Fig. 1.11a,b, 1.13a,b). It articulates proximally with the scapula and distally with the radius and ulna, the bones of the antebrachium or forearm. The head of the humerus is the large, smooth and rounded surface that articulates with the glenoid fossa of the scapula. Lateral to the head is the greater tuberosity, whereas the lesser tuberosity lies medially. Anteriorly between the tuberosities is the deep bicipital groove, along which passes the tendon of the

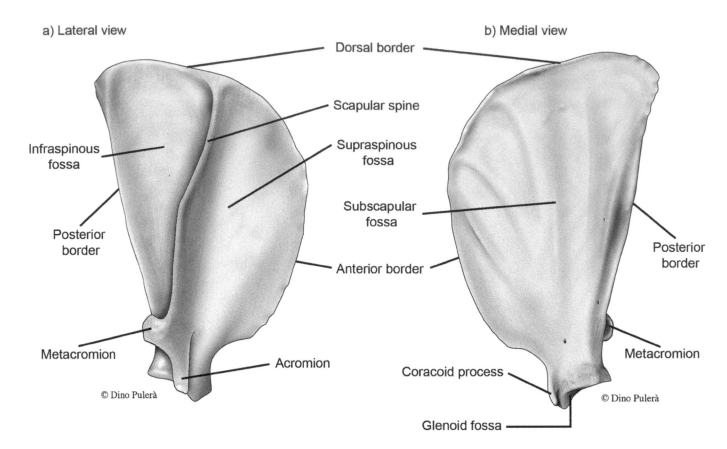

a) Lateral view

Dorsal border

b) Medial view

Scapular spine

Infraspinous
fossa

Supraspinous
fossa

Subscapular
fossa

Posterior
border

Posterior
border

Anterior border

Metacromion

Metacromion

Acromion

Coracoid process

© Dino Puerà

© Dino Puerà

Glenoid fossa

Appendix 1.10. The scapula (shoulder blade) of the cat in (a) lateral and (b) medial views.

Images © Dino Puerà reproduced by kind permission of Elsevier Inc.

biceps brachii muscle, which flexes the forearm. The pectoral ridge extends distally from the greater tuberosity on the anterior surface of the humeral shaft. The less prominent deltoid ridge extends distally and obliquely from the posterior part of the greater tuberosity. This ridge converges toward and meets the pectoral ridge about midway along the shaft. The features just described apply to the cat, as well as the dog, but the humerus of other mammals may differ. For example, in the horse and cow, the humerus is more massively built, with the head relatively posterior and the greater (particularly in the cow) and lesser tuberosities much more prominently developed, so much so that the head is obscured from anterior view (Appendix Fig. 1.11). Most notably, the deltoid ridge, especially in the horse, is greatly expanded into a crest that extends laterally away from the shaft. The bony protuberances just described are all sites of attachment of musculature and their (even relatively) increased size reflects a prominent musculature. For example, the deltoid ridge serves for the insertion of several muscles, such as the brachiocephlic, part of the superficial pectoral, deltoid, and teres minor, and the origin of part of the triceps. The latter has already been noted as a forearm extensor, whereas the others are involved in actions of the shoulder (such as flexing or extending and adducting or abducting the limb). The relevance of the increase in musculature is related to an investment in enhancing locomotion.

Distally, the humerus bears the spool-shaped condyle, which is actually composed of two articular surfaces. The smaller, lateral surface, the capitulum, articulates with the radius, while the larger, medial surface, the trochlea, articulates with the ulna. On the posterior surface, just proximal to the condyle, is the deep olecranon fossa that receives the olecranon process

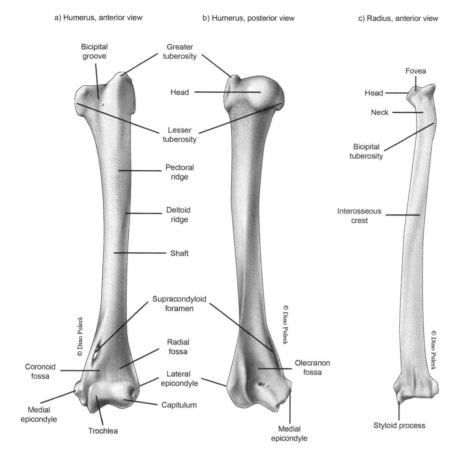

a) Humerus, anterior view   b) Humerus, posterior view   c) Radius, anterior view   d) Ulna, anterior view   e) Ulna, lateral view

of the ulna. The medial and lateral epicondyles are the rugose prominences that lie on either side of the trochlea. The oval passage lying proximal to the trochlea is the supracondyloid foramen.

The ulna is the longer of the two bones (Appendix Fig. 1.11c–e) in the forearm. It has a prominent proximal portion, but it tapers distally. Proximally the ulna articulates with the humerus and the radius. Distally it contacts the carpus or wrist. The trochlear notch is the deep semicircular surface for articulation with the trochlea of the humerus. The olecranon process extends proximal to the trochlear notch. It is the site of insertion of the tendon of the triceps brachii muscle, which extends the forearm, an action whose importance is further discussed in the main text (Chapter 7). The coronoid process extends anteriorly from the distal base of the trochlear notch. The radial notch, for articulation with the head of the radius, is a curved surface that lies laterally along the base of the trochlear notch. The ulna tapers distally and ends in the styloid process, which articulates with the lateral part of the carpus.

The radius is the other bone of the antebrachium. It is more slender proximally, but widens distally and articulates with the medial part of the carpus. As the radius articulates with the lateral part of the humerus and the medial part of the carpus, it crosses over the ulna when a mammal assumes a quadrupedal position (on all four limbs). In this position the manus is pronated (i.e., the palm faces the ground; the opposite position, with the palm facing up, is termed supinated). Proximally the radius consists of the head, which bears an oval, concave fovea that articulates with the capitulum of the humerus. The shape of the head allows the radius to rotate on

Appendix 1.11. Brachial (arm) and antebrachial (forearm) elements of the cat. Left humerus in (a) anterior and (b) posterior views; left radius in (c) anterior view; and left ulna in (d) anterior and (e) lateral views, respectively.

Images © Dino Pulerà reproduced by kind permission of Elsevier Inc.

a) Right manus, anterior view

b) Right pes, anterior view

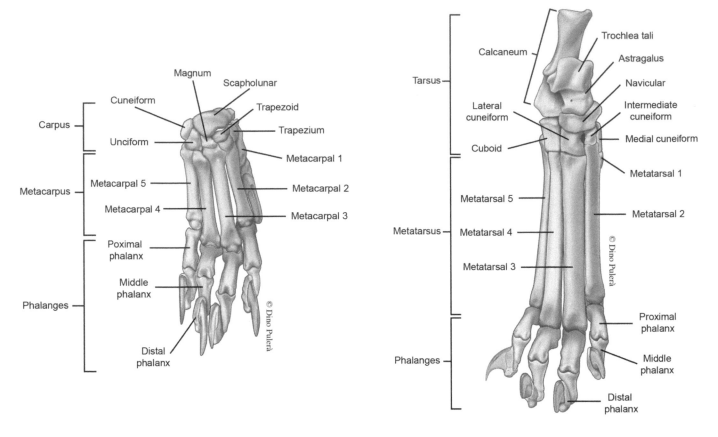

Appendix 1.12. The (a) right manus (hand; front paw) and (b) right pes (foot; hind paw) of the cat in anterior views.

Image © Dino Pulerà reproduced by kind permission of Elsevier Inc.

the capitulum. Immediately distal to the fovea, the head bears a smooth, narrow strip, the articular circumference, which articulates with the radial notch of the ulna. The neck of the radius is a short segment between the head and the bicipital tuberosity, onto which the tendon for the biceps brachii muscle inserts. The distal end of the radius has a small articular surface for the ulna medially that allows the radius to turn on the ulna. The distal surface has a large, concave surface that articulates with the scapholunar bone (in the cat) of the wrist. The short styloid process projects distally from the medial surface of the radius.

The manus (hand or front paw; Appendix Fig. 1.12a) includes the carpus (wrist), metacarpus, and phalanges in proximal to distal order. In the cat there are five digits (fingers), each of which is formed by a metacarpal and phalanges. The carpus consists of seven small, irregularly shaped bones arranged in two rows. In medial to lateral order, the proximal row includes the scapholunar and cuneiform, with the pisiform projecting from the ventral surface of the cuneiform. The distal row includes the trapezium, trapezoid, magnum, and unciform, which articulate distally with the metacarpals. The carpal and the metacarpal bones are arranged in an interlocking pattern that restricts motion. These bones are generally present in mammals, but some may display slightly different arrangements. For example, the scapholunar noted in the cat results from the fusion of two bones, the scaphoid and lunar, that remain separate in many other mammals. Fusion between other bones also occurs, and is usually a constant feature of a particular species.

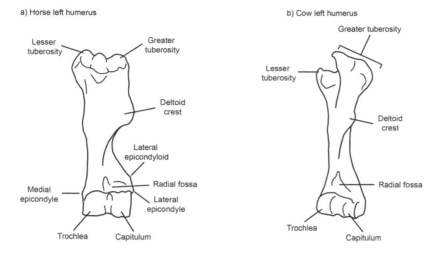

a) Horse left humerus

Lesser tuberosity

Greater tuberosity

Deltoid crest

Lateral epicondyloid

Medial epicondyle

Radial fossa

Lateral epicondyle

Trochlea

Capitulum

b) Cow left humerus

Greater tuberosity

Lesser tuberosity

Deltoid crest

Radial fossa

Trochlea

Capitulum

Appendix 1.13. Left forelimb elements of the horse and cow: humerus of (a) the horse and (b) the cow in anterior (from the front) views (modified from several sources); antebrachial (forearm) and manus (forefoot) elements of (c) the horse and (d) the cow in anterior views (modified from Getty, 1975, and France, 2009).

Illustrations by Sebastián Tambusso.

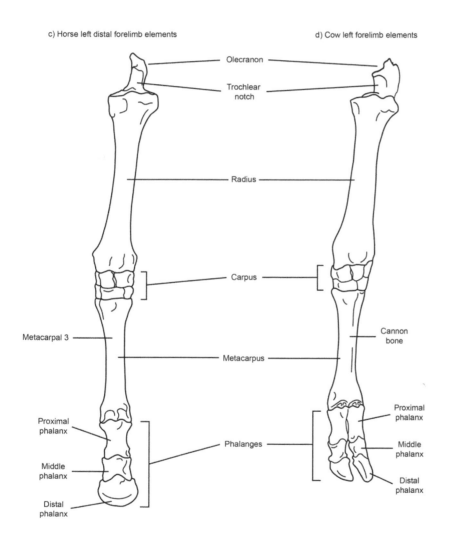

c) Horse left distal forelimb elements

d) Cow left forelimb elements

Olecranon

Trochlear notch

Radius

Carpus

Metacarpal 3

Cannon bone

Metacarpus

Proximal phalanx

Proximal phalanx

Phalanges

Middle phalanx

Middle phalanx

Distal phalanx

Distal phalanx

The five metacarpals are the bones in the palm of the manus. Metacarpal 1 is the most medial and shortest. Each metacarpal consists of a base proximally, a shaft, and a head distally. Digit 1 has a proximal phalanx and an ungual phalanx, which bears a large claw. Digits 2–5 each have proximal, intermediate, and ungual phalanges, with the latter also bearing large claws.

The descriptions above for the antebrachium and manus apply mainly to the cat. This pattern is repeated among many mammals, but other mammals differ considerably in anatomical details. In highly cursorial (or running) mammals, as is typical of grazing ungulates such as equids and ruminant artiodactyls, the limbs are modified to maximize speed. Among the changes are the reduction in the number of elements and the elongation of the distal limb segments (Appendix Fig. 1.13c,d). The first change reflects the need to restrict motion as much as possible to movement along the middle plane of a limb (that is, in its sagittal plane) to produce fore and aft motion—the less twisting among the limb elements, the better. The second change is related to increasing stride length: the longer the step, the more ground covered. The latter modification is addressed in connection with the hind limb, although the same effects are produced in the front limb.

With respect to limiting motion, we find that, in the antebrachium of the horse and cow, partial (though substantial) fusion occurs between the radius and ulna, which thus form a single function unit (Appendix Fig. 1.13c, d). This modification greatly restricts (and largely eliminates) rotation of the manus (it can't be supinated or pronated). In the horse, the ulna sits essentially behind radius and is strongly reduced: its shaft tapers abruptly, usually ending about midway along the radial shaft (the ulna's distal end usually persists, however, but fuses before birth to the end of the radius). Thus, it is the radius that articulates with the manus. In the cow the ulna, though reduced, is reasonably complete, but still largely fused to the radius. Though its distal end has a small articulation with the carpus, it is the radius through which all weight is borne and transferred to more distal elements.

In the manus of the horse, the typical carpals are present, but the digits, except for the middle or third, are markedly reduced or absent. The third metacarpal, generally termed the cannon bone by veterinarians, is a stout and elongated element. The second and fourth metacarpals are strongly reduced to distally tapered splints lying mainly against the posterior surface of the third, and the first and fifth metacarpals have been lost. Of the phalanges, only the third is retained. Its proximal phalanx is longer than the middle and distal, and the latter assumes a broadened hoof shape. In cows, and ruminant artiodactyls generally, the third and fourth metacarpals remain elongated and prominent, and are fused together to form a single element (also termed a cannon bone). A small remnant of the fifth metacarpal adheres to the proximal end of metacarpal 4, but does not make contact with the carpus. Metacarpals 1 and 2 are lost. Of the phalanges, the third and fourth are fully developed, each with the usual number of three phalanges. Only small portions of phalanges 2 and 4 are retained, and the first finger is entirely absent.

Hind limb

The pelvis or hip attaches the hind limb to the axial skeleton by attaching to the vertebral column. The hip consists of paired coxal bones that articulate with each other ventrally at the pelvic symphysis and with the sacrum of the vertebral column dorsally. Each coxal is composed largely from three bones, the ilium, ischium, and pubis, although a fourth center of ossification, the acetabular bone, makes a small contribution (Appendix Fig. 1.14).

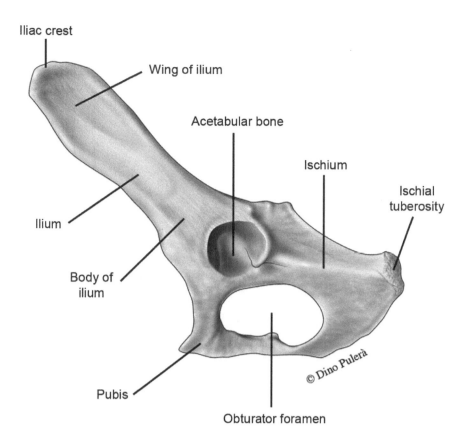

Iliac crest

Wing of ilium

Acetabular bone

Ischium

Ischial tuberosity

Ilium

Body of ilium

Pubis

Obturator foramen

© Dino Pulerà

Appendix 1.14. Left coxal bone of the cat in lateral view. The right and left coxal bones form the pelvis (or hip bones), which articulates with the sacrum and the hind limbs. Each coxal bone is a solid fusion of the three main bony elements: the ilium, ischium, and pubis.

Image © Dino Pulerà reproduced by kind permission of Elsevier Inc.

These bones are firmly fused together in the adult. The acetabulum is the deep socket that receives the head of the femur.

The ilium consists of a body, near the acetabulum, and an anterodorsally projecting wing. The iliac crest is the roughened, anterodorsal edge of the ilium. The ischium extends posteriorly from the acetabulum. Its enlarged termination is the ischial tuberosity. The pubis and the rest of the ischium are oriented ventromedially. The ischium and pubis of the each side of the body meet to form, respectively, the ischial and pubic symphyses, which together form the pelvic symphysis. The acetabular bone forms the thin, medial part of the acetabulum.

The femur is the bone of the thigh or proximal part of the hind limb (Appendix Fig. 1.15a,b). The hemispherical surface head of the femur fits into the acetabulum of the coxal. The neck, which supports the head, projects obliquely from the proximal end of the femur. The roughened greater trochanter lies laterally to the head and serves for attachment of hip musculature. The deep depression posteriorly between the trochanter and head is the trochanteric fossa. The lesser trochanter is on the posterior surface of the shaft, just distal to the head. The femur expands distally into the prominent and posteriorly projecting lateral and medial condyles. Each condyle bears a smooth, semicircular surface for articulation with the tibia. The intercondyloid fossa is the depression posteriorly between the condyles. The rugose areas for muscular attachment proximal to the condyles are the lateral and medial epicondyles. The patellar trochlea, for articulation with the patella, lies anteriorly between the condyles. It is a smooth, shallow trough oriented proximodistally. The patella, or kneecap, is a small, tear-shaped sesamoid bone (a bone that forms within the tendon of a muscle), with its apex directed distally. Its anterior surface is roughened.

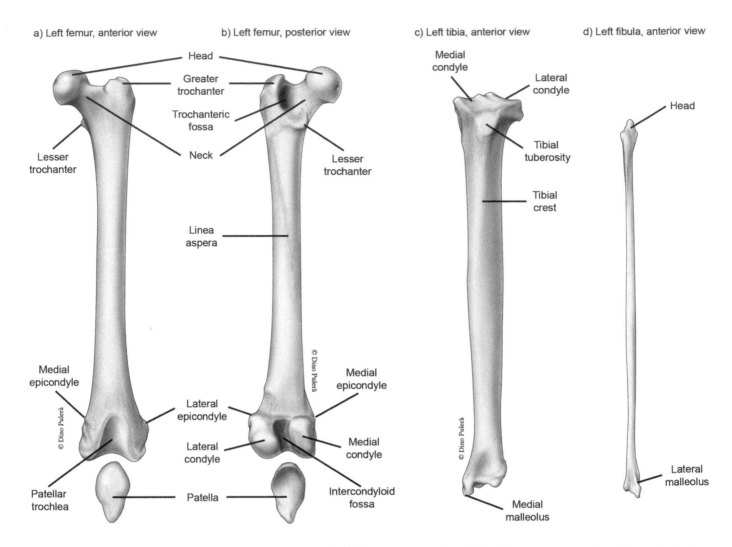

a) Left femur, anterior view
- Head
- Lesser trochanter
- Medial epicondyle
- Lateral epicondyle
- Lateral condyle
- Patellar trochlea

b) Left femur, posterior view
- Head
- Greater trochanter
- Trochanteric fossa
- Neck
- Lesser trochanter
- Linea aspera
- Medial epicondyle
- Medial condyle
- Patella
- Intercondyloid fossa

c) Left tibia, anterior view
- Medial condyle
- Lateral condyle
- Tibial tuberosity
- Tibial crest
- Medial malleolus

d) Left fibula, anterior view
- Head
- Lateral malleolus

© Dino Pulerà

Appendix 1.15. 1 Femoral (thigh) and crural (shin or lower leg) elements of the cat: left femur in (a) anterior and (b) posterior views; (c) left tibia in anterior view; (d) left fibula in anterior view.

Images © Dino Pulerà reproduced by kind permission of Elsevier Inc.

Appendix 1.16. Hind limb of a horse and a cow, anterior view. Left hind limb elements of the horse and cow: femur of (a) the horse and (b) the cow in anterior (from the front) views; crural (shin; lower leg) and pes (hind foot) elements of (c) the horse and (d) the cow in anterior views (modified from Getty, 1975, and France, 2009).

Illustrations by Sebastián Tambusso.

Posteriorly it bears a smooth, shallowly concave surface for articulation with the femur.

The next segment of the hind limb is termed the crus (that is, the region of the shin bone; Appendix Fig. 1.15c,d). The tibia is the larger and medial bone of the crus. Proximally, it bears lateral and medial condyles that articulate with the femur. The small, nearly oval facet for the head of the fibula is just distal to the lateral condyle. The popliteal notch lies between the condyles on the posterior surface of the tibia. A small muscle, the popliteus, lies in the notch and is a flexor of the knee joint. The tibial tuberosity, for insertion of the patellar ligament, lies anteriorly. The tibial crest continues distally from the tuberosity along the shaft.

The tibia has two articular surfaces distally. The large surface is the cochlea tibiae, which articulates with the astragalus, the tarsal bone with which the pes (or foot) largely articulates with the hind limb. This facet consists of two sulci separated by a median ridge, which restricts motion at the ankle largely to a fore and aft direction, producing flexion and extension. The small, nearly triangular facet for the fibula faces posterolaterally. The medial malleolus is the distal extension of the tibia's medial surface. It forms the medial protrusion of the ankle.

The fibula is the slender, lateral bone of the crus. Its head is expanded and bears a proximal facet that articulates with the tibia. The fibula widens

a) Horse left femur

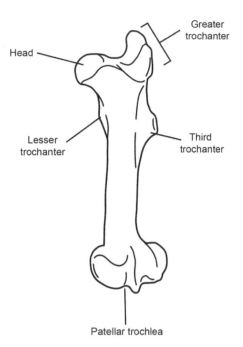

Head

Greater
trochanter

Lesser
trochanter

Third
trochanter

Patellar trochlea

b) Cow left femur

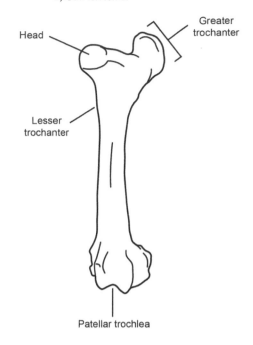

Head

Greater
trochanter

Lesser
trochanter

Patellar trochlea

c) Horse left distal hind limb elements

d) Cow left distal hind limb elements

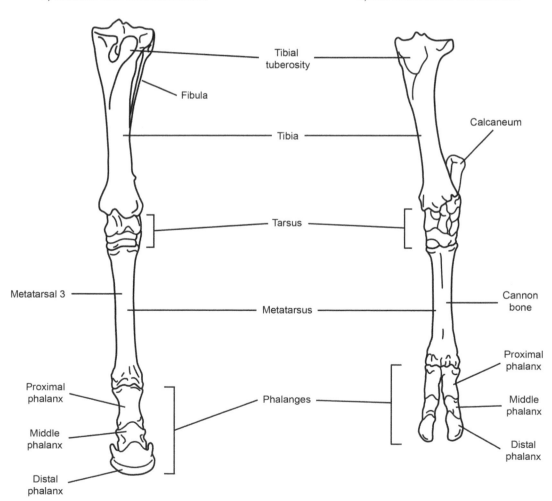

Tibial
tuberosity

Fibula

Tibia

Calcaneum

Tarsus

Metatarsal 3

Metatarsus

Cannon
bone

Proximal
phalanx

Phalanges

Proximal
phalanx

Middle
phalanx

Middle
phalanx

Distal
phalanx

Distal
phalanx

and bears two distal facets. The more proximal articulates with the tibia, whereas the distal facet articulates with the lateral part of the trochlea of the astragalus. The lateral malleolus projects distally from the end of the fibula.

The pes (or foot) includes of tarsals, metatarsals, and phalanges (Appendix Fig. 1.12b). Of the seven tarsals, the two most proximal, the astragalus and calcaneum, are larger than the others. The astragalus is the medial bone and articulates proximally with the tibia and fibula. The trochlea tali, the surface that articulates with the tibia, consists of medial and lateral keeled surfaces separated by a sulcus. The astragalus articulates ventrally with the calcaneum. The latter lies laterally, is about twice as long as the astragalus, and projects posteriorly as the heel. Distally the astragalus articulates with the navicular, and the calcaneum articulates with the cuboid. The navicular articulates distally with the lateral, intermediate, and medial cuneiform bones, and laterally with the cuboid. The articulations among the tarsals and metatarsals are arranged to produce interlocking joints that tend to restrict movement. Of the five metatarsals, the first is strongly reduced and articulates with the medial cuneiform. The phalanges for digit 1 have been lost during evolution in the cat. The other metatarsals are stout and elongated, each articulating with a series of three phalanges, the proximal, middle, and ungual phalanges.

We may note in the hind limb of horses and cows (Appendix Fig. 1.16) modifications similar to those discussed above for the forelimb. In the femur, the head is large, certainly, but the neck reduced, so that the head and shaft are more nearly aligned as a way to augment the effective length of the limb, which is useful in straight-ahead locomotion (Appendix Fig. 1.16a,b). The greater and lesser trochanters of the femur are greatly enlarged in both the horse and cow. These protuberances serve for the insertion of muscles such as the psoas major, iliacus, and parts of the gluteal (buttocks) musculature; these act as flexors or extensors of the hip joint and as adductors or abductors of the thigh, which fulfill requirements of changing direction during fast locomotion. A notable modification in the femur of the horse is the great development of a third trochanter, a large expansion along the lateral side of the femoral shaft. This structure serves for insertion of the gluteus maximus, which arises from the posterior surface of the ilium near or at the iliac crest and acts in extension and abduction of the femur with regard to the trunk of the body. When both right and left side muscles are contracted, abduction on one side is cancelled out by that on the other, and only the extension of the femora results. If the leg is fixed, the action of the muscles erects the body (as when a horse rears up), as occurs when horses fight or display strength. Indeed, horses are often depicted in just such a transitory (but rather prolonged) bipedal stance, as in a famous Italian car maker's symbol, the *cavallino rampante*. Cows and other bovines fight with their horns, and so need not be so agile in raising the body; and the third trochanter is predictably absent. It should be noted that rams do achieve a bipedal posture while fighting, as they first stand and let themselves fall back down in delivering a blow, with the head, to an opponent, but this is accomplished from a running rather than stationary position.

In the more distal parts of the limbs, the same pattern as in the forelimb may be observed (Appendix Fig. 1.16c,d). In the crus the fibula is much reduced, with its shaft tapering to a split that ends just past the midlength of

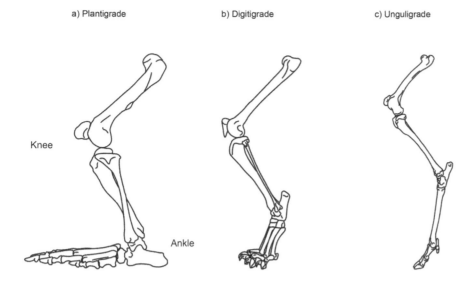

a) Plantigrade   b) Digitigrade   c) Unguligrade

Knee

Ankle

Appendix 1.17. The left hind limb of mammals in lateral view showing the posture of the pes (foot) and elongation of distal limb elements in (a) plantigrade, (b) digitigrade, and (c) unguligrade stances. The hind limb is illustrated here, but corresponding similarities are present in the forelimb as well. In plantigrady, the entire sole (plantar surface) of the pes comes into contact with the ground, and the three segments of the limb (femoral, crural, and pes) are of approximately equal length. In digitigrady, the posterior part of the pes is held off the ground, so that only the front parts of the fingers (or digits) strike the ground during locomotion. This effectively increases the length of the stride as well as reduces the amount of time the foot comes into contact with the ground, and both these factors contribute to increased speed. In unguligrady, raising of the pes is carried to an extreme, so only the distalmost parts of the fingers (i.e., the end of the toes, and so hooves, the equivalent of nails or claws; in the horse, only one toe is involved) strike the ground. The distal limb segments (crus and pes) are elongated relative to the proximal (i.e., femoral) segment. These features further increase cursorial (running) ability.

Illustrated by Sebastián Tambusso.

the tibia. The distal extremity is retained and fused to the tibia to form the lateral malleolus. The tarsus is followed by a much-reduced pes, at least in terms of number of bones. Metatarsal 3 is a long, stout bone (also termed the cannon bone), with metatarsals 2 and 4 retained as splints. Only phalanx 3, with the usual proximal, middle, and distal phalanges is present. In cows the fibula is even more reduced: the extremities remain, but the shaft has all but disappeared. The proximal extremity is fused to the tibia, while the distal remains as a small separate bone, forming the lateral malleolus, that articulates with the tibia. In the pes, metatarsals 3 and 4 fuse together (another cannon bone); a small nodular element, usually interpreted as the remnant of metatarsal 2 is also present. The phalanges are as in the manus.

As noted above, a second modification that occurs in cursorial mammals is elongation of the distal limb elements. The relative increase in length of the elements increases the length of the stride and thus speed. The degree of increase and position of the elements of the hind limb (modifications also occurring in the forelimb) may be appreciating in Appendix Fig. 1.17.

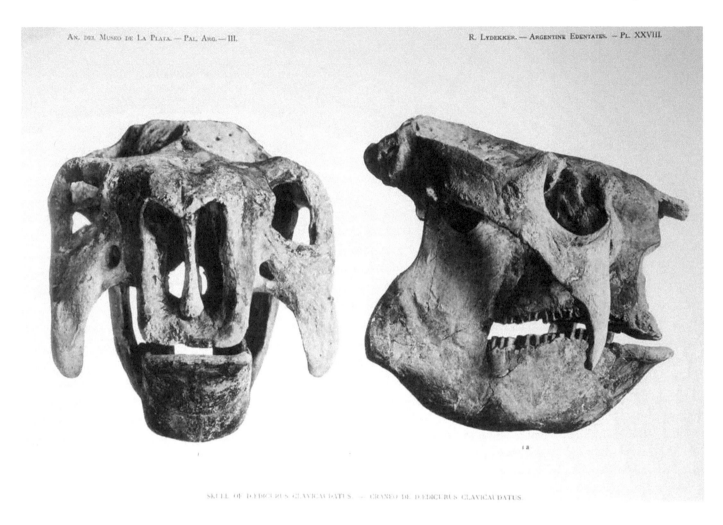

SKULL OF DOEDICURUS CLAVICAUDATUS. — CRANEO DE DOEDICURUS CLAVICAUDATUS.

Appendix 2.1. Skull and mandible of the glyptodont *Doedicurus.*

From Lydekker (1894).

The skeletal elements of xenarthrans reflect the clade's distinct character among mammals, provide support for uniting xenarthrans and appreciating their particular adaptations. McDonald (2003) provided a broad review of several of the distinctive skeletal features of xenarthrans. Given the importance of such features for our discussions on fossil (as well as living) forms, we discuss their presence or absence or degree of similarity to features present in other mammals.

Three major skull shapes may be recognized among living Xenarthra, reflecting its three main groups: armadillo-like, slothlike, and anteater-like. These three are easily recognizable among fossil xenarthrans, despite the vastly increased taxonomic diversity over extant forms. Clearly, xenarthran skulls share a close resemblance in bone number and pattern with the skulls of other mammals, except that of glyptodonts (Appendix Fig. 2.1), in which the skull has undergone a telescoping process to accommodate its center of mass near the neck joint (Fariña, 1985, 1988; Fariña and Vizcaíno, 2001).

Among the elements to consider is the septomaxilla (or os nariale). Although a small bone, its absence has been viewed as support for a more derived status among mammals. The septomaxilla is a small bone associated with the nasal region, particularly in basal tetrapods, and there has thus been much discussion over the presence in xenarthrans of a septomaxilla (which is present in monotremes but absent in other mammals). Based on an examination of fetal specimens of *Tamandua* and *Choloepus*, Zeller et al. (1993) concluded that the xenarthran septomaxilla was homologous with the central part of the septomaxilla of monotremes and several Mesozoic (that is, truly ancient) mammals, although they noted that in armadillos it has two additional components. On the other hand, Wible and Gaudin (2004) noted that it is probably not homologous with that element. Clearly, additional work is needed to fully evaluate the significance of this structure.

Among the features generally considered characteristics of xenarthrans is the reduction and loss of the zygomatic arch, which is regarded as part of a functional complex usually related to feeding upon social insects that also evolved in several other, though not closely related, mammalian groups. Indeed, absence of the arch in vermilinguas and its reduction in pangolins was among the evidence used to support a relationship between these xenarthrans and pangolins. However, among xenarthrans an incomplete or reduced zygomatic arch only occurs in pilosans (i.e., anteaters and many sloths), while in all of the cingulates (armadillos, pampatheres, and glyptodonts) the zygomatic arch is not only complete, but often strongly developed. In glyptodonts (Appendix Fig. 2.1), for instance, the zygomatic (bone in this case; Appendix 1) bears a strong ventral extension or flange

Cranial skeleton

that projects below level of the tooth row (similar projections occur in sloths, despite incompleteness of the arch). As the earliest xenarthrans were armadillo-like, and thus probably insectivorous, the argument that loss or reduction of the arch necessarily reflects such dietary habits is not at all firmly established. Further, an incomplete arch is present among the earliest known sloths (and is thus probably the ancestral state) and some authors have considered this evidence that the sloth ancestor was insectivorous, even though the earliest sloths have been interpreted as herbivores. A secondarily complete arch (i.e., in which the zygomatic bone and zygomatic process of the temporal bone come into contact) has evolved independently in some fossil sloths.

In the ear region, the ectotympanic is usually incomplete or semicircular (horseshoelike), a feature that has also been considered a primitive mammalian character (Van der Klaauw, 1931), although in some taxa this bone forms a circular ring. The floor of the middle ear is not usually covered by an ossified tympanic bulla.

Hyoid apparataus

In the hyoid apparatus, several of the elements that are usually independent in other mammals (Appendix 1) may be fused in xenarthrans. One of the most characteristic and widespread of these is the fusion of the thyrohyoid elements to the basihyoid to form a single "V-bone" (Pérez et al., 2010). Besides this fusion, many other xenarthrans have the typical form mentioned in Appendix 1, as noted by several authors (e.g., Pérez et al., 2010; De Iuliis et al., 2011), but in some cases fusion is even more extensive, as in *Megatherium americanum* and glyptodonts. In the latter, several of the elements (stylohyoid, epihyoid, and probably ceratohyoid) fuse to produce a rodlike element termed the sygmohyoid (or sygmohyal; Pérez et al., 2010). Other extreme modifications are present in vermilinguas: in Mymecophaga, the hyoid apparatus extends almost directly posterior to rather than beneath the skull, to which it is attached by a long and slender muscle. This arrangement permits the great mobility required to allow the extensive movements of this anteater's long and sticky tongue (Naples, 1999).

The teeth of vertebrates, particularly of mammals, play an important role in phylogenetic and functional studies (besides of, course, being important to the animals themselves). The vast majority of mammals display a dentition that is heterodont and diphyodont, with teeth formed by enamel, dentine, and cementum (Appendix 1).

Dentition

The dentition of xenarthrans, however, deviates in several ways from this general mammalian pattern. Indeed, a highly derived dentition was already present in the earliest members of the group. In nearly all xenarthrans the anterior teeth are strongly reduced or absent (with anteaters having no teeth at all) and the cheek teeth tend to be morphologically similar to each other. Indeed, in glyptodonts, pampatheres, and armadillos the cheek teeth are so homodont, i.e., so similar to each other, that it is not possible to distinguish premolars from molars. Although there is some variation in the cheek teeth of sloths, there is no clear and consistent distinction between teeth. The same may be said of other mammals, such as horses and cows, but recall that the criterion for distinguishing between premolars and molars is not form but replacement. The difference between xenarthrans and nearly all other mammals in this regard is that xenarthrans (with the sole exception of Dasypus and its extinct relative Propraopus) are monophyodont—they do not replace their teeth (i.e., they lack a deciduous dentition). Instead, their teeth have an open pulp cavity and grow continuously (juvenile teeth are initially conical but with continued growth and wear become parallel sided).

This means that we cannot establish correspondence (i.e., homology) between the teeth in xenarthrans and those in other mammals: the front teeth (implanted in the premaxilla in other mammals) are largely absent, and we cannot be sure that any of them represent the incisors of other mammals; among cheek teeth, the lack of replacement (combined with morphological similarity) means that we cannot determine which might represent the premolars and molars of other mammals. The lack of clear homologies between the teeth of xenarthrans and other mammals means that the terminology applied to the latter is unsuitable for the former and, at least in those forms—mainly sloths—where morphological differentiation is present, the teeth are instead referred to as incisiforms, caniniforms, and molariforms (Pujos and De Iuliis, 2007). In some groups such as the sloths the number of teeth is reduced to a maximum of five upper and four lower teeth while in the armadillo Priodontes the number of teeth ranges between 17 and 20.

Another characteristic feature of xenarthran teeth is the nearly universal lack of enamel (Vizcaíno, 2009). Enamel was reported for the early armadillo Utaetus by Simpson (1932), and its presence has been confirmed by Kaltoff (2011). As well, Martin (1916) noted that both juvenile and adult teeth of the living genus Dasypus feature a thin covering of enamel that is quickly worn away. Teeth are thus formed from different types of dentine, typically an inner softer core surrounded by a harder layer. The teeth do not usually bear any cusps, at least none that can be considered based on the ancestral tribosphenic pattern of mammals (Vizcaíno, 2009). In some sloths, cusplike eminences and ridges have been recognized on the occlusal surface of teeth, but this too has required a specialized terminology (Bargo et al., 2009; Pujos et al., 2011).

Vertebral column

Postcranial skeleton

As noted in Appendix 1, the vertebrae are usually independent elements in mammals, forming the vertebral column, the main axial support structure of the body. Xenarthrans display distinctive variations in the condition of these skeletal elements. Cingulates (armadillos, pampatheres and glyptodonts), for example, display fusion of cervical vertebrae, those of the neck region. Most commonly, fusion involves the axis, the second cervical element, and several vertebrae behind it to form a cervical tube. The only known departures from this condition are Utaetus (the earliest dasypodid for which the vertebral column is known), in which the axis is free, and the late Miocene or early Pliocene pampathere Vassallia, in which there

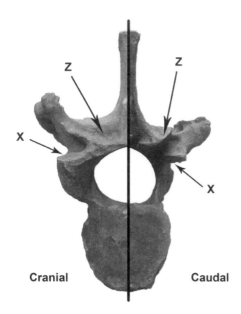

Cranial   **Caudal**

Appendix 2.2. Anterior (left side) and posterior (right side) views of a dorsal vertebra of the ground sloth *Lestodon*. The articular surfaces typical of mammals are indicated by z (that is, the corresponding facets of sequential vertebrae articulate here), whereas the additional (or xenarthrous) articular surfaces of xenarthrans are indicated by x.

Photos: Sebastián Tambusso.

appear to be fewer elements involved in the tube (De Iuliis and Edmund, 2002). On the other hand, sloths lack fusion of the cervical vertebrae. However, modern tree sloths depart from the standard mammalian number of seven cervical vertebrae, with *Bradypus* increasing the number to between eight to ten and *Choloepus* reducing it to five or six (Buchholtz and Stepien, 2009). All extinct ground sloths have seven cervicals.

Also, sloths reduce the number of lumbar vertebrae, in the lower back, to three (although Scott, 1903–1904, noted that four could be present in *Hapalops*), the lowest number among xenarthrans. Among other peculiarities, the centra in the lumbar and most of the thoracic vertebrae of both the living and extinct sloths are perforated by a pair of foramina that extended from the ventral surface to the base of the neural canal (de Burlet, 1922), which reflects one of the several peculiarities of the circulatory system in sloths.

The vertebrae also provide us with the one consistent morphological feature that has been used to unite fossil and living sloths, anteaters, armadillos and pampatheres into a single clade: the presence of additional or supplementary or xenarthral (meaning "strange articulations or joints") articulations (Appendix Fig. 2.2) between adjacent vertebrae in the more posterior presacral vertebrae (Gaudin, 1999b). Vaughan et al. (2000) noted these articulations supplement the typical articulations between the pre- and postzygapophyses (Appendix 1; Appendix Fig. 2.3). Presumably these additional articulations were also present in the juveniles, at least, of glyptodonts but the fusion of most of the thoracic vertebrae into a single tube and incorporation of the lumbar vertebrae into the synsacrum (a structure in which the sacrum is extended by incorporation of additional fused caudal or lumbar vertebrae) has precluded their recognition in this clade. Only one other placental mammals displays xenarthrous articulations, the Hero shrew *Scutisorex somereni*, but they are clearly distinct from those of xenarthrans (McDonald, 2003)

Glyptodonts achieve the greatest degree of fusion of among vertebral elements of any mammal. In addition to the fusion of the cervicals there is a "trivertebral" element formed by the second, third and fourth thoracic vertebrae (Gillette and Ray, 1981). The "trivertebral" element is followed by a dorsal tube, which is formed by the fusion of nine vertebrae. Glyptodonts depart not only from other placentals but all of the other xenarthrans with the fusion of thoracics and lumbar vertebrate into solid tubes. The lumbar vertebrae are incorporated into the synsacrum (Appendix Fig. 2.4). Finally, it is common in glyptodonts that the terminal caudal vertebrae are fused to each other and with the exoskeleton, forming a caudal tube that, in some species, reach very large size (Appendix Fig. 2.5).

Thoracic cage

The thoracic cage includes the thoracic vertebrae, ribs, and the sternum, or breastbone, which form a protective basket around the vital organs such as the lungs and heart, and participate in the mechanics of ventilating the lungs (by functioning to help expand and contract the thoracic cavity). This skeletal structure displays several more features distinctive to xenarthrans.

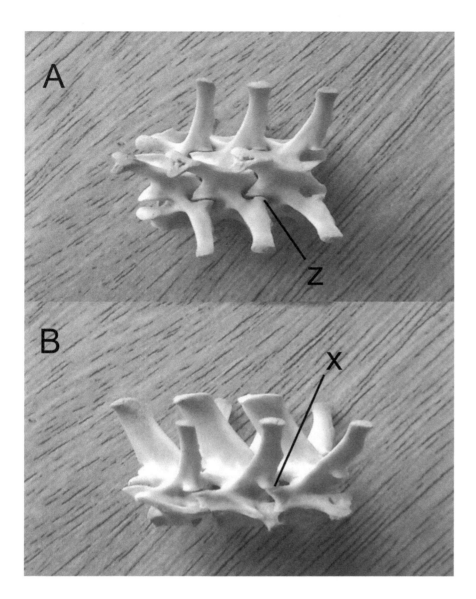

Appendix 2.3. The lumbar vertebrae of the armadillo *Dasypus* in (A) dorsal and (B) lateral views; anterior toward the right. The typical articulations, labeled z, are better seen in dorsal view (A). These articulations occur between facets borne by the zygapophyses. The xenarthrous articulations, denoted x, are observable in lateral view (B), which also indicates one of the typical articulations.

The ribs are curved and elongated bony, rodlike structures that articulate with the vertebrae. Specifically, these types of ribs are known as dorsal (or costal) ribs. In most mammals the tuberculum of the rib is a convex structure that articulates with a corresponding concave articular facet on the transverse process of a thoracic vertebra (Appendix 1). Sloth ribs can be readily distinguished in that the tuberculum is a concave surface and the articular facet on the transverse process of anterior thoracic vertebrae is a prominent convex structure.

Among vermilinguans, particularly *Cyclopes*, the ribs are expanded, resulting in an even more solid and extensive thoracic cage. Functionally, expanded ribs may increase the stability of the thorax, which, in turn, increases the stability of the vertebral column. This specialization in *Tamandua tetradactyla* (the lesser or collared anteater) and *Myrmecophaga tridactyla* (the giant anteater) probably relates to their fossorial habits; the trunk is stabilized during fossorial activity by xenarthrous processes in the lumbar region and by expanded ribs and robust intercostal muscles in the thorax. In strictly arboreal forms such as *Choloepus didactylus* (the two-toed sloth) and *Arctocebus calabarensis* (an African primate), lumbothoracic stability may be employed in methodical arboreal locomotion. *C. didactylus*,

Appendix 2.4. Pelvis of a reconstructed skeleton of the ground sloth *Lestodon* showing the synsacrum (reconstructed by Gustavo Lecuona). In the typical mammalian condition (see Appendix 1.8c), articulation between the hip bones and the vertebral column is essentially limited to the ilium of the coxal bone and the sacral vertebrae. The latter fuse to form a sacrum. In xenarthrans, fusion is more extensive and involves additional elements. Usually the transverse processes of the more posterior sacral and more anterior caudal vertebrae fuse to the ischium. This more complex and extensive vertebral fusion of elements is termed the synsacrum.

Photo courtesy of Ada Czerwonogora.

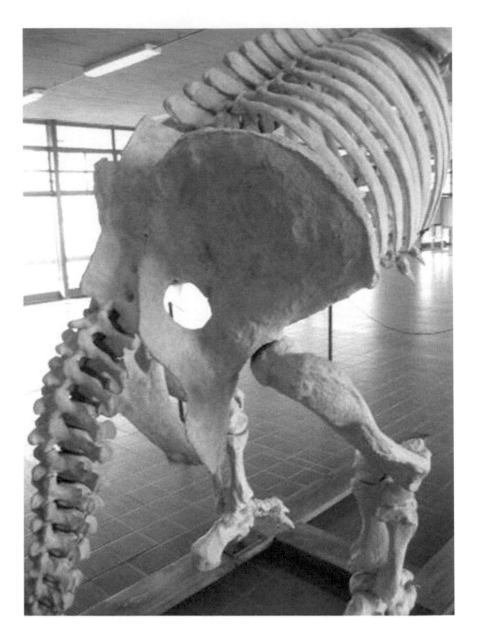

at least, is able to anchor itself with hind feet and tail and extend its trunk and forelimbs to reach an adjacent branch, a feat requiring unusual lumbothoracic stability.

A rib is connected to the sternum by another element (except for the most posterior ribs, which are attached only to the vertebrae). In all placental mammals this element is usually a cartilaginous structure, termed a costal cartilage. In xenarthrans these elements are strongly ossified as sternal ribs and the sternum has a prominent ventral portion that bears articular facets for articulation with the sternal ribs. None of the other mammals possess these sternal ribs with synovial articulations to the sternum. Also, there are often well-developed articular facets between adjacent sternal ribs. The costal cartilage may become ossified in many other mammals, but the presence of ossified sternal ribs that articulate both with the sternebrae and to each other via prominent articular surfaces are unique features of xenarthrans. This combination of features is as distinctive as the xenarthrous vertebrae (McDonald, 2003).

SKELETON AND CAUDAL TUBE OF PANOCHTHUS TUBERCULATUS. — ESQUELETO Y TUBO CAUDAL DE PANOCHTHUS TUBERCULATUS

Dermal armor

Dermal ossicles or osteoderms are common in many reptiles, including turtles and particularly tortoises where they may occur on the limbs, lizards such as the gila monster where they form a mosaic on the top of the skull, and crocodilians with their dorsal osteoderms. Xenarthra is the only mammalian clade whose members have evolved dermal armor, either as a solid carapace (Cingulata) or a mosaic of ossicles embedded in the skin (a few Cingulata and some Pilosa). The number of movable bands varies in the armadillos from one to eleven while the number of movable bands in pampatheres is three in all genera. The glyptodonts differ from the other cingulates in having a nearly solid carapace with few and, more typically, no movable bands. All of the cingulates have a cephalic shield of osteoderms on the skull. As well, the tail may be sheathed, particularly in glyptodonts, which generally have a variable number of moveable rings and a series of fused elements to form a caudal tube (Appendix Fig. 2.5).

Scapula

In all xenarthrans there is a secondary scapular spine along the posterior margin of the scapular blade that acts as the area of origin for two muscles, the teres minor (used during the retraction of the humerus) laterally and subscapularis minor medially. The spine is prominent in ground sloths but there is no posterior extension of the fossa as in the armadillos and giant anteater, as illustrated in Appendix Fig. 2.6. Its prominence in armadillos and the giant anteater, as well as in several other digging mammals, suggests

Appendix 2.5. A mounted skeleton of *Panochthus* with the carapace removed, revealing the condition of the vertebral column. As noted in the text, many of the individual vertebrae are fused to form a series of long tubular segments. In order of front to back, the cervical tube, just behind the skull, is followed by a free cervical vertebra (the sixth) and then by the trivertebral bone, formed by the posterior cervical and first two thoracic vertebrae. The remaining thoracics, usually numbering 10, are united into a compound dorsal tube. The lumbar tube is incorporated into the synsacrum. There are only two joints between the segments of the trunk region. In the tail, the vertebrae remain largely unfused, continuing within the caudal tube. Total length of skeleton is approximately 300 cm.

From Lydekker (1894).

Appendix 2.6. Scapula in the ground sloth *Lestodon*, as seen (a) in the skeleton and (b) in detail. In all xenarthrans, there is a secondary scapular spine along the posterior margin of the scapular blade acting as the area of origin for two muscles, the teres minor (used during the retraction of the humerus) laterally and subscapularis minor medially. The spine is prominent in ground sloths, but there is no posterior extension of the fossa as in armadillos: (c), as in this *Priodontes* from the exhibition at the Museo de la Plata.

Photograph by Sergio F. Vizcaíno by kind permission of Museo de La Plata, Argentina.

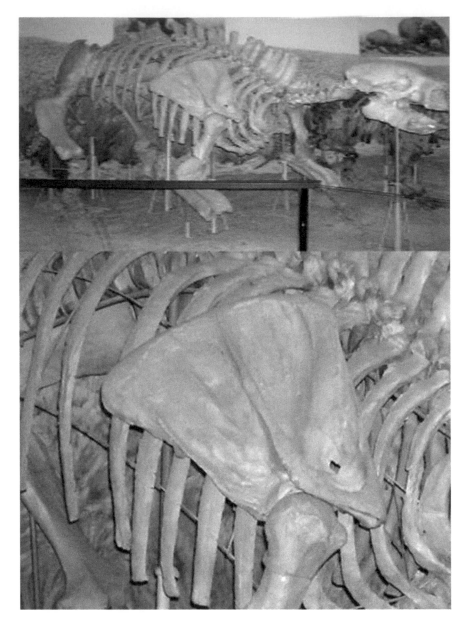

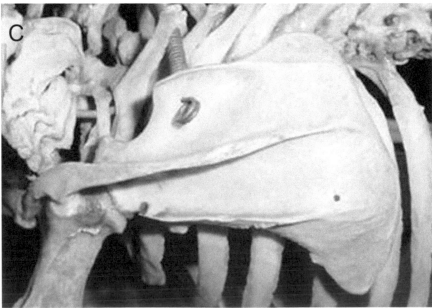

that its development is an adaptation related to digging. A feature characteristic of pilosans is the presence of a bony bar, the acromiocoracoid bridge, connecting the acromion and coracoid processes of the scapula (Appendix 1) that encloses an opening known as the coracoid foramen. The bridge is formed by a separate center of ossification considered by McDonald (2003) to be the procoracoid. Another feature related to digging habits is the long olecranon process present in the ulna of the armadillos. This enhances the leverage of the triceps muscle, which extends the forearm, yielding a powerful, force-optimizing movement able to comminute the soil (Vizcaíno et al., 1999).

Pelvis

Another of the distinctive features of the xenarthran skeleton is the fusion of the transverse processes of the posterior sacral vertebrae to the ischium and the inclusion of a variable number of the proximal caudal vertebrae to form a synsacrum (Appendix Fig. 2.4). As previously mentioned the lumbar vertebrae in glyptodonts fuse together into a single structure referred to as the lumbar tube, also incorporated into the synsacrum. This is interpreted as having a mechanical role because the whole structure becomes a single piece (Frechkop, 1950).

Femur

Xenarthrans generally possess a third trochanter (Appendix Fig. 2.7; Appendix 1), although it varies in its degree of development and may be secondarily lost. It is best developed in the cingulates. In armadillos and pampatheres it is a prominent structure, comparable in development to that observed in horses, positioned at midshaft, while in glyptodonts it is shifted distally, which improves the leverage of the gluteus maximus (Fariña, 1995). In sloths, the prominence of the third trochanter is obscured by an unusual (for mammals) expansion of the lateral margin of the femur, and it is absent in tree sloths.

Tibia, fibula, and calcaneum

The tibia and fibula are coossified in all of the cingulates, with fusion occuring both proximally and distally. This condition is more variable in sloths. For example, it may fuse at both ends, only proximally, or at neither end. Articulation between the fibula and the calcaneum (or heel bone), a primitive character, has been lost in all xenarthrans and the fibula articulates with the astragalus, or ankle bone, in a broad area of contact on its lateral side. In the two genera of tree sloths the distal end of the fibula is turned medially at a right angle to the shaft. The distal end is peglike and fits into a depression on the side of the astragalus (Appendix Fig. 2.8).

Hind foot

The hind foot, or pes, of xenarthrans displays a morphological range that not only parallels that of other mammals but also includes an arrangement

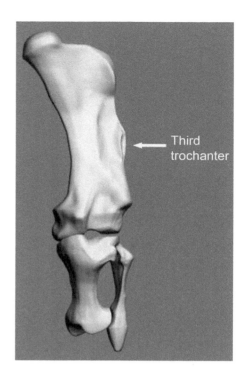

Third trochanter

Appendix 2.7. Femur (top bone) and tibia and fibula (bottom bones) of *Lestodon*. A third trochanter is generally present in all xenarthrans but varies in its degree of development; it may be secondarily lost. It is best developed in the cingulates. In armadillos and pampatheres, it is a prominent structure positioned at midshaft, while in glyptodonts it is shifted distally. In sloths, the third trochanter is not observed to be so prominent because of an unusual expansion of the lateral margin.

3D computer reconstruction by Sebastián Tambusso.

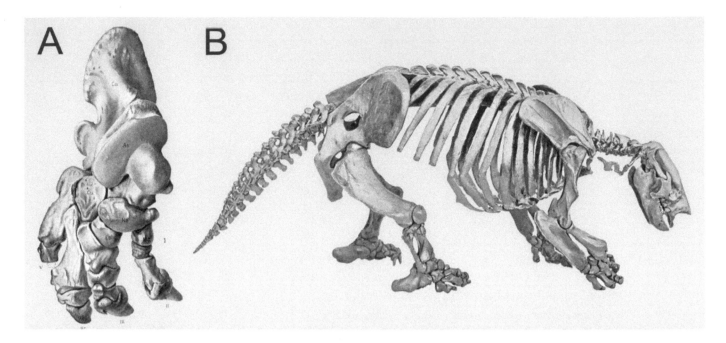

Appendix 2.8. *Paramylodon harlani.* The pes, isolated in (A), is shown in anatomically dorsal view. In some ground sloths, the pes is turned so that the sole faced almost entirely medially or inward, and the weight of the animal was borne by the outer margin, mainly digit 5 (V) and the heel or calcaneum (Ca). The astragalus (As), a large ankle bone, has a well-developed odontoid process, its smaller, peglike portion. Total length of pes is approximately 43 cm. (B) In the mount of the skeleton, the inwardly turned pes, a condition termed pedolateral, is clearly evident. Also, note that the manus is turned inward, so that weight rested mainly on the knuckles. It is likely that these positions allowed the large claws to be tucked in and protected during locomotion.

Images from Stock (1925).

that is unique to the clade. The pes can be plantigrade (anteaters), the manus and pes can be digitigrade (some armadillos). The larger anteaters support themselves on the dorsal surface of the unguals of the manus but they do not knuckle-walk in the manner of gorillas. In the larger glyptodonts, the manus and pes are graviportal (Vizcaíno et al., 2011a), a condition that describes in recent animals slow movement due to massiveness, which parallels features seen in proboscideans.

One of the truly distinctive character complexes unique to some of the extinct ground sloths is the rotation of the hind foot. As a result of this rotation the morphological sole of the foot faces medially and only the fifth metatarsal and the calcaneum contact the ground, an arrangement called pedolateral. This arrangement is prominent in several (nothrotheriids, scelidotheriines, and mylodontines), but not all sloths. As may be observed in the astragalus figured here (Fig. 2.8A), the medial trochlea of the astragalus forms a peglike odontoid process that fits into a complementary rounded socket on the medial side of the distal articular surface of the tibia is formed.

Palmar sesamoids

The presence of a large palmar sesamoid or extra bones derived from ossifications in the tendon of the flexor digitorum profundum is sporadic in xenarthrans, as in the North American glyptodont Glyptotherium (Gillette and Ray, 1981), the pampathere Pampatherium (Winge, 1915) and in different ground sloth clades (Winge, 1915; Stock, 1925; Cabrera, 1929a; McDonald, 1987; De Iuliis et al., 2011). While functionally it serves to increase the leverage of the tendon, its presence is not necessarily indicative of strictly fossorial habits because glyptodonts are not considered to have been active diggers and many sloths such as Megalonyx are interpreted as browsers. A similar sesamoid is present in the tendon of the flexor digitorum profundus of the pes. It has been described in some armadillos, such as Dasypus and in some ground sloths such as the scelidotheres and mylodontines.

Appendix 3. Equations Used to Estimate Body Masses Based on Dental and Skeletal Measurements and Their Respective Sources

Measurement	Equation	Source
Sum of humerus + femur circumference (H9 + F8)	mass = 0.000084 (H9 + F8)$^{2.73}$	Anderson et al. (1985)
Humerus length (H1)	log mass = $3.4026 \times$ log H1 -2.3707	Scott (1990)
Humerus length (H2)	log mass = $3.3951 \times$ log H2 -2.513	Scott (1990)
Condylar width (H3)	log mass = $2.7146 \times$ log H3 $+ 0.2594$	Scott (1990)
Trochlear width (H4)	log mass = $2.4815 \times$ log H4 $+ 0.4516$	Scott (1990)
Distal width (H5)	log mass = $2.5752 \times$ log H5 $+ 0.2863$	Scott (1990)
Transverse diameter (H7)	log mass = $2.485 \times$ log H7 $+ 1.0934$	Scott (1990)
Anteropost diameter (H8)	log mass = $2.4937 \times$ log H8 $+ 0.876$	Scott (1990)
Radius length (R1)	log mass = $2.8455 \times$ log R1 $- 1.8223$	Scott (1990)
Distal articular surface width (R2)	log mass = $2.5894 \times$ log R6 $+ 0.9092$	Scott (1990)
Distal articular surface height (R3)	log mass = $2.5894 \times$ log R6 $+ 0.9092$	Scott (1990)
Distal width (R4)	log mass = $2.5894 \times$ log R6 $+ 0.9092$	Scott (1990)
Maximum width (R5)	log mass = $2.5894 \times$ log R6 $+ 0.9092$	Scott (1990)
Transverse diameter (R6)	log mass = $2.5894 \times$ log R6 $+ 0.9092$	Scott (1990)
Anteroposterior diameter (R7)	log mass = $2.5038 \times$ log R7 $+ 1.4661$	Scott (1990)
Ulnar length (U1)	log mass = $2.9762 \times$ log U1 $- 2.3087$	Scott (1990)
Femur length (F1)	log mass = $3.4855 \times$ log F1 $- 2.9112$	Scott (1990)
Femur length (F2)	log mass = $2.6886 \times$ log F2 $- 0.2471$	Scott (1990)
Third trochanter, distal end (F3)	log mass = $2.9405 \times$ log F3 $- 0.087$	Scott (1990)
Trochlear width (F5)	log mass = $2.782 \times$ log F5 $- 0.0107$	Scott (1990)
Transverse diameter (F6)	log mass = $2.821 \times$ log F6 $+ 0.9062$	Scott (1990)
Anteropost diameter (F7)	log mass = $2.6016 \times$ log F7 $+ 0.9119$	Scott (1990)
Tibia length (T1)	log mass = $3.5808 \times$ log T1 $- 3.1732$	Scott (1990)
Proximal width (T2)	log mass = $2.8491 \times$ log T2 $- 0.2495$	Scott (1990)
Proximal anteroposterio diameter (T3)	log mass = $3.1568 \times$ log T3 $+ 0.137$	Scott (1990)
Distal transverse width (T4)	log mass = $2.6075 \times$ log T4 $+ 0.4247$	Scott (1990)
Distal anteroposterior width (T5)	log mass = $2.8949 \times$ log T5 $+ 0.642$	Scott (1990)
Transverse diameter (T6)	log mass = $2.7382 \times$ log T6 $+ 0.8761$	Scott (1990)
Anteropost diameter (T7)	log mass = $2.906 \times$ log T7 $+ 0.9909$	Scott (1990)
Occipital height (och)	log mass = log OCH $\times 2.783 - 0.42$	Janis (1990)
Basicranial length (bcl)	log mass = log BCL $\times 3.137 - 1.062$	Janis (1990)
Maseteric fossa length (mfl)	log mass = log MFL $\times 2.95 - 1.289$	Janis (1990)
Palatal width (paw)	log mass = log PAW $\times 3.27 - 0.196$	Janis (1990)
Muzzle width (mzw)	log mass = log MZW $\times 2.313 + 0.64$	Janis (1990)
Posterior skull length (psl)	log mass = log PSL $\times 2.758 - 0.973$	Janis (1990)
Mandibular angle height (dma)	log mass = log DMA $\times 2.448 - 0.331$	Janis (1990)
Posterior mandibular length (pjl)	log mass = log PJL $\times 2.412 + 0.031$	Janis (1990)
Width mandibular angle (wma)	log mass = log WMA $\times 2.803 - 0.352$	Janis (1990)
Lower molar row length (lmrl)	log mass = log LMRL $\times 3.265 - 0.536$	Janis (1990)
Lower premolar row length (lprl)	log mass = log LPRL $\times 2.673 + 0.438$	Janis (1990)
Anterior jaw length (ajl)	log mass = log AJL $\times 2.806 - 0.902$	Janis (1990)
Total skull length (tsl)	log mass = log TSL $\times 2.975 - 2.344$	Janis (1990)
Total jaw length (tjl)	log mass = log TJL $\times 2.884 - 1.952$	Janis (1990)
Second lower premolar length (SLPL)	log mass = log SLPL $\times 2.185 + 1.957$	Janis (1990)

Appendix 3, continued.

Measurement	Equation	Source
Second lower premolar width (SLPW)	log mass = log SLPW × 1.99 + 2.636	Janis (1990)
Third lower premolar length (TLPL)	log mass = log TLPL × 2.714 + 1.686	Janis (1990)
Third lower premolar width (TLPW)	log mass = log TLPW × 2.224 + 2.389	Janis (1990)
Fourth lower premolar length (FLPL)	log mass = log FLPL × 3.203 + 1.533	Janis (1990)
Fourth lower premolar width (FLPW)	log mass = log FLPW × 2.486 + 2.226	Janis (1990)
Fourth lower premolar area (FLPA)	log mass = log FLPA × 1.398 + 1.913	Janis (1990)
First lower molar length (FLML)	log mass = log FLML × 3.263 + 1.337	Janis (1990)
First lower molar width (FLMW)	log mass = log FLMW × 2.909 + 2.03	Janis (1990)
First lower molar area (FLMA)	log mass = log FLMA × 1.553 + 1.701	Janis (1990)
Second lower molar length (SLML)	log mass = log SLML × 3.201 + 1.13	Janis (1990)
Second lower molar width (SLMW)	log mass = log SLMW × 2.967 + 1.932	Janis (1990)
Second lower molar area (SLMA)	log mass = log SLMA × 1.563 + 1.541	Janis (1990)
Third lower molar length (TLML)	log mass = log TLML × 3.183 + 0.801	Janis (1990)
Third lower molar width (TLMW)	log mass = log TLMW × 2.933 + 1.991	Janis (1990)
Third lower molar area (TLMA)	log mass = log TLMA × 1.58 + 1.404	Janis (1990)
Second upper molar length (SUML)	log mass = log SUML × 3.184 + 1.091	Janis (1990)
Second upper molar width (SUMW)	log mass = log SUMW × 3.004 + 1.469	Janis (1990)
Second upper molar area (SUMA)	log mass = log SUMA × 1.568 + 1.277	Janis (1990)
Third upper molar length (M^3l)	log mass = log M^3l × 2.81 + 1.29	Damuth (1990)
Third upper molar width (M^3w)	log mass = log M^3w × 2.77 − 1.58	Damuth (1990)
Third upper molar area (M^3a)	log mass = log M^3a × 1.47 + 1.26	Damuth (1990)
Seventh lower molariform length	log mass = log 7LML × 3.201 + 1.13	Janis (1990)
Seventh lower molariform width	log mass = log 7LMW × 2.967 + 1.932	Janis (1990)
Seventh lower molariform area	log mass = log 7LMA × 1.563 + 1.541	Janis (1990)
Lower postcranial row length (pcrl)	log mass = log PCRL × 3.15 − 1.28	Janis (1990)
Lower postcranial row area (lpcta)	log mass = log LPCTA × 1.48 + 0.51	Janis (1990)
Upper postcrnial row length (pcru)	log mass = log PCRU × 3.07 − 1.1	Janis (1990)
Upper postcranial row area (upcta)	log mass = log UPCTA × 1.48 + 0.29	Janis (1990)
Shoulder height (eqn. *a*)	mass = (shoulder hght × 1.02 × 10^{-4})$^{3.11}$	Roth (1990, and references therein)
Shoulder height (eqn. *b*)	mass = (shoulder hght × 1.267 × 10^{-3})$^{2.631}$	Roth (1990, and references therein)
Shoulder height (eqn. *c*)	mass = (shoulder hght × 5.07 × 10^{-4})$^{2.803}$	Roth (1990, and references therein)
Shoulder height (eqn. *d*)	mass = (shoulder hght × 2.58 × 10^{-4})$^{2.917}$	Roth (1990, and references therein)
Shoulder height (eqn. *e*)	mass = (shoulder hght × 3.96 × 10^{-4})$^{2.890}$	Roth (1990, and references therein)
Shoulder height (eqn. *f*)	mass = (shoulder hght × 1.81 × 10^{-4})$^{2.97}$	Roth (1990, and references therein)
Shoulder height (eqn. *g*)	mass = (shoulder hght × 8.234 × 10^{-4})$^{2.711}$	Roth (1990, and references therein)
Shoulder height (eqn. *h*)	mass = (shoulder hght × 2.080 × 10^{-4})$^{2.934}$	Roth (1990, and references therein)
Shoulder height (eqn. *i*)	mass = (shoulder hght × 3.071 × 10^{-4})$^{2.917}$	Roth (1990, and references therein)
Shoulder height (eqn. *j*)	mass = (shoulder hght × 4.682 × 10^{-5})$^{3.263}$	Roth (1990, and references therein)
Shoulder height (eqn. *k*)	mass = (shoulder hght × 3.24 × 10^{-5})$^{3.356}$	Roth (1990, and references therein)
Shoulder height (eqn. *l*)	mass = (shoulder hght × 2.73 × 10^{-5})$^{3.387}$	Roth (1990, and references therein)
Humerus length (HL, all carnivores)	log mass = 2.93 × log HL − 5.11	Anyonge (1993)
Anteroposterior second moment of area (HIY, all carnivores)	log mass = 0.63 × log HIY − 0.61	Anyonge (1993)
Transverse second moment of area (HIX, all carnivores)	log mass = 0.6 × log HIX − 0.59	Anyonge (1993)
Cortical area humerus (HCA, all carnivores)	log mass = 1.18 × log HCA − 0.99	Anyonge (1993)
Humerus length (HL, felids)	log mass = 3.13 × log HL − 5.53	Anyonge (1993)
Anteroposterior second moment of area (HIY, felids)	log mass = 0.64 × log HIY − 0.61	Anyonge (1993)
Transverse second moment of area (HIX, felids)	log mass = 0.63 × log HIX − 0.64	Anyonge (1993)
Cortical area humerus (HCA, felids)	log mass = 1.25 × log HCA − 1.09	Anyonge (1993)
Femur length (FL, all carnivores)	log mass = 2.92 × log FL − 5.27	Anyonge (1993)
Anteroposterior second moment of area (FIY, all carn)	log mass = 0.67 × log FIY − 0.76	Anyonge (1993)
Transverse second moment of area (FIX, all carnivores)	log mass = 0.69 × log FIX − 0.77	Anyonge (1993)
Cortical area femur (FCA, all carnivores)	log mass = 1.25 × log FCA − 1.04	Anyonge (1993)
Articular area femur (FDA, all carnivores)	log mass = 1.31 × log FDA − 2.12	Anyonge (1993)
Femur length (FL, felids)	log mass = 3.2 × log FL − 5.9	Anyonge (1993)
Anteroposterior second moment of area (FIY, felids)	log mass = 0.69 × log FAP − 0.77	Anyonge (1993)

Appendix 3, continued.

Measurement	Equation	Source
Transverse second moment of area (FIX, felids)	log mass = 0.69 × log FT − 0.79	Anyonge (1993)
Cortical area (FCA, felids)	log mass = 1.31 × log CAF − 1.18	Anyonge (1993)
Articular area femur (FDA, felids)	log mass = 1.32 × log AAF − 2.16	Anyonge (1993)
m1 length (M1L, all carnivores)	log mass = 2.97 × log M1L − 2.27	Van Valkenburgh (1990)
Orbito-occiput length (OOL, all carnivores)	log mass = 3.44 × log OOL − 5.74	Van Valkenburgh (1990)
Skull length (SKL, all carnivores)	log mass = 3.13 × log SKL − 5.59	Van Valkenburgh (1990)
m1 length (M1L, felids)	log mass = 3.05 × log M1L − 2.15	Van Valkenburgh (1990)
Orbito-occiput length (OOL, felids)	log mass = 3.54 × log OOL − 5.86	Van Valkenburgh (1990)
Skull length (SKL, felids)	log mass = 3.11 × log SKL − 5.38	Van Valkenburgh (1990)
m1 length (M1L, ursids)	log mass = 0.49 × log M1L + 1.26	Van Valkenburgh (1990)
Orbito-occiput length (OOL, ursids)	log mass = 1.98 × log OOL − 2.38	Van Valkenburgh (1990)
Skull length (SKL, ursids)	log mass = 2.02 × log SKL − 2.8	Van Valkenburgh (1990)
m1 length (M1L, large carnivores)	log mass = 0.57 × log M1L + 1.45	Van Valkenburgh (1990)
Orbito-occiput length (OOL, large carnivores)	log mass = 1.51 × log OOL − 1.25	Van Valkenburgh (1990)
Skull length (SKL, large carnivores)	log mass = 1.56 × log SKL − 1.6	Van Valkenburgh (1990)
Head + body length (HBL, all carnivores)	log mass = 2.88 × log HBL − 7.24	Van Valkenburgh (1990)
Head + body length (HBL, felids)	log mass = 2.72 × log HBL − 6.83	Van Valkenburgh (1990)
Head + body length (HBL, large carnivores)	log mass = 2.46 × log HBL − 5.78	Van Valkenburgh (1990)
Head + body length (all ungulates)	log mass = 3.16 × log HBL − 5.12	Damuth (1990)

Appendix 4.1. Graphics showing the way in which forearm maximum (above) and mean (below) angular velocities (in rad s⁻¹) vary according to several possible lengths of the olecranon in meters.

Modified from Fariña and Blanco (1996).

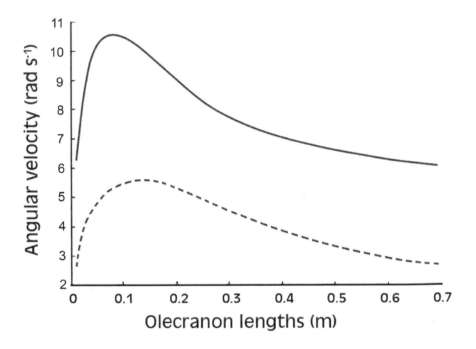

Appendix 4. Calculations

This section provides an example of the kinds of techniques at our disposal for using physics principles in helping us better understand the paleobiology of mammals. The mainly qualitative description of forearm speed and the energy of a forearm strike by *Megatherium americanum* is based on the following mathematical treatment of the subject.

Hill's (1938) equation, which relates several variables to muscle shortening, has the form

$$-\frac{dl}{dt} = \frac{(F_0 - F).b}{F + a} \text{ ,}$$

(Appendix 4.1)

where $-dl/dt$ is the velocity at which the muscle shortens, F_0 is the maximum isometric stress (i.e., the maximum force a muscle can exert per unit area of its section when it is not shortening), F is the actual force that the muscle exerts, and a and b are constants, obtained empirically, with dimensions of force and velocity, respectively.

F_0 was estimated from the maximal isometric stress of striated muscle and from an estimation of the active muscular section. This section was calculated from the volume divided by the fiber length. Both values were obtained from appropriate allometric equations for mammals (Alexander et al., 1981). Two data can be used as a rough check of the validity of such estimation: first, the fact that the observed area of insertion (0.02 m^2) is about 6 times smaller than the value obtained for the active section of the muscle (0.2 m^2), and second, that a 2500-kg African elephant *Loxodonta africana* had a 30.4-kg triceps with an active section of 0.176 m^2 (Alexander et al., 1979b). An intermediate value of a (i.e., $0.28\,F_0$) was taken from those given in the literature (Alexander, 1983). The value of b was obtained as follows. The maximum speed of shortening after Hill's equation, when $F = 0$, is $v_{max} = b\,F_0/a$ (Alexander, 1983). Also, this maximum speed is $v_{max} \leq 3.3\,f\,l_0$ (Weis-Fogh and Alexander, 1977), where f is the repetition frequency of fast striated muscle, and l_0 is the initial fiber length, estimated as 0.29 m. The frequency, f, was calculated based on observations of the few quadrupedal tracks assigned to *Megatherium americanum* in Pehuén Có (see Chapters 5 and 7), and assuming the fast striated musculature of both pairs of limbs had the same frequency. Thus,

$$f = \frac{1}{t_\lambda} \text{ ,}$$

(Appendix 4.2)

where t_1 is the time in which the stride length l is completed. Therefore, assuming that the animal is moving at a constant speed u,

$$t_\lambda = \frac{u}{\lambda}.$$

(Appendix 4.3)

The speed u can be calculated after the equation

$$\frac{\lambda}{h} = 2.3 \left(\frac{u^2}{g\,h} \right)^{0.3},$$

(Appendix 4.4)

where h is the height of the hip from the ground, and g the acceleration of free fall (Alexander, 1976). In the case of those footprints assigned to *Megatherium*, the stride length l was between 1.5 and 1.8 m and the hip height h is about 1.25 m. Rearranging equations Appendix 4.2, Appendix 4.3 and Appendix 4.4 and plotting those values, the frequency turned out to be between 1 and 1.3 Hz. However, the animal was unlikely to have been moving at cruising (let alone maximum) speed, as the substratum was soft mud (Aramayo and de Bianco, 1987). Therefore, a value of 1.5 Hz was chosen as a more suitable estimate of the maximum frequency, and consequently the respective maximum strain rate is 5 muscle lengths s^{-1}. This is higher than, but on the same order of magnitude as, the value 2.73 muscle lengths s^{-1}, obtained by extrapolating an allometric equation for fast contracting, type IIB muscle (Rome et al., 1990), and is also coherent with other results for similar muscles (Calow and Alexander, 1973; Winters and Starck, 1985). Thus, the value of v_{max} so obtained was 1.4 m s^{-1}. Therefore, the appropriate value of b is 0.4 m s^{-1}.

The force F was calculated by taking moments at the ulnar–humerus articulation (the effect of the acceleration of gravity was considered negligible):

$$F.s_1 \sin\theta = I.\frac{d\omega}{dt},$$

(Appendix 4.5)

where s_1 is the olecranon length, q is the angle formed by the forearm and the arm, I is the moment of inertia of the forearm (as described in the text; see Eq. Appendix 4.6), and dw/dt is the angular acceleration of the forearm.

I is defined for a rigid body consisting of N point masses m_i with distances r_i to the rotation axis, as the sum of the point-mass moments of inertia or

$$I = \sum_{i=1}^{N} m_i\, r_i^2$$

(Appendix 4.6)

From the general equation of circular movement,

$$\omega \sin\theta\, s_1 = \frac{dl}{dt}.$$

(Appendix 4.7)

The equations were rearranged and F, as given by Eq. Appendix 4.5, and dl/dt, as given in Eq. Appendix 4.7, were replaced in Eq. Appendix 4.1. Thus, the following differential equation of motion was obtained:

$$\ddot{\theta} = \frac{(F_0 \, b - a \, s_1 \, \dot{\theta} \sin\theta) \, s_1 \, \sin\theta}{(b + s_1 \, \dot{\theta} \sin\theta) \, I} \quad , \qquad\qquad (\text{Appendix } 4.8)$$

where angular velocity is represented by $\dot{\theta}$, angular acceleration $\ddot{\theta}$, and the other symbols as already indicated above. This equation was solved numerically using second and third order Runge-Kutta-Fehlberg formulas of 1^{0-3} accuracy for an initial angle $q = 30°$ and an initial velocity of o rad s^{-1}. Those results were not particularly sensitive to variations of the parameters used, except for the constant b in Hill's equation, which we estimated based on footprints attributed to *Megatherium americanum*.

The olecranon length (s_1), as noted in the text, is the distance of the point of application of the muscular force of the triceps muscle to the rotational axis, which is the articulation at the elbow joint. In Eq. Appendix 4.8, a range of possible, realistic values of olecranon length was considered, to see how the speed of rotation, or, to say it properly, the angular velocity, varied with that length. The graph obtained (Appendix Fig. 4.1) presents two angular velocities, the maximal, important because this value squared is related with the energy delivered by the blow and hence its destructive power, and the mean velocity, which has to do with how fast the blows reaches the intended object. Both velocities increase steeply until they reach an optimum, after which they tend to decrease with a lower slope.

In the range selected, mean velocity is more uniform, whereas maximal velocity shows a greater variation. The estimated optimal olecranon length for improving maximal velocity is 7.8 cm, and the actual is 12 cm, still within a range of velocity close to the optimum. In the case of mean velocity, the estimated optimal olecranon length is 12.9 cm. This figure is also close to the observed length, and also lies in the region of high angular velocity. Interestingly, the maximal angular velocity (10.8 rad s^1) of the forearm was achieved when the angle q is 180°, i.e., when the forearm was completely extended. The time of complete forearm extension is 0.57 s.

The kinetic energy of the forearm is defined by the equation $E_k = \frac{1}{2} I w^2$ (symbols as above), and according to Fariña and Blanco's (1996) calculations equals about 2.7 kJ, although new, more refined calculations lowered this figure to about 2 kJ.

References

Abel, O. 1912. Grundzüge der Palaeobiologie der Wirbeltiere. E. Schweizerbart'sche Verlagsbuchhandlung, Stuttgart, 708 pp.

Abrantes, É. A. L., Ávilla, L. S., and Vizcaíno, S. F. 2005. Paleobiologia e paleoecologia de *Pampatherium humboldti* (Lund, 1839) (Mammalia: Cingulata: Dasypodidae). Boletim de Resumos do II Congresso Latino-Americano de Paleontologia de Vertebrados, Rio de Janeiro 1:16–17.

Aceñolaza, F. G., and Herbst, R. (eds). 2000. El Neógeno de Argentina. Correlación Geológica 14:1–289.

Aerts, P., Ker, R. F., De Clercq, D., Isley, D. W., and Alexander, R. McN. 1995. The mechanical properties of the human heel pad: a paradox resolved. Journal of Biomechanics 28:1299–1308.

Agnolin, F. L., and Chimento, N. R. 2011. Afrotherian affinities for endemic South American "ungulates." Mammalian Biology 76:101–108.

Akersten, W. A. 1985. Canine function in *Smilodon* (Mammalia; Felidae; Machairodontidae). Contributions in Science 365:22. Natural History Museum of Los Angeles County, Los Angeles.

Akersten, W. A., and McDonald, H. G. 1991. *Nothrotheriops* from the Pleistocene of Oklahoma and paleogeography of the genus. The Southwestern Naturalist 36:178–185.

Alberdi, M. T. 1987. La Familia Equidae, Gray 1821 (Perissodactyla, Mammalia) en el Pleistoceno de Sudamérica. IV Congreso Latinoamericano de Paleontología, Bolivia 1:484–499.

Alberdi, M. T., and Prado, J. L. 1993. Review of the genus *Hippidion* Owen 1869 (Mammalia: Perissodactyla) from the Pleistocene of South America. Zoological Journal of the Linnean Society 108:1–22.

———. 1995. Los mastodontes de América del Sur; in Evolución biológica y climática de la Región Pampeana durante los últimos cinco millones de años. Un ensayo de correlación con el Mediterráneo occidental; in Alberdi, M. T., Leone, G., and Tonni, E. P. (eds.), Monografías del Museo Nacional de Ciencias Naturales, CSIC, España 12:277–292.

———. 2004. Los caballos fósiles de América del Sur. Una historia de 3 millones de años. Universidad Nacional del Centro de la Provincia de Buenos Aires, INQUAPA Serie monográfica (Argentina) 3. 269 pp.

Alberdi, M. T., Leone, G., and Tonni, E. P. (eds.) 1995a. Evolución biológica y climática de la región pampeana durante los últimos cinco millones de años. Monografías del Museo Nacional de Madrid 12:1–423.

Alberdi, M. T., Prado, J. L., and Ortiz-Jaureguizar, E. 1995b. Patterns of body size changes in fossil and living Equini (Perissodactyla). Biological Journal of the Linnean Society 54:349–370.

Alberdi, M. T., Menegaz, A. N., Prado, J. L., and Tonni, E. P. 1989. La Fauna local Quequén Salado-Indio Rico (Pleistoceno Tardío) de la Provincia de Buenos Aires, Argentina. Aspectos paleoambientales y estratigráficos. Ameghiniana 25:225–236.

Alberdi, M. T., Prado, J. L., López, P., Labarca R., and Martínez, I. 2007. *Hippidion saldiasi* Roth 1899 (Mammalia, Perissodactyla) en el Pleistoceno tardío de Calama, norte de Chile. Revista Chilena de Historia Natural 80:157–171.

Albino, A. 2005. A late Quaternary lizard assemblage from the southern Pampean region of Argentina. Journal of Paleontology 25:185–191.

Alexander, R. McN. 1976. Estimates of speeds of dinosaurs. Nature 261:129–130.

———. 1981. Factors of safety in the structure of animals. Science Progress 67:109–130.

———. 1982. Locomotion of Animals. Blackie, Glasgow, 163 pp.

———. 1983. Animal Mechanics. 2nd ed. Oxford, Blackwell, 301 pp.

———. 1985. Mechanics of posture and gait of some large dinosaurs. Zoological Journal of the Linnean Society 83:1–25.

———. 1989. Dynamics of Dinosaurs and other extinct giants. Columbia University Press, New York, 167 pp.

———. 1991. Optimum timing of muscle activation for simple models of throwing. Journal of Theoretical Biology 150:349–372.

———. 1992. The work that muscles can do. Nature 357:360–361.

———. 1995. Leg design and jumping technique for humans, other vertebrates and insects. Philosophical Transactions of the Royal Society of London 347:235–248.

———. 1996. *Tyrannosaurus* on the run. Nature 379:121.

———. 1999. Energy for animal life. Oxford University Press, Oxford, 165 pp.

———. 2003. Evolution enhanced: a rodent as big as a buffalo. Science 301:1678–1679.

Alexander, R. McN., Jayes, A. S., Maloiy, G. M. O., and Wathuta, E. M. 1979a. Allometry of the limb bones of mammals from shrews (*Sorex*) to elephant (*Loxodonta*). Journal of Zoology (London) 189:305–314.

Alexander, R. McN., Maloiy, G. M. O., Hunter, B., Jayes, A. S., and Nturibi, J. 1979b. Mechanical stresses in fast locomotion of buffalo (*Syncerus caffer*) and elephant (*Loxodonta africana*). Journal of Zoology (London) 189:135–144.

Alexander, R. McN., Jayes, A. S., Maloiy, G. M. O., and Wathuta, E. M. 1981. Allometry of the leg muscles of mammals. Journal of Zoology (London) 194:539–552.

Alexander, R. McN., Fariña, R. A., and Vizcaíno, S. F. 1999. Tail blow energy and carapace fractures in a large glyptodont (Mammalia, Edentata). Zoological Journal of the Linnean Society 126:41–49.

Allison, P. A., and Bottjer, D. J. (eds.) 2010. Taphonomy: Process and Bias through Time. 2nd ed. Topics in Geobiology 32. Springer, Dordrecht, 599 pp.

———. 2001. A multispecies overkill simulation of the end-Pleistocene megafaunal mass extinction. Science 292:1893–1896.

Alvarez, L. W., Alvarez, W., Asaro, F., and Michel, H. V. 1980. Extraterrestrial cause for the Cretaceous–Tertiary extinction. Science 208:1095–1108.

Akersten, W. A. 1985. Canine function in *Smilodon* (Mammalia;. Felidae; Machairodontinae). Contributions to Science of the Natural History Museum of Los Angeles County 356:1–22

Ameghino, F. 1884. Filogenia. Principios de clasificación transformista basados sobre leyes naturales y proporciones matemáticas. Buenos Aires and Paris, 512 pp.

———. 1887. Observaciones generales sobre el orden de mamíferos extinguidos llamados toxodontes (Toxodontia) y sinopsis de los géneros y especies hasta ahora conocidos. Anales del Museo de La Plata (entrega especial):1–66.

———. 1889. Contribución al conocimiento de los mamíferos fósiles de la República Argentina. Actas de la Academia Nacional de Ciencias de Córdoba 6:xxxii + 1028 pp., and atlas of 98 plates.

———. 1891a. Los monos fósiles del Eoceno de la República Argentina. Revista Argentina de Historia Natural 1:383–397.

———. 1891b. Mamíferos y aves fósiles argentinas. Especies nuevas, adiciones y correcciones. Revista Argentina de Historia Natural 1:240–259.

———. 1894. Sur les oiseaux fossiles de Patagonie; et la faune mammalogique des couches à *Pyrotherium*. Boletín del Instituto Geográfico Argentino 15:501–660.

———. 1898. An existing ground-sloth in Patagonia. Natural Science 13:324–326.

———. 1906. Les formations sedimentaires du Crétacé supérieur et du Tertiaire de Patagonie. Anales del Museo Nacional de Buenos Aires 8:1–568.

———. 1907. Les toxodontes à cornes. Anales del Museo Nacional de Historia Natural de Buenos Aires 16:49–91

———. 1908. Las formaciones sedimentarias de la región litoral de Mar del Plata y

Chapadmalán. Anales del Museo Nacional de Historia Natural de Buenos Aires 10:343–428.

———. 1909. Le Diprothomo platensis: un pré-curseur de l'homme du Pliocene inférieur de Buenos Aires. Anales del Museo Nacional de Historia Natural de Buenos Aires 19:107–209.

———. 1915. Taquigrafía Ameghino. Publicada en el N° 155, año XV, III cuatrimestre de la Revista del Centro Estudiantes de Ingeniería. Buenos Aires, Ediciones del Centro de Estudiantes de Ingeniería, 35 pp.

Anderson, J. F., Hall-Martin, A., and Russell D. A. 1985. Long bone circumference and weight in mammals, birds and dinosaurs. Journal of Zoology (London) 207:53–61.

Anderson, R. P., and Handley, C. O. 2001. A new species of three-toed sloth (Mammalia: Xenarthra) from Panamá, with a review of the genus Bradypus. Proceedings of the Biological Society of Washington 114:1–33.

Anderson, T. F., Arthur, M. A. 1983. Stable isotopes of oxygen and carbon and their application to sedimentologic and paleoenvironmental problems; in Arthur, M. A. (organizer), Stable Isotopes in Sedimentary Geology. Society of Economic Paleontologists and Mineralogists Short Course 10:1–151.

Andrews, J. T. 2006. The Laurentide ice sheet: a review of history and processes; pp. 201–208 in Knight, P. G. (ed.), Glaciology and Earth's Changing Environment. Blackwell, London.

Anonymous. 1873. Cartoon Portraits and Biographical Sketches of Men of the Day. Illustrated by Frederick Waddy. Tinsley Brothers, London, 64 pp.

Anthony, H. E. 1926. Mammals of Porto Rico, living and extinct. Rodentia and Edentata. Scientific Survey of Porto Rico and West Islands. Publication of the New York Academy of Sciences 9: 97–241.

Anyonge, W. 1993. Body mass in large extant and extinct carnivores. Journal of Zoology (London) 231:339–350.

———. 1996. Locomotor behaviour in Plio-Pleistocene sabre-tooth cats: a biomechanical analysis. Journal of Zoology (London) 238:395–413.

Aramayo, S. 1998. Nuevos restos de Proscelidodon sp. (Edentata, Mylodontidae) del yacimiento de Monte Hermoso (Plioceno Inferior a Medio) Provincia de Buenos Aries, Argentina. Estudio morfológico functional. Actas Segundas Jornadas Geológicas Bonaerenses 1998:99–107.

Aramayo, S., and Manera de Bianco, T. 1987. Hallazgo de una icnofauna continental (Pleistoceno tardío) en la localidad de Pehuén-Có (Partido de Coronel Rosales), Provincia de Buenos Aires, República Argentina. Parte I: Edentata, Litopterna, Proboscidea. Anales del IV Congreso Latinoamericano de Paleontología, Bolivia 1:516–531.

———. 1996. Edad y nuevos hallazgos de icnitas de mamíferos y aves en el yacimiento paleoicnológico de Pehuén-Có (Pleistoceno tardío), Provincia de Buenos Aires, Argentina. Asociación Paleontológica Argentina. Publicación Especial 4. 1° Reunión Argentina de Icnología, pp. 47–57.

Arbour, V. M. 2009. Estimating impact forces of tail club strikes by ankylosaurid dinosaurs. Public Library of Science ONE 4:e6738.

Arnason, U., Gullberg, A., and Janke, A. 1997 Phylogenetic analyses of mitochondrial DNA suggest a sister group relationship between Xenarthra (Edentata) and Ferungulates. Molecular Biology and Evolution 14:762–768.

Arribas, A., and Palmqvist, P. 1999. On the ecological connection between sabre-tooths and hominids: faunal dispersal events in the lower Pleistocene and a review of the evidence for the first human arrival in Europe. Journal of Archaeological Science 26:571–585.

Arribas, A., Palmqvist, P., Pérez-Claros, J. A., Castilla, R., Vizcaíno, S. F., and Fariña, R. A. 2001. New evidence on the interaction between humans and megafauna in South American. Publicaciones del Seminario de Paleontología de Zaragoza 5:228–238.

Arroyo-Cabrales, J., Polaco, O. J., Laurito, C., Johnson, E., Alberdi, M. T., Zamora, A. L. V. 2007. The proboscideans (Mammalia) from Mesoamerica. Quaternary International 169–170:17–23.

Asara J. M., Schweitzer, M. H. Phillips, M. P. Freimark, L. M., and Cantley, L. C. 2007. Protein sequences from mastodon (Mammut americanum) and dinosaur (Tyrannosaurus rex) revealed by mass spectrometry. Science 316:280–284.

Ausich, W. I., and Lane, N. G. 1999. Life of the Past. 4th ed. Prentice-Hall, Upper Saddle River, N.J., 321 pp.

Ávila, L. S., and Bergqvist, L. P. 2005. Sobre o Hábito Locomotor de Carodnia vieirai Paula-Couto, 1952 (Mammalia: Xenungulata) On the Locomotory Pattern of Carodnia vierai Paula-Couto, 1952 (Mammalia: Xenungulata). Anuário do Instituto de Geociências—UFRJ 28:185–186.

Babot, M. J., Powell, J. E., and de Muizon, C. 2002. Callistoe vincei, a new Proborhyaenidae (Borhyaenoidea, Metatheria, Mammalia) from the Early Eocene of Argentina. Geobios 35:615–629.

Barsness, L. 1985. Heads, Hides and Horns: The Complete Buffalo Book. Texas Christian University Press, Fort Worth, 233 pp.

Bartoli, G., Sarnthein, M. Weinelt, M. Erlenkeuser, H. Garbe-Schönberg, D., and Lea, D. W. 2005. Final closure of Panama and the onset of Northern Hemisphere glaciation. Earth and Planetary Science Letters 237:33–44.

Baker, P. A., Rigsby, C. A., Seltzer, G. O., Fritz, S. C., Lowenstein, T. K., Bacher, N. P., and Veliz, C. 2001. Tropical climate changes at millennial and orbital timescales on the Bolivian Altiplano. Nature 409:698–701.

Bakker, R. T. 1986. The Dinosaur Heresies: New Theories Unlocking the Mystery of the Dinosaurs and Their Extinction. William Morrow, New York, 165 pp.

———. 1987. The return of the dancing dinosaurs; pp. 38–69 in Czerkas, S. M., and Olson, E. C. (eds.), Dinosaurs Past and Present. Vol. 1. University of Washington Press, Seattle.

Ballreich, R., and Kuhlow, A. 1986. Biomechanik des Kugelstosses; pp. 89–109 in Ballreich, R., and Kuhlow, A. (eds.), Biomechanik des Sportarten. Stuttgart: Enke.

Barnett, R., Shapiro, B., Barnes, I., Ho, S. Y. W., Burger, J., Yamaguchi, N., Higham, T. F. G., Wheeler, H. T., Rosendahl, W., Sher, A. V., Sotnikova, M., Kuznetsova, T., Baryshnikov, G. F., Martin, L. D., Harington, C. R., Burns, J. A., and

Cooper, A. 2009. Phylogeography of lions (Panthera leo ssp.) reveals three distinct taxa and a late Pleistocene reduction in genetic diversity. Molecular Ecology 18:1668–1677.

Barreda, V., and Palazzesi, L. 2007. Patagonian vegetation turnovers during the Paleogene–Early Neogene: origin of arid-adapted floras. Botanical Review 73:31–50.

Barreda, V., Anzótegui, L. M., Prieto, A. M., Aceñolaza, P., Bianchi, M, Borromei, A. M., Brea, M., Caccavari, M., Cuadrado, G. A., Garralla, S., Grill, S., Guerstein, R., Lutz, A. I., Mancini, M. V., Mautino, L. R., Ottone, E. G., Quattrocchio, M. E., Romero, E. J., Zamaloa, M. C., and Zucol, A. 2007. Diversificación y cambios de las angiospermas durante el Neógeno en Argentina. Asociación Paleontológica Argentina. Ameghiniana (Publicación Especial de la Associación Paleontológica Argentina 50° aniversario)11:173–191.

Barreda, V. D., Palazzesi, L., Tellería, M. C., Katinas, L., Crisci, J. V., Bremer, K., Passalia, M. G., Corsolini, R., Rodríguez Brizuela, R., and Bechis, F. 2010. Eocene Patagonia fossils of the daisy family. Science 329:1621.

Bargo, M. S. 2001a. The ground sloth Megatherium americanum (Xenarthra, Tardigrada, Megatheriidae): skull shape, bite forces, and diet; pp. 173–192 in Vizcaíno, S. F., Fariña, R. A. and Janis, C. (eds.), Acta Palaeontologica Polonica, Special Issue on Biomechanics and Palaeobiology 46.

———. 2001b. El aparato masticatorio de los perezosos terrestres (Xenarthra, Tardigrada) del Pleistoceno de Argentina. Morfometría y biomecánica. Ph.D. dissertation, Universidad Nacional de La Plata, República Argentina, 400 pp.

Bargo, M. S., and Vizcaíno, S. F. 2008. Paleobiology of Pleistocene ground sloths (Xenarthra, Tardigrada): biomechanics, morphogeometry and ecomorphology applied to the masticatory apparatus. Ameghiniana 45:175–196.

Bargo, M. S., Vizcaíno S. F., Archuby, F. M., and Blanco, R. E. 2000. Limb bone proportions, strength and digging in some Lujanian (late Pleistocene-early Holocene) mylodontid ground sloths (Mammalia, Xenarthra). Journal of Vertebrate Paleontology 20:601–610.

Bargo, M. S., De Iuliis, G., and Vizcaíno, S. F. 2006a. Hypsodonty in Pleistocene ground sloths. Acta Palaeontologica. Polonica 51:53–61.

Bargo, M. S., Toledo, N., and Vizcaíno, S. F. 2006b. Muzzle of South American Pleistocene ground sloths (Xenarthra, Tardigrada). Journal of Morphology 267:248–263

Bargo, M. S., Vizcaíno, S. F., and Kay, R. F. 2009. Predominance of orthal masticatory movements in the early Miocene Eucholaeops (Mammalia, Xenarthra, Tardigrada, Megalonychidae) and other megatherioid sloths. Journal of Vertebrate Paleontology 29:870–888.

Barnosky, A. D., Koch, P. L., Feranec, R. S., Wing, S. L., and Shabel, A. B. 2004. Assessing the causes of late Pleistocene extinctions on the continents. Science 306:70–75.

Bauer, A., and McCarthy, C. 2010. Darwin's pet Galápagos tortoise, Chelonoidis darwini, rediscovered. Chelonian Conservation and Biology 9:270–276.

Behling, H., Arz, H., Pätzold J., and Wefer, G. 2000. Late Quaternary vegetational and climatic dynamics in northeastern Brazil, inferences from marine core GeoB 3104-1. Quaternary Science Review 19:981–994.

Behrensmeyer, A. K., and Dechant Boaz, D. E. 1980. The recent bones of Amboseli Park, Kenya, in relation to East African Paleoecology; pp. 72–93 in Behrensmeyer, A. K., and Hill, A. P., (eds.), Fossils in the Making. University of Chicago Press, Chicago.

Behrensmeyer, A. K., Todd, N. E., Potts, R., and McBrinn, G. 1977. Late Pliocene Faunal Turnover in the Turkana Basin of Kenya and Ethiopia. Science 278:1589–1594.

Benton, M. J. 2003. When Life Nearly Died: The Greatest Mass Extinction of All Time. Thames and Hudson, London, 336 pp.

————. 2005. Vertebrate Palaeontology. 3rd ed. Blackwell, Malden, Mass., 455 pp.

Bergqvist, L. P. 2010. Deciduous premolars of Paleocene litopterns of São José de Itoboraí Basin, Rio de Janeiro, Brazil. Journal of Paleontology 84:858–867.

Bergqvist, L. P., and Maciel, L. 1994. Icnofósseis de mamiferos (crotovinas) na planicie costeira do Rio Grande do Sul, Brasil. Annais da Academia Brasileira de Ciências 52:359–377.

Berman, W. D. 1994. Los carnívoros continentales (Mammalia, Carnivora) del Cenozoico en la provincia de Buenos Aires. Ph.D. dissertation, Facultad de Ciencias Naturales y Museo, Universidad Nacional de La Plata, 412 pp.

Bertels-Psotka, A. 2000. Ostrácodos (Arthropoda, Crustacea) de hábitat límnico de la Formación Collón Cura, provincia del Chubut, Argentina. Ameghiniana 37:39–45.

Bertram, B. C. R. 1979. Serengeti predators and their social systems; pp. 221–248 in Sinclair, A. R. E., and Norton Griffiths, M. (eds.), Serengeti. Dynamics of an Ecosystem. Chicago University Press, Chicago.

Biewener, A. A. 1983. Allometry of quadrupedal locomotion: the scaling of duty factor, bone curvature and limb orientation to body size. Journal of Experimental Biology 105:147–171.

————. 1989a. Scaling body support in mammals: limb posture and muscle mechanics. Science 245:45–48.

————. 1989b. Mammalian terrestrial locomotion and size. Mechanical design principles define limits. Bioscience 39:776–783.

————. 1990. Biomechanics of mammalian terrestrial locomotion. Science 250:1097–1103.

Biknevicius, A. R., and Van Valkenburgh, B. 1996. Design for killing: craniodental adaptations of predators; in Gittleman, J. L. (ed.), Carnivore Behavior, Ecology, and Evolution 2:393–428. Cornell University, Ithaca, N.Y.

Billet, G. 2010. New observations on the skull of Pyrotherium (Pyrotheria, Mammalia) and new phylogenetic hypotheses on South American ungulates. Journal of Mammalian Evolution 17:21–59.

Billet, G., Patterson, B., and Muizon, C de. 2009. Craniodental anatomy of late Oligocene archaeohyracids (Notoungulata, Mammalia) from Bolivia and Argentina and new phylogenetic hypotheses. Zoological Journal of the Linnean Society 155:458–509

Bininda-Emonds, O. R. P., Cardillo, M., Jones, K. E., MacPhee, R. D. E., Beck, R. M. D., Grenyer, R., Price, S. A., Vos, R. A., Gittleman J. L., and Purvis, A. 2007. The delayed rise of present-day mammals. Nature 446:507–512.

Birabén, M. 1968. Germán Burmeister, su vida, su obra. Ediciones Culturales Argentinas, Buenos Aires, 98 pp.

Blackburn, T. M., and Gaston, K. J. 1996. Abundance–body size relationships: the area you census tells you more. Oikos 75:303–309.

Blanco, R. E., and Czerwonogora A. 2003. The gait of Megatherium Cuvier 1796 (Mammalia, Xenarthra, Tardigrada); in Fariña, R. A., Vizcaíno, S. F., and Storch, G. (eds.), Morphological Studies in Fossil and Extant Xenarthra (Mammalia), Special Issue 101. Senckenbergiana Biologica 83:61–68.

Blanco, R. E., and Rinderknecht, A. 2008. Estimation of hearing capabilities of Pleistocene ground sloths (Mammalia, Xenarthra) from middle-ear anatomy. Journal of Vertebrate Paleontology 28:274–276.

Blanco, R. E., Jones, W. W., and Rinderknecht, A. 2009. The sweet spot of a biological hammer: the centre of percussion of glyptodont (Mammalia: Xenarthra) tail clubs. Proceedings of the Royal Society B 276:3971–3978.

Blodgett, T. A. Lenters, J. D., and Isacks, B. L. 1997. Constraints on the Origin of Paleolake Expansions in the Central Andes. Earth Interactions 1:1–28.

Bobe, R., and Eck, K. G. G. 2001. Responses of African Bovids to Pliocene Climatic Change. Paleobiology Memoirs 27, Supp. 2:1–47.

Boisserie J.-R., Likius A., Vignaud P., and Brunet M. 2005a. A new late Miocene hippopotamid from Toros-Menalla, Chad. Journal of Vertebrate Paleontology 25:665–673.

Boisserie, J.-R., Lihoreau, F., and Brunet, M. 2005b. The position of Hippopotamidae within Cetartiodactyla. Proceedings of the National Academy of Sciences of the United States of America 102:1537–1541.

————. 2005c. Origins of Hippopotamidae (Mammalia, Cetartiodactyla): towards resolution. Zoologica Scripta 34(2):119–143.

Bon, C., Caudy, N., de Dieuleveult, M., Fosse, P. Philippe, M., Maksud, F., Beraud-Colomb, E., Bouzaid, E., Kefi, R., Laugier, C., Rousseau, B., Casane, D., van der Plicht, J., and Elalouf, J.-M. 2008. Deciphering the complete mitochondrial genome and phylogeny of the extinct cave bear in the Paleolithic painted cave of Chauvet. Proceedings of the National Academy of Sciences of the United States of America 105:17447–17452.

Bonaparte, J. F. 1986. A new and unusual Late Cretaceous mammal from Patagonia. Journal of Vertebrate Paleontology 6:264–270.

————. 1990. New Late Cretaceous mammals from the Los Alamitos Formation, northern Patagonia, and their significance. National Geographic Research 6:63–93.

Bonaparte, J. F., and Morales, J. 1997. Un primitivo Notonychopidae (Litopterna) del Paleoceno Inferior de Punta Peligro, Chubut, Argentina. Estudios Geológicos 53:263–274.

Bond, M. 1999. Quaternary native ungulates of Southern South America. A synthesis. Quaternary of South America and Antarctic Peninsula 12:177–205.

Bond, M., and García, M. 2002. Nuevos restos de toxodonte (Mammalia, Notoungulata) en estratos de la Formación Chucal, Mioceno, Altiplano de Arica, norte de Chile. Revista Geológica de Chile 29:81–91.

Bond, M., Cerdeño, E. P., and López, G. 1995. Los ungulados nativos de América de Sur; in Alberdi, M. T., Leone, G. and Tonni, E. P. (eds.), Evolución climática y biológica de la región Pampeana durante los últimos cinco millones de años. Un ensayo de correlación con el Mediterráneo occidental. Monografías del Museo de Ciencias Naturales, Consejo Superior de Investigaciones Científicas, Madrid 12:259–275.

Bondesio, P. 1986. Lista sistemática de los vertebrados terrestres del Cenozoico de Argentina. 4° Congreso Argentino de Paleontología y Bioestratigrafía (Mendoza), Actas 2:187–190.

Borrero, L. A. 2001. El poblamiento de la Patagonia. Toldos, milodones y volcanes. Emecé, Buenos Aires, 195 pp.

————. 2003. Taphonomy of the Tres Arroyos 1 Rockshelter, Tierra del Fuego, Chile. Quaternary International 109–110:87–93.

————. 2008. Extinction of Pleistocene megamammals in South America: the lost evidence. Quaternary International 185:69–74.

————. 2009. The elusive evidence: the archeological record of the South American extinct megafauna; pp. 146–168 in Haynes, G. (ed.), American Megafaunal Extinctions at the End of the Pleistocene. Springer, Dordrecht.

Boule, M., and Thévenin A. 1920. Mammifères fossiles de Tarija. Librairie H. le Soudier, Paris, 256 pp.

Bozinovic, F., and Martínez del Río, C. 1996. Animals eat what they should not: why do they reject our foraging models? Revista Chilena de Historia Natural 69:15–20.

Brandoni, D., Soibelzon, E., and Scarano, A. 2008. On Megatherium gallardoi (Mammalia, Xenarthra, Megatheriidae) and the Megatheriinae from the Ensenadan (lower to middle Pleistocene) of the Pampean region, Argentina. Geodiversitas 30:793–804.

Breda, M., and Marchetti, M. 2005. Systematical and biochronological review of Plio-Pleistocene Alceini (Cervidae; Mammalia) from Eurasia. Quaternary Science Review 24:775–805.

Britton, S. W. 1941a. Form and function in the sloth. Quarterly Review of Biology 16:13–34.

————. 1941b. Form and function in the sloth. Quarterly Review of Biology 16:190–207.

Brown, J. H. 1995. Macroecology. University of Chicago Press, Chicago, 269 pp.

Brown, M. F., and Cook, R. G. (eds.) 2006. Animal spatial cognition: comparative, neural, and computational approaches. http://www.pigeon .psy.tufts.edu/asc/.

Bru de Ramón, J. B. 1784–1786. Colección de láminas que representan a los animales y monstruos del real Gabinete de Historia Natural. 2 vols. Imprenta de Andrés de Sotos, Madrid, vi + 76 pp.

Buchholtz, E. A., and Stepien, C. C. 2009. Anatomical transformation in mammals: developmental origin of aberrant cervical anatomy in tree sloths. Evolution and Development 11:69–79.

Buchmann, F. S., Lopes, R. P., and Caron F. 2010. Paleotoca do Município de Cristal, RS, Registro da atividade fossorial de mamíferos gigantes extintos no sul do Brasil; pp. 1–10 in Winge, M., Schobbenhaus, C., Souza, C. R.

G., Fernandes, A. C. S., Berbert-Born, M., Sallun filho, W., and Queiroz, E. T. (eds.), Sítios Geológicos e Paleontológicos do Brasil. http://www.unb.br/ig/sigep/sitio048/sitio048.pdf.

Burmeister, H. 1863–1864. Observaciones sobre las diferentes especies de Glyptodon en el Museo Público de Buenos Aires. Revista Farmacéutica de la Sociedad de Farmacia Nacional Argentina. Año 5, tomo III, pp. 271–280 1° de octubre de 1863; Año 6°, Tomo III, pp. 327–336 1° de enero de 1864. Translated to English as "Observations of the various species of Glyptodon in the Public Museum of Buenos Ayres. Translated by C. Carter Blake, with corrections and addenda by the author. The Annals: a Magazin of Natural History (3) XIV, pp. 81–97. London.

———. 1864. Sumario sobre la fundación y los progresos del Museo Público de Buenos Aires. Anales del Museo Público de Buenos Aires 1:1–11.

———. 1866. Suplementos a la lista de mamíferos fósiles del terreno diluviano. Actas de la Sociedad Paleontológica de Buenos Aires 1866:299.

———. 1866–1867. Fauna argentina. I. Mamíferos fósiles. Lista de los mamíferos fósiles del terreno diluviano. Anales del Museo Público de Buenos Aires 1:87–300.

———. 1871. Descripción del género Hoplophorus. Monografía de los Gliptodontes en el Museo Público de Buenos Aires. Anales del Museo Público de Buenos Aires 2:157–226.

———. 1874. Monografía de los glyptodontes en el Museo Público de Buenos Aires. Anales del Museo Nacional de Buenos Aires 2: vi + 1–412, pl. 42.

———. 1879. Description physique de la République Argentine d'aprés des observations personnelles et étrangères. III. Paul-Emile Coni, Buenos Aires; 556 pp.

Burmeister, C. G. C. 1885. Examen crítico de los mamíferos y reptiles fósiles denominados por Augusto Bravard y mencionados en su obra precedente. Anales del Museo Público de Buenos Aires 3:95–174.

Burmeister, C. G. C. 1888. Relación de un viaje a la Gobernación del Chubut. Anales del Museo Público de Buenos Aires 3:175–252

Burness, G. P., Diamond, J., and Flannery, T. F. 2001. Dinosaurs dragons, and dwarfs: the evolution of maximal body size. Proceedings of the National Academy of Sciences of the United States of America 98:518–523.

Burney, D. A., and Flannery, T. F. 2005. Fifty millennia of catastrophic extinctions after human contact. Trends in Ecology and Evolution 20:395–401.

Cabrera, Á. 1926. Sobre la alimentación del megaterio. Boletín de la Real Sociedad Española de Historia Natural 26:338–391.

———. 1928. Los orígenes de la fauna argentina. Gaea 3:146–160.

———. 1929a. Sobre la estructura de la mano y del pie en el megaterio. Anales de la Sociedad Científica Argentina 107:425–443.

———. 1929b. Una revisión de los mastodontes argentinos. Revista del Museo de La Plata 32:61–144.

———. 1929c. A propósito de la biología de los Xenartros. Conferencias y Reseñas Científicas

de la Real Sociedad Española de Historia Natural 4:123–126.

———. 1932. Sobre los camélidos fósiles y actuales de la América Austral. Revista del Museo de La Plata 33:89–117.

———. 1934. Los yaguares vivientes y extinguidos de la América austral. Notas Preliminares del Museo de La Plata (Argentina) 2:34–50.

———. 1935. Sobre la osteología de Palaeolama. Anales Museo Argentino de Ciencias Naturales, Paleontología de vertebrados 66:283–312.

———. 1936. Las especies del género Glossotherium. Notas del Museo de La Plata Paleontología 1:193–206.

Cabrera, A., and Kraglievich, L. 1931. Diagnosis previas de los ungulados del Arroyo Chasicó. Notas Preliminares Museo de La Plata 1:107–113

Cadbury, D. 2000. The Dinosuar Hunters: A True Story of Scientific Rivalry and the Discovery of the Prehistoric World. Fourth Estate, London, 384 pp.

Calow, L. J., and Alexander, R. McN. 1973. A mechanical analysis of a hind leg of a frog (Rana temporaria). Journal of Zoology (London) 171:293–321.

Camacho, H. H. 2006. [Biography of Lucas Kraglievich (1886–1932)]. Galería de paleontólogos. http://www.ugr.es/~mlamolda/galeria/biografia/kraglievich.html.

Campbell, N. A. 1996. Biology. 4th ed. Benjamin/Cummings, Menlo Park, Calif., 542 pp.

Campbell, Jr., K. E., Frailey, C. D., and Romero-Pittman, L. 2009. In defense of Amahuacatherium (Proboscidea: Gomphotheriidae). Neues Jahrbuch für Geologie und Paläontologie Abhandlungen 252:113–128.

Campbell, Jr., K. E., Prothero, D. R., Romero-Pittman, L., Hertel, F., and Rivera, N. 2010. Amazonian magnetostratigraphy: dating the first pulse of the Great American Faunal Interchange. Journal of South American Earth Sciences 29:619–626.

Candela, A. M., and Morrone, J. J. 2003. Biogeografía de puercoespines neotropicales (Rodentia: Hystricognathi): integrando datos fósiles y actuales a través de un enfoque panbiogeográfico. Ameghiniana 40:361–378.

Cantú, M., and A. Becker. 1988. Holoceno del arroyo Spernanzoni, Dpto. Río Cuarto, Prov. Córdoba, Argentina. Abstracts International Symposium Holocene in South America, p. 24.

Cao, Y., Janke, A., Waddell, P. J., Westerman, M., Takenaka, O., Murata, S., Okada, N., Pääbo, S., and Hasegawa, M. 1998. Conflict among individual mitochondrial proteins in resolving the phylogeny of eutherian orders. Journal of Molecular Evolution 47:307–322.

Carrano, M. 1998. Locomotion in non-avian dinosaurs: integrating data from hindlimb kinematics, in vivo strains, and bone morphology. Paleobiology 24:450–469.

Carbone, C., and Gittleman, J. L. 2002. A common rule for the scaling of carnivore density. Science 295:2273–2276.

Carbone, C., Maddox, T., Funston, P. J., Mills, M. G., Grether, G. F., and Van Valkenburgh, B. 2009. Parallels between playbacks and Pleistocene tar seeps suggest sociality in an extinct sabretooth cat, Smilodon. Biological Letters 23:81–85

Cardozo, A. 1975. Origen y filogenia de los Camélidos Sudamericanos. Academia Nacional de Ciencias de Bolivia, La Paz, 116 pp.

Carlini, A. A., Scillato-Yané, G. J., Vizcaíno, S. F., and Dozo, M. T. 1992. Un singular Myrmecophagidae (Xenarthra, Vermilingua) de Edad Colhuehuapense (Oligoceno tardío, Mioceno temprano) de Patagonia, Argentina. Ameghiniana 29:176.

Cartelle, C. 1980. Estudo comparativo do rádio e esqueleto da mão de "Glossotherium (Ocnotherium) giganteum" Lund 1842. Anais da Academia Brasileira de Ciências 52:359–377.

———. 1991. Um novo Mylodontinae (Edentata, Xenarthra) do Pleistoceno final da região intertropical brasileira. Anais da Academia Brasileira de Ciências 63:161–170.

———. 1992. Edentata e megamamíferos herbívoros extintos da toca dos ossos (Ourolândia, BA). Ph.D. dissertation, Programa de Pós-Graduação em Morfologia, Universidade Federal de Minas Gerais, Belo Horizonte, Brazil, 516 pp.

———. 1994a. Mamíferos extintos. Ciência Hoje da Criança 39:2–6.

———. 1994b. Tempo Passado. Mammíferos do Pleistoceno em Minas Gerais. Editora Palco, Belo Horizonte, 131 pp.

———. 1999. Pleistocene mammals of the Cerrado and Caatinga of Brazil; pp. 27–46 in Eisenberg, J. F., and Redford, K. H. (eds.), Mammals of the Neotropics: The Central Neotropics. Vol. 3, Ecuador, Peru, Bolivia, Brazil. University of Chicago Press, Chicago.

———. 2002. Peter W. Lund, a naturalist of several sciences. Lundiana 3:83–85.

Cartelle, C., and Bohórquez, G. A. 1982. "Eremotherium Laurillardi" Lund 1842, determinação específica e dimorfismo sexual. Iheringia 7:45–63.

———. 1983. "Pampatherium paulacoutoi" Uma Nova Especie de Tatu Gigante da Bahia. Revista Brasileira de Zoologia 2:229–254,

———. 1986. Presença de ossículos dérmicos em Eremotherium laurillardi (Lund) Cartelle & Bohórquez, 1982. (Edentata, Megatheriidae). Iheringia, Série Geológica 11:3–8.

Cartelle, C., and Lessa, G. 1988. Descrição de um novo gênero e espécie de Macraucheniidae (Mammalia, Litopterna) do Pleistoceno do Brasil. Paula Coutiana 3:3–26.

Cartelle, C., and De Iuliis, G. 1995. Eremotherium laurillardi: the Panamerican Late Pleistocene megatheriid sloth. Journal of Vertebrate Paleontology 15:830–841.

———. 2006. Eremotherium laurillardi (Lund) (Xenarthra, Megatheriidae) the Panamerican giant ground sloth: taxonomic aspects of the ontogeny of skull and dentition. Journal of Systematic Palaeontology 4:199–209.

Cartelle, C., De Iuliis, G., and Pujos, F. 2008. A new species of Megalonychidae (Mammalia, Xenarthra) from the Quaternary of Poço Azul (Bahia, Brazil). Comptes Rendus Palevol 7:335–346.

Cartelle, C., De Iuliis, G., and Ferreira, R. L. 2009. Systematic revision of tropical Brazilian scelidotheriine sloths (Xenarthra, Mylodontoidea). Journal of Vertebrate Paleontology 29:555–566.

Casinos, A. 1996. Bidedalism and quadrupedalism in Megatherium: an attempt at biomechanichal reconstruction. Lethaia 29:87–96.

Cassini, G. H., and Vizcaíno, S. F. 2012. An Approach to the Biomechanics of the Masticatory Apparatus of Early Miocene (Santacrucian Age) South American Ungulates (Astrapotheria, Litopterna, and Notoungulata): Moment Arm Estimation Based on 3D Landmarks. Journal of Mammalian Evolution 19:9–25.

Cassini, G. H., Mendoza, M., Vizcaíno, S.F., and Bargo, M. S. 2011. Inferring habitat and feeding behaviour of early Miocene notoungulates from Patagonia. Lethaia 44:153–165.

Castellanos, A. 1931. "La librería del Glyptodon" de Ameghino. Cultura, Órgano de la Biblioteca Popular Bernardino Rivadavia 3:4–9.

———. 1932. Nuevos géneros de gliptodontes en relación con su filogenia. Physis 11:92–100.

———. 1940. A propósito de los géneros Plohophorus, Nopachthus y Panochthus (2a. parte). Publicaciones de la Facultad de Ciencias Matemáticas, Físico-Químicas y Naturales Aplicadas a la Industria, Universidad Nacional del Litoral. Serie Técnico-Científica 20:279–418.

———. 1959. Trascendencia de la obra de Florentino Ameghino. Revista de la Facultad de Ciencias Naturales de Salta 1:35–56.

Cayssials, R. 1979. Interpretación de los recursos básicos de suelos para su uso, manejo y conservación a nivel nacional. Boletín Técnico N° 1, Ministerio de Ganadería, Agricultura y Pesca, Montevideo, Uruguay.

Chapman, R. E. 1990. Conventional Procrustes approaches; pp. 251–267 in Rohlf, F. J., and Bookstein, F. L. (eds.), Proceedings of the Michigan Morphometrics Workshop. Special Publication 2. University of Michigan Press, Ann Arbor.

Charrier, R., and Vicente, J. C. 1970. Liminary and geosynclinal Andes: major orogenic phases and synchronical evolution of the Southern Andes. Upper Mantle Symposyum, Buenos Aires 2:451–470

Chatwin, B. 1977. In Patagonia. Jonathan Cape, London, 240 pp.

Chimento, N. R., and Agnolin, F. L. 2010. Afinidades entre los Afrotheria y los "ungulados" nativos sudamericanos (Astrapotheria, Notoungulata, Pyrotheria y Xenungulata). Resúmenes XXIII Jornadas Argentinas de Mastozoología, p. 35.

Christiansen, P. 1998 (2000). Strength indicator values of theropod long bones, with comments on limb proportions and cursorial potential; in Pérez-Moreno, B., Holtz, Jr., T. R., Sanz, J. L., and Moratalla, J. J., eds. Theropod Paleobiology Special Volume. Gaia 15:241–255.

———. 1999. What size were Arctodus simus and Ursus spelaeus (Carnivora: Ursidae)? Annales Zoologica Fennici 36:93–102.

———. 2008. Phylogeny of the great cats (Felidae: Pantherinae), and the influence of fossil taxa and missing characters. Cladistics 24:977–992.

Christiansen, P., and Fariña, R. A. 2003. Mass estimation of two fossil ground sloths (Xenarthra; Mylodontidae) in Fariña, R. A., Vizcaíno, S. F., and Storch, G. (eds.), Morphological Studies in Fossil and Extant Xenarthra (Mammalia). Senckenbergiana Biologica 83:95–101.

———. 2004. Mass prediction in theropod dinosaurs. Historical Biology 16:85–92.

Churcher, C. S., Morgan, A. V., and Carter, L. D. 1993. Arctodus simus from the Alaskan Arctic slope. Canadian Journal of Earth Science 30:1007–10013.

Cifelli, R. L. 1985. Biostratigraphy of the Casamayoran, Early Eocene, of Patagonia. American Museum Novitates 2820:1–26.

———. 1993. The phylogeny of the native South American ungulates; pp. 195–216 in Szalay, F. S., Novacek, M. J., and McKenna, M. C., (eds.), Mammal Phylogeny. Vol. 2, Placentals. Springer-Verlag, New York.

Cione, A. L., and López Arbarello, A. 1995. Los peces fósiles del Cenozoico tardío de la región pampeana; in Alberdi, M. T., Leone, G., and Tonni, E. P. (eds.), Evolución climática y biológica de los últimos cinco millones de años. Monografías del Museo Nacional de Madrid 12:129–142.

Cione, A. L., and Tonni, E. P. 1995a. Chronostratigraphy and "land mammal-ages": the Uquian problem. Journal of Paleontology 69:135–159.

———. 1995b. Bioestratigrafía y cronología del Cenozoico superior de la región pampeana; in Alberdi, M. T. Leone G., and Tonni, E. P. (eds.), Evolución climática y biológica de los últimos cinco millones de años. Monografías del Museo Nacional de Madrid 12:47–74.

———. 1995c. El estratotipo de los pisos Montehermosense y Chapadmalalense (Plioceno) del esquema cronológico sudamericano. Ameghiniana 32:369–374.

———. 1999. Biostratigraphy and chronological scale of uppermost Cenozoic in the Pampean area, Argentina; in Tonni E. P., and Cione, A. L. (eds.), Quaternary vertebrate paleontology in South America. Quaternary of South America and Antarctic Peninsula 12:23–51.

———. 2001. Correlation of Pliocene to Holocene southern South American and European Vertebrate-Bearing units. Bolletino della Società Paleontologica Italiana 40:167–173.

———. 2005. Bioestratigrafía basada en mamíferos del Cenozoico superior de la provincia de Buenos Aires, Argentina; in de Barrio, R. E., Etcheverry, R. O., Caballé, M. F., and Llambías, E. (eds.), Geología y Recursos Minerales de la Provincia de Buenos Aires. 16° Congreso Geológico Argentino, Relatorio 11:183–200.

Cione, A. L., Tonni, E. P., and Soibelzon, L. H. 2003. The broken zig-zag: Late Cenozoic large mammal and turtle extinction in South America. Revista del Museo Argentino de Ciencias Naturales "Bernardino Rivadavia" 5:1–19.

Cione, A. L., Tonni E. P., Bargo S., Bond M., Candela A. M., Carlini A. A., Deschamps C. M., Dozo M. T., Esteban G., Goin F. J., Montalvo C. I., Nasif N., Noriega J. I., Ortiz Jaureguizar E., Pascual R., Prado J. L., Reguero M. A., Scillato-Yané G. J., Soibelzon L., Verzi D. H., Vieytes C. E., Vizcaíno S. F., and Vucetich M. G. 2007. Mamíferos continentales del Mioceno tardío a la actualidad en la Argentina: cincuenta años de estudios Publicación Especial de la Asociación Paleontólogica Argentina 11:257–278.

Cione, A. L., Tonni, E. P., and Soibelzon, L. 2009. Did humans cause the Late Pleistocene–Early Holocene mammalian extinctions in South America in a context of shrinking open areas? pp. 125–144 in Haynes, G. (ed.), American Megafaunal Extinctions at the End of the Pleistocene. Springer, Dordrecht.

Clapperton, C. 1983. The glaciation of the Andes. Quaternary Science Review 2:83–155.

Claramunt, S., and Rinderknecht, A. 2005. A new fossil furnariid from the Pleistocene of Uruguay, with remarks on nasal type, cranial kinetics, and relationships of the extinct genus Pseudoseisuropsis. The Condor 107:114–127.

Clark, W. B. 2006. A Medieval Book of Beasts: The Second-Family Bestiary: Commentary, Art, Text, and Translation. Boydell Press, Woodbridge, 480 pp.

Clift, W. 1835. Notice on the Megatherium brought from Buenos Ayres by Woodbine Parish, Esq., F. R. S. Transactions of the Geological Society s2–3:437–450, pl. XLIV–XLVI.

Coates, A. G., and Obando, J. A. 1996. The geologic evolution of the Central American Isthmus; pp. 21–56 in Jackson, J. B. C., Budd, A. F., and Coates, A. G. (eds.), Evolution and Environment in Tropical America. University of Chicago Press, Chicago.

Coates, A. G., Collins, L. S., Aubry M.-P., and Berggren,W. A. 2004. The geology of the Darien, Panama, and the late Miocene–Pliocene collision of the Panama arc with northwestern South America. Geological Society of America Bulletin 116:1327–1344.

Coe, M. J., Dilcher, D. L., Farlow, J. O., Jarzen, D. M., and Russell, D. A. 1987. Dinosaurs and land plants; pp. 225–258 in Friis, E. M., Chaloner, W. G., and Crane, P. R. (eds.), The Origins of Angiosperms and their Biological Consequences. Cambridge University Press, Cambridge, U.K.

Colinvaux, P. 1980. Why big fierce animals are rare. Penguin, Middlesex, 224 pp.

Coltorti, M., Abbazzi, L., Ferreti, M. P., Iacumin, P., Paredes Ríos, F., Pellegrini, M., Pieruccini, P., Rustioni, M., Tito, G., and Rook, L. 2007. Last Glacial mammals in South America: a new scenario from the Tarija Basin (Bolivia). Naturwissenschaften 94:288–299.

Coltrain, J. B., Harris, J. M., Cerling, T. E., Ehleringer, J. R., Dearing, M., Ward, J., and Allen, J. 2004. Rancho La Brea stable isotope biogeochemistry and its implications for the palaeoecology of the late Pleistocene, coastal southern California. Palaeogeography, Palaeoclimatology, Palaeoecology 205:199–219.

Coombs, M. C. 1983. Large mammalian clawed herbivores: a comparative study. Transactions of the American Philosophical Society 73:1–96.

Costa, R. L., and Greaves, W. S. 1981. Experimentally produced tooth wear facets and the direction of jaw motion. Journal of Paleontology 55:635–638.

Creel, S. 2001. Four factors modifying the effect of competition on carnivore population dynamics as illustrated by African wild dogs. Conservation Biology 15:271–274.

Creel, S., Spong, G., and Creel, N. 2001. Interspecific competition and the population biology of extinction-prone carnivores; pp. 35–60 in Gittleman, J. L., Funk, S. M., Macdonald, D. W., and Wayne, R. K. (eds.), Carnivore Conservation. Cambridge University Press, Cambridge.

Crepet, W. L., and Feldman, G. D. 1991. The earliest remains of grasses in the fossil record. American Journal of Botany 78:1010–1014.

Croft, D. A. 1999. Placentals: South American ungulates; pp. 890–906 in Singer, R. (ed.),

Encyclopedia of Paleontology. Fitzroy-Dearborn, Chicago.

————. 2001. Cenozoic environmental change in South America as indicated by mammalian body size distributions (cenograms). Diversity and Distributions 7:271–287.

————. 2007. The Middle Miocene (Laventan) Quebrada Honda fauna, Southern Bolivia and a description of its notoungulates. Palaeontology 50:277–303.

Cuenca Anaya, J. 1995. El aparato locomotor de los escelidoterios (Edentata, Mammalia) y su paleobiología. Colección "Estudis" 6 452 pp., Adjuntament de València, València.

Cuvier, G. 1796a. Notice sur de squelette d'une très grande espèce de quadrupède inconnue jusqu'á présent, trouvé au Paraguay, et déposé au Cabinet dªHistoire Naturelle de Madrid. Magasin encyclopédique, ou Journal des Sciences, des lettres et des Arts 2(1):303–310.

————. 1796b. Mémoire sur les espèces d'éléphants tant vivantes que fossiles, lu á la séance publique de l'Institut nationale le 15 germinal, an IV. Magasin Encyclopédique, ou Journal des Sciences, des Lettres et des Arts 2(3):440–445.

————. 1804. Sur le Megatherium. Annales du Muséum d'Histoire Naturelle 5.

————. 1806. Sur différent dents du genre des mastodontes, mais d'espèces moindres que celles de l'Ohio, trouvées en plusieurs lieux des dux continents. Annales du Muséum national d' Histoire Naturelle 3:401–424.

————. 1812. Recherches sur les ossemens fossiles: où l'on rétablit les caractères de plusieurs animaux dont les révolutions du globe ont détruit les espèces. G. Dufour et E. d'Ocagne, Paris, 232 pp.

————. 1824. Recherches sur les ossements fossiles. 2nd ed. G. Dufour et E. D'Ocagne, Paris; pages not continuously numbered.

————. 1825. Discours sur les révolutions de la surface du globe, et sur les changemens qu'elles ont produits dans le règne animal 3rd ed. G. Dufour et E. d'Ocagne, Paris, 514.

————. 1834. Rozprava o přewratech kůry zemnj, a o proměnách w žiwočstwu gimi způsobených, w ohledu přjrodopisném a děgopisném. Knjžecj arcibiskupská knihtiskárna, Praha, 318 pp.

Czelusniak, J., Goodman, M., Koop, B. F., Tagle, D. A., Shoshani, J., Braunitzer, G., Kleinschmidt, T. K., DeJong, W. W., and Matsuda, G. 1990. Perspectives from amino acid and nucleotide sequences on cladistic relationships among higher taxa of Eutheria; pp. 545–572 in Genoways, H. H. (ed.), Current Mammalogy. Vol. 2. Plenum Press, New York.

Czerwonogora, A., and Fariña, R. A. In press. How many Pleistocene species of Lestodon (Mammalia, Xenarthra, Tardigrada)? Journal of Systematic Palaeontology.

Czerwonogora, A., Fariña, R. A., and Tonni, E. P. 2011. Diet and isotopes of Late Pleistocene ground sloths: first results for Lestodon and Glossotherium (Xenarthra, Tardigrada). Neues Jahrbuch für Geologie und Paläontologie. Abhandlungen Band 262:257–266.

Damuth, J. 1981a. Population density and body size in mammals. Nature 290:699–700.

————. 1981b. Home range, home range overlap and energy use among animals. Biological Journal of the Linnean Society 15:185–193.

————. 1982. Analysis of the preservation of community structure in assemblages of fossil mammals. Paleobiology 8:434–446.

————. 1987. Interspecific allometry of population density in mammals and other animals: the independence of body mass and population energy use. Biological Journal of the Linnean Society 331:193–246.

————. 1990. Problems in estimating body masses of archaic ungulates using dental measurements; pp. 229–253 in Damuth, J., and MacFadden, B. J. (eds.), Body Size in Mammalian Paleobiology: Estimation and Biological Implications. Cambridge University Press, Cambridge.

————. 1991. Of size and abundance. Nature 351:268–269.

————. 1993. Cope's rule, the island rule and the scaling of mammalian population density. Nature 365:748–750.

Damuth, J., and MacFadden, B. J. (eds.) 1990. Body Size in Mammalian Paleobiology: Estimation and Biological Implications. Cambridge University Press, Cambridge, 397 pp.

Dana, J. D. 1896. Manual of Geology. Treating of the principles of the science with special reference to American geological history. 4th ed. American Book Company, New York, 728 pp.

Dantas, M. A. T., and Zucon, M. H. 2007. Occurrence of Catonyx cuvieri (Lund 1839) (Tardigrada, Scelidotheriinae) in late Pleistocene-Holocene of Brazil. Revista Brasileira de Paleontologia 10:129–132.

Darwin, C. 1837. The correspondence of Charles Darwin, vol. 2:1837–1843. Bulletin Géologique.

————. 1839. Narrative of the surveying voyages of His Majesty's Ships Adventure and Beagle between the years 1826 and 1836, describing their examination of the southern shores of South America, and the Beagle's circumnavigation of the globe. Vol. 3, Journal and Remarks, 1832–1836. Henry Colburn Press, London, 615 pp.

————. 1859. The Origin of Species by Means of Natural Selection, or The Preservation of Favoured Races in the Struggle for Life. John Murray, London, 432 pp.

de Burlet, H. M. 1922. Ueber durchbohrte wirbelkorper fossiler und recenter Edentaten. Morphologisches Jahrbuch 51:555–584.

De Clercq, D., Aerts, P., and Kunnen, M. 1994. The mechanical characteristics of the human heel pad during foot strike in running: in vivo cineradiographic study. Journal of Biomechanics 27:1213–1222.

Dechaseaux, C. 1958. Encéphales de Xenarthres fossiles; in Piveteau, J. (ed.), Traité de Paléontologie 6:637–640. Masson et Cie, Paris.

De Esteban-Trivigno, S., Mendoza, M., and De Renzi, M. 2008. Body mass estimation in Xenarthra: a predictive equation suitable for all quadrupedal terrestrial placentals? Journal of Morphology 269:1276–1293.

De Iuliis, G. 1994. Relationships of the Megatheriidae, Nothrotheriinae and Planopsinae: some skeletal characteristics and their importance for phylogeny. Journal of Vertebrate Paleontology 14:577–591.

————. 1996. A systematic review of the Megatheriinae (Mammalia: Xenarthra: Megatheriidae). Ph.D. dissertation, University of Toronto, Toronto, 781 pp.

————. 2006. On the taxonomic status of Megatherium sundti. Philippi 1893 (Mammalia: Xenarthra: Megatheriidae). Ameghiniana 43:161–169.

De Iuliis, G., and Cartelle, C. 1993. The medial carpal and metacarpal elements of Eremotherium and Megatherium (Xenarthra: Mammalia). Journal of Vertebrate Paleontology 13:525–533.

————. 1999. A new giant megatheriine ground sloth (Mammalia: Xenarthra: Megatheriidae) from the late Blancan to early Irvingtonian of Florida. Zoological Jounal of the Linnean Society 127:495–515.

De Iuliis G., and Edmund, A. G. 2002. Vassallia maxima Castellanos 1946 (Mammalia: Xenarthra: Pampatheriidae), from Puerta del Corral Quemado (late Miocene to early Pliocene), Catamarca Province, Argentina. Smithsonian Contributions to Paleobiology 93:49–64.

De Iuliis, G., and Pulerà, D. 2010. The Dissection of Vertebrates: A Laboratory Manual. 2nd ed. Academic Press, Amsterdam, 332 pp.

De Iuliis, G., Bargo, M. S., and Vizcaíno, S. F. 2000. Variation in skull morphology and mastication in the fossil giant armadillos Pampatherium spp., and allied genera (Mammalia: Xenarthra: Pampatheriidae), with comments on their systematics and distribution. Journal of Vertebrate Paleontology 20:743–754.

De Iuliis, G., Pujos, F., and Cartelle, C. 2009. A new ground sloth (Mammalia: Xenarthra) from the Quaternary of Brazil. Comptes Rendus Palevol 8:705–715.

De Iuliis, G., Gaudin, T. J., and Vicars, M. J. 2011. A new genus and species of nothrotheriid sloth (Xenarthra, Tardigrada, Nothrotheriidae) from the Late Miocene (Huayquerian) of Peru. Palaeontology 54:171–205.

de la Fuente, M. 1999. A review of the Pleistocene reptiles of Argentina: taxonomic and palaeoenvironmental considerations. Quaternary of South America and the Antarctic Peninsula 12:109–136.

D'Elía, G., Luna, L. González, E. M., and Patterson, B. D. 2005. On the Sigmodontinae radiation (Rodentia, Cricetidae): an appraisal of the phylogenetic position of Rhagomys. Molecular Phylogenetics and Evolution 38:558–564.

DeJong, W. W., Leunissen, J. A. M., and Wistow, G. J. 1993. Eye lens crystallins and the phylogeny of placental orders: evidence for a macroscelid–paenungulate clade?; pp. 81–102 in Szalay, F. S., Novacek, M. J., and McKenna, M. C. (eds.), Mammal Phylogeny: Placentals. Springer-Verlag, New York.

Delsuc, F., and Douzery, E. J. P. 2008. Recent advances and future prospects in xenarthran molecular phylogenetics; pp. 11–23 in Vizcaíno S. F., and Loughry, W. J. (eds.), The Biology of the Xenarthra. University Press of Florida, Gainesville.

Delsuc, F., Catzeflis, F. M., Stanhope, M. J., and Douzery, E. J. P. 2001. The evolution of armadillos, anteaters and sloths depicted by nuclear and mitochondrial phylogenies: implications for the status of the enigmatic fossil *Eurotamandua*. Proceedings of the Royal Society B 268:1605–1615.

Delsuc, F., Scally, M., Madsen, O., Stanhope, M. J., De Jong, W. W., Catzeflis, F. M., Springer, M. S., and Douzery, E. J. P. 2002. Molecular phylogeny of living xenarthrans and the impact of character and taxon sampling on the placental tree rooting. Molecular Biology and Evolution 19:1656–1671.

Desmond, A. 1982. Archetypes and Ancestors: Paleontology in Victorian London 1850–1875. University of Chicago Press, Chicago, 287 pp.

Dillehay, T. D. 1999. The Late Pleistocene cultures of South America. Evolutionary Anthropology 7:206–216.

———. 2000. The Settlement of the Americas: A New Prehistory. Basic Books, New York, 371 pp.

Dimery, N. J., Alexanderm, R. McN., and Deyst, K. A. 1985. Mechanics of the ligamentum nuchae of some artiodactyls. Journal of Zoology (London), London 206:341–351.

Ding, S.-Y. 1979. A new edentate from the Paleocene of Guang-dong. Vertebrata Palasiatica 12:62–64.

D'Orbigny, A. 1842. Voyage dans l'Amérique Méridionale. (Le Brésil, La République orientale de l'Uruguay, la République argentine, la Patagonie, la République du Chili, la République de Bolivia, la République du Pérou), exécuté pendant les années 1826, 1827, 1828, 1829, 1830, 1831, 1832 et 1833. Pitois-Levrault et Cie, Paris and Strasbourg, 152 pp.

Dyke, A. S., and Prest, V. K. 1987. Late Wisconsinan and Holocene history of the Laurentide ice sheet. Geographie Physique et Quaternaire 41:237–264.

East, R. 1984. Rainfall, soil nutrient status and biomass of large African savanna mammals. African Journal of Ecology 22:245–270.

Economos, A. 1981. The largest land mammal. Journal of Theoretical Biology 89:211–214.

Edmund, A. G. 1985. The fossil giant armadillos of North America (Pampatheriinae, Xenarthra = Edentata); pp. 83–93 in Montgomery, G. G. (ed.), The Evolution and Ecology of Armadillos, Sloths, and Vermilinguas. Smithsonian Institution Press, Washington, D.C.

———. 1996. A review of Pleistocene giant armadillos (Mammalia, Xenarthra, Pampatheriidae); pp. 300–321 in Stewart, K. M., and Seymour, K. L. (eds.), Palaeoecology and Palaeoenvironments of Late Cenozoic Mammals. University of Toronto Press, Toronto.

Edwards, M. E., Anderson, P. M., Brubaker, L. B., Ager, T. A., Andreev, A. A., Bigelow, N. H., Cwynar, L. C., Eisner, W. R., Harrison, S. P., Hu, F. S., Jolly, D., Lozhkin, A. V., MacDonald, G. M., Mock, C. J., Ritchie, J. C., Sher, A. V., Spear, R. W., Williams, J. W., and Yu, G. 2000. Plant-based biomes for Beringia 18,000 6,000, and 014C yr B. P. Journal of Biogeography 27:521–554.

Eizirik, E., Murphy, W. J., and O'Brien, S. J. 2001. Molecular dating and biogeography of the early placental mammal radiation. The Journal of Heredity 92:212–219.

Elissamburu, A. 2010. Estudio biomecánico y morfofuncional del esqueleto apendicular de *Homalodotherium* Flower 1873 (Mammalia, Notoungulata). Ameghiniana 47:25–43.

Elissamburu, A., and Vizcaíno, S. F. 2004. Limb proportions and adaptations in caviomorph rodents (Rodentia: Caviomorpha). Journal of Zoology (London) 262:145–159.

———. 2005. Diferenciación morfométrica del húmero y fémur de las especies de *Paedotherium* (Mammalia, Notoungulata) del Plioceno y Pleistoceno temprano. Ameghiniana 42:159–166.

Engelmann, G. 1985. The phylogeny of the Xenarthra; pp. 51–64 in Montgomery, G. G. (ed.), The Evolution and Ecology of Armadillos, Sloths, and Vermilinguas. Smithsonian Institution Press, Washington, D.C.

Esteban, G. I. 1996. Revisión de los Mylodontinae cuaternarios (Edentata–Tardigrada) de Argentina, Bolivia y Uruguay. Sistemática, filogenia, paleobiología, paleozoogeografía y paleoecología. Ph.D. dissertation, Facultad de Ciencias Naturales e Instituto Miguel Lillo, Universidad Nacional de Tucumán, 314 pp.

Evans, A. H. 1899. Birds. Illustrated by George Edward Lodge. The Cambridge Natural History Society/MacMillan, London, 635 pp.

Ewer, R. F. 1973. The Carnivores. Cornell University Press, Ithaca, N.Y., 494 pp.

Faith, J. T., and Surovell, T. A. 2009. Synchronous extinction of North America's Pleistocene mammals. Proceedings of the National Academy of Sciences of the United States of America 106:20641–20645.

Falkner, T. 1774. A description of Patagonia and the adjoining parts of South America; containing an account of the soil, produce, animals, vales, mountains, rivers, lakes &c. of those countries; the religion, government, policy, customs, dress, arms, and language of the Indian inhabitants; and some particulars relating to Falkland's Islands. C. Pugh-T. Lewis, London-Hereford, 144 pp.

Fariña, R. A. 1985. Some functional aspects of mastication in Glyptodontidae. Fortschritte der Zoologie 30:277–280.

———. 1988. Nuevas observaciones sobre la biomecánica masticatoria en Glyptodontidae. Boletín de la Sociedad Zoológica del Uruguay 4:5–9.

———. 1995. Limb bone strength and habits in large glyptodonts. Lethaia 28:189–196.

———. 1996. Trophic relationships among Lujanian mammals. Evolutionary Theory 11:125–134.

———. 2002a. *Megatherium*, el pelado: sobre la apariencia de los grandes perezosos (Mammalia; Xenarthra) cuaternarios. Ameghiniana 39:241–244.

———. 2002b. Taphonomy and Palaeoecology of the South American giant mammals; pp. 97–113 in De Renzi, M., Pardo Alonso, M. V., Belinchón, M., Peñalver, E., Montoya, P., and Márquez-Aliaga, A. (eds.), Current Topics in Taphonomy and Fossilization. Ayuntamiento de Valencia, Valencia.

Fariña, R. A., and Álvarez, F. 1994. La postura de *Toxodon*: una nueva reconstrucción. Acta Geologica Leopoldensia 39:565–571.

Fariña, R. A., and Blanco, R. E. 1996. *Megatherium*, the stabber. Proceedings of the Royal Society B 263:1725–1729.

Fariña, R. A., and Castilla, R. 2007. Earliest evidence for human–megafauna interaction in the Americas; in Corona, M. E., and Arroyo-Cabrales, J. (eds.), Human and Faunal Relationships Reviewed: An Archaeozoological Approach. BAR S1627:31–33. Archaeopress, Oxford.

Fariña, R. A., and Parietti, M. 1983. Uso del método RFTRA en la comparación de la morfología craneana en Edentata. Resúmenes y Comunicaciones de las 3as. Jornadas de Ciencias Naturales:106–108.

Fariña, R. A., and Vizcaíno, S. F. 1995. Hace Sólo Diez Mil Años. Fin de Siglo, Montevideo, 128 pp.

———. 1996. Hábitos locomotores del armadillo pleistoceno *Propraopus grandis* (Mammalia, Dasypodidae): una comparación con formas actuales. Actas del VI Congreso Argentino de Paleontología y Bioestratigrafía, pp. 123–127.

———. 1997. Allometry of the bones of living and extinct armadillos (Dasypoda). Zeitschrift für Säugetierkunde 62:65–70.

———. 1999. A century after Ameghino: the palaeobiology of the large Quaternary mammals of South America revisited. Quaternary of South America and the Antarctic Peninsula 12:255–277.

———. 2001. Carved teeth and strange jaws: how glyptodonts masticated; in Vizcaíno, S. F., Fariña, R. A., and Janis, C. (eds.), Acta Palaeontologica Polonica, Special Issue on Biomechanics and Palaeobiology 46:87–102.

Fariña, R. A., Vizcaíno, S. F., and Blanco, E. 1997. Scaling of the indicator of athletic capability in fossil and extant land tetrapods. Journal of theoretical Biology 185:441–446.

Fariña, R. A., Vizcaíno, S. F., and Bargo, M. S. 1998. Body mass estimations in Lujanian (Late Pleistocene-Early Holocene of South America) mammal megafauna. Mastozoología Neotropical 5:87–108.

Fariña, R. A., Blanco, R. E., and Christiansen, P. 2005. Swerving as the escape strategy of *Macrauchenia patachonica* (Mammalia; Litopterna). Ameghiniana 42:751–760.

Farlow, J. O. 1993. On the rareness of big, fierce animals: speculation about the body sizes, population densities, and geographic ranges of predatory ammals and large carnivorous dinosaurs. American Journal of Science 293-A:167–199.

Farlow, J. O., and Brett-Surman, M. K. (eds.) 1997. The Complete Dinosaur. Indiana University Press, Bloomington, 752 pp.

Farlow, J. O., Brinkman, D. L., Abler, W. L., and Currie, P. J. 1991. Size, shape, and serration density of theropod dinosaur lateral teeth. Modern Geology 16:161–198.

Farlow, J. O., Dodson, P., and Chinsamy, A. 1995. Dinosaur biology. Annual Review of Ecology and Systematics 26:445–471.

Farrand, W. R. 1961. Frozen mammoths and modern geology: the death of the giants can be explained as a hazard of tundra life, without evoking catastrophic events. Science 133:729–735.

Ferigolo, J. 1985. Evolutionary trends of the histological pattern in the teeth of Edentata (Xenarthra). Archives oral Biology 30:71–82.

———. 1999. Late Pleistocene South America land-mammal extinctions: the infection hypothesis. Quaternary of South America and Antarctic Peninsula 12 279–299.

Ferigolo, J., and Berman, W. D. 1993. Dental paleopathology and paleodiet of *Arctotherium bonariensis* (Mammalia, Carnivora, Ursidae) from the Ensenadan (Lower to Middle Pleistocene), Buenos Aires Province, Argentina. Boletim de Resumos 13 Congresso Brasileiro de Paleontologia, São Leopoldo.

Fernicola, J. C. 2008. Nuevos aportes para la sistemática de los Glyptodontia Ameghino 1889 (Mammalia/Xenarthra/Cingulata). Ameghiniana 48:553–574.

Fernicola, J. C., Vizcaíno, S. F., Fariña, R. A. 2008. The evolution of armored xenarthrans and a phylogeny of the glyptodonts; pp. 79–85 in Vizcaíno, S. F., and Loughry, W. J. (eds.), Biology of the Xenarthra. University of Florida Press, Gainesville.

Fernicola, J. C., De Iuliis, G., and Vizcaíno, S. F. 2009. The fossil mammals collected by Charles Darwin in South America during his travels on board the HMS Beagle. Revista de la Asociación Geológica Argentina 64:147–159.

Ferretti, M. P. 2008. A review of South American gomphotheres. New Mexico Natural History and Science Museum Bulletin 44:381–391.

———. 2010. Anatomy of Haplomastodon chimborazi (Mammalia, Proboscidea) from the late Pleistocene of Ecuador and its bearing on the phylogeny and systematics of South American gomphotheres. Geodiversitas 32:663–721.

Ficcarelli, G., Borselli, V., Herrera, G., Moreno Espinosa, M., and Torre, D. 1995. Taxonomic remarks on the South American mastodons referred to Haplomastodon and Cuvieronius. Geobios 28:745–756.

Fiedel, S., and Haynes, G. 2004. A premature burial: comments on Grayson and Meltzer's "Requiem for overkill." Journal of Archaeological Science 31:121–131.

Figueirido, B., and Soibelzon, L. H. 2009. Inferring palaeoecology in extinct tremarctine bears (Carnivora, Ursidae) using geometric morphometrics. Lethaia 43:209–222.

Figueirido, B., Palmqvist, P., and Pérez-Claros, J. A. 2008. Ecomorphological correlates of craniodental variation in bears and paleobiological implications for extinct taxa: an approach based on geometric morphometrics. Journal of Zoology (London) 227:70–80.

Figueirido, B., Pérez-Claros, J. C., Torregrosa, V., Martín-Serra, A., and Palmqvist, P. 2010. Demythologizing Arctodus simus, the "short-faced" long-legged and predaceous bear that never was. Journal of Vertebrate Paleontology 30:262–275.

Firestone, R. B., West, A., Kennett, J. P., Becker, L., Bunch, T. E., Revay, Z. S., Schultz, P. H., Belgya, T., Kennett, D. J., Erlandson, J. M., Dickenson, O. J., Goodyear, A. C., Harris, R. S., Howard, G. A., Kloosterman, J. B., Lechler, P., Mayewski, P. A., Montgomery, J., Poreda, R., Darrah, T., Que Hee, S. S., Smith, A. R., Stich, A., Toping, W., Wiike, J. H., and Wolbach, W. S. 2007. Evidence for an extraterrestrial impact 12,900 years ago that contributed to the megafaunal extinctions and the Younger Dryas cooling. Proceedings of the National Academy of Sciences of the United States of America 104:16016–16021.

Fischer de Waldheim, G. 1814. Zoognosia Tabulis synopticis illustrata. Vol. 3. Nicolai S. Vsevolozsky, Moscow, 694 pp.

Fleagle, J. G. 1999. Primate Adaptation and Evolution. 2nd ed. Academic Press, San Diego, 596 pp.

Flessa, K. W., Erben, H. K., Hallam, A., Hsü, K. J., Hüssner, H. M., Jablonski, D., Raup, D. M., Sepkoski, Jr., J. J., Soulé, M. E., Sousa, W., Stinnesbeck,W., and Vermeij, G. J. 1986. Causes and consequences of extinctions; pp. 235–257 in Raup, D. M., and Jablonski, D. (eds.), Patterns and Processes in the History of Life. Springer Verlag. Berlin.

Flynn, J. J., and Swisher, III, C. 1995. Cenozoic South American Land Mammal Ages: correlation to global geochronologies. (SEPM) Society for Sedimentary Geology, Special Publication 54:317–333.

Flynn, J. J., and Wesley-Hunt, G. D. 2005. Carnivora; pp. 175–198 in Rose, K. D., and Archibald J. D. (eds.), The Rise of Placental Mammals: Origins and Relationships of the Major Extant Clades. Johns Hopkins University Press, Baltimore.

Flynn, J. J., and Wyss, A. R. 1999. New marsupials from the Eocene–Oligocene transition of the Andean Main Range, Chile. Journal of Vertebrate Paleontology 19:533–549.

Flynn, J. J., Croft, D. A., Charrier, R., Hérail, G., and Wyss, A. R. 2002. The first Cenozoic mammal fauna from the Chilean altiplano. Journal of Vertebrate Paleontology 22:200–206.

Flynn, J. J., Kowallis, B. J., Núñez, C., Carranza-Castañeda, Ó., Miller, W. E, Swisher, C. C., III, and Lindsay, E. H. 2005. Geochronology of Hemphillian–Blancan aged strata, Guanajuato, Mexico, and implications for timing of the Great American Biotic Interchange. Journal of Geology 113:287–307.

France, D. L. 2009. Human and nonhuman bone identification: a colour atlas. CRC Press, Boca Raton, 584 pp.

France, C. A. M., Zelanko, P. M., Kaufman. A. J., and Holtz, T. R. 2007. Carbon and nitrogen isotopic analysis of Pleistocene mammals from the Saltville Quarry (Virginia, USA): implications for trophic relationships. Palaeogeography, Palaeoclimatology, Palaeoecology 249:271–282

Frechkop, S. 1949. Explication biologique fournie par les tatous, d'un des caractères distinctifs des Xenarthres et d'un caractre adaptatif analogue chez les Pangolins. Bulletin de l'Institut Royal des Sciences naturelles de Belgique 25:1–12.

———. 1950. Locomotion et la structure des tatus et des pangolins. Annales de la Société Royale Zoologique de Belgique 80:5–8.

Frenguelli, J. 1928. Observaciones geológicas en la región costanera sur de la Provincia de Buenos Aires. Anales de la Universidad Nacional del Litoral, Facultad de Ciencias de la Educación 3:101–130.

Friedman, W. E. 2008. The meaning of Darwin's "abominable mystery." American Journal of Botany 96:5–21,

Fuller, T. K., and Sievert, P. R. 2001. Carnivore demography and the consequences of changes in prey availability; pp. 163–178 in Gittleman, J. L., Funk, S. M., Macdonald, D. W., and Wayne, R. K. (eds.), Carnivore Conservation. Cambridge University Press, Cambridge.

Gallardo, Á. 1912 (ed.). Anales del Museo Nacional de Historia Natural de Buenos Aires, 22: 1–462.

Gardner, A. L. 2007. Mammals of South America. Vol. 1, Marsupials, Xenarthrans, Shrews, and Bats. University of Chicago Press, Chicago, 669 pp.

Garland, Jr., T., and Janis, C. M. 1993. Does metatarsal/tarsal ratio predict maximal running speed in cursorial mammals? Journal of Zoology (London) 229:133–151.

Gaudin, T. J. 1993. The phylogeny of the Tardigrada (Xenarthra, Mammalia) and the evolution of locomotor function in the Xenarthra. Ph.D. dissertation, University of Chicago, Chicago, 226 pp.

———. 1995. The ear region of edentates and the phylogeny of the Tardigrada (Mammalia, Xenarthra). Journal of Vertebrate Paleontology 15:672–705.

———. 1999a. Xenarthra; pp. 1347–1353 in Singer, R. S. (ed.), Encyclopedia of Paleontology. Vol. 2. Fitzroy Dearborn Publishers, Chicago.

———. 1999b. The morphology of xenarthrous vertebrae (Mammalia, Xenarthra). Fieldiana (Geology), n. s. 41:1–38.

———. 2003. Phylogeny of the Xenarthra (Mammalia); in Fariña, R. A., Vizcaíno, S. F., and Storch, G. (eds.), Morphological Studies in Fossil and Extant Xenarthra (Mammalia). Senckenbergiana Biologica 83:5–17.

Gaudin, T. J., and Branham, D. G. 1998. The phylogeny of the Myrmecophagidae (Mammalia, Xenarthra, Vermilingua) and relationship of Eurotamandua to the Vermilingua. Journal of Mammalian Evolution 5:237–265.

Gaudin, T. J., and McDonald, H. G. 2008. Morphology-based investigations of the phylogenetic relationships among extant and fossil xenarthrans; pp. 24–36 in Vizcaíno, S. F., and Loughry, W. J. (eds.), The Biology of the Xenarthra. University of Florida Press, Gainesville.

Gaudin, T. J., and Wible, J. R. 1999. The entotympanic of pangolins and the phylogeny of the Pholidota. Journal of Mammalian Evolution 6:39–65.

———. 2006. The phylogeny of living and extinct armadillos (Mammalia, Xenarthra, Cingulata): a craniodental analysis; pp. 153–198 in Carrano, M. T., Gaudin, T. J., Blob, W. R., and Wible, J. R. (eds.), Amniote Paleobiology: Perspectives on the Evolution of Mammals, Birds and Reptiles. University of Chicago Press, Chicago.

Gayet, M. 2001. A review of some problems associated with the occurrences of fossil vertebrates in South America. Journal of South American Earth Sciences 14:131–145.

Geikie, A. 1905. The Founders of Geology. 2nd ed. Macmillan, New York, 486 pp.

Gelfo, J. N., and Pascual, R. 2001. Peligrotherium tropicalis (Mammalia, Dryolestida) from the early Paleocene of Patagonia, a survival from a Mesozoic Gondwanan radiation. Geodiversitas 23:369–379.

Gelfo, J. N., López, G. M., and Bond, M. 2008. A new Xenungulata (Mammalia) from the Paleocene of Patagonia, Argentina. Journal of Paleontology 82:329–335.

Geraads, D. 2006. The late Pliocene locality of Ahl al Oughlam, Morocco: vertebrate fauna and interpretation. Transactions of the Royal Society of South Africa 61:97–101.

Gervais, P. 1874. Lestodon trigonidens et Valgipes deformis. Journal de Zoologie 3:162–164.

Gervais, H., and Ameghino, F. 1880. Los Mamíferos de la América del Sud. Igon Hermanos, Buenos Aires, 225 pp.

Getty, R. 1975. Sisson and Grossman's The Anatomy of the Domestic Animals. 5th ed, vol. 1. W. B. Saunders, Philadelphia, 1211 pp.

Gheerbrant, E., and Tassy, P. 2009. L'origine et l'évolution des éléphants. Comptes Rendus Palevol 8:281–294.

Gillette, D. D., and Ray, C. E. 1981. Glyptodonts of North American. Smithsonian Contributions to Paleobiology 40:1–255.

Gilmore, D., Fittipaldi Duarte, D., and Peres da Costa, C. 2008. The physiology of two- and three-toed sloths; pp. 130–142 in Vizcaíno, S. F., and Loughry, W. J. (eds.), The Biology of the Xenarthra. University Press of Florida, Gainesville.

Grayson, D. K. 1984. Archaeological associations with extinct Pleistocene mammals in North America. Journal of Archaeological Science 11:213–221.

Griffiths, D. 1998. Sampling effort, regression method, and the shape and the slope of size-abundance relations. Journal of Animal Ecology 67:795–804.

Gobetz, K. E., and Martin, L. D. 2001. An exceptionally large short-faced bear (Arctodus simus) from the late Pleistocene(?)/early Holocene of Kansas. Current Research in the Pleistocene 18:97–98.

Goin, F. J. 1995. Los Marsupiales; pp. 165–179 in Alberdi, M. T., Leone, G., and Tonni, E. P. (eds.), Evolución biológica y climática de la Región Pampeana durante los últimos cinco millones de años. Un ensayo de correlación con el Mediterráneo occidental. Museo Nacional de Ciencias Naturales, Consejo Superior de Investigaciones Científicas, Madrid, 423 pp.

———. 1997. New clues for understanding Neogene marsupial radiations; pp. 187–206 in Kay, R. F., Madden, R. H., Cifelli, R. L., and Flynn, J. J. (eds.), Vertebrate Paleontology in the Neotropics: The Miocene Fauna of La Venta, Colombia. Smithsonian Institution Press, Washington, D.C.

Goloboff, P. A., Catalano, S. A., Mirande, J. M., Szumika, C. A., Ariasa, S. J., Källersjö, M., and Farris, J. S. 2009. Phylogenetic analysis of 73 060 taxa corroborates major eukaryotic groups. Cladistics 25:211–230.

Gorman, M. L., Mills, M. G., Raath, J. P., and Speakman, J. R. 1998. High hunting costs make African wild dogs vulnerable to kleptoparasitism by hyaenas. Nature 391:479–481.

Gould, S. J. 1983. Hen's Teeth and Horse's Toes: Further Reflections in Natural History. W. W. Norton, New York, 413 pp.

———. 1985. The Flamingo's Smile: Reflections in Natural History. W. W. Norton, New York, 479 pp.

———. 2000. The Lying Stones of Marrakech: Penultimate Reflections in Natural History. Harmony Books, New York, 372 pp.

Gradstein, F. M., Ogg, J. G., Smith, A. G., Bleeker, W., and Lourens, L. J. 2004. A new geologic time scale, with special reference to Precambrian and Neogene. Episodes 27:83–100.

Grassé, P.-P. 1955. Ordre des Édentés; pp. 1182–1266 in Grassé, P.-P. (ed.), Traité de Zoologie. Vol. 17, Mammifères. Masson et Cie, Paris, 1170 pp.

Gray, J. E. 1869. Catalogue of carnivorous, pachydermatous and edentate Mammalia in the British Museum. Taylor & Francis, London, 398 pp.

Grayson, D. K. 2001. The archaeological record of human impacts on animal populations. Journal of World Prehistory 15:1–68.

———. 2007. Deciphering North American Pleistocene extinctions. Journal of Anthropological Research 63:185–213.

Grayson, D. K., and Meltzer, D. J. 2003. Requiem for North American overkill. Journal of Archaeological Science 30:585–593.

Greaves, W. S. 1973. The inference of jaw motion from tooth wear facets. Journal of Paleontology 47:1000–1001.

Greenwood, A. D., Castresana, J., Feldmaier-Fuchs, G., and Pääbo, S. 2001. A molecular phylogeny of two extinct sloths. Molecular Phylogenetics and Evolution 18:94–103.

Griffin, D. R. 1992. Animal Minds. Chicago: University of Chicago Press, 311 pp.

Gröcke, D. R. 1997. Stable-isotope studies on the collagen and hydroxylapatite components of fossils: palaeoecological implications. Lethaia 30:65–78.

Groves, C. P. 2007. Family Cervidae; pp. 249–256 in Prothero, D. R., and Foss, S. E. (eds.), The Evolution of Artiodactyls. John Hopkins University Press, Baltimore.

Guérin, C., and Faure, M. 1999. Palaeolama (Hemiauchenia) niedae nov. sp, nouveau Camelidae du Nordeste Brasilien, et sa place parmi les Lamini d'Amérique du Sud, Geobios 32:629–659.

———. 2000. La véritable nature de Megatherium laurillardi Lund 1842 (Mammalia, Xenarthra): un nain parmi les géants. Geobios 33:475–488.

———. 2004a. Scelidodon piauiense nov. sp., nouveau Mylodontidae, Scelitheriinae (Mammalia, Xenarthra) du Quaternaire de la region du parc national Serra da Capibara (Piauí, Brésil). Comptes Rendus Palévol 3:3–42.

———. 2004b. Macrauchenia patachonica Owen (Mammalia, Litopterna) de la région de São Raimundo Nonato (Piauí, Nordeste brésilien) et la diversité des Macraucheniidae pléistocènes. Geobios 37:516–535.

Guilday, J. E. 1984. Pleistocene extinction and environmental change: case study of the Appalachians; pp. 250–258 in Martin, P. S., and Klein, R. G. (eds.), Quaternary Extinctions: A Prehistoric Revolution. University of Arizona Press, Tucson.

Guimarães, P. R., Galetti, M., and Jordano, P. 2008. Seed dispersal anachronisms: rethinking the fruits extinct megafauna ate. PLOS One 3:e1745.

Guthrie, R. 2003. Rapid body size decline in Alaskan Pleistocene horses before extinction. Nature 426:169–171.

———. 2006. New carbon dates link climatic change with human colonization and Pleistocene extinctions. Nature 441:207–209.

Gutiérrez, M. A., Martínez, G. A., Bargo, M. S., and Vizcaíno, S. F. 2010. Supervivencia diferencial de mamíferos de gran tamaño en la región pampeana en el Holoceno temprano y su relación con aspectos paleobiológicos; pp. 231–241 in Gutiérrez, M. A., De Nigris, M., Fernández, P. M., Giardina, M., Gil, A. F., Izeta, A., Neme, G., and Yacobaccio, H. D.

(eds.), Zooarqueología a principios del siglo XXI: Aportes Teóricos, Metodológicos y Casos de Estudio. Ediciones El Espinillo, Buenos Aires.

Hallam, A. 1983. Great Geological Controversies. Oxford University Press, Oxford, 182 pp.

Haller, M. J. 2004. En memoria de John Bell Hatcher (1861–1904) y las expediciones de la Universidad de Princeton a la Patagonia de 1896 a 1899. Revista de la Asociación Geológica Argentina 59:523–524.

Handy, J. P., and Biggot, J. D. 1979. Population change in lions and other predators; pp. 249–262 in Sinclair, A. R. E., and Griffiths, M. N. (eds.), Serengeti. Dynamics of an Ecosystem. University of Chicago Press, Chicago.

Haq, B. U., Hardenbol, J., and Vail, P. R. 1987. Chronology of fluctuating sea levels since the Triassic. Science 235:1156–1167.

Hassig, D. 1995. Medieval Bestiaries: Text, Image, Ideology. Cambridge University Press, New York, 300 pp.

Hatcher, J. B. 1899. Scientific Books: The Mysterious Mammal of Patagonia, Grypotherium Domesticum. Science 10:814–815

———. 1985 (1903). Bone Hunters in Patagonia: Narrative of the Expedition. Ox Bow Press, Woodbridge, Conn., 209 pp.

Hauthal, R., Roth, S., and Lehmann-Nitsche, R. 1899. El mamífero misterioso de la Patagonia "Grypotherium domesticum." Revista del Museo de La Plata 9:411–412.

Hawking, S. 1988. A Brief History of Time: From the Big Bang to Black Holes. Bantam Books, New York, 198 pp.

Head, J. J., Bloch, J. I., Hastings, A. K., Bourque, J. R., Cadena, E. A., Herrera, F. A., Polly, P. D., and Jaramillo, C. A. 2009. Giant boid snake from the paleocene neotropics reveals hotter past equatorial temperatures. Nature 457:715–718.

Hellman, H. 1998. Great Feuds in Science: Ten of the Liveliest Disputes Ever. John Wiley and Sons, New York, 240 pp.

Henderson, D. M. 1999. Estimating the masses and centers of mass of extinct animals by 3–D mathematical slicing. Paleobiology 25:88–106.

Herbert, S. 1980. The Red Notebook of Charles Darwin. Bulletin of the British Museum (Natural History) Historical Series 7:1–164

Hiiemäe, K. M. 1978. Mammalian mastication: a review of the activity of the jaw muscles and the movements they produce in chewing; pp. 359–398 in Butler, P. M., and Joysey, K. A. (eds.), Development, Function and Evolution of Teeth. London, Academic Press.

Hiiemäe, K., and Crompton, A. 1985. Mastication, food transport and swallowing; pp. 262–290 in Hildebrand, M., Bramble, D., Liem, K., and Wake, D. (eds.), Functional Vertebrate Morphology. Belknap Press, Harvard University Press, Cambridge.

Hill, A. V. 1938. The heat of shortening and the dynamic constants of muscles. Proceedings of the Royal Society B 126:136–195.

Hinojosa, L. F. 2005. Cambios climáticos y vegetacionales inferidos a partir de paleofloras cenozoicas del sur de Sudamérica. Revista Geológica de Chile 32:95–115.

Hinojosa, L. F., and Villagrán, C. 1997. Historia de los bosques del sur de Sudamérica, I: antecedentes paleobotánicos, geológicos y climáticos del Terciario del cono sur de

América. Revista Chilena de Historia Natural 70:225–239.

Hoffmann, J. A. J. 1975. Climatic Atlas of South America. WMO, UNESCO and Cartographica, Budapest, 27 pp.

Hoffstetter, R. 1952. Les mammifères pléistocènes de la République de l'Équateur. Mémoires de la Société Géologique de France, nouvelle série 66:1–391.

———. 1954a. Les Gravigrades cuirassés du Déséadien de Patagonie (note préliminaire). Mammalia 18:159–169.

———. 1954b. Phylogenie des Édentés Xenarthres. Bulletin du Museum National del Histoire Naturelle 26:433–438.

———. 1958. Édentés Xénarthres; pp. 536–636 in Piveteau, J. (ed.), Traité de Paléontologie. Vol. 6. Masson et Companie, Paris.

———. 1963. La faune pléistocène de Tarija (Bolivie): note préliminaire. Bulletin du Muséum National d'Histoire Naturelle 35:194–203.

———. 1968. Sur la répartition géographique des Macraucheniidae (Mammifères, Litopternes) au Pléistocène. Comptes Rendus sommaires des séances de la Société géologique de France 3:85–86.

———. 1969. Remarques sur la phylogénie et la classification des Édentés Xénarthres (Mammifères) actuels et fossiles. Bulletin du Museum National del Histoire Naturelle 41:91–103.

———. 1982. Les Édentés Xénarthres, un groupe singulier de la faune néotropicale (origines, affinités, radiations adaptatives, migrations et extinctions); pp. 385–443 in Gallitelli, E. M. (ed.), Paleontology, Essential of Historical Geology. Proceedings of the First International Meeting of Paleontology, Venice, 1981.

———. 1986. High Andean mammalian faunas during the Plio-Pleistocene; pp. 218–245 in Vuilleumier, F., and Monasterio, N. (eds.), High Altitude Tropical Biogeography. Oxford University Press, Oxford.

Hoffstetter, R., and Paskoff, R. 1966. Présence des genres *Machrauchenia* et *Hippidion* dans la faune Pleistocène du Chili. Bulletin du Museum national d' Histoire naturelle (2nd ser.) 38:476–490.

Hofmann, R. R., and Stewart, D. R. M. 1972. Grazer or browser: a classification based on the stomach-structure and feeding habits of east African ruminants. Mammalia 36:226–240.

Höner, O. P., Wachter, B., East, M. L., Runyoro, V. A., and Hofer, H. 2005. The effect of prey abundance and foraging tactics on the population dynamics of a social, territorial carnivore, the spotted hyena. Oikos 108:544–554.

Hooghiemstra, H. 1984. Vegetational and climatic history of the high plain of Bogotá, Colombia: a continuous record of the last 3.5 million years. Dissertationes Botanicae 79:42–67.

———. 2002. The dynamic rainforest ecosystem on geological, Quaternary and human time scales; pp. 7–19 in Verweij, P. (ed.), Understanding and Capturing the Multiple Values of Tropical Forest. Tropenbos International, Wageningen.

Hooker, J. J., and Dashzeveg, D. 2004. The origin of chalicotheres (Perissodactyla, Mammalia). Palaeontology 47:1363–1386

Horovitz, I. 2003. The type skeleton of *Ernanodon antelios* is not a single specimen. Journal of Vertebrate Paleontology 23:706–708

———. 2004. Eutherian mammal systematics and the origins of South American ungulates; in Dawson, M., and Lillegraven. J. (eds.), Fanfare for an Uncommon Vertebrate Paleontologist: Papers on Vertebrate Evolution in Honor of Malcolm Carnegie McKenna. Bulletin of the Carnegie Museum of Natural History 36:63–79.

Horovitz, I., Sánchez-Villagra, M. R., and Aguilera, O. A. 2006. The fossil record of *Phoberomys pattersoni* Mones 1980 (Mammalia, Rodentia) from Urumaco (Late Miocene, Venezuela), with an analysis of its phylogenetic relationships. Journal of Systematic Palaentology 4:293–306

Hutton, J. 1788. Theory of the Earth; or an investigation of the laws observable in the composition, dissolution, and restoration of land upon the Globe. Transactions of the Royal Society of Edinburgh 1:209–304.

Janis, C. M. 1988. An estimation of tooth volume and hypsodonty indices in ungulate mammals and the correlation of these factors with dietary preferences; pp. 367–387 in Russel, D. E., Santorio, J. P., and Signogneu-Russel, D. (eds.), Teeth Revised: Proceedings of the VII International Symposium on Dental Morphology. Muséum national de Histoire Naturelle Memoir série C, Paris.

———. 1995. Correlations between craniodental morphology and feeding behavior in ungulates: reciprocal illumination between living and fossil taxa; pp. 76–98 in Thomason, J. J. (ed.), Functional Morphology in Vertebrate Paleontology. Cambridge University Press, Cambridge.

Honeycutt, R. L., Frabotta, L. J., and Rowe, D. L. 2007. Rodent evolution, phylogenetics and biogeography; pp. 8–23 in Wolff, J. O., and Sherman, P. W. (eds.), Rodent Societies: An Ecological and Evolutionary Perspective. University of Chicago Press, Chicago.

Horner, J. R., and Padian, K. 2004. Age and growth dynamics of *Tyrannosaurus rex*. Proceedings of the Royal Society B 271:1875–1880.

Hörnes, M. 1853. Mitteilung an Prof. Bronn. Gerichtet: Wien 3. Okt. Neues Jahrbuch der Mineralogie, Geognosie, Geologi und Petrefaktenkunde, 1853:806–810.

Höss, M., Dilling, A., Currant, A., and Pääbo, S. 1996. Molecular phylogeny of the extinct ground sloth *Mylodon darwini*. Proceedings of the National Academy of Science 93:181–185.

Huchon, D. E., and Douzery, J. P. 2001. From the Old World to the New World: a molecular chronicle of the phylogeny and biogeography of hystricognath rodents. Molecular Phylogenetics and Evolution 20:238–251.

Houle, A. 1999. The origin of platyrrhines: an evaluation of the Antarctic scenario and the floating island model. American Journal of Physical Anthropology 109:541–559.

Howard, J. 1982. Darwin. Oxford University Press, Oxford, 101 pp.

Howland, H. C. 1974. Optimal strategies for predator avoidance: the relative importance of speed and manoeuvrability. Journal of theoretical Biology 47:333–350.

Hubbe, A., Hubbe, M., and Neves, W. 2007. Early Holocene survival of megafauna in South America. Journal of Biogeography 34:1642–1646.

Hulbert, Jr., R. C., Baskin, J. A., Ray, C. E. and Tessman, N. 2001. Mammalia 3, Carnivorans; pp. 188–225 in Hulbert, Jr., R. C. (ed.), The Fossil Vertebrates of Florida, University Press of Florida, Gainesville.

Hunter, J.P., and Janis, C. M. 2006. Garden of Eden or fool's paradise? Phylogeny, dispersal, and the southern continent hypothesis of placental mammal origins. Paleobiology 32:339–344.

Hutchinson, H. N. 1893. Extinct monsters. A popular account of some of the larger forms of ancient animal life, with illustrations by J. Smit and others. Chapman & Hall, London, xxii + 270 pp.

Hutchinson, J. R., and Garcia, M. 2002. *Tyrannosaurus* was not a fast runner. Nature 415:1018–1021.

Hutton, J. 1795. Theory of the Earth. With proofs and illustrations. William Creech, Edinburgh, 567 pp.

Huxley, T. H. 1863. On the osteology of the genus *Glyptodon*. Proceedings of the Royal Society of London 13:108.

Huxley, J., and Kettlewell, H. B. D. 1965. Charles Darwin and His World. Thames and Hudson, London, 144 pp.

Imbellone, P., and Teruggi, M. 1988. Sedimentación crotovínica en secuencias cuaternarias bonaerenses. II Reunión Argentina de Sedimentología, 125–129.

Imbellone, P., Teruggi, M., and Mormeneo, L. 1990. Crotovinas en sedimentos Cuaternarios del partido de La Plata. International Symposium on loess. CADINQUA, pp. 166–172. Mar del Plata.

International Commission on Zoological Nomenclature. 1999. International Code of Zoological Nomenclature. 4th ed. International Trust for Zoological Nomenclature, Natural History Museum, London, 306 pp.

Iriondo, M. 1990. Map of the South American plains: its present state; pp. 297–308 in Rabassa J. (ed.), Quaternary of South America and the Antarctic Peninsula. Balkema, Rotterdam.

Iriondo, M. H., and García, N. O. 1993. Climatic variations in the Argentine plains during the last 18,000 years. Palaeogeography, Palaeoclimatology, Palaeoecology 101:209–220.

Iturralde-Vinent, M. A., and MacPhee, R. D. E. 1999. Paleogeography of the Caribbean region: implications for Cenozoic biogeography. Bulletin of the American Museum of Natural History 238:1–95.

Janis, C. M. 1976. The evolutionary strategy of the Equidae, and the origin of rumen and caecal digestion. Evolution 30:757–774.

———. 1990. Correlation of cranial and dental variables with body size in ungulates and macropodoids; pp. 255–299 in Damuth J., and MacFadden, B. J. (eds.), Body Size in Mammalian Paleobiology: Estimation and Biological Implications. Cambridge University Press, Cambridge.

Janis, C. M., and Ehrhardt, D. 1988. Correlation of the muzzle width and relative incisor width with dietary preference in ungulates. Zoological Journal of the Linnean Society 92:267–284.

Janis, C.M., and Constable, E. 1993. Can ungulate craniodental features determine digestive physiology? Journal of Vertebrate Paleontology 13:43A.

Janis, C., and Fortelius, M. 1988. On the means whereby mammals achieve increased functional durability of their dentitions, with special reference to limiting factors. Biological Reviews (Cambridge) 63:197–230.

Janzen, D. H., and Martin, P. S. 1982. Neotropical anachronisms: the fruits the gomphotheres ate. Science 215:19–27.

Jaramillo, C. A., Bayona, G., Pardo-Trujillo, A., Rueda, M., Torres, V., Harrington, G. J., and Mora, G. 2007. The palynology of the Cerrejón Formation (Upper Paleocene) of Northern Colombia. Palynology 31:153–189.

Jarman, P. J. 1974. The social organisation of antelope in relation to their ecology. Behaviour 48:215–267.

Jaslow, C. R. 1990. Mechanical properties of cranial sutures. Journal of Biomechanics 23:313–321.

Jefferys, T. 1775. The Russian discoveries, from the map published by the Imperial Academy of St. Petersburg; in The American Atlas, or A geographical description of the whole continent of America by the late Mr. Thomas Jefferys, geographer to the king, and others. Engraved on forty-eight copper plates. R. Sayer, J. Bennet, London, 237 pp.

Jenkins, Jr., F. A. 1970. Anatomy and function of expanded ribs in certain Edentates and Primates. Journal of Mammalogy 51:288–301.

Jerison, H. J. 1973. The evolution of the brain and intelligence. Academic Press, New York, 482 pp.

Johnson, C. N. 2002. Determinants of loss of mammal species during the late Quaternary "megafauna" extinctions: life history and ecology, but not body size. Proceedings of the Royal Society B 269:2221–2228.

———. 2009. Ecological consequences of Late Quaternary extinctions of megafauna. Proceedings of the Royal Society B 276:2509–2519.

Johnson, S. C., and Madden, R. H. 1997. Uruguaytheriinae Astrapotheres of Tropical South America; pp. 355–381 in Kay, R. F., Madden, R. H., Cifelli, R. L., and Flynn, J. J. (eds.), Vertebrate Paleontology in the Neotropics: The Miocene Fauna of La Venta, Colombia. Smithsonian Institution Press, Washington, D.C.

Jolley, D., Gilmour, I., Gurov, E., Kelley, S., and Watson, J. 2010. Two large meteorite impacts at the Cretaceous–Paleogene boundary. Geology 38:835–838

Kalthoff, D. C. 2011. Microstructure of dental hard tissues in fossil and Recent xenarthrans (Mammalia: Folivora and Cingulata). Journal of Morphology 272:641–661.

Kalthoff, D., and Tütken, D. C. 2007. Stable isotope composition of extant xenarthran teeth and their potencial fr the reconstruction of the diet of fosil xenarthrans (Mammalia). 8th International Congress of Vertebrate Morphology (Paris), Abstracts, 62.

Kandel, R. 2003. Water from Heaven: The Story of Water from the Big Bang to the Rise of Civilization and Beyond. Columbia University Press, New York, 312 pp.

Kay, R. F., Johanson, D., and Meldrum, J. 1998. A new Pitheciin primate from the Miocene of Argentina. American Journal of Primatology 45:317–336.

Kay, R. F., Madden, R. H., Vucetich, M. G., Carlini, A. A., Mazzoni, M. M., Re, G. H., Heizler, M., and Sandeman, H. 1999. Revised age of the Casamayoran South American land mammal "age": climatic and biotic implications. Proceedings of the National Academy of Sciences of the United States of America 96:13235–13240.

Keller, G., Abramovich, S., Berner, Z., and Adatte, T. 2009. Biotic effects of the Chicxulub impact, K–T catastrophe and sea level change in Texas. Palaeogeography, Palaeoclimatology, Palaeoecology 271:52–68.

Kennett, D. J., Kennett, J. P., West, A., West, G. J., Bunch, T. E., Culleton, B. J., Erlandson, J. M., Que Hee, S. S., Johnson, J. R., Mercer, C., Shen, F., Sellers, M., Stafford, Jr., T. W., Stich, A., Weaver, J. C., Wittke, J. H., and Wolbach, W. S. 2009. Shock-synthesized hexagonal diamonds in Younger Dryas boundary sediments. Proceedings of the National Academy of Sciences of the United States of America 106:12623–12628.

Keynes, R. D. 2001. Charles Darwin's Beagle Diary. Cambridge University Press, Cambridge, 464 pp.

Kleiber, M. 1932. Body size and metabolism. Hilgardia 6:315–353.

Kirby, M. X., Jones D. S., and MacFadden, B. J. 2008. Lower Miocene stratigraphy along the Panama Canal and its bearing on the Central American peninsula. PLoS One 3:1–14.

Kirby, M. X., Jones, D. S., and MacFadden. 2011. Lower Miocene stratigraphy along the Panama Canal and its bearing on the Central American Peninsula. PloS One 3:e2791.

Koepfli, K.-P., Gompper, M. E., Eizirik, E., Ho, C.-C., Linden, L., Maldonado, J. E., and Wayne, R. K. 2007. Phylogeny of the Procyonidae (Mammalia: Carvnivora): molecules, morphology and the Great American Interchange. Molecular Phylogenetics and Evolution 43:1076–1095.

Kraglievich, L. (1928): "Mylodon Darwin" Owen, es la especie genotipo de "Mylodon" Ow. Revista de la Sociedad Argentina de Ciencias Naturales, Physis 9:169–185.

Kraglievich, L. 1930. La Formación Friaseana del río Frías, río Fénix, Laguna Blanca, etc. y su fauna de mamíferos. Physis 10:127–161.

———. 1932. Nuevos apuntes para la geología y paleontología uruguayas. Anales del Museo de Historia Natural de Montevideo 2:1–65.

———. 1940. Obras de geología y paleontología. Vols. 1–3. Impresiones Oficiales, La Plata.

Krause, D. W., and Bonaparte, J. F. 1990. The Gondwanatheria, a new suborder of Multituberculata from South America. Journal of Vertebrate Paleontology 10:31A.

Kurten, B. 1973. Pleistocene Jaguars in North America. Commentationes Biologicae Societas Scientiarum Fennica 62:1–23.

Kurtén, B., and Werdelin, L. 1990. Relationships between North and South American Smilodon. Journal of Vertebrate Paleontology 10:158–169.

Lacourse, T., and Gajewski, K. 2000. Late Quaternary Vegetation History of Sulphur Lake, Southwest Yukon Territory, Canada. Arctic 53:27–35

Lambeck, K., Esat, T. M., and Potter, E.-K. 2002. Links between climate and sea levels for the past three million years. Nature 419:199–206.

Lambert, W. D. 1992. The feeding habits of the shovel-tusked gomphotheres (Mammalia, Proboscidea, Gomphotheriidae): evidence from tusk wear patterns. Paleobiology 18:132–147

Lambert, W. D., and Shoshani, J. 1998. Proboscidea; pp. 606–621 in Janis, C. M., Scott, K. M., and Jacobs, L. L. (eds.), Evolution of Tertiary Mammals of North America. Vol. 1, Terrestrial Carnivores, Ungulates, and Ungulatelike Mammals. Cambridge University Press, Cambridge.

Lankester, E. R. 1905. Extinct Animals. Archibald Constable, London, 332 pp.

Laporte, L. F. 2000. George Gaylord Simpson: Paleontologist and Evolutionist. Columbia University Press, New York, 332 pp.

Lavocat, R. 1958. Litopterna; pp. 31–58 in Piveteau, J. (ed.), Traité de Paléontologie. Vol. 6. Masson et Cie, Paris.

———. 1969. La systématique des rongeurs hystricomorphes et la dérive des continents. Compte Rendus de la Académie des Sciences Série D 269:1496–1497.

Legendre, S. 1986. Analysis of mammalian communities from the late Eocene and Oligocene of Southern France. Palaeovertebrata 16:191–212.

———. 1987. Les communautés de mammifères du. Paléogène (Éocène supérieur et Oligocéne): structures et milieux. Münchner Geowissenschaftliche Abhandlungen (A) 10:301–312.

Leidy, J. 1855. A Memoir of the Extinct Sloth Tribe of North America. Smithsonian Contributions to Knowledge 7:1–68.

Lemoine, V. 1891. Étude d'ensemble sur les dents des mammifères fossiles des environs de Reims, Bulletin de la Société géologique de France 19:263–290.

Lessa, E. P., and Fariña, R. A. 1996. Reassessment of extinction patterns among the late Pleistocene mammals of South America. Palaeontology 39:651–662.

Lessa, E. P., Van Valkenburgh, B., and Fariña, R. A. 1997. Testing hypotheses of differential mammalian extinctions subsequent to the Great American Biotic Interchange. Palaeogeography, Palaeoclimatology, Palaeoecology 135:157–162.

Li, H. M., and Zhou, Z. K. 2007. Fossil nothofagaceous leaves from the Eocene of western Antarctica and their bearing on the origin, dispersal and systematics of Nothofagus. Science in China 50:1525–1535.

Lindstedt, S. L., Miller, B. J., and Buskirk, S. W. 1986. Home range, time, and body size in mammals. Ecology 67:413–418

Lister, A. M. 1993. Evolution of mammoths and moose: the Holarctic perspective; pp. 178–204 in Barnosky, A. D. (ed.), Morphological Change in Quaternary Mammals of North America. Cambridge University Press, Cambridge.

Liu, F.-G. R., Miyamoto, M. M., Freire, N. P., Ong, P. Q., Tennant, M. R., Young, T. S., and Gugel, K. F. 2001. Molecular and morphological supertrees for eutherian (placental) mammals. Science 291:1786–1789.

Lloyd, J., and Mitchinson, J. 2008. The Book of General Ignorance. Faber and Faber, London, 299 pp.

Longo, G. 2007. The Tunguska event; pp. 303–330 in Bobrowsky, P. T., and Rickman, H. (eds.), Comet/Asteroid Impacts and Human Society: An Interdisciplinary Approach. Springer Verlag, Berlin.

Lönnberg, E. 1899. On some remains of "Neo-mylodon listai" Ameghino, brought home by Swedish Expedition of Tierra del Fuego; in Nordenskjöld, O. (ed.), Wissenschaftliche Ergebnisse der Schwedischen Expedition nach den Magellansländern 1895–1897, unter Leitung von Dr Otto Nordenskjöld 2:149–170. P. A. Norstedt & Söner, Stockholm.

Loomis, F. B. 1913. Hunting Extinct Animals in the Patagonian Pampas. Dodd, Mead, New York, 141 pp.

López Aranguren, D. J. 1930. Camélidos fósiles argentinos. Anales de la Sociedad Científica Argentina 59:15–56.

López Piñero, J. M. 1989. Juan Bautista Bru y la difusión por Cuvier de su obra paleontológica. Arbor 527–528:79–100.

Lucas, F. A. 1901. Animals of the past. McClure, Phillips, New York, 258 pp.

Ludvigsen, R. 1979a. A trilobite zonation of Middle Ordovician rocks, southwestern District of MacKenzie. Bulletin of the Geological Survey of Canada 312:1–99.

———. 1979b. Fossils of Ontario, Part 1: The Trilobites. Royal Ontario Museum Life Sciences Miscellaneous Publications, Toronto, 96 pp.

Lull, R. S. 1929. A remarkable ground sloth. Memoirs of the Peabody Museum of Yale University 3:1–39.

Lund, P. W. 1839. Blik paa Brasiliens dyreverden för sidote jordomvæltning. Anden afhandling: Pattedyrene. Det Kongelige Danske Videnskabernes Selskabs Naturvidenskabelige og Matematiske Afhandlinger 8:61–144.

———. 1842. Memórias sobre a Paleontologia Brasileira, Revistas e comentadas por Carlos de Paula Couto. Ministério da Educação e Saúde, Instituto Nacional do Livro, Rio de Janeiro, 589 pp.

———. 1844. Carta escripta de Lagôa Santa a 21 de abril de 1844. Revista do Instituto Histórico e Geográfico Brasileiro 6:334–342.

———. 1846. Meddlelse af det Udbytte de i 1844 undersøgte Knoglehuler Have afgivet til kundskaben om Brasiliens Dyreverden før sidste Jordomvæltning. Det Kongelige Danske Videnskaberne Selskabs Naturvidenskabelige og Matematiske Afhandlinger 12:57–94.

Lund, T. F. 1920–1922. Bakkehus og Solbjerg—Træk af et nyt Livssyns Udvikling i Norden I–III. Nordisk forlag Gyldendal.

Luo, Z.-X., Ji, Q., Wible, J. R., and Yuan, C-X. 2003. An early Cretaceous tribosphenic mammal and metatherian evolution. Science 302:1934–1939.

Lydekker, R. 1887. Catalogue of the fossil Mammalia in the British Museum (Natural History). Part 5. The Trustees of the British Museum (Natural History), XXXV, London, 345 pp.

———. 1893. Contributions to a knowledge of the fossil vertebrates of Argentina., 3. A study of the extinct ungulates of Argentina. Anales del Museo de La Plata 2:1–91.

———. 1894. Contributions to a knowledge of the fossil vertebrates of Argentina. Part II (2). The extinct edentates of Argentina. Anales del Museo de La Plata 3:1–118.

Lyell, C. (ed.) 1830–1833. Principles of geology: being inquiry how far the former changes of the earth's surface are referable to causes now in operation. Vol. 3. John Murray, London, 398 pp.

Macalister, A. 1869. On the myology of Brad-ypus tridactylus; with remarks on the general anatomy of the Edentata. Annual Magazine of Natural History 4:51–67.

Macdonald, D. W. 1983. The ecology of carnivore social behaviour. Nature 301:379–384.

MacFadden, B. J. 1992. Fossil Horses: Systematics, Paleobiology, and Evolution of the Family Equidae. Cambridge University Press, Cambridge, 369 pp.

———. 2000. Cenozoic mammalian herbivores from the Americas: reconstructing ancient diets and terrestrial communities. Annual Review on Ecology and Systematics 31:33–59.

MacFadden, B. J., and Shockey, B. J. 1997. Ancient feeding ecology and niche differentiation of Pleistocene mammalian herbivores from Tarija, Bolivia: morphological and isotopic evidence. Paleobiology 23:77–100.

MacFadden, B. J., Wang, Y., Cerling, T. E., and Anaya, F. 1994. South American fossil mammals and carbon isotopes: a 25 million-year sequence from the Bolivian Andes. Palaeogeography, Palaeoclimatology, Palaeoecology 107:257–268

MacFadden, B. J., Cerling, T. E., and Prado, J. L. 1996. Cenozoic Terrestrial Ecosystem in Argentina Evidence from Carbon isotopes of Fossil Mammal Teeth. Palaios 11:319–327.

MacFadden, B. J., Solounias, N., and Cerling, T. E. 1999. Ancient diets, ecology, and extinction of 5-millionyear-old horses from Florida. Science 283:824–827.

MacPhee, R. D. E., and Iturralde-Vinent, M. 2005. The interpretation of Caribbean paleogeography: reply to Hedges; in Alcover, J. A., and Bover, P. (eds.), Proceedings of the International Symposium "Insular Vertebrate Evolution: the Palaeontological Approach." Monografies de la Societat d'Història Natural de les Balears 12:175–184.

MacPhee, R. D. E., and Marx, P. A. 1997. The 40,000-year plague: humans, hyperdisease, and first-contact extinctions; pp. 169–217 in Goodman, S. M., and Patterson, B. D. (eds.), Natural Change and Human Impact in Madagascar. Smithsonian Institution Press, Washington, D.C.

Macrini, T. E., Muizon, C. de, Cifelli, R. L., and Rowe, T. E. 2007. Digital cranial endocast of Pucadelphys andinus, a Paleocene metatherian. Journal of Vertebrate Paleontology 27:99–107.

Madden, R., Guerrero, J., Kay, R. F., Flynn, J. J., Swisher, III, C. C., and Walton, A. H. 1997. The Laventan Stage and Age; pp. 499–519 in Kay, R. F., Madden, R. H. Cifelli, R. L., and John, J. F. (eds.), Vertebrate Paleontology in the Neotropics: The Miocene Fauna of La Venta, Colombia. Smithsonian Institution Press, Washington, D.C.

Mader, B. J. 1998. Brontotheriidae; pp. 525–536 in Janis, C. M., Scott, K. M., and Jacobs, L. L. (eds.), Evolution of Tertiary Mammals of North America. Vol. 1, Terrestrial Carnivores, Ungulates, and Ungulatelike Mammals. Cambridge University Press, Cambridge.

Madsen, O., Scally, M., Douady, C. J., Kao, D. J., DeBry, R. W., Adkins, R., Amrine, H. M., Stanhope, M. J., DeJong, W. W., and Springer, M. S. 2001. Parallel adaptive radiations in two major clades of placental mammals. Nature 409:610–614.

Maley, J. 2004. Le bassin du lac Tchad au Quaternaire récent: formations sédimentaires, paléoenvironnements et préhistoire; pp. 179–217 in Renault-Miskovsky, J., and Semah, A. M. (eds.), La question des Paléotchad., L'Evolution de la Végétation Depuis Deux Millions d'Années. Publication Errance, Paris.

Marcus, L. F., and Berger, R. 1984. The significance of radiocarbon dates for Rancho La Brea; pp. 159–183 in Martin, P. S., and R. G. Klein, (eds.), Quaternary Extinctions; A Prehistoric Revolution. University of Arizona Press, Tucson.

Margalef, R. 1980. Ecología. Omega, Barcelona, 951 pp.

Marinho, T. S., and Carvalho, I. S. 2009. An armadillo-like sphagesaurid crocodyliform from the Late Cretaceous of Brazil. Journal of South American Earth Sciences 27:36–41

Marcon, G. T. G. 2007. Contribuição ao estudo dos proboscídea (Mammalia, Gomphotheriidae) do Quaternário do estado do Rio Grande do Sul, Brasil. M.Sc. dissertation, Instituto de Geociências, Programa de Pos-graduação em Geociências, Universidade Federal do Rio Grande do Sul, Porto Alegres, RS, 113 pp.

Marquet, P. A., and Cofré, H. 1999. Large temporal and spatial scales in the structure of mammalian assemblages in South America: a macroecological approach. Oikos 85:299–309.

Markgraf, V. 1985. Late Pleistocene faunal extinctions in southern Patagonia. Science 228:1110–1112.

———. 1989. Palaeoclimates in Central and South America since 18,000 BP based on pollen and lake-level records. Quaternary Science Review 8:1–24.

——— (ed.) 2001. Interhemispheric Climate Linkages. Academic Press, San Diego, 454 pp.

Marshall, L. G., and Cifelli, R. L. 1990. Analysis of changing diversity patterns in Cenozoic land mammal age faunas, South America. Palaeovertebrata 19:169–210.

Marshall, L. G., and Muizon, C. de. 1988. The dawn of the age of mammals in South America. National Geographic Research 4:23–55.

Marshall, L. G., Webb, S. D., Sepkoski, Jr., J. J., and Raup, D. M. 1982. Mammalian evolution and the great American interchange. Science 215:1351–1357.

Martin, B. E. 1916. Tooth development in Dasypus novemcinctus. Journal of Morphology 27:647–691.

Martin, P. S. 1984. Prehistoric overkill: the global model; pp. 354–403 in Martin, P. S., and Klein, R. G. (eds.), Quaternary Extinctions: A Prehistoric Revolution. University of Arizona Press, Tucson.

Martin, P. S., and Klein, R. K. (eds.) 1984. Quaternary Extinctions: A Prehistoric Revolution. University of Arizona Press, Tucson, 892 pp.

Martinic, M. 1996. La cueva del milodon: historia de los hallazgos y otros sucesos. Relación de los estudios realizados a lo largo de un siglo (1895–1995). Anales del Instituto de la Patagonia, Chile. Serie Ciencias de la Tierra 24:43–80.

Matthew, W. D. 1906. The osteology of Sinopa, a creodont mammal of the Middle Eocene. Proceedings of the U.S. National Museum 30:203–234.

———. 1910. The phylogeny of the Felidae. Bulletin of the American Museum of Natural History 28:289–316.

Mayle, F. E. 2004. Assessment of the Neotropical dry forest refugia hypothesis in the light of palaeoecological data and vegetation model simulations. Journal of Quaternary Science 19:713–720.

———. 2006. The Late Quaternary biogeographical history of South American seasonally dry tropical forests: insights from palaeo-ecological data; pp. 395–416 in Pennington, R. T., Lewis, G. P., and Ratter, J. A. (eds.), Neotropical Savannas and Seasonally Dry Forests: Plant Diversity, Biogeography and Conservation. Systematics Association Special Volume 69. CRC Press, Taylor and Francis, Boca Raton.

Maynard Smith, J., and Savage, R. J. G. 1959. The mechanics of mammalian jaws. School Sciences Review 141:289–301.

Mazzetta, G., Christiansen, P., and Fariña, R. A. 2004. Gigantic and bizarres: body mass of southern South American Cretaceous dinosaurs. Historical Biology 16:71–83.

McBee, K., and Baker, R. J.1982. *Dasypus novemcinctus*. Mammalian Species 162:1–4.

McDonald, H. G. 1987. A sistematic review of the Plio-Pleistocene Scelidotherine Ground Sloths (Mammalia, Xenarthra; Mylodontidae). Ph. D. Thesis, University of Toronto, Toronto.

———. 1995. Gravigrade xenarthrans from the early Pleistocene Leisey Shell Pit 1A, Hillsborough County, Florida. Bulletin of Florida Museum of Natural History 37:345–373.

———. 2003. Xenarthran skeletal anatomy: primitive or derived? (Mammalia, Xenarthra); in Fariña, R. A., Vizcaíno, S. F., and Storch, G. (eds.), Morphological Studies in Fossil and Extant Xenarthra (Mammalia). Senckenbergiana Biologica 83:5–17.

———. 2005. Paleoecology of extinct xenarthrans and the Great American Biotic Interchange. Bulletin of the Florida Musuem of Natural History 45:319–340.

———. 2006. Sexual dimorphism in the skull of Harlan's ground sloth, Contributions on Science, Natural history Museum of Los Angeles County 510:1–9.

———. 2007. Biomechanical inferences of locomotion in ground sloths: integrating morphological and track data. New Mexico Museum of Natural History and Science Bulletin 42:201–208.

McDonald, H. G., and De Iuliis, G. 2008. Fossil history of sloths; pp. 39–55 in Vizcaíno, S. F., and Loughry, W. J. (eds.), The Biology of the Xenarthra. University Press of Florida, Gainesville.

McDonald, H. G., and Naples, V. L. 2008. Xenarthra: pp. 147–160 in Janis, C. M. Gunnell, G. E., and Uhen, M. D. (eds.), Evolution of Tertiary Mammals of North America. Vol. 2. Cambridge University Press, Cambridge.

McDonald, H. G., and Perea, D. 2002. The large Scelidothere *Catonyx tarijensis* (Xenarthra, Mylodontidae) from the Pleistocene of Uruguay. Journal of Vertebrate Paleontology 22:677–683.

McDonald, H. G., Jefferson, G. T., and Force, C. 1996. Pleistocene distribution of the ground sloth *Nothrotheriops shastensis* (Xenarthra, Megalonychidae). 1996 Desert Research Symposyum, San Bernardino County Museum Association Quarterly 43:151–152.

McDonald, H. G., Harington, C. R., and De Iuliis, G. 2000. The ground sloth *Megalonyx* from Pleistocene deposits of the Old Crow Basin, Yukon, Canada. Arctic 53:213–220.

McDonald, H. G., Vizcaíno, S. F., and Bargo, M. S. 2008. Skeletal anatomy and the fossil history of the Vermilingua; pp. 64–78 in Vizcaíno, S. F., and Loughry, W. J. (eds.), The Biology of the Xenarthra. University Press of Florida, Gainesville.

McGowan, C. 1994. Diatoms to dinosaurs. The size and scale of living things. Island Press, Washington, 288 pp.

———. 2001. The Dragon Seekers. Perseus Publishing, Cambridge, 254 pp.

McKenna, M. C. 1975. Toward a phylogenetic classification of the Mammalia; pp. 21–46 in Luckett W. P., and Szalay, F. S. (eds.), Phylogeny of the Primates. Plenum Press, New York.

McKenna, M. C., and Bell, S. K. 1997. Classification of Mammals above the Species Level. Columbia University Press, New York, 631 pp.

McNab, B. K. 1979. Climatic adaptation in the energetics of heteromyid rodents. Comparative Biochemistry and Physiology 62:813–820.

———. 1980. Energetics and the limits to a temperate distribution in armadillos. Journal of Mammalogy 61:606–627.

———. 1985. Energetics, population biology, and distribution of xenarthrans, living and extinct; pp. 219–232 in Montgomery, G. G. (ed.), The Evolution and Ecology of Armadillos, Sloths and Vermilinguas. Smithsonian Institution Press, Washington, D.C.

———. 2008. An analysis of the factors that influence the level and scaling of mammalian BMR. Comparative Biochemichemistry and Physiology A 151:5–28.

———. 2009. Resources and energetics determined dinosaur maximal size. Proceedings of the National Academy of Sciences of the United States of America 106:12184–12188.

McNaughton, S. J., Oesterheld, M., Frank, D. A., and Williams, K. J. 1989. Ecosystem-level patterns of primary productivity and herbivory in terrestrial habitats. Nature 341:142–144.

Meachen-Samuels, J., and Van Valkenburgh, B. 2009a. Forelimb indicators of prey-size preference in the Felidae. Journal of Morphology 270:729–744.

———. 2009b. Craniodental indicators of prey size preference in the Felidae. Biological Journal of the Linnean Society 96:784–799.

Meltzer, D. J. 2008. First Peoples in a New World: Colonizing Ice Age America. University of California Press, Berkeley, 446 pp.

Méndez, J., and Keys, A. 1960. Density and composition of mammalian muscle. Metabolism 9:184–188.

Méndez Alzola, R. 1950. Estudio sobre la obra científica de Larrañaga. Su iconografía paleomastozoológica. Memoria de las exposiciones de Montevideo 4–28 de julio 1948. Anales de la Universidad. 60:49–129.

Mendoza, M., and Palmqvist, P. 2008. Hypsodonty in ungulates: an adaptation for grass consumption or for foraging in open habitat? Journal of Zoology 274:134–142.

Mendoza, M., Janis, C. M., and Palmqvist, P. 2002. Characterizing complex craniodental patterns related to feeding behaviour in ungulates: a multivariate approach. Journal of Zoology (London) 258:223–246.

Menegaz, A., and Ortiz Jaureguizar, E. 1995. Los artiodáctilos; in Alberdi, M., Leone, G. y Tonni, E. P. (eds.), Evolución biológica y climática de la Región Pampeana durante los últimos cinco millones de años. Un ensayo de correlación con el Mediterráneo occidental. Monografías del Museo Nacional de Ciencias Naturales CSIC 12:311–337.

Menéndez, C. A. 1971. Estudio palinológico del Pérmico de Bajo de Véliz, provincia de San Luis. Revista del Museo Argentino de Ciencias Naturales "B. Rivadavia." Paleontología 1:263–306.

Merino, M. L., Milne, N., and Vizcaíno, S. F. 2005. A morphometric study of deer (Mammalia, Cervidae) crania from Argentina using three dimensional landmarks. Acta Theriologica 50:91–108.

Messineo, P., Politis, G., and Rivas, M. I. 2004. Cazadores tempranos y megamamíferos tardíos en la región pampeana: el sitio Campo Laborde. XX Congreso Nacional de Arqueología Argentina, Río Cuarto.

Milankovitch, M. 1941. Kanon der Erdbestrahlungen und seine Anwendung auf das Eiszeitenproblem. Eiszeitenproblem, Belgrad, Königlich Serbische Akademie; Editions spéciales, CXXXIII, Séction des Sciences Mathématiques et Naturelles 33:1–633.

Millen, V. 2008. The largest among the smallest: the body mass of the giant rodent *Josephoartigasia monesi*. Proceedings of the Royal Society B 275:1953–1955.

Mills, M. G., and Gorman, M. L. 1997. Factors affecting the density and distribution of wild dogs in the Kruger National Park. Conservation Biology 11:1397–1406.

Milne, N., Vizcaíno S. F., and Fernicola, J. C. 2009. A 3D geometric morphometric analysis of digging ability in the extant and fossil cingulate humerus. Journal of Zoology (London), 278:48–56.

Minoprio, J. L. 1947. Fósiles de la Formación Divisadero Largo. Anales de la Sociedad Científica Argentina 144:365–378

Miotti, L. 1992. Paleoindian occupation at Piedra Museo Locality, Santa Cruz Province, Argentina. Current Research in the Pleistocene 9:27–30.

Montgomery, G. G. (ed). 1985. The Evolution and Ecology of Armadillos, Sloths and Vermilinguas. Smithsonian Institution Press, Washington, D.C., 451 pp.

Montuire, S., and Desclaux, E. 1997. Palaeoecological analysis of mammalian faunas and environmental evolution in the south of France during the Pleistocene. Boreas 26:355–365.

Montuire, S., and Marcolini, F. 2002. Palaeoenvironmental significance of the mammalian faunas of Italy since the Pliocene. Journal of Quaternary Science 17:87–96.

Moore, D. M. 1978. Post-glacial vegetation in the south Patagonian territory of the giant ground sloth, *Mylodon*. Botanical Journal of the Linnaean Society 77:177–202.

Moore, W. J. 1981. The Mammalian Skull. Cambridge University Press, Cambridge, 369 pp.

Moreno, F. P., and Woodward, A. S. 1899. On a portion of mammalian skin named *Neomylodon listai*, from a cavern near Consuelo Cove Last Hope Inlet, Patagonia. Proceedings of the Zoological Society of London 67:144–156.

Morgan, G. S. 2005. The Great American Biotic Interchange in Florida. Bulletin of the Florida Museum of Natural History 45:271–311.

Morgan, M. E., Badgley, C. Gunnell, G. F. Gingerich, P. D. Kappelman, J. W., and Maas, M. C. 1995. Comparative paleoecology of Paleogene and Neogene mammalian faunas: body-size structure. Palaeogeography, Palaeoclimatology, Palaeoecology 115:287–317.

Mothé, D., Ávilla, L. S., and Winck, G. R. 2010. Population structure of the gomphothere Stegomastodon waringi (Mammalia: Proboscidea: Gomphotheriidae) from the Pleistocene of Brazil. Anais Academia Brasileira de Ciências 82:983–996.

Muizon C. de, and Cifelli, R. L. 2000. The "condylarths" (archaic Ungulata, Mammalia) from the early Palaeocene of Tiupampa (Bolivia): implications on the origin of the South American ungulates. Geodiversitas 22:47–150.

Muñiz, F. J. 1845. Descripción del Muñi-felis bonarensis. Gaceta Mercantil (6603). Buenos Aires 9 de octubre de 1845.

———. 1916. Escritos Científicos. Ensayos de Domingo F. Sarmiento y juicios críticos de Bartolomé Mitre y Florentino Ameghino. La Cultura Argentina, Buenos Aires, 279 pp.

Murray, P. F., and Vickers-Rich, P. 2004. Magnificent Mihirungs: The Colossal Flightless Birds of the Australian Dreamtime. Indiana University Press, Bloomington, 410 pp.

Murphy, W. J., Eizirik, E., Johnson, W. E., Zhang, Y. P., Ryder, O. A., and O'Brien, S. J. 2001a. Molecular phylogenetics and the origins of placental mammals. Nature 409:614–618.

Murphy, W. J., Eizirik, E., O'Brien, S. J., Madsen, O., Scally, M., Douady, C. J., Teeling, E. C., Ryder, O. A., Stanhope, M. J., De Jong, W. W., and Springer, M. S. 2001b. Resolution of the early placental mammal radiation using Bayesian phylogenetics. Science 294:2348–2351.

Musters, G. C. 1873. At home with the Patagonians. A year's wandering over untrodden ground from the straits of Magellan to the Rio Negro. John Murray, London, 756 pp.

Musters, P. T. A. 2001. The Musters: A Family Gathering. 2nd ed. Oakham, Rutland.

Muybridge, E. 1899. Animals in motion. An electro photographic investigation of consecutive phases of animal progressive moivements. Chapman and Hall, London, 277 pp.

Nabel, P. E., Cione, A., and Tonni, E. P. 2000. Environmental changes in the Pampean area of Argentina at the Matuyama-Brunhes (C1r–C1n) Chrons boundary. Palaeogeography, Palaeoclimatology, Palaeoecology 162:403–412.

Nagel, D., Sabine, S., Benesch, A., and Scholz, J. 2003. Functional morphology and fur patterns in Recent and fossils Panthera species. Scripta Zoologica 126:227–241.

Nagy, K. A. 2005. Review. Field metabolic rate and body size. Journal of Experimental Biology 208:1621–1625.

Nagy, K. A., Girard, I. A., and Brown, T. K. 1999. Energetics of free-ranging mammals, reptiles, and birds. Annual Review of Nutrition 19:247–277.

Naish, D. 2008. Dude, where's my astrapothere? Science Blogs, February 3. http://scienceblogs .com/tetrapodzoology/2008/02/dude_wheres_ my_astrapothere.php.

Naples, V. L. 1982. Cranial osteology and function in the tree sloths, Bradypus and Choloepus. American Museum Novitates 2739:1–41.

———. 1985. Form and function of the masticatory musculature in the tree sloths, Bradypus and Choloepus. Journal of Morphology 183:25–50.

———. 1987. Reconstruction of cranial morphology and analysis of function in the Pleistocene ground sloth Nothrotheriops shastense (Mammalia, Megatheriidae). Contributions in Science, N° 389, pp. 1–23. Los Angeles County Museum of Natural History.

———. 1989. The feeding mechanism in the Pleistocene ground sloth, Glossotherium. Contributions in Science, N° 415, pp. 1–21. Los Angeles County Museum of Natural History.

———. 1999. Morphology, evolution and function of feeding in the giant anteater (Myrmecophaga tridactyla). Journal of Zoology (London) 249:19–41.

Nascimento E. R. do, Cozzuol, M. A., and Sant'Anna Filho, M. J. 2010. O primeiro registro de um proboscídeo não gonfoteriídeo na América do Sul, um exemplo da diversidade "escondida" no passado da Amazônia. Boletim de Resumos, 7 Simpósio Brasileiro de Paleontologia de Vertebrados, p. 97.

Nee, S., Read, A. F., Greenwood, J. J. D., and Harvey, P. H. 1991. The relationship between abundance and body size in British birds. Nature 351:312–313.

Newman, C. C., and R. J. Baker. 1942. Armadillo eats young rabbit. Journal of Mammalogy 23:450.

Nodot, L. 1857. Description d' un nouveau genre d' édenté fósil renfermant plusieurs espèces voisines du Glyptodon, etc. Memoires de l' Académie Impériale des Sciences, Arts et Belles-Lettres de Dijon 2:1–172.

Noe-Nygaard, N. 1989. Man-made trace fossils on bones. Human Evolution 4:461–491.

Novoa, A. and Levine, A. 2010. From Man to Ape: Darwinism in Argentina, 1870–1920. University of Chicago Press, Chicago, 281 pp.

Nowak, R. L. 1999. Walker's Mammals of the World. 6th ed. Johns Hopkins University Press, Baltimore, 1936 pp.

Oken, L. 1933. Allgemeine Naturgeschichte für alle Stände. Hoffmann'sche Verlags, Stuttgart, 1432 pp.

Olson, E. C. 1991. George Gaylord Simpson: June 16, 1902–October 6, 1984. Biographical memoirs. Proceedings of the National Academy of Sciences of the United States of America 60:331–353

Onna, A. F. 2000. Estrategias de visualización y legitimación de los primeros paleontólogos en el Río de la Plata durante la primera mitad del siglo XIX. Francisco Javier Muñiz y Teodoro Miguel Vilardebó; pp. 53–70 in Montserrat, M. (ed.), La ciencia en la Argentina entre siglos. Textos, contextos e instituciones. Ediciones Manantial, Buenos Aires.

Osborn, H. F. 1910. The Age of Mammals in Europe, Asia, and North America. Macmillan, New York, 635 pp.

———. 1924. Additional generic and specific stages in the evolution of the Proboscidea. American Museum Novitates 154:1–5.

Ortiz Jaureguizar, E. 1986. Evolución de las comunidades de mamíferos cenozoicos sudamericanos: un estudio basado en técnicas de análisis multivariado. 4° Congreso Argentino de Paleontología y Bioestratigrafía (Mendoza), Actas 2:191–207.

———. 1988. Evolución de las comunidades de mamíferos cenozoicos sudamericanos: un análisis cuali-cuantitativo basado en el registro argentino. Unpublished PhD thesis. Universidad Nacional de La Plata, Facultad de Ciencias Naturales y Museo, pp. 350.

Ortiz Jaureguizar, E., and Cladera, G. 2006. Paleoenvironmental evolution of southern South America during the Cenozoic. Journal of Arid Environments 66:489–532.

Ortiz Jaureguizar, E., and Pascual, R. 1989. South American Land-Mammal faunas during the Cretaceous–Tertiary transition: evolutionary biogeography; A231–A251 in Spalletti, L. (ed.), Contribuciones de los Simposios sobre el Cretácico de América Latina. Parte A: Eventos y Registro Sedimentario. GSGP-IGCP 242. La Paz.

Ortiz Jaureguizar, E., and Posadas, P. E. 1999. Desde el Cretácico tardío a la Actualidad: los cambios composicionales en la fauna de mamíferos de América del Sur a la luz del PAE ("Parsimony Analysis of Endemicity"). 2° Reunión Argentina de Cladismo y Biogeografía (Buenos Aires), Actas: p. 22.

———. 2000. Cambios composicionales en la fauna de mamíferos de América del Sur durante el lapso Cretácico tardío-Paleoceno: un análisis aplicando un algoritmo de simplicidad. 15° Jornadas Argentinas de Mastozoología (La Plata), Actas: p. 91.

Ortiz Jaureguizar, E., Cladera, G., and Giallombardo, A. 1999. Relaciones de similitud entre las faunas del lapso Cretácico superior-Paleoceno superior en América del Sur. Temas Geológico-Mineros. Instituto Tecnológico Geominero de España 26:280–283.

Owen, R. 1838. Fossil Mammalia (1); in Darwin, C. R. (ed.), The Zoology of the Voyage of HMS Beagle. Smith, Elder, London, 40 pp.

———. 1839. Description of a tooth and part of the skeleton of the Glyptodon, a large quadruped of the edentate order, to which belongs the tessellated bony armour figurated by Mr. Clift in his memoir on the remains of the Megatherium, brought to England by Sir Woodbine Parish, F.G.S. Proceedings of the Geological Society of London 3:108–113.

———. 1840. Part 1. Fossil Mammalia; in Darwin, C. R. (ed.), The Zoology of the Voyage of HMS Beagle. Smith, Elder, London, 64 pp.

———. 1842. Description of the Skeleton of an Extinct Giant Sloth, Mylodon robustus, Owen; with observations on the osteology, natural affinities, and probable habits of the Megatheroid quadrupeds in general. John Van Voorst, London, 176 pp.

———. 1845a. Account of various portions of the Glyptodon, an extinct quadruped, allied to the armadillo, and recently obtained from the Tertiary deposits in the neighbourhood of Buenos Ayres. Quarterly Journal of the Geological Society of London 1:257–262

———. 1845b. Descriptive and illustrated catalogue of the fossil organic remains of Mammalia and Aves contained in the Museum of the Royal College of Surgeons of London. Richard & John E. Taylor, London, 391 pp.

————. 1846. Notices of some fossil mammalia of South America. Report of the British Association for the Advancement of Science 16:65–67

————. 1848. On the Archetype and Homologies of the Vertebrate Skeleton. J. van Voorst, London, 203 pp.

————. 1851. On the Megatherium (Megatherium Americanum, Cuvier and Blumenbach). Part I. Philosophical Transactions of the Royal Society of London 141:719–764.

————. 1855. On the Megatherium (Megatherium Americanum, Cuvier and Blumenbach). Part II. Vertebrae of the Trunk. Philosophical Transactions of the Royal Society of London 145:359–388.

————. 1856. On the Megatherium (Megatherium Americanum, Cuvier and Blumenbach). Part III: The skull. Philosophical Transactions of the Royal Society of London 146:571–589.

————. 1858. On the Megatherium (Megatherium Americanum, Cuvier and Blumenbach). Part IV. Bones of the anterior extremities. Philosophical Transactions of the Royal Society of London 148:261–278.

————. 1860. On the Megatherium (Megatherium Americanum, Cuvier and Blumenbach). Part V. Bones of the posterior extremities. Philosophical Transactions of the Royal Society of London 149:809–829.

Owen-Smith, N., and Mills, M. G. L. 2008a. Shifting prey selection generates contrasting herbivore dynamics within a large-mammal predator–prey web. Ecology 89:1120–1133.

————. 2008b. Predator–prey size relationships in an African large-mammal food web. Journal of Animal Ecology 77:173–183.

Owen-Smith, R. N. 1987. Pleistocene extinctions: the pivotal role of megaherbivores. Paleobiology 13:351–362.

————. 1988. Megaherbivores: The Influence of Very Large Body Size on Ecology. Cambridge University Press, Cambridge, 369 pp.

Packard G. C., and Birchard, G. F. 2008. Traditional allometric analysis fails to provide a valid predictive model for mammalian metabolic rates. Journal of Experimental Biology 211:3581–3587.

Packer, C., Hilborn, R. H., Mosser, A., Kissui, B., Borner, M., Hopcraft, G., Wilmshurst, J., Mduma, S., and Sinclair, S. R. E. 2005. Ecological change, group territoriality, and population dynamics in Serengeti lions. Science 307:390–393.

Paine, R. T. 1966. Food web complexity and species diversity. American Naturalist 100:65–75.

Palazzesi, L., and Barreda V. 2007. Major vegetation trends in the Tertiary of Patagonia (Argentina): a qualitative paleoclimatic approach based on palynological evidence. Flora 202:328–337.

Paley, W. 1802. Natural Theology, or Evidences of the Existence and Attributes of the Deity collected from the Appearances of Nature. J. Faulder, London, 421 pp.

Palmqvist, P., and Arribas, A. 2001. Taphonomic decoding of the paleobiological information locked in a lower Pleistocene assemblage of large mammals. Paleobiology 27:512–530.

Palmqvist, P., Gröcke, D. R., Arribas, A., and Fariña, R. A. 2003. Paleoecological reconstruction of a lower Pleistocene large mammals community using biogeochemical (13C,15N,18O, Sr:Zn) and ecomorphological approaches. Paleobiology 29:204–228.

Palomares, F., and Caro, T. M. 1999. Interspecific killing among mammalian carnivores. The American Naturalist 153:493–508.

Palombo, M. R., and C. Giovinazzo. 2006. What do cenograms tell us about mammalian palaeoecology? The example of Plio-Pleistocene Italian faunas. Courier Forschungs 59 Institut Senckenberg 256:215–235.

Palombo, M. R., Raia, P., and Giovinazzo, C. 2005. Early–Middle Pleistocene structural changes in mammalian communities from the Italian peninsula; pp. 251–262 in Head, M. J., and Gibbard, P. L. (eds.), Early–Middle Pleistocene Transitions: The Land–Ocean Evidence. Geological Society of London Special Publication 247.

Panario, D., and Gutiérrez, O. 1999. The continental Uruguayan Cenozoic: an overview. Quaternary International 62:75–84.

Panario, D., and May, H. 1994. Estudio comparativo de la sucesión ecológica de la flora pratense en dos sitios de la región basáltica, suelo superficial y suelo profundo, en condiciones de exclusión y pastoreo; pp. 55–67 in Anonymous (ed.), Contribución de los estudios edafológicos al conocimiento de la vegetación en la República Oriental del Uruguay. Boletín Técnico No. 13, Ministerio de Ganadería, Agricultura y Pesca, Montevideo.

Pardiñas, U. F. J., D'Elia, G., and Ortiz, P. E. 2002. Sigmodontinos fósiles (Rodentia, Muroidea, Sigmodontinae) de América del Sur: estado actual de su conocimiento y prospectiva. Mastozoología Neotropical 9:209–252

Pardiñas, U. F. J., Abba, A. M., and Merino, M. I. 2004. Micromamíferos (Didelphimorphia y Rodentia) del sudoeste de la provincia de Buenos Aires (Argentina): taxonomía y distribución. Mastozoología Neotropical 11:211–232.

Parish, W. 1839. Buenos Ayres and the provinces of the Rio de la Plata: their present state, trade, and debt with some account from original documents of the progress of geographical discovery in those parts of South America during the last sixty years. John Murray, London, 415 pp.

Pascual, R. 1984a. La sucesión de las Edades-mamífero, de los climas y del diastrofismo sudamericano durante el Cenozoico: fenómenos concurrentes. Anales de la Academia Nacional de Ciencias Exactas, Físicas y Naturales 36:15–37.

————. 1984b. Late Tertiary mammals of southern South America as indicators of climatic deterioration. Quaternary of South America and Antarctic Peninsula 2:1–30.

Pascual, R., and Bondesio, P. 1982. Un roedor Cardiatheriinae (Hydrochoeridae) de la edad Huayqueriense (Mioceno tardío) de La Pampa. Sumario de los ambientes terrestres en la Argentina durante el Mioceno. Ameghiniana 19:19–35.

Pascual, R., and Ortiz Jaureguizar, E. 1990. Evolving climates and mammal fauna in Cenozoic South America. Journal of Human Evolution 19:23–60.

————. 1992. Evolutionary pattern of land mammal faunas during the Late Cretaceous and Paleocene in South America: a comparison with the North American pattern. Annales Zoologici Fennici 28:245–252.

————. 2007. The Gondwanan and South American episodes: two major and unrelated moments in the history of the South American mammals. Journal of Mammalian Evolution 14:75–137.

Pascual, R., Hinojosa, E., Gondar, D., and Tonni, E. 1965. Las Edades del Cenozoico mamalífero de la Argentina, con especial atención a aquéllas del territorio bonaerense. Anales de la Comisión de Investigación Científica de Buenos Aires 6:165–193.

Pascual, R., Vucetich, M. G., Scillato-Yané, G. J., and Bond, M. 1985. Main pathways of mammalian diversification in South America; in Stehli, F. G., and Webb, S. D. (eds.), The Great American Interchange. Series Topics in Geobiology 4:219–247.

Pascual, R., Goin, F. J., and Carlini, A. A. 1994. New data on the Groeberiidae: unique late Eocene–early Oligocene South American marsupials. Journal of Vertebrate Paleontology 14:247–259.

Pascual, R., Ortiz Jaureguizar, E., and Prado, J. L. 1996. Land mammals: paradigm for Cenozoic South American geobiotic evolution; in A. Arratia (ed.), Contribution of Southern South America to Vertebrate Paleontology, Munich, Geowissenschaftlich Abhandlungen (A) 30:265–319.

Pascual, R., Goin, F. J., Krause, D. W., Ortiz-Jaureguizar, E., and Carlini, A. A. 1999. The first gnathic remains of Sudamerica: implications for gondwanathere relationships. Journal of Vertebrate Paleontology 19:373–382.

Pascual, R., Goin, F. J., Balarino, L., and Sauthier, D. E. U. 2002. New data on the Paleocene monotreme Monotrematum sudamericanum, and the convergent evolution of triangulate molars. Acta Palaeontologica Polonica 47:487–492.

Pasquali, R. C., and Tonni, E. T. 2008. Los Hallazgos de Mamíferos Fósiles Durante el Período Colonial en el Actual Territorio de la Argentina; in Aceñolaza, E. G. (ed.), Los geólogos y la geología en la historia argentina, INSUGEO, Tucumán. Serie Correlación Geológica 24:35–42.

Patterson, B. 1977. A primitive pyrothere (Mammalia, Notoungulata) from the early Tertiary of Northwestern Venezuela. Fieldiana: Geology 33:397–421.

Patterson, B., and Pascual, R. 1968. The fossil mammal fauna of South America. The Quarterly Review of Biology 43:409–451.

————. 1972. The fossil mammal fauna of South America; pp. 247–309 in Keast, A., Erk, F. C., and Glass, B. (eds.), Evolution, Mammals, and Southern Continents. State University of New York Press, Albany.

Patterson, B., Segall, W., and Turnbull, W. D. 1989. The ear region in xenarthrans (=Edentata: Mammalia), part I. Cingulates. Fieldiana Geology (n.s.) 18:1–46.

Patterson, B., Segall, W., Turnbull, W. D., and Gaudin, T. J. 1992. The ear region in xenarthrans (=Edentata: Mammalia), part 2. Pilosa (sloths, anteaters), palaeanodonts, and a miscellany. Fieldiana Geology n.s. 24:1–79.

Paul, G. S. 1988. Predatory Dinosaurs of the World: A Complete Illustrated Guide. Simon and Shuster, New York, 464 pp.

Paula Couto, C. de. 1947. Contribuição para o estudo de *Hoplophorus euphractus* Lund, 1839. Summa Brasiliensis Geologiae 1:1–14.

———. 1950. Memórias da paleontologia Brasileira. Instituto Nacional do Livro, Rio de Janeiro, 589 pp.

———. 1957. Sôbre um gliptodonte do Brasil. Boletim Divisão de Geologia e Mineralogia, 165:1–37.

———. 1979. Tratado de paleomastozoologia. Academia Brasileira de Ciências, Rio de Janeiro, 590 pp.

PC-MATLAB for MS-DOS Personal Computers. 1990. MathWorks, Natick, Mass.

Pennington, R. T., Prado, D. E., and Pendry, C. A. 2000. Neotropical seasonally dry forests and Quaternary vegetation changes. Journal of Biogeography 27:261–273.

Penny, D., Hasegawa, M., Waddell, P. J., and Hendy, M. D. 1999. Mammalian evolution: timing and implications from using the LogDeterminant transform for proteins of differing amino acid composition. Systematic Biology 48:76–93.

Pérez, L. M., Toledo, N., De Iuliis, G., Bargo, M. S., and Vizcaíno, S. F. 2010. Morphology and function of the hyoid apparatus of fossil Xenarthrans (Mammalia). Journal of Morphology 271:1119–1133.

Pérez-Barbería, F. J., and Gordon, I. J. 1998. Factors affecting food comminution during chewing in ruminants: a review. Biological Journal of the Linnean Society 63:233–256.

Peters, R. H. 1983. The Ecological Implications of Body Size. Cambridge University Press, Cambridge, 329 pp.

Pictet, F.-J., 1844–1846. Traité élémentaire de paléontologie ou histoire naturelle des animaux fossiles considérés dans leurs rapports zoologiques et géologiques, 2. Langlois & Leclerq, Paris, 407 pp.

Poinar, H. N., Hofreiter, M., Spaulding, W. G., Martin, P. S., Stankiewicz, B. A., Bland, H., Evershed, R. P., Possnert, G., and Pääbo, S. 1998. Molecular coproscopy: dung and diet of the extinct ground sloth *Nothrotheriops shastensis*. Science 281:402–406.

Politis, G., and Gutiérrez, M. 1998. Gliptodontes y cazadores-recolectores de la región pampeana (Argentina). Latin American Antiquity 9:111–134.

Politis, G., Johnson, E., Gutiérrez M., and Hartwell, T. 2003. Survival of the Pleistocene fauna: new radiocarbon dates on organic sediments from La Moderna (Pampean Region, Argentina); pp. 45–50 in Miotti, L., Salemme, M., and Flegenheimer, N. (eds.), Where the South Winds Blow: Ancient Evidence for Paleo South Americans. CSFA and Texas A&M University Press, College Station.

Politis, G. G., Messineo, P. G., and Kaufmann, C. A. 2004. El poblamiento temprano de las llanuras pampeanas de Argentina y Uruguay. Complutum 15:207–224.

Pope, K. O., Ocampo, A. C., Kinsland, G. L., and Smith, R. 1996. Surface expression of the Chicxulub crater. Geology 24:527–530.

Porpino, K. O., Fernicola, J. C., and Bergqvist, L. P. 2010. Revisiting the Intertropical Brazilian Species *Hoplophorus euphractus* (Cingulata, Glyptodontoidea) and the Phylogenetic

Affinities of *Hoplophorus*. Journal of Vertebrate Paleontology 30:911–927.

Poux, C., Chevret. P., Huchon, D., de Jong, W. W., Douzery, E. J. 2006. Arrival and diversification of caviomorph rodents and platyrrhine primates in South America. Systematic Biology 55:228–244.

Prado, J., and Alberdi, M. T. 1994. A quantitative review of the horse *Equus* from South America. Palaeontology 37:459–481.

———. 1996. A cladistic analysis of the horses of the tribe Equini. Palaeontology 39:663–680.

Prado, D. E., and Gibbs, P. E. 1993. Patterns of species distributions in the dry seasonal forests of South America. Annales of the Missouri Botanical Garden 80:902–927.

Prado, J. L., Menegaz, A. Z., Tonni, E. P., and Salemme, M. C. 1987. Los mamíferos de la Fauna local Paso Otero (Pleistoceno Tardío), Provincia de Buenos Aires. Aspectos paleoambientales y bioestratigráficos. Ameghiniana 24:217–233.

Prámparo, M. B., Quattrocchio, M., Gandolfo, M. A. Zamaloa, M. del C., and Romero, E. 2007. Historia evolutiva de las angiospermas (Cretácico-Paleógeno) en Argentina a través de los registros paleoflorísticos Asociación Paleontológica Argentina. Publicación Especial 11. Ameghiniana 50° aniversario:157–172.

Prasad, V., Stroemberg, C. A. E., Alimohammadian, H., Sahni, A. 2005. Dinosaur coprolites and the early evolution of grasses and grazers. Science 310:1177–1180.

Prasad, G. V. R., Verma, O., Sahni, A., Parmar, V., and Khosla., A. 2007. A Cretaceous hoofed mammal from India. Science 318:937.

Prasad, A. B., Allard, M. W., NISC Comparative Sequencing Program, and Green, E. D. 2008. Confirming the phylogeny of mammals by use of large comparative sequence data sets. Molecular Biology and Evolution 25:1795–1808.

Prevosti, F. J. 2006. New materials of Pleistocene cats (Carnivora, Felidae) from southern South America, with comments on biogeography and the fossil record. Geobios 39:679–694.

———. 2010. Phylogeny of the large extinct South American canids (Mammalia, Carnivora, Canidae) using a "total evidence" approach. Cladistics 26:456–481.

Prevosti, F. J., and Soibelzon, L. H. 2012. Evolution of the South American carnivores (Mammalia, Carnivora): a paleontological perspective; pp. 102–122 in Patterson, B.D. and Costa, L.P. (eds.), Bones, Clones, and Biomes: an 80-million Year History of Modern Neotropical Mammals. University of Chicago Press, Chicago.

Prevosti, F. J., and Vizcaíno, S. F. 2006. Paleoecology of the large carnivore guild from the late Pleistocene of Argentina. Acta Palaeontologica Polonica 51:407–422.

Prevosti, F. J., Turazzini, G. F., and Chemisquy, M. A. 2010. Morfología craneana en tigres dientes de sable: alometría, función y filogenia. Ameghiniana 47:239–256.

Prieto, A. R. 1996. Late Quaternary vegetational and climatic changes in the Pampa grassland of Argentina. Quaternary Research 45:73–88.

Prior, H., Schwarz, A., and Güntürkün, O. 2008. Mirror-induced behavior in the magpie (*Pica pica*): evidence of self-recognition. PLoS Biology 6:e202

Prothero, D. R. 1994. The Eocene–Oligocene Transition: Paradise Lost. Columbia University Press, New York, 291 pp.

———. 2006. After the Dinosaurs: The Age of Mammals. Indiana University Press, Bloomington, 362 pp.

Pujos, F. 2000. *Scelidodon chilensis* (Mammalia, Xenarthra) du Pleistocene terminal de "Pampa de los Fósiles" (Nord-Pérou). Quaternaire 11:197–206.

———. 2002. Contribution à la connaissance des Tardigrades (Mammalia: Xenarthra) du Pléistocène péruvien: systématique, phylogénie, anatomie fonctionnelle et extinction. Ph.D. dissertation, Muséum National d'Histoire Naturelle, Paris, France, 513 pp.

———. 2006. *Megatherium celendinense* sp. nov. from the pleistocene of the peruvian Andes and the phylogenetic relationships of megatheriines. Palaeontology 49:305–306.

Pujos, F., and De Iuliis, G. 2007. Late Oligocene Megatherioidea fauna (Mammalia: Xenarthra) from Salla-Luribay (Bolivia): new data on basal sloth radiation and Cingulata-Tardigrada split. Journal of Vertebrate Paleontology 27:132–144.

Pujos, F., De Iuliis, G., Argot, C., and Werdelin, L. 2007. A peculiar climbing Megalonychidae from the Pleistocene of Peru and its implications for sloth history. Zoological Journal of the Linnean Society 149:179–235.

Pujos, F., De Iuliis, G., and Mamani Quispe, B. 2011. *Hiskatherium saintandrei* gen. et sp. nov.: an unusual sloth from the Santacrucian of Quebrada Honda (Bolivia) and an overview of middle Miocene, small megatherioids. Journal of Paleontology 31:1131–1149.

Quattrocchio, M., Deschamps, C., Martínez, D., Grill, S., and Zavala, C. 1988. Caracterización paleontológica y paleoambiental de sedimentos cuaternarios, Arroyo Napostá Grande, Provincia de Buenos Aires. Actas 2as. Jornadas Geológicas Bonaerenses, pp. 37–46.

Quintana, C. A. 1992. Estructura interna de una paleocueva, posiblemente de un Dasypodidae (Mammalia, Edentata), del Pleistoceno de Mar del Plata (Provincia de Buenos Aires, Argentina). Ameghiniana 29:87–91.

Ramírez Rozzi, F., and Podgorny I. 2001. La metamorfosis del megaterio. Ciencia Hoy. Buenos Aires 11:12–19,

Rate Inflation. (N.d.) UK—consumer price index (CPI) history. http://www.rateinflation.com/consumer-price-index/uk-historical-cpi.php.

Redford, K. H. 1985. Food habits of armadillos (Xenarthra: Dasypodidae); pp. 429–437 in Montgomery, G. G. (ed.), The Evolution and Ecology of Armadillos, Sloths, and Vermilinguas, Smithsonian Institution, Washington, D.C.

Reguero, M. A., and Gasparini, Z. 2006. Late Cretaceous–Early Tertiary marine and terrestrial vertebrates from James Ross Basin, Antarctic Peninsula: a review; pp. 55–76 in Rabassa, J., and Borla, M. L. (eds.), Antarctic Peninsula and Tierra del Fuego: 100 years of Swedish–Argentine Scientific Cooperation at the End of the World. Taylor and Francis, London.

Reguero, M.A., Dozo, M.T., and Cerdeño, E. 2007. A poorly known rodentlike mammal (Pachyrukhinae, Hegetotheriidae, Notoungulata) from the Deseadan (Late Oligocene)

of Argentina: paleoecology, biogeography, and radiation of the rodentlike ungulates in South America. Journal of Paleontology 81:1301–1307.

Reig, O. 1981. Teoría del origen y desarrollo de la fauna de mamíferos de América del Sur. Monographie Naturae 1:1–162.

Reinhardt, J. 1875. De i Brasiliens knogelhuler fundne Glyptodon-levninger og en ny, til de gravigrade edentater hørende slaegt. Videnskabelige Meddelelser fra den naturhistoriske Forening i Kjøbenhavn 1875:165–236.

———. 1879. Beskrivelse af Hovedskallen af et Kæmpedovendyr, Grypotherium darwinii. Videnskabernes Selskabs Skrifter. 5 Række. Naturvidenskabelig og Mathematisk Afdeling 12:351–381.

Reiss, M. 1988. Scaling of home range size: body size, metabolic needs and ecology. Trends in Ecology and Evolution 3:85–86.

Rich, T. H. 1990. Monotremes, placentals and marsupials: the record in Australia and its biases; pp. 893–1004 in Vickers-Rich, P., Monaghan, J. M., Baird, R. F., and Rich, T. H. (eds.), Vertebrate Palaeontology of Australasia. Monash University Publications Committee, Singapore.

Richards, R. L., C. S. Churcher, and W. D. Turnbull. 1996. Distribution and size variation in North American short-faced bears, Arctodus simus; pp. 191–246 in Stewart, K. M., and Seymour, K. L. (eds.), Palaeoecology and Palaeoenvironments of Late Cenozoic Mammals: Tributes to the Career of C. S. (Rufus) Churcher. University of Toronto Press, Toronto.

Ridewood, W. G. 1901. On the structure of the hairs of Mylodon listai and other South American edentates. Quarterly Journal of Microscopical Science (Ser. 2) 44:393–410.

Riggs, E. S. 1935. A skeleton of Astrapotherium. Geological Series of Field Museum of Natural History 6:167–177.

———. 1937. Mounted skeleton of Homalodotherium. Field Museum of Natural History, Geological Series 6:233–243.

Rincón, A. D. 2003. Los mamíferos fósiles del pleistoceno de la Cueva del Zumbador (fa. 116), Estado Falcón, Venezuela. Boletín de la Sociedad Venezolana de Espeleología 37:18–26.

———. 2006. A first record of the Pleistocene saber-toothed cat Smilodon populator Lund 1842 (Carnivora: Felidae: Machairodontinae) from Venezuela. Ameghiniana 43:499–501.

Rinderknecht, A., and Blanco, R. E. 2008. The largest fossil rodent. Proceedings of the Royal Society B 275:923–928.

Rinderknecht, A., and Claramunt, S. 2000. Primer registro de Colaptes Vigors 1826, para el Pleistoceno de Uruguay (Aves: Piciformes, Picidae). Comunicaciones Paleontológicas del Museo de Historia Natural de Montevideo 32:157–160.

Ringuelet, R. A. 1957. Restos de probables huevos de Nematodos en el estiércol del edentado extinguido Mylodon listai (Ameghino 1889). Ameghiniana 1:15–16.

Robinson, D. 2001. 15N as an integrator of the nitrogen cycle. Trends in Ecology and Evolution 16:153–162.

Robinson, T. J., and Matthee, T. A. 2005. Phylogeny and evolutionary origins of the Leporidae: a review of cytogenetics, molecular analyses and a supermatrix analysis. Mammal Review 35:231–247.

Roca, A. L., Bar-Gal, G. K., Eizirik, E., Helgen, K. M., Maria, R., Springer, M. S., O'Brien, S. J., and Murphy, W. J. 2004. Mesozoic origin for West Indian insectivores. Nature 429:649–651.

Rodríguez, J. 1999. Use of Cenograms in mammalian paleoecology. A critical review. Lethaia 32:331–347.

———. 2000. Comment on "Mammalian faunas as indicators of environmental and climatic changes in Spain during the Pliocene–Quaternary transition" (Montuire 1999). Quaternary Research 54:433–435.

Rome, L. C., Sosnicki, A., and Goble, D. 1990. Maximum velocity of shortening of three fiber types from horse soleus muscle: implications for scaling. Journal of Physiology 431 173–185.

Romero, E. 1978. Paleoecología y paleofitogeografía de las tafofloras del Cenofítico de Argentina y áreas vecinas. Ameghiniana 15: 209–227

———. 1986. Paleogene phytogeography and climatology of South America. Annals of the Missouri Botanical Garden 73:449–61.

Rose, K. D. 1999. Eurotamandua and Palaeanodonta: convergent or related? Paläontologische Zeitschrift 73:395–401.

———. 2006. The Beginning of the Age of Mammals. Johns Hopkins University Press, Baltimore, 428 pp.

Rose, K. D., and Emry, R. J. 1993. Relationships of Xenarthra, Pholidota, and fossil "Edentates"; pp. 81–102 in Sazlay, F. S., Novacek, M. J., and McKenna, M. C. (eds.), Mammal Phylogeny (Placentals). Springer, New York.

Roth, S. 1899. El mamífero misterioso de la Patagonia Grypotherium domesticum. II. Descripción de los restos encontrados en la Caverna de Última Esperanza. Revista del Museo de La Plata 9:421–453.

Roth, V. L. 1990. Insular dwarf elephants: a case study in body mass estimation and ecological inference; pp. 151–179 in Damuth, J., and MacFadden, B. J. (eds.), Body Size in Mammalian Paleobiology: Estimation and Biological Implications. Cambridge University Press, Cambridge.

Rudwick, M. J. S. 1985. The Meaning of Fossils: Episodes in the History of Palaeontology. 2nd ed. University of Chicago Press, Chicago, 287 pp.

———. 1997. Georges Cuvier, Fossil Bones, and Geological Catastrophes: New Translations and Interpretations of the Primary Texts. University of Chicago Press, Chicago, 301 pp.

Rupke, N. A. 1993. 1994 Richard Owen's Vertebrate Archetype. Isis 84:231–251.

———. 1994. Richard Owen: Victorian Naturalist. Yale University Press, New Haven, 462 pp.

Rusconi, C. 1930. Las especies fósiles argentinas de pecaríes (Tayassuidae) y sus relaciones con las del Brasil y Norteamérica. Anales del Museo Nacional de Historia Natural "Bernardino Rivadavia" 36:121–241.

Russell, D. 2009. Islands in the Cosmos: The Evolution of Life on Land. Indiana University Press, Bloomington, 480 pp.

Sacco, T., and Van Valkenburgh, B. 2004. Ecomorphological indicators of feeding behaviour in the bears (Carnivora: Ursidae). Journal of Zoology (London) 263:41–54.

Saint-André, P.-A., and De Iuliis, G. 2001. The smallest and most ancient representative of the genus Megatherium Cuvier 1796 (Xenarthra, Tardigrada, Megatheriidae), from the Pliocene of the Bolivian Altiplano. Geodiversitas 23:625–645.

Sánchez, B., Prado, J. L., and Alberdi, M. T. 2004. Feeding ecology, dispersal, and extinction of South American Pleistocene gomphotheres (Gomphotheriidae, Proboscidea). Paleobiology 30:146–163.

Sánchez-Villagra, M. R., Aguilera, O., and Horovitz, I. 2003. The Anatomy of the World's Largest Extinct Rodent. Science 30:1708–1710.

Sarmiento, D. F. 1885. Vida y escritos del coronel doctor Francisco Javier Muñiz. Félix Lajouane, Buenos Aires, 369 pp.

Savage, R. J. G. 1973. Megistotherium, new genus gigantic hyaenodont from Miocene of Gegel Zelten Lybia. Bulletin of the British Museum (Natural History) Geology 22:485–511.

Savin, S. M., Douglas, R. G., and Stehli, F. G. 1975. Tertiary marine paleotemperatures. Geological Society of America Bulletin 86:1499–1510.

Schaller, G. B. 1972. The Serengeti Lion. University of Chicago Press, Chicago, 480 pp.

Schmidt-Nielsen, K. 1984. Scaling: Why Is Animal Size So Important? Cambridge University Press, Cambridge, 241 pp.

Scholander, P. F. 1955. Evolution of climatic adaptation in homeotherms. Evolution 9:15–26.

Schönenberger, J., and Conti, E. 2003. Molecular phylogeny and floral evolution of Penaeaceae, Oliniaceae, Rhynchocalycaceae, and Alzateaceae (Myrtales). American Journal of Botany 90:293–309.

Schubert, B. W., Graham, R. W., McDonald, H. G., Grimm, E. C., and Stafford, Jr., T. W. 2004. Latest Pleistocene paleoecology of Jefferson's ground sloth (Megalonyx jeffersonii) and elk-moose (Cervalces scotti) in northern Illinois. Quaternary Research 61:231–240.

Schubert, B. W., Hulbert, Jr., R. C., MacFadden, B. J., Searle, M., and Searle S. 2010. Giant short-faced bears (Arctodus simus) in Pleistocene Florida USA, a substantial range extension. Journal of Paleontology 84:79–87.

Schultz, P. H., Zárate, M. A., Hames, W., Camilión, C., and King, J. 1998. A 3.3-Ma impact in Argentina and possible consequences. Science 282:2061–2063.

Schweitzer, M. H., Wittmeyer, J. L. Horner, J. R., and Toporski, J. K. 2005. Soft-tissue vessels and cellular preservation in Tyrannosaurus rex. Science 307:1952–1955.

Schweitzer M. H, Wittmeyer, J. L., and Horner, J. R. 2007. Soft tissue and cellular preservation in vertebrate skeletal elements from the Cretaceous to the present. Proceedings of the Royal Society B 274:183–187.

Scillato-Yané, G. J. 1976. Sobre algunos restos de Mylodon (?) listai (Edentata, Tardigrada) procedentes de la cueva "Las Buitreras" (Provincia de Santa Cruz, Argentina). Relaciones de la Sociedad Argentina de Antropología 10:309–312.

———. 1982. Los Dasypodidae (Mammalia, Edentata) del Plioceno y Pleistoceno de Argentina. Ph.D. dissertation, Nc 406, Facultad de Ciencias Naturales y Museo, Universidad Nacional de La Plata, 244 pp.

Scillato-Yané, G. J., and Carlini, A. A. 1998. Nuevos Xenarthra del Friasense (Mioceno Medio) de Argentina. Studia Geologica Salamanticensia 34:43–67.

Scillato-Yané, G. J., and Pascual, R. 1985. Un peculiar Xenarthra del Paleoceno medio de Patagonia (Argentina). Su importancia en la sistemática de los Paratheria. Ameghiniana 21:173–176.

Scillato-Yané, G. J., Carlini, A. A., Tonni, E. P., and Noriega, J. I. 2005. Paleobiogeography of the late Pleistocene pampatheres of South America. Journal of South American Earth Sciences 20:131–138

Scott, K. M. 1990. Postcranial dimensions of ungulates as predictors of body mass; pp. 301–335 in Damuth, J., and MacFadden, B. J. (eds.), Body Size in Mammalian Paleobiology: Estimation and Biological Implications. Cambridge University Press, Cambridge.

Scott, W. B. 1903–1904. Mammalia of the Santa Cruz beds. I. Edentata; 1–364 in Scott, W. B (ed.), Reports of the Princeton University Expeditions to Patagonia, 1896–1899. Princeton University Press, Princeton.

———. 1912. Mammalia of the Santa Cruz Beds. Vol. 6. Part 2–3. Toxodontia. Report Princeton University Expedition to Patagonia 6:111–300

———. 1913. A History of Land Mammals in the Western Hemisphere. Macmillan, New York, 693 pp.

———. 1937. The Astrapotheria. Proceedings of the American Philosophical Society 77:300–393.

Scott, K. P. 1990. Postcranial dimensions of ungulates as predictors of body mass; pp. 301–335 in Damuth, J., and MacFadden, B. J. (eds.), Body Size in Mammalian Paleobiology: Estimation and Biological Implications. Cambridge University Press, Cambridge.

Sefve, I. 1924. Macrauchenia patachonica. Bulletin of the Geological Institute of the University of Upsala 19:1–21.

Senter, P. 2005. Function in the stunted forelimbs of Mononykus olecranus (Theropoda), a dinosaurian anteater. Paleobiology 31:373–381.

Seymour, K. 1993. Size change in North American Quaternary jaguars; pp. 343–372 in Martin, R. A., and Barnosky, A. D. (eds.), Morphological Change in Quaternary Mammals of North America. Cambridge University Press, Cambridge.

Shipman, P. 1981. Life History of a Fossil: An Introduction to Taphonomy and Paleoecology. Harvard University Press, Cambridge, Mass., 222 pp.

Shockey, B. J., Croft, D. A., and Anaya, F. 2007. Analysis of function in the absence of extant functional homologues: a case study using mesotheriid notoungulates (Mammalia). Paleobiology 33:227–247

Shoshani, J. 1996. Para- or monophyly of the gomphotheres and their position within Proboscidea; 149–177 in Shoshani, J., and Tassy, P. (eds.), The Proboscidea: Evolution and Palaeoecology of Elephants and Their Relatives. Oxford University Press, Oxford.

Sicher, H. 1944. Masticatory apparatus of the sloths. Fieldiana: Zoology 29:161–168.

Silva, M., and Downing, J. A. 1995. The allometric scaling of density and body mass: a nonlinear relationship for terrestrial mammals. The American Naturalist 145:704–727.

Silva, M., Brimacombe, M., and Downing, J. A. 2001. Effects of body mass, climate, geography, and census area on population density of terrestrial mammals. Global Ecology and Biogeography 10:469–485.

Silverman, H., and Isbell, W. H. 2008. Handbook of South American Archaeology. Springer, New York.

Simpson, G. G. 1928. A Catalogue of the Mesozoic Mammalia in the Geological Department of the British Museum. Trustees of the British Museum, London, 215 pp.

———. 1929. American Mesozoic Mammalia. Memories of the Peabody Museum Yale University 3:1–235.

———. 1932. Enamel on the teeth of an Eocene edentate. American Museum Novitates 567:1–4.

———. 1933. Stratigraphic nomenclature of the Early Tertiary of Central Patagonia. American Museum Novitates 644:1–13.

———. 1934. Attending Marvels: A Patagonian Journal. University of Chicago Press, 310 pp.

———. 1935. Ocurrence and relationships of the Río Chico fauna of Patagonia. American Museum Natural History Novitates 818:1–21.

———. 1936. Structure of a primitive notoungulate cranium. American Museum Novitates 824:1–31.

———. 1944. Tempo and Mode in Evolution. Columbia University Press, New York, 237 pp.

———. 1945. The principles of classification and a classification of mammals. Bulletin of the American Museum of Natural History 85:1–350

———. 1948. The beginning of the age of mammals in South America. Part 1: Introduction. Systematics: Marsupialia, Edentata, Condylarthra, Litopterna and Notioprogonia. Bulletin of the American Museum of Natural History 91:1–232.

———. 1950. History of the Fauna of Latin America. American Scientist 38:361–389.

———. 1953a. The Major Features of Evolution. Columbia University Press, New York, 434 pp.

———. 1953b. Evolution and Geography. Condon Lectures, Eugene, Ore., 64 pp.

———. 1961. The Principles of Animal Taxonomy. Columbia University Press, New York, 247 pp.

———. 1967. The beginning of the age of mammals in South America. Part 2. Systematics: Notoungulata, concluded (Typotheria, Hegetotheria, Toxodonta, Notoungulata incertae sedis), Astrapotheria, Trigonostylopoidea, Pyrotheria, Xenungulata, Mammalia incertae sedis. Bulletin of the American Museum of Natural History 137:1–259.

———. 1980. Splendid Isolation, the Curious History of South American Mammals. New Haven, Yale University Press, 266 pp.

———. 1984. Discoverers of the Lost World: An Account of Some of Those Who Brought Back to Life South American Mammals Long Buried in the Abyss of Time. Yale University Press, New Haven, 222 pp.

Simpson, G. G., and Paula Couto, C. 1957. Os Mastodontes do Brasil. Instituto Brasileiro de Bibliografia e Documentação 2:1–20.

Simpson, G. G., and Roe, A. 1939. Quantitative Zoology: Numerical Concepts and Methods in the Study of Recent and Fossil Animals. McGraw-Hill, New York, 414 pp.

Sinclair, W. J. 1906. Marsupialia of the Santa Cruz beds; in Scott, W. B. (ed.), Reports of the Princeton University Expeditions to Patagonia, 1896–1899. Princeton University Press, Princeton 4:333–460.

———. 1909. Mammalia of the Santa Cruz Beds. Part I, Typotheria; in Scott, W. B. (ed.), Reports of the Princeton University Expeditions to Patagonia, 1896–1899. Princeton University Press, Princeton 6:1–110.

Smallwood, K. S. 1997. Interpreting puma (Puma concolor) population estimates for theory and management. Environmental Conservation 24:283–289.

———. 1999. Scale domains of abundance amongst species of mammalian Carnivora. Environmental Conservation 26:102–111.

———. 2001. The allometry of density within the space used by populations of mammalian Carnivora. Canadian Journal of Science 79:1634–1640.

Smallwood, K. S., and Schonewald, C. 1998. Study design and interpretation of mammalian carnivore density estimates. Oecologia 113:474–491.

Smith, R. J. 1993. Categories of allometry: body size versus biomechanics. Journal of Human Evolution 24:173–182.

Smith-Woodward, A. 1900. On some remains of Grypotherium (Neomylodon) listai and associated mammals from a cavern near Consuelo Cave, Last Hope Inlet, Patagonia. Proceedings of the Zoological Society of London 1900:64–79.

Soibelzon, L. H. 2002. Los Ursidae (Carnivora, Fissipedia) fósiles de la República Argentina. Aspectos Sistemáticos y Paleoecológicos. Ph.D. dissertation, Facultad de Ciencias Naturales y Museo, Universidad Nacional de La Plata, 239 pp.

———. 2004. Revisión sistemática de los Tremarctinae (Carnivora, Ursidae) fósiles de América del Sur. Revista del Museo Argentino de Ciencias Naturales "Bernardino Rivadavia" 6:107–133.

Soibelzon, L., and Prevosti, F. 2007. Los carnívoros (Carnivora, Mammalia) terrestres del Cuaternario de América del Sur; pp. 49–68 in Pons, G. X., and Vicens, D. (eds.), Geomorphologia Litoral i Quaternari. Homenatge a D. Joan Cuerda Barceló. Monografies de la Societat d'Història Natural de les Balears, Palma de Mallorca.

Soibelzon, L., and Schubert, B. W. 2011. The largest known bear, Arctotherium angustidens, from the early Pleistocene Pampean region of Argentina: with a discussion of size and diet trends in bears. Journal of Paleontology 85:69–75.

Soibelzon, L., and Tarantini, V. B. 2009. Estimación de la masa corporal de las especies de osos fósiles y actuales (Ursidae, Tremarctinae) de América del Sur. Revista del Museo Argentino de Ciencias Naturales, nueva serie 11:243–254.

Soibelzon L., Isla, F. I., and Dondas, A. 2001. Primer registro de tres individuos asociados de Arctotherium latidens (Ursidae: Tremarctinae). Ameghiniana 38:40R–41R.

Soibelzon, L. H., Romero, M. R., Huziel Aguilar, D., and Tartarini, V. B. 2008. A new finding in the

Blancan of the El Salvador indicates the probable dating of the Tremarctinae's (Ursidae, Carnivora) entrance to South America. Neues Jahrbuch für Geologie und Paläontologie 250:1–8.

Solounias, N., and Moelleken, S. M. 1993. Dietary adaptation of some extinct ruminants determined by premaxillary shape. Journal of Mammalogy 74:1059–1071.

Solounias, N., Teaford, M., and Walker, A. 1988. Interpreting the diet of extinct ruminants: the case of a non-browsing giraffid. Paleobiology 14:287–300.

Sorkin, B. 2008. A biomechanical constraint on body mass in terrestrial mammalian predators. Lethaia 41:333–347

Spencer, L. M. 1995. Morphological correlates of dietary resource partitioning in the African Bovidae. Journal of Mammalogy 76:448–471.

Springer, M. S., Burk, A., Kavanagh, J. R., Waddell, V. G., and Stanhope, M. J. 1997. The interphotoreceptor retinoid binding protein gene in therian mammals: implications for higher level relationships and evidence for loss of function in the marsupial mole. Proceedings of the National Academy of Sciences of the United States of America 94:13754–13759.

Springer, M. S., Stanhope, M. J., Madsen, O., and De Jong, W. W. 2004. Molecules consolidate the placental mammal tree. Trends in Ecology and Evolution 19:430–438.

Stangerup, H. 1987. The Road to Lagoa Santa. Marion Boyars, London, 284 pp.

Stanhope, M. J., Waddell, V. G., Madsen, O., DeJong, W., Hedges, S. B., Cleven, G. C., Kao, D., and Springer, M. S. 1998. Molecular evidence for multiple origins of Insectivora and for a new order of endemic African insectivore mammals. Proceedings of the National Academy of Science 95:9967–9972.

Stanley, S. M. 1974. Relative growth of the titanothere horn: a new approach to an old problem. Evolution 28:447–457.

Steadman, D. W., Martin, P. S., MacPhee, R. D. E., Jull, A. J. T., McDonald, H. G., Woods, C. A., Iturralde-Vinent, M., and Hodgins, G. W. L. 2005. Asynchronous extinction of late Quaternary sloths on continents and islands. Proceedings of the National Academy of Sciences of the United States of America 102:11763–11768.

Steele, J., and Politis, G. 2009. AMS 14C dating of early human occupation of southern South America. Journal of Archaeological Science 36:419–429.

Stehli, F. G., and Webb, S. D. (eds.) 1985. The Great American Biotic Interchange. Plenum Press, New York, pp. 532.

Stehlin, H. G. 1910. Remarques sur les faunules de Mammifères des couches éocènes et oligocènes du Bassin de Paris. Bulletin de la Société Géologique de France 4:488–520.

Stevens, P. F. (N.d.) Angiosperm phylogeny website. http://www.mobot.org/mobot/research/apweb/.

Stock, C. 1925. Cenozoic gravigrade edentates of western North America, with special reference to the Pleistocene Megalonychidae and Mylodontidae of Rancho La Brea. Carnegie Institute of Washington Publication 331:1–206.

———. 1956. Rancho La Brea: a record of Pleistocene life in California. Natural History Museum of Los Angeles County Science Series 20:1–81.

Storch, G. 1981. Eurotamandua joresi, ein Myrmecophagidae aus dem Eozän der "Grube Messel" bei Darmstadt (Mammalia, Xenarthra). Seckenbergiana lethaea 61:247–289.

———. 2003. Fossil Old World "edentates" (Mammalia); in Fariña, R. A., Vizcaíno, S. F., and Storch, G. (eds.), Morphological Studies in Fossil and Extant Xenarthra (Mammalia). Senckenbergiana Biologica 83:51–60.

Storch, G., and Haubold, H. 1989. Additions to the Geiseltal mammalian faunas, middle Eocene: Didelphidae, Nyctitheriidae, Myrmecophagidae. Palaeovertebrata 19:95–114.

Stewart, L. 1994. A Guide to Palms and Cycads of the World. Angus and Robertson, Sydney, Australia, 246 pp.

Stuart, A. J., Sulerzhitsky, L. D., Orlova, L. A., Kuzmin Y. V., and Lister, A. M. 2002. The latest woolly mammoths (Mammuthus primigenius Blumenbach) in Europe and Asia: a review of the current evidence Quaternary Science Reviews 21:1559–1569.

Sun, G., Ji, Q., Dilcher, D. L. Zheng, S. Nixon, K. C., and Wang, X. 2002. Archaefructaceae, a new basal angiosperm family. Science 296:899–904.

Sutcliffe, A. J. 1977. Further notes on bones and antlers chewed by deer and other ungulates. Deer 4:73–82.

Swihart, R. K., Slade, N. A., and Bergstrom, B. J. 1988. Relating body size to the rate of home range use in mammals. Ecology 69:393–399.

Szalay, F. S., and Schrenk, F. 1998. The middle Eocene Eurotamandua and a Darwinian phylogenetic analysis of "edentates." Kaupia. Darmstädter Beiträge zur Naturgeschichte 7:97–186.

Taboada, G. S., Duque, W. S., and Franco, S. D. 2007. Mamíferos terrestres autóctonos de Cuba, vivientes y extinguidos. Museo Nacional de Historia Natural, Havana, 465 pp.

Tambussi, C., Noriega, J. I., and Tonni, E. P. 1993. Late Cenozoic birds of Buenos Aires Province (Argentina): an attempt to document quatitative faunal changes. Palaeogeography, Palaeoclimatology, Palaeoecology 101:117–129.

Tattersall, I. 2002. The Monkey in the Mirror: Essays on the Science of What Makes Us Human. Harcourt, San Diego, 217 pp.

Therrien, F. 2005. Feeding behaviour and bite force of sabretoothed predators. Zoological Journal of the Linnean Society 145:393–426.

Thomas, O. 1887. On the homologies and succession of the teeth in the Dasypodidae, with an attempt to trace the history of the evolution of mammalian teeth in general. Philosophical Transactions of the Royal Society of London B 1887:443–462.

Thomas, K. 2008. The Legacy of the Mastodon. The Golden Age of Fossils in America. Yale University Press, New Haven, 386 pp.

Thompson, L. G. 2000. Ice core evidence for climate change in the tropics: implications for our future. Quaternary Science Review 19:19–35.

Thompson, L. G., Davis, M. E., Mosley-Thompson, E., Sowers, T. A., Henderson, K. A., Zagorodnov, V. S., Lin, P.-N., Mikhalenko, V. N., Campen, R. K., Bolzan, J. F., Cole-Dai, J., and Francou, B. 1998. A 25,000–year tropical climate history from Bolivian ice cores. Science 282:1858–1864.

Toledo, M. J. 2005. Secuencias pleistocenas "lujanenses" en su sección tipo: primeras dataciones C14 e implicancias estratigráficas, arqeológicas e históricas, Luján–Jáuregui, provincia de Buenos Aires. Revista de la Asociación Geológica Argentina 60:417–424.

Tonni, E. P. 1985a. Mamíferos del Holoceno del Partido de Lobería, Pcia. de Bs. As. Aspectos paleoambientales y bioestratigráficos del Holoceno del Sector Oriental de Tandilla y Area Interserrana. Ameghiniana 22:283–288.

———. 1985b. The Quaternary climate in the Buenos Aires province through the mammals. Actas 1° Reunión latinoamericana sobre la importancia de los fenómenos periglaciales y 3° Reunión del Grupo Periglacial Argentino 3:114–121.

———. 1990. Mamíferos del Holoceno en la Provincia de Buenos Aires. Paula-Coutiana 4:3–21.

Tonni, E. P., and Cione, A. L. (eds.) 1999. Quaternary Vertebrate Paleontology in South America. Quaternary of South America and Antarctic Peninsula 12.

Tonni E. P., and Laza, J. 1980. Las aves de la Fauna local Paso Otero (Pleistoceno Tardío) de la provincia de Buenos Aires. Su significación ecológica, climática y zoogeográfica. Ameghiniana 17:313–322.

Tonni, E. P., Prado, J. L., Menegaz, A. N., and Salemme, M. C. 1985. La unidad mamífero (Fauna) Lujanense. Proyección de la estratigrafía mamaliana al cuaternario de la región pampeana. Ameghiniana 22:255–261.

Tonni, E. P., Cione, A. L., and Figini, A. J. 1999. Predominance of arid climates indicated by mammals in the pampas of Argentina during the late Pleistocene and Holocene. Palaeogeography, Palaeoclimatology, Palaeoecology 147:257–281.

Tonni, E. P., Huarte, R. A., Carbonari, J. E., and Figini, A. J. 2003. New radiocarbon chronology for the Guerrero Member of the Luján Formation (Buenos Aires, Argentina): palaeoclimatic significance. Quaternary International 109–110:45–48.

Toohey, L. 1959. The species of Nimravus (Carnivora, Felidae). Bulletin of the American Museum of Natural History 118:71–112.

Troncoso, A., and Romero, E. J. 1998. Evolución de las comunidades florísticas en el extremo sur de Sudamérica durante el Cenofítico; 149–172 in Fortunato, R., and Bacigalupo, N. (eds.), Proceedings of the Congreso Latinoamericano de Botánica, No. 6. Monographs in Systematic Botany from the Missouri Botanical Garden.

Tseng, Z. J., Takeuchi, G. K., and Wang, X. 2010. Discovery of the upper dentition of Barbourofelis whitfordi (Nimravidae, Carnivora) and an evaluation of the genus in California. Journal of Vertebrate Paleontology 30:244–254.

Tsubamoto, T., Egi, N. Takai, M., Sein, C., and Maung, M. 2005. Middle Eocene ungulate mammals from Myanmar: a review with description of new specimens. Acta Palaeontologica Polonica 50:117–138.

Turnbull, W. D. 1970. Mammalian masticatory apparatus. Fieldiana Geology 18:149–356.

Turner, A., and Antón, M. 1997. The big cats and their fossil relatives: an illustrated guide to their evolution and natural history. Columbia University Press, New York, 256.

Udvardy, M. 1975. A classification of the biogeographical provinces of the world.

IUCN Occasional Paper 18. IUCN, Morges, Switzerland.

Valverde, J. A. 1964. Remarques sur la structure et l'évolution des communautés de vertébrés terrestres. I. Structure d'une commumauté. II. Rapports entre prédateurs et proies. La Terre et la Vie 111:121–154.

————. 1967. Estructura de una comunidad de vertabrados terrestres. Monografías de la Estación Biológica de Doñana 1:1–129.

Van der Hammen, T. 1974. The Pleistocene changes of vegetation and climate in tropical South America. Journal of Biogeography 1:3–26.

Van der Klaauw, C. J. 1931. On the auditory bulla in some fossil mammals, with a general introduction to this region of the skull. Bulletin of the American Museum of Natural History 62:1–352.

Van Dijk, M. A. M., Paradis, E., Catzefelis, F., and DeJong, W. W. 1999. The virtues of gaps: Xenarthran (Edentate) monophyly supported by a unique deletion in a-Crystallin. Systematic Biology 48:94–106.

Van Valkenburgh, B. 1990. Skeletal and dental predictors of body mass in carnivores; pp. 181–205 in Damuth, J., and MacFadden, B. J. (eds.), Body Size in Mammalian Paleobiology: Estimation and Biological Implications. Cambridge University Press, Cambridge.

————. 1991. Iterative evolution of hypercarnivory in canids (Mammalia: Carnivora): evolutionary interactions among sympatric predators. Paleobiology 17:340–362.

Vaughan, T. A., Ryan, J. M., and Czaplewski, N. J. 2000. Mammalogy. 4th ed. Saunders College Publishing, Orlando, Fla., vii + 565 pp.

Verzi D. H., Deschamps, C. M., and Tonni, E. P. 2004. Biostratigraphic and paleoclimatic meaning of the Middle Pleistocene South American rodent Ctenomys kraglievichi (Caviomorpha, Octodontidae). Palaeogeography, Palaeoclimatology, Palaeocology 212:315–329.

Vicq-d'Azyr, F. 1792. Système anatomique. Quadrupèdes. Encyclopédie méthodique. Vol. 2, Discours sur l'anatomie simple et comparée, avec l'exposition du plan que j'ai suivi dans la rédaction de cet ouvrage. Panckoucke, Paris, Plomteux, Liège, 632 pp.

Vilardebó, T. M., and Berro, B. P. 1838. Informe presentado á la Comision de Biblioteca y Museo por los miembros de ella, D. Bernardo Berro y D. Teodoro M. Vilardebó, sobre el reciente descubrimiento de un animal fosil en el Partido de la Piedra Sola Departamento del Canelon. El Universal (31 March 1838) 2551:2; (2 April 1838) 2552:2; (3 April 1838) 2553:2–3; (5 April 1838) 2555:2–3.

Villagrán, C. 1993. Una interpretación climática del registro palinológico del último ciclo glacial–postglacia en Sudamérica. Bulletin du Institut français d'Études Andines 22:243–258.

Villagrán, C., and Hinojosa, L. F. 1997. Historia de los bosques del sur de Sudamérica, II: análisis fitogeográfico. Revista Chilena de Historia Natural 70:241–267.

Vizcaíno, S. F. 1995. Identificación específica de las "mulitas," genero Dasypus L. (Mammalia, Dasypodidae), del noroeste Argentino. Descripción de una nueva especie. Mastozoología Neotropical 2:5–13.

————. 2000. Vegetation partitioning among Lujanian (Late Pleistocene-Early Holocene) armored herbivores in the pampean region. Current Research in the Pleistocene 17:135–137.

————. 2009. The teeth of the "toothless." Novelties and key innovations in the evolution of xenarthrans (Mammalia, Xenarthra). Paleobiology 35:343–366.

Vizcaíno, S. F., and Bargo, M. S. 1998. The masticatory apparatus of Eutatus (Mammalia, Cingulata) and some allied genera. Evolution and paleobiology. Paleobiology 24:371–383.

Vizcaíno, S. F., and De Iuliis, G. 2003. Evidence for advanced carnivory in fossil armadillos (Mammalia: Xenarthra: Dasypodidae). Paleobiology 29:123–138.

Vizcaíno, S. F., and Fariña, R. A. 1997. Diet and locomotion of the armadillo Peltephilus: a new view. Lethaia 30:79–86.

————. 1999. On the flight capabilities and distribution of the giant Miocene bird Argentavis magnificens (Teratornithidae). Lethaia 32:271–278.

Vizcaíno S. F., and Loughry, W. J. (eds.). 2008. The Biology of the Xenarthra. University Press of Florida, Gainesville, 370 pp.

Vizcaíno S. F., and Milne, N. 2002. Structure and function in armadillo limbs (Mammalia: Xenarthra: Dasypodidae). Journal of Zoology, London 257:117–127.

Vizcaíno, S. F., De Iuliis, G., and Bargo, M. S. 1998. Skull shape, masticatory apparatus, and diet of Vassallia and Holmesina (Mammalia: Xenarthra: Pampatheriidae). When anatomy constrains destiny. Journal of Mammalian Evolution 5:293–321.

Vizcaíno, S. F., Fariña R. A., and Mazzetta, G. 1999. Ulnar dimensions and fossoriality in armadillos and other South American mammals. Acta Theriologica 44:309–320.

Vizcaíno, S. F., Palmqvist, P., Fariña, R. A. 2000. ¿Hay un límite para el tamaño corporal en las aves voladoras? Encuentros en Biología 64:6–8.

Vizcaíno, S. F., Zárate, M., Bargo, M. S., and Dondas, A. 2001. Pleistocene caves in the Mar del Plata area (Buenos Aires Province, Argentina) and their probable builders; in Vizcaíno, S. F., Fariña, R. A., and Janis, C., (eds.), Acta Palaeontologica Polonica. Special Issue on Biomechanics and Palaeobiology 46:157–169.

Vizcaíno S. F., Milne, N., and Bargo, M. S. 2003. Limb reconstruction of Eutatus seguini (Mammalia: Dasypodidae). Paleobiological implications. Ameghiniana 40:89–101.

Vizcaíno, S. F., Fariña, R. A. Bargo, M. S., and De Iuliis, G. 2004a. Functional and phylogenetical assessment of the masticatory adaptations in Cingulata (Mammalia, Xenarthra). Ameghiniana 41:651–664.

Vizcaíno, S. F., Fariña, R. A., Zárate, M. A., Bargo, M. S., and Schultz, P. 2004b. Palaeoecological implications of the mid-Pliocene faunal turnover in the Pampean region. Palaeogeography, Palaeoclimatology, Palaeoecology 213:101–113.

Vizcaíno, S. F., Bargo, M. S., and Cassini, G. H. 2006. Dental occlusal surface area in relation to food habits and other biologic features in fossil xenarthrans. Ameghiniana 4:11–26.

Vizcaíno, S. F., Bargo, M. S., and Fariña, R. A. 2008. Form, function and paleobiology in xenarthrans; pp. 86–99 Vizcaíno, S. F., and Loughry, W. J. (eds.), The Biology of the Xenarthra. University Press of Florida, Gainesville.

Vizcaíno, S. F., Blanco, R. E., Bender, J. B., and Milne, N. 2011a. Proportions and function of the limbs of glyptodonts. Lethaia 44:93–101.

Vizcaíno, S. F., Cassini, G. H. Fernicola, J. C., and Bargo, M. S. 2011. Evaluating habitats and feeding habits through ecomorphological features in glyptodonts (Mammalia, Xenarthra). Ameghiniana. 48(3): 305–309.

Vizcaíno, S. F., G. H. Cassini, N. Toledo, and M. S. Bargo. 2012. On the evolution of large size in mammalian herbivores of Cenozoic faunas of South America. Pp. 76–101. In: Bones, clones, and biomes: the history and geography of Recent Neotropical mammals, B. D. Patterson & L. P. Costa, eds., University of Chicago Press, Chicago.

Voorhies, M. R. 1969. Taphonomy and population dynamics of an early Pliocene vertebrate fauna, Knox County, Nebraska. University of Wyoming Contributions in Geology, Special Paper no. 1, 69 pp

Vucetich, J. A., and Creel, S. 1999. Ecological Interactions, social organization, and extinction risk in African wild dogs. Conservation Biology 13:1172–1182.

Vucetich, M. G., Vieytes, E. C., Verzi, D. H., Noriega, J. I., and Tonni, E. P. 2005. Unexpected primitive rodents in the Quaternary of Argentina. Journal of South American Earth Sciences 20:57–64.

Vucetich, M. G., Reguero M. A., Bond, M., Candela, A. M., Carlini, A. A., Deschamps, C. M., Gelfo, J. N., Goin, F. J., López, G. M., Ortiz Jaureguizar, E., Pascual, R., Scillato-Yané, G. J., and Vieytes E. C. 2007. Mamíferos continentales del Paleógeno argentino: las investigaciones de los últimos cincuenta años. de la Asociación Paleontológica Argentina Publicación Especial 11:239–255.

Vucetich, M. G., Vieytes, E. C., Pérez, M. E., and Carlini, A. A. 2010. The rodents from La Cantera and the early evolution of caviomorphs in South America; pp. 193–205 in Madden, R. H. Carlini, A. A., Vucetich, M. G., and Kay, R. F. (eds.), The Paleontology of Gran Barranca. Cambridge University Press, Cambridge.

Wallace, D. R. 2004. Beasts of Eden: walking whales, dawn horses, and other enigmas of mammal evolution. University of California Press, Berkeley, 340 pp.

Waters, M. R., Forman, S. L., Jennings, T. A., Nordt, L. C., Driese, S. G., Feinberg, J. M., Keene, J. L., Halligan, J., Lindquist, A., Pierson, J., Hallmark, C. T., Collins, M. B., and Wiederhold, J. E. 2011. The Buttermilk Creek Complex and the origins of Clovis at the Debra L. Friedkin site, Texas. Science 331:1599–1603.

Wayne, R. K., Geffen, E., and Vila, B. C. 2004. Population genetics; pp. 55–84 in Macdonald, D. W., and Sillero-Zubiri, C. (eds.), Biology and Conservation of Wild Canids. Oxford University Press, New York.

Webb, S. D. 1974. Pleistocene llamas of Florida, with a brief review of the Lamini; pp. 170–213 in Webb, S. D. (ed.), Pleistocene Mammals of Florida. University Press of Florida, Gainesville.

————. 1976. Mammalian faunal dynamics of the Great American interchange. Paleobiology 2:216–234.

————. 1978. A history of savanna vertebrates in the New World. Part 2, South America and the Great Interchage. Annual Review of Ecology an Systematics 9:393–426.

————. 1985a. The interrelationships of tree sloths and ground sloths; pp. 105–112 in Montgomery, G. G. (ed.), The Evolution and Ecology of Armadillos, Sloths, and Vermilinguas. Smithsonian Institution Press, Washington, D.C.

————. 1985b. Late Cenozoic mammal dispersals between the Americas; pp. 357–386 in Stehli, F. G., and ————. (eds), The Great American Biotic Interchange. Plenum, New York.

Weis-Fogh, T., and Alexander, R. McN. 1977. The sustainable power output obtainable from striated muscle; pp. 511–525 in Pedley, T. J. (ed.), Scale Effects in Animal Locomotion. Academic Press, London.

Weiss, C. S. 1830. Über das südliche Ende des Gebirgszuges von Brasilien in der Provinz S. Pedro do Sul und der Banda oriental oder dem Staate von Monte Video; nach den Sammlungen des Herrn Fr. Sellow (nach einem Vortrag in der Akademie der Wissenschaften am 9. August 1827 und 5. Juni 1828); in Abhandlungen der Königlichen Akademie der Wissenschaften zu Berlin 1827. Dümmler, Berlin.

Wetzel, R. 1985. The identification and distribution of recent Xenarthra (=Edentata); pp. 5–21 in Montgomery, G. G. (ed.), The Evolution and Ecology of Armadillos, Sloths, and Vermilinguas. Smithsonian Institution Press, Washington, D.C.

Wetzel, R. M., Dubois, R. E., Martin, R. L., and Myers, P. 1975. Catagonus, an "extinct" peccary, alive in Paraguay. Science 89:379–38

White, J. L. 1997. Locomotor adaptations in Miocene xenarthrans; pp. 246–264 in Kay, R. F., Madden, R. H., Cifelli, R. L., and Flynn, J. J. (eds.), Vertebrate Paleontology in the Neotropics: The Miocene Fauna of La Venta, Colombia. Smithsonian Institution Press, Washington, D.C.

White, T. H. (ed.) 1960. The Book of Beasts, Being a Translation from Latin Bestiary of the XII Century [reprint]. Putnam, New York, 296 pp.

White, J. L., and MacPhee, R. D. E. 2001. The sloths of the West Indies: a systematic and phylogenetic review; pp. 201–235 in Woods, C. A., and Sergile, F. E. (eds.), Biogeography of the West Indies: Patterns and Perspectives. 2nd ed. CRC Press, New York.

Wible, J. R., and Gaudin, T. J. 2004. On the cranial osteology of the yellow armadillo Euphractus sexcinctus (Dasypodidae, Xenarthra, Placentalia). Annals of Carnegie Museum 73:117–196.

Windle, B. C. A., and Parsons, F. G. 1899. On the myology of Edentata. Proceedings of the Zoological Society of London 1:314–339.

Winge, H. 1915. Jordfundne og nulevende gumlere (Edentata) fra Lagoa Santa, Minas Geraes, Brasilien. Med udsigt over gumlernes indbyrdes slætskab. E Museo Lundi 3:1–321.

————. 1941. The interrelationships of the mammalian genera, Rodentia, Carnivora, Primates (translated from Danish by E. Deichmann and G. M. Allen). Reitzels Forlag, Copenhagen, 376 pp.

Winters, J. M., and Starck, L. 1985. Analysis of fundamental human movements patterns through the use of in-depth antagonistic muscle models. Transactions on Biomedical Engineering 32:826–839.

Wood, A. E. 1985. The relationships, origin, and dispersal of hystricognath rodents; pp. 475–513 in Luckett, W. P., and Hartenberger, J.-R. (eds.), Evolutionary Relationships among Rodents: A Multidisciplinary Approach. Plenum Press, New York.

Woodburne, M. O. 2010. The Great American Biotic Interchange: dispersals, tectonics, climate, sea level and holding pens. Journal of Mammalian Evolution 17:245–264.

Wroe, S., and Field, J. 2007. A review of the evidence for a human role in the extinction of Australian megafauna and an alternative interpretation; in Cupper, M. L., and Gallagher, S. J. (eds.), Climate Change or Human Impact? Australia's Megafaunal Extinction. Selwyn

Symposium of the GSA Victoria Division, September 2007. Geological Society of Australia Extended Abstracts 79:5–10.

Wroe, S., Field, J., and Grayson, D. K. 2006. Megafaunal extinction: climate, humans and assumptions. Trends in Ecology and Evolution 21:61–62.

Wyss, A. R., Flynn, J. J., Norell, M. A., Swisher, III, C. C., Charrier, R., Novacek, M. J., and McKenna, M. C. 1993. South America's earliest rodent and recognition of a new interval of mammalian evolution. Nature 365:434–437.

Yoganandan, N., Pintar, F. A., Sances, A., Walsh, P. R., and Ewing, C. I. 1995. Biomechanics of skull fracture. Journal of Neurotrauma 12:659–668.

Zachos, J., Pagani, M., Sloan, L., Thomas, E., and Billups, K. 2001. Trends, rhythms, and aberrations in global climate 65 Ma to present. Science 292:686–693.

Zanno, L. E., Gillette, D. D., Albright, L. B., and Titus, A. L. 2009. A new North American therizinosaurid and the role of herbivory in "predatory" dinosaur evolution. Proceedings of the Royal Society B 276:3505–3511.

Zárate, M. A. 2003. Loess of southern South America. Quaternary Science Reviews 22:1987–2006.

Zárate, M. A., Bargo, M. S., Vizcaíno, S. F., Dondas, A., and Scaglia, O. 1998. Estructuras biogénicas en el Cenozoico tardío de Mar del Plata (Argentina) atribuibles a grandes mamíferos. Revista de la Asociación Argentina de Sedimentología 5:95–103.

Zárate, M. A., Schultz, P., Blasi, A., Heil, C., King J., and Hames, W. 2007. Geology and geochronology of type Chasicoan (late Miocene) mammal-bearing deposits of Buenos Aires (Argentina). Journal of South American Earth Sciences 23:81–90.

Zeller, U., Wible, J. R., and Elsner, M. 1993. New ontogenetic evidence on the septomaxilla of Tamandua and Choloepus (Mammalia, Xenarthra), with a reevaluation of the homology of the mammalian septomaxilla. Journal of Mammalian Evolution 1:31–46.

Index

caimans, 98
calcium phosphate, 14
calculations, 396–399
Callistoe, 138
Callistoe vincei, 89, *89*
Callitrichidae, 319, 321 [t]
Calomys, 322 [t]
Calvin-Benson cycle, 287
Camacho, H. H., 59
camels and Camelidae, *112,* **192–193**; characteristics of, 116–117, *117;* classification of, 110, 111, 116; diets of, 280; distinguishing features of, 116–117; at Estancia La Moderna excavation, 167; and extinction events, 319; fossil record of, 167; and humans, 339; of the Lujanian Age, 280; migration of (GABI), 144 [c], 147, 150, 322 [t], 325; and Pehuén-Có tracksites, 165; at Piedra Museo site, 168; predation on, *283;* size of, 117, 281 [t]; TYPES OF: *Camelopardalis, 112; Camelus,* 117; *Camelus bactrianus,* 116; *Eulamaops parallelus,* 280; *Hemiauchenia,* 322 [t] (see also *Lama*)
Campbell, K. E., Jr., 145, 200
Campo Laborde site, 344
Cañadón Vaca, 89
canids and Canidae: classification of, 125; diets of, 281–282; metabolic rates of, 295, *295;* migration of (GABI), 144 [c], 150, 165, 321 [t], 325; and Pehuén-Có tracksites, 165; at Piedra Museo site, 168; predation of, *282;* size of, 281 [t], *282;* success of, 329; THROUGH TIME: the Lujanian Age, 280, 281–282; the Pleistocene, 165; TYPES OF, 321 [t]; *Canis,* 321 [t]; *Cerdocyon,* 321 [t]; *Chrysocyon,* 321 [t]; *C. nehringi,* 280, 281–282, *282, 283; Protocyon,* 321 [t]; *Pseudalopex,* 321 [t]; *Theriodictis,* 321 [t] (see also *Dusicyon avus*)
Cape buffalo (African buffalo), 111
Capriolinae, 118–119
capybaras, **195–196**; diets of, 305; origins of, 139; predation on, *283;* size of, 3, 240; skulls of, *196.* See also *Neochoerus*
carbon dating, 28
carbon dioxide (CO2), 78, 85
Caribbean Islands, 146
Caribbean Plate, 135
caribou, 118 [b]
carnassials, 124, 125
carnivorans and Carnivora, 123–133; classification of, 172; diets of, 124; and kleptoparasitism, *99,* 266, 298, 300, 304–310, *307;* and large-mammal impact on vegetation, 345; of the Lujanian Age, 280–281; metabolic rates of, 295, *295,* 299; migration of (GABI), 98, 147, 321–322 [t], 323–324, 325; and northern-superiority theory, 323–324; opportunism in, 305, 306, 308, 310; of the Pleistocene, 165; and population densities, 282, 294; size of, 125, 280–281, 281 [t], *282;* skulls of, *128, 132;* success of, 325, 329; teeth of, *123,* 124, 125; terms, 123–124; types of, 125, 321 [t] (see also canids and Canidae; felids and Felidae; mustelids and Mustelidae; procyonids and Procyonidae; ursids and Ursidae). See also predation and predators
carnivores: dentition of, 188–189; digestive physiology of, 304; diversity of, 298, 313; of the early Pleistocene, 158; and habitats, 157–158; and herbivore diversity, 298, 313; in the Marplatan Age, 101; niche of, 300; and northern-superiority theory, 319–320; and population densities, 294; size of predators, 280–281, *283;* size of prey, 281, *283;* skeletal anatomy of,

358, 363 [b], 364–366; teeth of, 365; terms, 123–124. *See also* predation and predators
Carodnia, 87 [b]
Caroloameghinia, 89, 229 [b]
carrion eaters, 304–310
Carson City, Nevada tracksites, 254–255
Cartelle, Cástor, 65 [b]; on burrow-digging, 250–251; on glyptodonts, 228; on megalonychids, 220; on pampatheres, 221; portrait, *67;* on scelidotheres, 205, 206; on sloths, 65 [b], 216, 218
Carus, Carl Gustav, 45
Casamayoran Age, 88–89
Casinos, A., 213, 239, 256 [b]
Castellanos, Alfredo, 54, 186, 230 [b], 289
Catagonus, 322 [t]
Catagonus wagneri, 319 [b]
catarrhines, 142
catastrophism theory, 35, 48
Cathartidae, 164
Catonyx, **205–206**
cats: migration of (GABI), 147, 150; SKELETAL ANATOMY OF, *354;* cranium, *123, 356,* 357, 358, 359, 363, 364; forelimbs, 371, 372, *372, 373,* 374, *374,* 376; hind limbs, *377, 378,* 380; ribs, *370,* 370–371; sternum, 371; teeth, 365–366; vertebral column, 368–370, *369. See also* felids and Felidae
cattle, 111, 246 [t]
Caudiverbera amazonensis, 88
Cavendish, Thomas, 92
caves, 47, 204–205, 208, 253–254, 260
Caviidae, 144 [c], 321 [t]
caviomorphs and Caviomorpha, 139–143, 147, 164
Cebidae, 319, 321 [t]
Cebus, 321 [t]
cellulose, 15
cenograms, 289–292, *291, 293*
Cenozoic: climatic conditions of, 76–77, *77,* 79; flora evolution during, 78–80; mammals, predaceous, 124; MEGACYCLES: Neocenozoic Megacycle, 82, 92–101; Paleocenozoic Megacycle, 82, 83–91
Central America: flora of, 162; and migrations (GABI), 146, 149; and seaway closure, *134,* 135–136; taxonomic representation of, 145 [c]. *See also* Isthmus of Panama
Central American Seaway, 135–136
Ceratotherium, 243 [b]
Ceraurus maewestoides, 229 [b]
Cerdocyon, 321 [t]
Cerrejón Formation, 84
Cerro Azul Formation, 164
Cervalces, 119
Cervalces scotti, 118 [b], 119
cervids and Cervidae: characteristics of, 117–118; diets of, 280, 284; of the Lujanian Age, 280; migration of (GABI), 144 [c], 322 [t]; size of, 281 [t]; success of, 329; TYPES OF: *Blastocerus,* 322 [t]; *Mazama,* 322 [t]; *Ozotoceros,* 322 [t]; *Ozotoceros bezoarticus,* 284; Pudu, 322 [t] (see also *Morenelaphus*)
cetaceans, 110–111, 172
cetartiodactyls and Cetartiodactyla, 110–111, 172
Chaetophractus villosus, 281
chalicotheres (or ancylopods), 119, 122, 175
Chalicotherium, 122
Chapadmalalan Age (formerly "Chapalmalalan"), 99–101, 150, *316*
Chapalmalania, 99–100, 321 [t]
Chapalmatherium, 196

Chasicoan Age, 97
Chasicó Creek, Buenos Aries Province, Argentina, 97
Chatwin, Bruce, 167
Chelidae, 163
chelonians, 84, 89
Chelonoidis, 163
Chenopodiaceae, 80
chenopodiines, 159
Chile: and Colloncuran Age fauna, 96; early humans in, 167, 344; Monte Verde, 167, 336–337, 338; Tagua-Tagua, 167; and Tinguirirican Age fauna, 91; Última Esperanza caves in, 166–167, 254
Chimento, N. R., 174
China, 328
chitin, 15
Choloepus, 180, *181,* 303, 321 [t], 386
Cholomys, 322 [t]
Christiansen, P., 130, 207, 211, 238, 242
Christian VIII, 48
chronostratigraphy, 24, 29 [b]
Chrysocyon, 321 [t]
Chubut Province, Argentina, 92, 97
Ciconiidae, 164
Cifelli, R. L., 173
cingulates and Cingulata, 181–182; classification of, 177, *178,* 181–182, *183;* diets of, 181–182; digestive physiology of, 301, 303; diversification of, 150; extinction of, 323; of the Lujanian Age, 283, 284; migration of (GABI), 147; size of, 239, *316;* skeletal anatomy of, 383, 385–386, 389; teeth of, *220. See also* armadillos; glyptodonts and Glyptodontidae; pampatheres and Pampatheriidae
Cione, A. L.: on broken-zigzag hypothesis of extinction, 342–343; on Chapadmalalan Age, 99; on extinction, 344; and fauna characteristics, 83; on glacial cycles, 279; on human occupation, 338; on the Lujanian Age, 151; on Marplatan Age, 100
civets (viverrids), 125
Cladera, G., 76–77, 84, 156
clades, 22
classes (in classification system), 20
classification systems, 9–10, 18–23, *19,* 144–145 [c]
Claussen, Peter, 47
Clift, William, 43
climate, *plate 4;* aeolian (wind) activity, 277; Andes' impact on, 77, 78, 277, 330; astronomical influences on, 151–154, *153,* 155; and biomes, 76–77, *157;* effect of glacial/interglacial periods on, 77, *77,* 79, 96, 342 (see also glaciation and interglacial periods); and extinction events, 334–335, 339–344; and faunistic replacement, 101; and flora, 78–80, 159–160; and habitats, 156–159; and land mass, 311; and migrations (GABI), 330; and niches, 329–331; and primary productivity, 296–297; temperate regions, 330, *331;* temperature estimates, 154–155, 159; THROUGH TIME: Lujanian Age, 150–151, 277, 279, 311; Neocenozoic Megacycle fauna, 92–101; Paleocenozoic Megacycle fauna, 83–91; Pleistocene, 156, 163, 335. *See also* glaciation and interglacial periods
Clovis people, 335–339
Coates, A. G., 136
Cochabamban Cycle, *81,* 83–85
Coe, M. J., 303
Cofré, H., 289
Colburn, A. E., 56

of, 239 [t], 240, 281 [t]; strength of, 260–263, 261 [t], *262, 264; Xenorhinotherium,* 158, 201
macroecology, 289
Madden, R., 96, *96*
Madtsoia bai, 89
Magallanes Region, 166
Major Features of Evolution, The (Simpson), 62
Malayan sun bears, 129
Mammalia class, 23
mammoths: 105–110; characteristics of, 106, 108; classification of, 105, 106; diets of, 110; extinction of, 343; habitats of, 108; and kill sites, 339; preservation of Siberian specimens, 15; size of, 109; TYPES OF: *Mammuthus,* 106; *M. columbi,* 109, 110; *M. imperator,* 109; *M. jeffersonii,* 109; *M. meridionalis,* 109; *M. primigenius,* 108–109, 110; *M. trogontherii,* 109, 110
mammutids and Mammutidae. *See* mastodons
manatees, 172
mandibles: anatomy of, 357, 363–366, *382;* and mastication, 268, 357; rarity of, in fossil record, 221; of specific species, *44, 123, 124, 126, 138, 140, 196, 198, 219, 232, 267, 308*
Manera, Teresa, 165
Manera de Bianco, T., 165, 166, 255, 256 [b]
Mantell, Gideon, 45
mantle of the earth, 75
mara, 281
Marchetti, M., 119
Mar del Plata cliffs, 99, 100, 249, *249*
Markgraf, V., 159
marmosets, 139
Marplatan Age, 100–101, 150
Marquet, P. A., 289
marsupials and Marsupialia: as ancient SA inhabitants, 137–138; and carnivore niche, 325; carnivorous marsupials, 319, 325; characteristics of, 137; diets of, 292, *293;* extinction of marsupial species, 319, 323, 328; migration of (GABI), 321 [t]; *Necrolestes patagonensis, 97;* SABERTOOTH MARSUPIALS: *Achlysictis,* 98–99, *99,* 127–128, *139; Anachlysictis gracilis, 97;* size of, 164; THROUGH TIME: Casamayoran Age, 88, 89; Divisaderan Age, 91; Itaboraian Age, 85; Laventan Age, 97; Mayoan Age, 97; Mustersan Age, 91; Paleocenozoic Megacycle, 83, 91; Pleistocene, 164; Riochican Age, 88; Santacrucian Age, 91, 94; Tinguirirican Age, *92*
Martin, P. S., 160, 161–162 [b], 335, 336, 339
Martínez del Río, C., 306
Martinic, M., 208
masticatory musculature of sloths, 265–272
mastodons, 105–110; American mastodons, 106, 107, 110 [b]; hunting of, 339, 344; Jefferson's interest in, 110 [b]; and kill sites, 339; migration of (GABI), 147; in Museo de La Plata mural, *274,* 276; and Pehuén-Có; predation on, *283;* size of, 110; term "mastodons," 107; tracksites, 165; TYPES OF: *Mammut,* 106, 107; *M. americanum,* 106, 107, *107,* 110. See also *"Stegomastodon"*
mastodonts, 167
Matthew, W. D., 61
Mauritius, 340
Mayoan Age, 97
Mayo River, Chubut Province, Argentina, 97
Mazama, 322 [t]
McDonald, H. G.: on cingulates, 181; on Isthmus of Panama, 149; on scelidotheres, 266; on sexual dimorphism, 238; on skeletal anatomy, 383, 391; on sloths, 150, 238, 254–255, 272; on vermilingua, 180; on xenarthrans, 177

McKenna, M. C., 178
McNab, B. K., 212, 293–294, 304, 308
McNaughton, S. J., 299–300
Meadowcroft Rock Shelter, Pennsylvania, 336
megafaunal dispersal syndrome, 160, 161–162 [b]
Megalocnus, 146, 218
megalonychids and Megalonychidae: classification of, *181;* extant species of, 319; fossil record of, *219;* migration of (GABI), 144 [c], 146, 321 [t], 325; of the Mustersan Age, 90; ranges of, 183; size of, 182; taxonomic representation of, 147; TYPES OF: *Ahytherium,* 183, **218–220,** 219; *Ahytherium aureum,* 220; *Australonyx,* 183; *Australonyx aquae,* 220; *Choloepus,* 180, 321 [t]; *Diabolotherium,* 219; *Megalonychops,* 219; *Megalonyx,* 146, 321 [t]; *Meizonyx,* 150, 321 [t]; *Pliometanastes,* 146, 321 [t]; *Pliomorphus,* 219. See also sloths (Tardigrada)
Megalonyx: diet of, 289; fossil record of, *148;* migration of (GABI), 146, 321 [t], 325; *M. jeffersonii,* 289; range of, 147; skeletal anatomy of, 392
megatheres, 256 [b]
megatheriines and Megatheriidae, **213–218;** of the Colloncuran Age, 95, 96; distinguishing features of, 215–216; distribution of, 214, 216; fossil record of, *217;* habitats of, 158; migration of (GABI), 144 [c], 321 [t], 325; and sexual dimorphism, 216, 218; size of, 182, 183, *217;* taxonomic representation of, 144 [c], 147; TYPES OF: *M. altiplanicum,* 214; *M. gallardoi,* 215; *M. medinae,* 214; *M. sundti,* 214; *M. tarijense,* 214 (see also *Eremotherium; Megatherium americanum*)
Megatherium americanum, **213–216;** athleticism of, 246 [t]; bipedality of, 254–256, *255;* Bru's mounting and illustration of, 32, *33,* 35; and Cuvier, 31–33, 35; diet of, 266, 268, 271–272, 280, 286, 304–310, *307;* digestive physiology of, 301, 303–304; distinguishing features of, 213–214, *214;* at Estancia La Moderna excavation, 167, 168; extinction of, 344; forearms and claws of, used aggressively, 256–259, *257, 258;* fossil record of, *257, 258,* 344; furry coat of, 259–260, *259;* and *Glossotherium,* 266; habitats of, 158; kleptoparasitism of, 305–310, *307;* and Larrañaga, 39, *40;* locomotion of, 165–166, 254–256, *255,* 256 [b]; of the Lujanian Age, 280; masticatory apparatus of, 268, 270 [t], 270–271, 284; metabolic rates of, 296 [t]; in Museo de La Plata mural, *274,* 276; and Owen, 39, 42, *44;* predation on, *283;* size of, 183, 239, 239 [t], 280, 281 [t]; skeletal anatomy of, 384; skull and mandible of, *44, 308;* specimen from Luján, 31–32, 51; tracksites of, 165–166, 254–256, *255*
Megatylopus, 117
Megistotherium, 132–133
Meiolania, 184
Meizonyx, 150, 321 [t]
Meltzer, D. J., 339, 340, 343, 346
Memórias sobre a Paleontologia Brasileira (Paula Couto), 64
Mendoza, M., 271, 285, 288
Mendoza Province, Argentina, 98
Mephitidae, 165
Meridiungulata, 86–87 [b], 173–174
Merychippus, 119, 120
Mesocenozoic Supercycle, *plate 1, 81;* Panaraucanian Cycle, 82, 95–100; Patagonian Cycle, 82, 92–95
Mesohippus, 119, 120

mesonychids, 132–133
metabolic rates: of carnivorans, 295, *295,* 299; and digestive physiology, 300–304; and extinctions, 327; and field energy expenditures (FEEs), 308–309, 309 [b]; formulas for, 293–294; and phylogenetic heritage, 300–304; and population densities, 292–296, 296 [t], 297; and primary productivity, 304–305; and size of species, 292–295, *294,* 296 [t], 297 [t], 327
Metatheria, 179
Mexico, 158
miacids, 124
miacoids, 124
migrations: and glaciation and interglacial periods, 104–105; and Isthmus of Panama, 110, 136, 146–147; to North America, 104–105, 137, 145–146, 147–148, 321–322 [t], 325; from North America, 110, 136–137, 142, 147–148, 318–331, 321 [t], 325; north–south disparity in, *145,* 148–149, 318; seasonal migrations, 312. *See also* Great American Biotic Interchange (GABI)
Milankovitch, Milutin, 151–152, *151,* 154
Minoprio, J. L., 91
Miocene Age, 86
mioclaenids, 173, 174
Miotti, L., 168
Mitchinson, J., 3
mixed feeders, 157–158, 272, 284–286
Mixotoxodon, 158, **203,** 321 [t], 323, 325
moa, 43, 340
modern synthesis, 62
Moelleken, S. M., 284–285
Moeritherium, 106
moles, 172
monkeys: migration of (GABI), 139, 141, 142–143; THROUGH TIME: Colhuehuapian Age, 93–94; Colloncuran Age, 96; Deseadan Age, 92; Santacrucian Age, 94; TYPES OF: *Branisella boliviana,* 68, 92; *Dolichocephalus gaimanensis,* 93–94; *Homunculus patagonicus,* 94; howler monkeys, 139; New World monkeys, 142; Old World monkeys, 142; platyrrhine monkeys, 93–94, 95, 139, 141, 142–143
monophyletic groups, 22
Monotrematum, 84, *85*
Monotrematum sudamericanum, 83
monotremes, 82, 85
Montehermosan Age, 99
Monte Hermoso Formation, 164
Monte Verde, Chile, 167, 336–337, 338
Moore, D. M., 272
moose, 118 [b], 118–119
Morenelaphus: diets of, 280; metabolic rates of, 296 [t]; migration of (GABI), 322 [t]; *M. lujanensis,* 280, 281 [t], *283,* 296 [t]; predation on, *283*
Moreno, Francisco P., 54, 70, 94, 208
Moropus, 122
morphological diversity, 38
Mothé, D., 200
mouflon, *112*
mouse opossum, 148
Muizon, Christian de, 69, 173
Multituberculata, 83
mummification, 15
Muñiz, Francisco Javier, 39–41; influence of, 50; portrait, *41;* on *Smilodon,* 187, 189, 190 [b]
murids, 141, 144
muroids and Muroidea, 322 [t], 324
Museo Argentino de Ciencias Naturales, 41, 49
Museo de Ciencias Naturales, 59

paleoecology, 275–314; background of field, 275–276; depicted in Museo de La Plata mural, *274*, 276; and herbivore/carnivore composition, 279–282; in LGM–Holocene interval, 278–279; of the Lujanian Age, 277–278; paleoautecology, 275; and paleogeographic reconstruction, 278, *278*; paleosynecology, 275

paleomammalogists, 31–71; Ameghinos, Carlos, 50–55, *53*; Ameghinos, Florentino, 50–55, *51*; Burmeister, Hermann, 48–50, *48*; Cuvier, Georges, 31–35, *32*; Darwin, Charles, 35–38, *36*; Hatcher, John Bell, 55–59, *56*; Hoffstetter, Robert, 66–69, *68*; Kraglievich, Lucas, 59–60, *59*; Larrañaga, Dámaso Antonio, 38–39, *38*; Lund, Peter Wilhelm, 46–48, *47*; Muñiz, Francisco Javier, 39–41, *41*; Owen, Richard, 42–46, *43*; Pascual, Rosendo, 69–71, *70*; Paula Couto, Carlos de, 63–66, *64*; Simpson, George Gaylord, 60–62, *61*. *See also* full entries for many of these paleomammalogists

paleoneurology, 332–333, *334*

paleontology, etymology of term, 11

Paleoparaná River, 311

Paley, W., 37

Palmae, 78

Palmqvist, P., 271

palynological records, 156

pampatheres and Pampatheriidae, 183–184, **220–223**; burrow-digging of, 222–223, *223*, 252; diets of, 223, 280; digestive physiology of, 301, 303; of the Lujanian Age, 280, 283; metabolic rates of, 301; migration of (GABI), 144 [c], 147, 321 [t], 325; osteoderms of, 220–221, *221*, *222*; paleobiology of, 222–223; predation on, *283*; size of, 183, 220, 281 [t]; skeletal anatomy of, 384; taxonomy and systematics of, *178*, *183*, 220–221; TYPES OF, 221, 321 [t] (see also *Holmesina*; *Pampatherium*)

Pampatherium, **220–223**; burrow-digging of, *222*, 252; diets of, 223, 280, 305; of the Lujanian Age, 280; metabolic rates of, 296 [t]; migration of (GABI), 321 [t], 325; predation on, *283*; size of, 239, 281 [t]; TYPES OF: *P. humboldti*, 221; *P. mexicanum*, 221; *P. paulacoutoi*, 221; *P. typum*, 220, 221, 222, 281 [t], *283*, 296 [t]

Pampean region: climatic conditions of, 158, 223, 277; flora of, 278–279; fossil record of, 164, 199; in LGM–Holocene interval, 278–279; in Lujanian Age, 277–278

Pampian Subcycle, *81*, 95, 100–101, 150

Panama, 158

Panama Canal, 136. *See also* Isthmus of Panama

Panameriungulata, 173

Panaraucanian Cycle, *81*, 82, 95–100

Pancasamayoran Subcycle, *81*, 88–91

pandas, 124, 129

Pangaea, 74

pangolins, 172, 177, 178

Panochthidae: athleticism of, 242; size of, 239 [t]; tails of, 246–247, *247*. See also *Hoplophorus*; *Neosclerocalyptus*; *Panochthus*

Panochthus, **226**; at Arroyo del Vizcaíno site, 169; athleticism of, 242; carapace and tail of, *227*; caudal tube of, *232*; characteristics of, 232; classification of, *186*; diets of, 280; and *Hoplophorus*, 228–229; of the Lujanian Age, 280; metabolic rates of, 296 [t]; in Museo de La Plata mural, *274*, 276; osteoderms of, *226*; predation on, *283*; *P. tuberculatus*, 239 [t], 243 [b], 280, 281 [t], *283*, 296 [t]; size of, 239, 281 [t]; skeletal anatomy of, *389*; tail of, 246, *247*

Panpampian Cycle, *81*, 82, 100–101, 150

Panrlochican Subcycle, *81*, 85–88

Pansantacrucian Subcycle, *81*, 93–95

Panthera atrox, 129, 187, 238

Panthera leo, 129, 187, 261 [t]

Panthera onca: diets of, 282; distribution of, 129; of the Lujanian Age, 280; metabolic rates of, 295, *295*; predation of, 281 [t], *282*, *283*; size of, 129, 280, 281 [t], *282*; strength of, 261 [t]; systematics of, 187

Panthera pardus, 129, 261 [t]

Panthera spelaea, 129, 187

Panthera tigris altaica, 261 [t]

Panthera tigris tigris, 261 [t]

Panthera vereshchagini, 129

Paraceratherium, 105, 122

Parahippus, 119

páramos ecosystem, 156

Paramylodon: migration of (GABI), 146, 321 [t], 325; *P. harlani*, 238, *392*

Paraná River, 48, 98, 141, 155–156, 311

Paraná River Basin, *312*

paraphyletic groups, 23

Paratheria, 179

Pardiñas, U. F. J., 164

Parish, Woodbine, 39, 185, 186

Parodi, Lorenzo, 54

parrots, 164

Pascual, Rosendo, 69–71; on ages, 82; on diversity of fossil record, 80; and Neocenozoic Megacycle, 92; on pampatheres and glyptodonts, 184; on Patagonian biota, 84; portrait, *70*; and Romer, 71; on sloths, 180

Patagonia: Ameghino's work in, 51, 53; climatic conditions of, 77, 89, 92, 93–94, 96, 160; fossil record of, 80; Hatcher's expeditions in, 55, 56; as isolated biogeographic province, 84; and Patagonian Cycle, *81*, 82, 92–95; and Prepatagonian Cycle, 88, 89; vegetation trends of, 78–79

Patterson, B., 180, 184

Paula Couto, Carlos de, 63–66, *64*; and Cartelle, 65 [b]; on glyptodonts, 186, 227, 231; on gomphotheres, 197, 199; and Lund, 48; on megalonychids, 219–220; *Memórias sobre a Paleontologia Brasileira*, 64; mistaken taxonomic identification of, 206; and São José of Itaboraí, 85; and Simpson, 64; *Tratado de Paleomastozoologia*, 65

Peabody Museum, Yale University, 56

Pecari, 322 [t]; diet of, 280; metabolic rates of, 110, 296 [t]; *P. tajacu*, 280, 281 [t], 296 [t]; size of, 281 [t]

peccaries, 145, 147

pecorans, 111

Pehuén-Có tracksites, 165–166, 244–245, 254–255

Peligran Age, 83–84

Peltephilus, *183*

perciforms, 162

periods (geochronologic unit), 29 [b]

periptychids, 173

perissodactyls, 119–122, *120*; basal perissodactyls, 121; diets of, 280; digestive physiology of, 300; extinction of, 325; herbivory of, 280; of the Lujanian Age, 280; migration of (GABI), 322 [t], 328; and phylogenetic scheme, 172, 174; size of, 238, 281 [t]; skeletal anatomy of, *122*; TYPES OF, 119 (see also equids and Equidae; tapirs and Tapiridae)

permineralization, 15

Peru, 68, 84, 98, 158

Peters, R. H., 294, 306

Peterson, O. A., 56

petrification, 14–15

Phalacrocoracidae, 164

phenacodonts, 173

Phiomia, 106

Pholidota, 180

phorusrhacids, 93, 94, 97

Phorusrhacos, *93*

Phospatherium, 106

Phrynops, 163

Phylogenie des Édentés Xenarthres (Hoffstetter), 68

phylogeny, 22–23, 37–38, 172–174

Phylogeny (Ameghinos), 54

phylum (in classification system), 20

physics of megafauna, 235–272; aggression in sloths, 256–259, *257*, *258*; and Archimedes' principle, 237 [b]; athleticism of megafauna, 240–246, 241 [t], *242*, 243 [t], *245*, 246 [t]; bipedality of *Megatherium*, 254–256, *255*; digging of sloths, 248–254, *249*, *250*, *251*, *252*; furry coat of *Megatherium*, *259*, 259–260; masticatory function and geometry of sloth skulls, 265–272, *267*, *269*, 270 [t]; posture of *Toxodon*, 263–265, *264*, *265*; size of megafauna, 236–240, 239 [t]; speed of megatheres, 256 [b]; strength of a *Macrauchenia*, 260–263, 261 [t], *262*; tails of glyptodonts, 246–248, *247*

Picidae, 164

Pictet, F.-J., 184

Piedra Museo site, 168

pigs, 110

pilosans and Pilosa, 177, *178*, 383

pinnipeds, 125

placentals and Placentalia, 178–179

Plaina, 321 [t]

plate tectonics, 73–76, *74*, *75*, 135

Platybelodon, 107–108

Platygonus, 322 [t]

platypus, 83

platyrrhine monkeys: of the Colhuehuapian Age, 93–94; of the Colloncuran Age, 95; migration of (GABI), 139, 141, 142–143

Plaxhaplous canaliculatus, 280, 281 [t], 296 [t]

Pleistocene, 150–169; climate fluctuations of, 156, 163, 335; communities of, 159–160; extinction event of, 133, *145*, 150, 318, 325–326 (*see also* extinction); habitats of, 156–159; and Luján, Buenos Aires Province, Argentina, 51; migrations of (GABI), 150; sites and specimens of, 165–169; small mammals of, 164–165; stratigraphy of, 150–156; vertebrates of, 150–156, 162–164

Pleistocene Pampean Formation, 49

plesiomorphy, 22

Pleurolestodon lettsomi, 211

Pliocene, 156

Pliohippus, 120, 121

Pliometanastes, 146, 321 [t]

Pliomorphus, 219

plovers, 164

Podgorny, I., 39

Podkamennaya Tunguska River, Central Siberia, 346

Podocarpaceae, 79, 80

polar bears, 125, 129–130

Politis, G., 167, 338, 344

pollen and pollen fossil record, *12*, 159–160, 311

polychrotids, 163

polydolopimorphians, 85

polyphyletic groups, 23

RICHARD A. FARIÑA is Professor of Paleontology at the Facultad de
Ciencias, Universidad de la República, Montevideo, Uruguay. He has also
published science books for a general readership and collaborated in TV
documentaries on the subjects of his expertise.

SERGIO F. VIZCAÍNO is Professor of Vertebrate Zoology at the Universidad
Nacional de La Plata and researcher of the Consejo Nacional de
Investigaciones Científicas y Técnicas working at the Museo de La Plata,
Argentina. His research focuses on the paleobiology of South American
fossil vertebrates, mostly mammals. He has participated in numerous field
work seasons in Argentina and Antarctica. He was the President of the
Asociación Paleontologica Argentina.

GERRY DE IULIIS is affiliated with the Department of Ecology and
Evolutionary Biology at the University of Toronto.